TOWARDS HYDROGEN INFRASTRUCTURE

TOWARDS HYDROGEN INFRASTRUCTURE

Advances and Challenges in Preparing for the Hydrogen Economy

Edited by

DEEPSHIKHA JAISWAL-NAGAR

*Associate Professor, School of Physics,
Indian Institute of Science Education and
Research Thiruvananthapuram (IISER-TVM), India*

VINEY DIXIT

*Postdoctoral Fellow, School of Physics,
Indian Institute of Science Education and
Research Thiruvananthapuram (IISER-TVM), India*

SHEILA DEVASAHAYAM

*Senior Lecturer, WASM: Minerals,
Energy and Chemical Engineering,
Curtin University, Australia*

ELSEVIER

Elsevier
Radarweg 29, PO Box 211, 1000 AE Amsterdam, Netherlands
The Boulevard, Langford Lane, Kidlington, Oxford OX5 1GB, United Kingdom
50 Hampshire Street, 5th Floor, Cambridge, MA 02139, United States

Notices
Knowledge and best practice in this field are constantly changing. As new research and experience broaden our understanding, changes in research methods, professional practices, or medical treatment may become necessary.

Practitioners and researchers must always rely on their own experience and knowledge in evaluating and using any information, methods, compounds, or experiments described herein. In using such information or methods they should be mindful of their own safety and the safety of others, including parties for whom they have a professional responsibility.

To the fullest extent of the law, neither the Publisher nor the authors, contributors, or editors, assume any liability for any injury and/or damage to persons or property as a matter of products liability, negligence or otherwise, or from any use or operation of any methods, products, instructions, or ideas contained in the material herein.

ISBN: 978-0-323-95553-9

For Information on all Elsevier publications visit our website at
https://www.elsevier.com/books-and-journals

Publisher: Megan R. Ball
Acquisitions Editor: Peter Adamson
Editorial Project Manager: Ali Afzal-khan
Production Project Manager: Surya Narayanan Jayachandran
Cover Designer: Greg Harris

Typeset by Aptara, New Delhi, India

Contents

SECTION 1 Hydrogen economy and international hydrogen strategy

1.1 Hydrogen economy and international hydrogen strategies   3
Joydev Manna

1.2 Potential market failures inhibiting the development of a green hydrogen export industry   39
Lee V. White, David Gourlay and Emma Aisbett

1.3 Global hydrogen economy and hydrogen strategy overview   59
Anuraag Nallapaneni and Soham Kshirsagar

SECTION 2 Hydrogen production

2.1 Recent progress and developments in photocatalytic overall water splitting   77

Simon Joyson Galbao, Sujana Chandrappa and Dharmapura H. K. Murthy

SECTION 3 Hydrogen storage

3.1 Current state and challenges for hydrogen storage technologies   101

Zainul Abdin, Chunguang Tang, Yun Liu and Kylie Catchpole

3.2 Hydrogen storage in high entropy alloys   **133**

Abhishek Kumar, Nilay Krishna Mukhopadhyay and Thakur Prasad Yadav

3.3 Hydrogen storage technology   **165**

Peng Lv

SECTION 4 Hydrogen transportation and distribution

4.1 Hydrogen transportation and distribution   **187**

Arash Aghakhani, Nawshad Haque, Cesare Saccani, Marco Pellegrini
and Alessandro Guzzini

SECTION 6 Power to gas 'pathway'

6.1 Potential of hydrogen in powering mobility and grid sectors   **349**
Biswajit Ghosh

SECTION 7 Potential use of hydrogen for vehicular and power grid application

7.1 Total cost of ownership analysis of fuel cell electric vehicles in India 379
Ujwal R. Sontakke, Santosh Jaju and Dhiraj K. Mahajan

7.2 Overview on application of hydrogen   **401**
Joydev Manna

7.3 Fuel cell applications: Portable-domestic-distributed-mobility   431

M Shaneeth

SECTION 8   Current and future of hydrogen economy

8.1 Current and future of the hydrogen economy   465

Anandh Subramaniam

Contributors

Zainul Abdin
School of Engineering, The Australian National University, Canberra, ACT, Australia

Mahvash Afzal
Department of Mechanical Engineering, Islamic University of Science and Technology, Kashmir, Jammu and Kashmir, India

Arash Aghakhani
Department of Industrial Engineering, University of Bologna, Bologna, Italy; Commonwealth Scientific and Industrial Research Organization (CSIRO energy), Melbourne, Australia

Emma Aisbett
College of Law, Australian National University, Canberra, ACT, Australia

Akash Kumar Burolia
Chemical Engineering Department, Indian Institute of Technology Kharagpur, Kharagpur, West Bengal, India

Kylie Catchpole
School of Engineering, The Australian National University, Canberra, ACT, Australia

Sujana Chandrappa
Department of Chemistry, Manipal Academy of Higher Education, Manipal Institute of Technology, Manipal, Karnataka, India

Mustafa Erkovan
Instituto de Engenharia de Sistemas E Computadores—Microsistemas e Nanotecnologias (INESC MN), Lisbon, Portugal; Instituto Superior Tecnico (IST), Universidade de Lisboa, Lisbon, Portugal

Simon Joyson Galbao
Department of Chemistry, Manipal Academy of Higher Education, Manipal Institute of Technology, Manipal, Karnataka, India

Biswajit Ghosh
The Neotia University, 24 Paragons (S), West Bengal, India

David Gourlay
Zero Carbon Energy for the Asia-Pacific Grand Challenge, Australian National University, Canberra, ACT, Australia

Alessandro Guzzini
Department of Industrial Engineering, University of Bologna, Bologna, Italy

Nawshad Haque
Commonwealth Scientific and Industrial Research Organization (CSIRO energy), Melbourne, Australia

Santosh Jaju
Department of Mechanical Engineering, G. H. Raisoni College of Engineering, Nagpur, Maharashtra, India

Necmettin Kilinc
Department of Physics, Faculty of Science & Arts, Inonu University, Malatya, Türkiye

Soham Kshirsagar
World Resources Institute, New Delhi, India

Abhishek Kumar
Hydrogen Energy Centre, Department of Physics, Institute of Science, Banaras Hindu University, Varanasi, Uttar Pradesh, India

Sanjay Kumar
Department of Physics, Faculty of Science, University of South Bohemia, České Budějovice, Czech Republic

Yun Liu
Research School of Chemistry, The Australian National University, Canberra, ACT, Australia

Peng Lv
Engineering Research Center of Nuclear Technology Application, Ministry of Education, East China University of Technology, Nanchang, China; School of Water Resources and Environmental Engineering, East China University of Technology, Nanchang, China

Dhiraj K. Mahajan
Ropar Mechanics of Materials Laboratory, Department of Mechanical Engineering, Indian Institute of Technology Ropar, Rupnagar, Punjab, India

Ananya Mandal
Chemical Engineering Department, Indian Institute of Technology Kharagpur, Kharagpur, West Bengal, India

Joydev Manna
Hydrogen Energy Division, National Institute of Solar Energy, Gwal Pahari, Gurugram, Haryana, India

Nilay Krishna Mukhopadhyay
Department of Metallurgical Engineering, Indian Institute of Technology, Banaras Hindu University, Varanasi, Uttar Pradesh, India

Dharmapura H. K. Murthy
Department of Chemistry, Manipal Academy of Higher Education, Manipal Institute of Technology, Manipal, Karnataka, India; Center for Renewable Energy, Manipal Academy of Higher Education, Manipal Institute of Technology, Manipal, Karnataka, India

Anuraag Nallapaneni
World Resources Institute, New Delhi, India

Anilia Nanan-Surujbally
Department of Physics, The University of the West Indies, St. Augustine, Trinidad & Tobago

Swati Neogi
Chemical Engineering Department, Indian Institute of Technology Kharagpur, Kharagpur, West Bengal, India

Dinesh Pathak
Department of Physics, The University of the West Indies, St Augustine, Trinidad & Tobago

Marco Pellegrini
Department of Industrial Engineering, University of Bologna, Bologna, Italy

Cesare Saccani
Department of Industrial Engineering, University of Bologna, Bologna, Italy

M Shaneeth
Vikram Sarabhai Space Centre, ISRO, Thiruvananathapuram, Kerala, India

Pranjali Sharma
Chemical Engineering Department, Indian Institute of Technology Kharagpur, Kharagpur, West Bengal, India

Davinder Pal Sharma
Department of Physics, The University of the West Indies, St Augustine, Trinidad & Tobago

Orhan Sisman
Center for Functional and Surface Functionalized Glass, Alexander Dubcek University of Trencin, Trencin, Slovakia

Ujwal R. Sontakke
Ropar Mechanics of Materials Laboratory, Department of Mechanical Engineering, Indian Institute of Technology Ropar, Rupnagar, Punjab, India

Anandh Subramaniam
Materials Science and Engineering (MSE), Centre for Environmental Science and Engineering (CESE), Sustainable Energy Engineering (SEE), Indian Institute of Technology, Kanpur, Uttar Pradesh, India

Chunguang Tang
Research School of Chemistry, The Australian National University, Canberra, ACT, Australia

Lee V. White
School of Regulation and Global Governance, Australian National University, Canberra, ACT, Australia

Thakur Prasad Yadav
Hydrogen Energy Centre, Department of Physics, Institute of Science, Banaras Hindu University, Varanasi, Uttar Pradesh, India; Department of Physics, Faculty of Science, University of Allahabad, Prayagraj, Uttar Pradesh, India

Preface

The Sustainable Development Goals (SDG) are a call for action by all countries, to promote prosperity while protecting the planet. Of the 17 SDGs, number 13 calls for urgent action to combat climate change and its impacts. Goal 7 calls for affordable and clean energy and Goal 8 for promoting sustained, inclusive, and sustainable economic growth, full and productive employment, and decent work for all through a transition to renewable energy. Goal 9 calls for developing resilient infrastructure. International Governments at the 26th UN Climate Change Conference of the Parties (COP26) in Glasgow in 2021, put forward new 2030 emissions targets to keep global warming to no more than 1.5°C, as called for in the Paris Agreement; that emissions need to be reduced by 45% by 2030 and reach net zero by 2050. There was also an agreement to phase down coal power and to limit methane emissions by 30 percent by 2030, compared to 2020 levels. To achieve these objectives, the consensus reached among the most developed countries now includes a commitment to hydrogen as a key alternative energy source, including Australia, the United Kingdom, Japan, Korea, Germany, China, the United States, and the United Arab Emirates.

There are numerous ways in which hydrogen can contribute to decarbonizing the economy and achieving the goal of net zero emissions. Not only can hydrogen provide a practical solution for storing energy generated from renewable sources, but it can also serve as an important source of combined heat and power to replace natural gas. Furthermore, hydrogen is being viewed as a crucial option for transportation in sectors that are difficult to decarbonize, such as heavy vehicles, trains, and shipping. This is because hydrogen offers a clean alternative to fossil fuels, resulting in zero emissions.

It is our pleasure to introduce this comprehensive book on "Hydrogen Infrastructure: Towards a Sustainable Energy Future." As editors, we are excited to present a collection of expert contributions that provide a thorough exploration of the development and challenges of the hydrogen infrastructure.

Hydrogen is an extremely small molecule with some unique properties like very low density, high diffusivity, higher ignition power than diesel and petrol, and a broad flammability range than other hydrocarbon fuels. Therefore, the hydrogen economy faces significant challenges regarding hydrogen storage, production, distribution, and safety.

This book addresses the critical issues surrounding the hydrogen economy, including the potential market failures that inhibit the development of a green hydrogen export industry, and the global overview of the hydrogen economy and strategies. The latest developments in photocatalytic overall water splitting for hydrogen production, as well as the current state and challenges of hydrogen storage technologies are explored in depth.

The book also provides an in-depth analysis of hydrogen transportation and distribution, including commercially available resources for physical hydrogen storage and distribution, and a systems perspective on metal hydride hydrogen storage. Moreover, the crucial issue of hydrogen safety and standards is discussed, including the development of hydrogen sensors for safety applications and national and international document standards on hydrogen energy and fuel cells.

The potential of hydrogen in powering mobility and grid sectors is discussed in detail, along with an overview of the application of hydrogen for vehicular and power grid applications, including a total cost of ownership analysis of fuel cell electric vehicles in India, and the use of fuel cell applications in portable houses and distributed mobility.

Finally, the book concludes with a thought-provoking discussion of the current and future of the hydrogen economy, highlighting the opportunities and challenges ahead.

We are grateful to the distinguished experts who contributed their insights and perspectives to this book, and we hope that readers will find this book informative and engaging. It is our sincere hope that this book will contribute to a better understanding of the challenges and opportunities associated with the development of a sustainable hydrogen infrastructure, and inspire the future generation of researchers, engineers, and policymakers in this exciting field.

Acknowledgment

The editors would like to acknowledge the Elsevier editorial team for their guidance and suggestions. Dr. Deepshikha Jaiswal-Nagar and Dr. Viney Dixit would like to thank the Department of Science and Technology (DST), Ministry of Science and Technology, Government of India and the Indian Institute of Science Education and Research Thiruvananthapuram (IISER-TVM), Kerala, India for providing them with funding and a working environment. Dr. Sheila Devasahayam wishes to acknowledge Western Australian School of Mines (WASM), Kalgoorlie, Australia and Curtin University, Perth, Australia for their support. Dr. Dixit and Dr. Jaiswal-Nagar would like to express their gratitude to their families for their constant motivation and moral support.

Hydrogen economy and international hydrogen strategy

Hydrogen economy and international hydrogen strategies

Joydev Manna
Hydrogen Energy Division, National Institute of Solar Energy, Gwal Pahari, Gurugram, Haryana, India

1.1.1 Introduction

As a consequence of the continuous increase in the world's population, huge development in the industrial sectors, and change in human lifestyle to consume more energy, the demand for energy will continue to rise. Till 2030, the world's primary energy demand is expected to increase by more than 50%.

Despite a growing proportion of renewable energy (RE) resources, the majority of the world's energy needs are currently met by fossil fuels. Fig. 1.1.1 shows the increase in energy demand for the last few years till 2020. As can be seen from the figure, energy demand has been increasing every year, and around 580 exajoule of primary energy was consumed during 2019. Global primary energy consumption increased annually with a growth rate of around 2% in the last few decades and it is expected to be increased with an annual growth of about 2.5% in coming years. The drop in primary energy demand in 2020 is mainly due to the less fuel consumption during the COVID-19 lockdown. However, it has been noted that energy demand is back after the COVID-19 lockdown and above 2019 pre-COVID-19 levels. It is expected that global energy demand will increase by 4.6% in 2021, more than offsetting the contraction in 2020.

Fossil fuels make up around 80% of the world's primary energy supply today among the numerous primary energy sources. Crude oil fulfills more than 33% of global primary energy demand most of which is used in the transport sector. As the margin between production and demand of oil increased in the last couple of years it caused a rapid rise in oil prices. Fig. 1.1.2 shows the primary energy consumption for different parts of the world in terms of various primary energy sources. Among various end-users, the industrial sector has been consuming the highest amount of primary energy resources since the 1950s. Besides, the commercial and transport

Towards Hydrogen Infrastructure: Advances and Challenges in
Preparing for the Hydrogen Economy.
DOI: https://doi.org/10.1016/B978-0-323-95553-9.00009-1

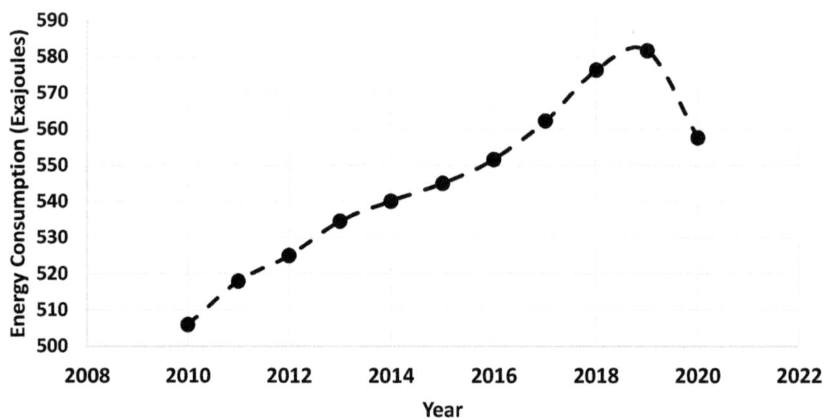

Figure 1.1.1 Global energy consumption during 2010–2020. *(From: Statista, Global Primary Energy Consumption [15]).*

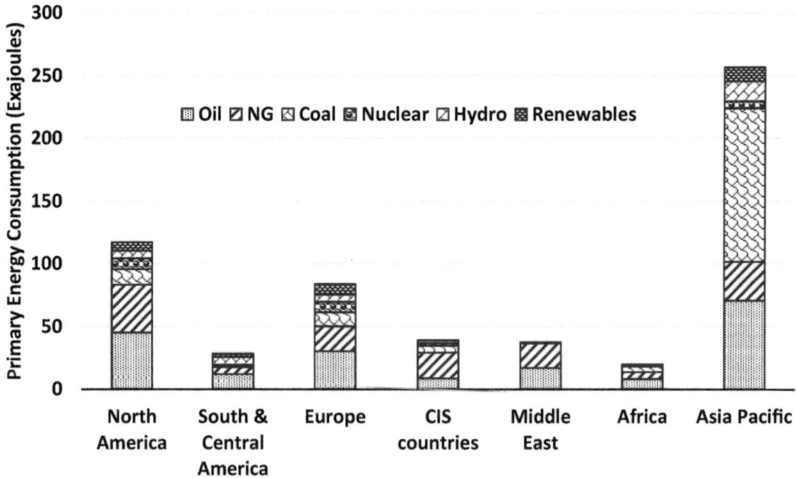

Figure 1.1.2 Energy consumption for a different part of the world in terms of various primary energy sources. *(From: IEA [19]).*

sectors are in the second and third places for primary energy consumption followed by the residential sector in the fourth position (Fig. 1.1.3). It is assumed that the transport sector will come out as the highest consumption sector in coming years. The drop in primary energy consumption is mostly seen in the transport sector in 2020 due to the global COVID-19 lockdown. The demand for energy in the transport sector is observed to be growing in

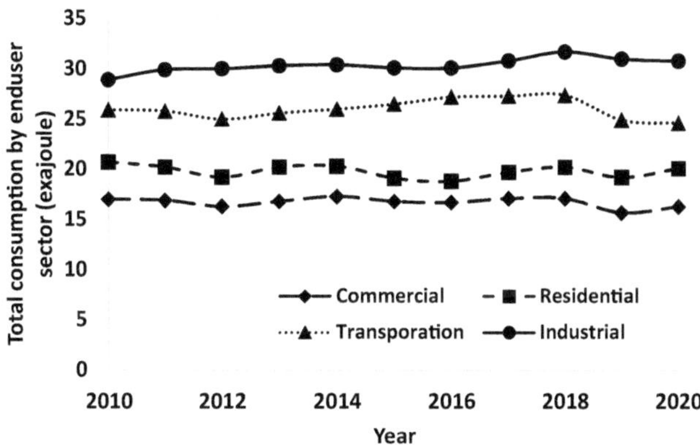

Figure 1.1.3 Energy consumption by end-user sector during 2010–2020. *(From: EIA, Monthly energy consumption [25]).*

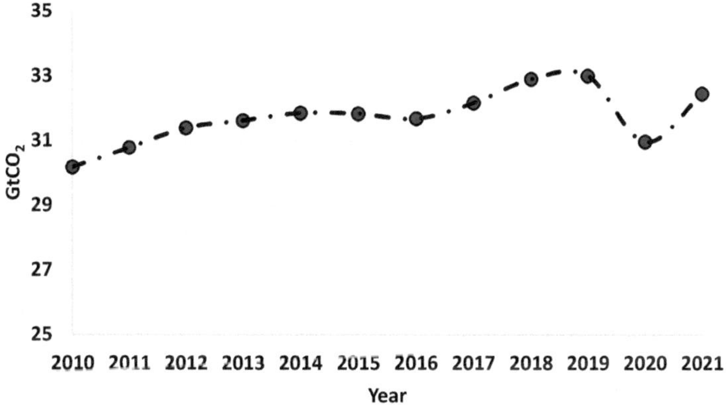

Figure 1.1.4 Carbon dioxide emission in the world during 2010–2021. *(From: IEA, Global Energy Review [19]).*

the current year and it is expected that it will surplus the pre-COVID-19 energy demand. Consequently, global energy-related carbon dioxide emissions are increasing annually. As the demand for fossil fuels was grown in the recent year (2021), demand for coal alone is anticipated to rise by 60% higher than that for all RE sources combined which will cause to rise in greenhouse gas (GHG) emissions by almost 5% (around 1500 Mt).

Emerging and developing countries like China and India now release more than two-thirds of global carbon dioxide emissions [9]. Fig. 1.1.4 shows the CO_2 emission trends during the last decades. It has been predicted that CO_2 emission by China will increase by 500 Mt CO_2 during 2021. Amongst all fossil fuels, coal is expected to dominate in carbon dioxide emissions mainly due to its use in the power sector. India's CO_2 emission will increase by 200 Mt in 2021. Nonetheless, it is important to note that per capita CO_2 emissions in India continue to be 60% lower than those in the rest of the world and two-thirds lower than those in Europe.

The ever-increasing energy demand, a huge amount of CO_2 emission by extensive use of fossil fuels, threats of increasing global average temperature by 1.5°C, and the issue of local air pollution made policymakers look for alternate energy resources. It has also been envisaged that the depleted fossil fuel resources will not be able to deal with the increasing world primary energy demand. According to projections, the total oil production will need to double by 2030 in order to meet the growing global energy demand. Another concern for the policymakers is energy security due to the import dependence on crude oil from the Middle East region which area is facing economic and geopolitical turmoil for the last few decades. Energy security is always a strong pillar of energy policy, and recent climate change and GHG emission restrictions are raising concerns for every country to develop a cost-effective energy policy ensuring energy security as well as minimum GHG emissions. Therefore, scientists, policymakers, and energy implementation agencies are looking for alternative RE resources to replace the conventional fossil fuel-based energy ecosystem.

Extensive research and development have been carried out during the last few decades throughout the world to decarbonize the energy sector [11]. To resolve the fossil fuel energy consumption in transport sectors, different types of innovation and alternate fuel options are tested out. Development of lightweight vehicle construction, improvement in the conventional internal combustion engines (ICEs), hybridization as well as dieselization of transport vehicles are thought to be short-term procedures to improve the fuel economy of vehicles. However, for a permanent and long-term solution, it has been strategized to develop alternative fuels such as biofuels, electricity, methanol, and hydrogen. In the biofuel category, ethanol and biodiesel are the most produced and used worldwide. Corn, sugar cane, and other crops can be used to make ethanol, while animal and vegetable oils can be used to make biodiesel. Biodiesel and ethanol fuels are facing development as well as implementation issues mainly due to feedstock unavailability, proper supply chain of feedstocks, and fewer installed production facilities. Methanol is a comparatively new fuel for transport applications. Transport vehicles may need material modifications and methanol mixing calibration in order to embrace methanol. However, methanol is mostly produced from fossil fuels (compressed natural gas [CNG] and coal), and using green methanol, its economic viability will considerably be influenced by the cost of renewable electricity as well as the cost of green hydrogen. On a well-to-wheel basis, using electricity in the transportation sector is the most energy-efficient option. However, a breakthrough in battery technology is needed to increase its storage capacity as well as minimize the charging time and cost to adopt widely. In the case of hydrogen, it can be utilized to power electric vehicles (EVs) using fuel cells or in vehicles with ICEs. In an ICE, hydrogen is used in a similar way to petrol, diesel, or CNG engines and fired to produce thermal energy which eventually converts to mechanical energy. In fuel cells, hydrogen is reacted with oxygen to generate electricity as well as water vapor [12]. The generated electricity is supplied to an electric motor to power the wheels. However, the wide adoption of hydrogen in transport sectors has many barriers such as efficient and large-scale production of green hydrogen, safe and scalable hydrogen storage methods, efficient and affordable hydrogen vehicles as well as adequate hydrogen refueling infrastructure. These issues are taken up and tackled by researchers for the last few decades and it is believed that this is the time to spread and implement the hydrogen economy in our society.

Hydrogen can also be utilized in other sectors such as industrial applications, household applications, electricity production, etc. It is considered one of the promising choices as a secondary energy carrier as it is the most abundant element on the earth and can be produced using any primary energy sources, and it is a green energy carrier as it can be produced using RE resources and does not emit GHGs after its final use. The dream of using green hydrogen in every sector seems possible nowadays only due to an increase in installed RE capacity and the reduction in solar photovoltaics (SPV) and wind electricity costs during the last few years.

It has been reported that RE uses increased by 3% in 2020 while demand for all other fuels decreased. RE resources are primarily used for electricity generation. The share of RE in global electricity production has increased to 29% in 2020. China, the EU, the United States, and India are leading in the implementation of RE for electricity generation [2]. As the RE sector is growing, the cost of RE electricity from wind and SPV is also declining and is now comparable with conventional grid electricity. Therefore, it is envisaged that the excess electricity generated during the daytime could be used to produce green hydrogen and store it for future use. The contribution

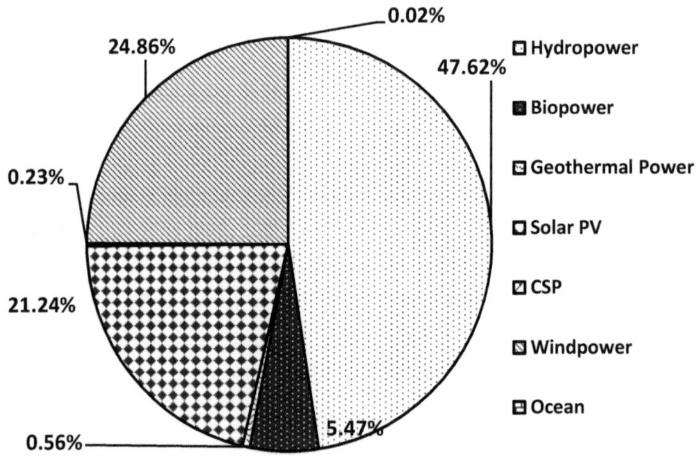

Figure 1.1.5 Global renewable energy share during 2018. *(From: EIA, 2013; Ref. [7]).*

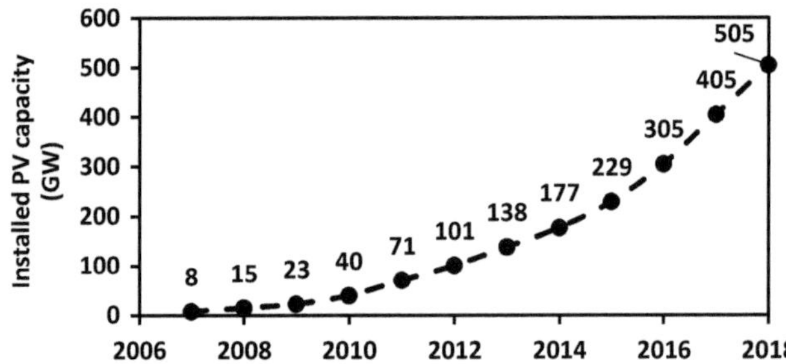

Figure 1.1.6 Growth of solar photovoltaics capacity globally. *(From: EIA, 2013; Ref. [7]).*

of renewable power is growing worldwide and e-mobility using renewable electricity would almost make the transport sector a zero-emission sector. Fig. 1.1.5 shows the break-up of the global renewable power generating capacity of 2377.3 GW in 2018. Most of the renewable power capacities are contributed by hydro and wind. Recently solar energy-based power generating capacity, both thermal and photovoltaic, is increasing rapidly in recent years. Worldwide SPV installations have increased significantly during the past few years. As of 2018, globally 505 GW of SPV power was installed. The growth of solar SPV is shown in Fig. 1.1.6. Following section discusses how the available RE capacity could be used to produce green hydrogen and a RE-based green hydrogen economy could successfully be implemented.

1.1.2 Concept of the hydrogen economy

The production, use, storing, and handling of hydrogen is not an unknown process, and it has been extensively used in industrial sectors, especially in the petroleum and fertilizer industry. Hydrogen was first utilized to carry people in a hydrogen balloon, and it was subsequently used in several nations as town gas for street lighting until the 1960s.

The idea and conceptual basis of a hydrogen economy was formulated after the oil crises of the 1970s [4]. Interest in hydrogen has revived again after the successful development of fuel cell technologies [3]. A tremendous effort in terms of research and development projects has been put to prove the feasibility of hydrogen as an energy carrier since then.

With its ability to operate in the transportation, heating, industrial, and electricity sectors, hydrogen is regarded as a practical alternative fuel that can balance electricity as a zero-carbon energy carrier that is simple to store and transport. This makes it possible to create a more secure energy system with less reliance on fossil fuels. Due to a downward trend in electricity generation costs and an increase in the percentage of RE in various countries' energy mixes, electricity is proving to be relatively simple to decarbonize, whereas hydrogen lags far behind and has been making up ground in recent years. Fig. 1.1.7 demonstrates the versatility of hydrogen as an energy carrier, in terms of its production from different feedstocks ranging from fossil fuels to biomass and electricity, distribution and end-use for electricity generation, as a transport fuel, in industry and buildings. This figure also depicts a diagram of the so-called hydrogen economy.

As discussed earlier, the use of hydrogen is not a new concept, and currently, hydrogen gas is widely used in industrial sectors mainly in refineries, ammonia, methanol, and other chemical production [13]. Since 1975, the demand for hydrogen has increased more than fourfold, reaching more than 70 MMT in 2018. Around 38.2 MMT of hydrogen was used in refineries and 31 MMT was consumed in ammonia synthesis in 2018. The rest of the hydrogen was used in other chemical syntheses such as methanol production, polymers (ethylene, propylene), and solvents (benzene, toluene) production as well as some other applications such as in glass, semiconductor, synthetic fuel, and food processing industries. It has been estimated that hydrogen demand in refineries will grow by 7% and reach 41 MMT by 2030. The hydrogen used in these industries is primarily generated from fossil fuels. Hence, the primary aim of the policymakers is to replace the existing

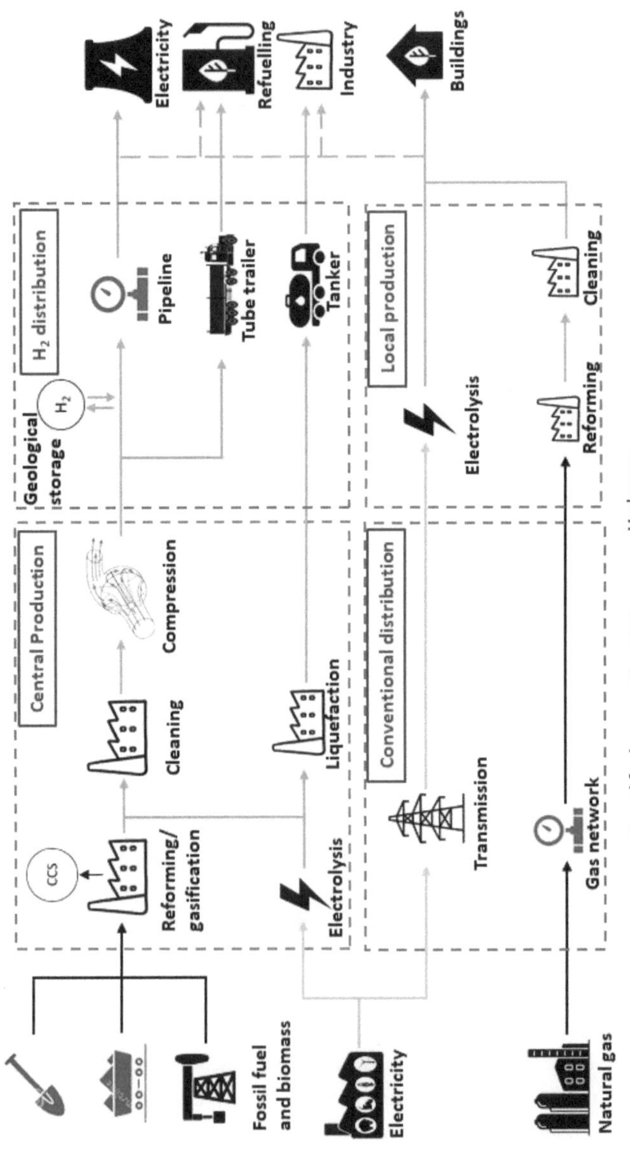

Figure 1.1.7 Schematic diagram to show the various parts of the hydrogen economy [20].

hydrogen production methods with green hydrogen production and reduce fossil fuel consumption as well as GHG emissions.

Hard-to-abate industrial sectors such as iron, steel plants, cement production, etc. generate a huge quantity of hydrogen mixed with other gases such as coke oven gas, which are largely used within plants. These gases are produced by the reformation of fossil fuels. Efforts are underway to make use of hydrogen in the iron and steel industry as a reducing agent in place of coal. Commercial steel and cement plants making use of green hydrogen are likely to become operational within a decade.

The potential of hydrogen in the transport sector is huge and different types of vehicles can be powered by either fuel cell (FC) or hydrogen-powered internal combustion engine (HICE) technologies. Currently, EVs and hydrogen-fueled vehicles (HFVs) are the only options with no GHG emissions at the use point as the energy carriers used for their operation can be generated using renewable sources. HFVs offer a similar driving range and lower refueling time as available with fossil fuel-based ICE. Nevertheless, the high cost of hydrogen and FCs and technical/infrastructure challenges associated with hydrogen production, transportation, and storage are inhibiting the introduction of HFVs and their rapid market deployment. Hydrogen could also be used in the maritime sector. Around 2.5% of global energy-related CO_2 emissions are caused by maritime freight. The rail sector is mostly run by electricity and alternative fuels could be used instead of diesel in those portions of the tracks where electrification is difficult. Germany is already using two hydrogen FC trains which can travel 800 km/day on a single refueling, and they intend to run 14 such trains in coming years. Austria, France, Japan, and the United Kingdom have plans to deploy hydrogen trains. Recently in 2023, India also announced its plan to adopt 35 hydrogen-powered trains on its railway lines. For the operation of material handling equipment like forklifts, hydrogen has already emerged as an attractive option in the United States. Hydrogen could also be used in the future for other rail yards and logistics hub machinery. The aviation sector is responsible for around 2.8% of global CO_2 emissions (in 2017) and it will increase further. Alternative fuels such as biofuels and hydrogen are among the best options to avoid increases in emissions from this sector. Although feasibility tests and demonstrations have been performed for using hydrogen in small planes, significant research and development (R&D) on refueling, storage infrastructure, and aircraft design are needed to use 100% hydrogen-fueled engines in the aviation sector. Hydrogen-based liquid fuel could be used in the current aircraft with little or no modifications.

In this scenario, the transportation sector is not the only target; it is also thought that stationary applications using hydrogen and fuel cells might be employed to generate electricity for small appliances and homes [6]. The use of hydrogen in the power sector for the production of electricity using fuel cells, traditional ICE, hydrogen-fired gas turbines, and combined-cycle gas turbines has enormous potential. A gas mixture containing 3–5% hydrogen could be used in the gas turbine technology now in use. A successful demonstration of using a maximum of 30% hydrogen gas mixture in a modified gas turbine was performed by Mitsubishi Hitachi Power Systems (MHPS). The technological development towards using hydrogen for power generation are on a positive track as reported by major power equipment manufacturers such as MHPS, GE Power, Siemens Energy, and Ansaldo Energia. The power industry is highly confident that gas turbines entirely run-on hydrogen could be developed by 2030. FCs are versatile and can also be used for stationary power generation. In the last decade, many stationary FCs have been installed for power generation, and as of now total global installed capacity of FC is around 1.6 GW. These are mostly installed in Japan, Germany, Korea, and the United States. Japan aims to install 1 GW FC by 2030. Korea targets to install 1.5 GW FC by 2022 and 15 GW by 2040. Hydrogen is an excellent energy storage medium with variable renewable power generation systems based on solar and wind power. A combination of RE power sources, electrolyzer for hydrogen generation, hydrogen storage tanks and FCs could be a better option for power generation, depending on needs as a microgrid.

Natural gas (NG) is currently used for heating in buildings and a huge amount of NG is consumed every year. Hydrogen could be blended into existing NG networks for heating buildings. Local district energy networks could also use hydrogen for cooling or heating their building networks. In fact, the complete use of hydrogen in household applications such as cooking, heating, cooling, etc. could significantly reduce the environmental impact, caused by the current use of fossil fuels.

Therefore, the hydrogen economy does not only deal with the production, storage, and delivery of hydrogen, it also includes the use of hydrogen in various sectors wherever fossil fuels are used. However, it is debatable whether it is viable to use green hydrogen in almost all sectors or not and it needs to be properly evaluated based on the availability of RE resources, risk involvement, and most importantly economic analysis. It is also true that a huge amount of investments are required prior to implementing a hydrogen economy in terms of infrastructure and efficient technology development.

Table 1.1.1 Properties of different fuels [8].

Property	Hydrogen	Gasoline	Natural gas	Liquefied petroleum gas
Density (kg/m^3)	0.084	4.4	0.651	1.87
Density relative to air	0.07	4	0.55	1.52
Molecular weight	2	107	16	44
Diffusion coefficient in air (cm^2/s)	0.61	0.05	0.16	0.12
Ignition energy (mJ)	0.02	0.24	0.29	0.26
Ignition temperature (°C)	585	228–471	540	410–580
Explosive energy (MJ/m^3)	9	407	32	93
Flame temperature (°C)	2045	2197	1875	1970
Flame speed (cm/s)	346	42	43	47

Although the hydrogen economy provides great hope for decarbonizing society, there are many unresolved issues and challenges before implementing the hydrogen economy in a complete sense. This chapter will provide an overview of different hydrogen production methods, storage methods, and their application in various sectors. It will also discuss hydrogen safety-related issues and different strategies developed by various countries to implement a hydrogen economy on a short-term as well as long-term basis.

1.1.3 Properties of hydrogen

The lightest element is hydrogen. It is a tasteless, odorless, colorless, and nontoxic gas that is present in the air at amounts of around 100 parts per million (ppm; 0.01%) and makes up 75% of all ordinary matter in the universe. Table 1.1.1 compares the fuel characteristics of hydrogen with those of other commonly used fuels. The distinct combustion properties of hydrogen are what allow HICE to burn cleanly and function effectively. Its wide range of flammability, low ignition energy, short quenching distance, high autoignition temperature, the fast flame speed at stoichiometric ratios, high diffusivity, and extremely low density are among the qualities that make it useful as a combustible fuel. The relative heating value of hydrogen and other fossil fuels is shown in Fig. 1.1.8 [5]. Compared to methane and gasoline, hydrogen has a far wider flammability range of 4–75% by volume with air. Because of this, hydrogen may burn in an engine at a variety of fuel–air ratios, including lean ratios, resulting in complete combustion, improved

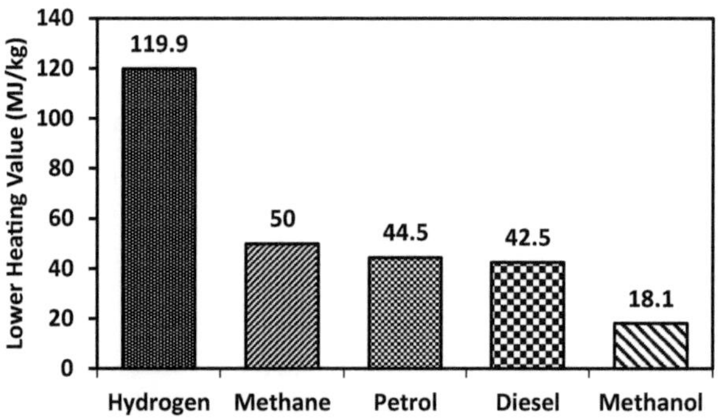

Figure 1.1.8 Heating values of commonly used fuels.

fuel efficiency, and a decrease in the number of pollutants like nitrogen oxide (NOx) due to reduced combustion temperature. As hydrogen has a very low ignition energy (0.02 mJ), it can quickly ignite lean mixtures and power engines. The low ignition energy has the unfortunate side effect of making hot gases and hot spots on the cylinder's potential sources of ignition, leading to issues with premature ignition and flashback. Due to hydrogen's broad range of flammability, a hot spot can ignite practically any mixture. The quenching distance of hydrogen is rather short—about three times that of petrol. The likelihood of backfire may rise with a shorter quenching distance. Because hydrogen has a comparatively high autoignition temperature, a hydrogen engine can use a higher compression ratio than a hydrocarbon engine, which improves the engine's thermal efficiency. Hydrogen has a rapid flame. It is also true that in the event of a hydrogen leakage, it spreads quickly due to hydrogen's extremely high diffusivity in the air.

1.1.4 Overview of hydrogen production methods

A variety of carbonaceous feedstocks and/or water can be converted into hydrogen utilizing a variety of processes. Gasification and reforming processes such as steam methane reforming and partial oxidation/auto-thermal reforming can transform coal, NG, petroleum fractions, and biomass into hydrogen. Different hydrogen production methods used in industries are depicted in Fig. 1.1.9. Globally, about 95% of the hydrogen requirement is produced by fossil fuels. Depending on the hydrogen production methods and amount of carbon oxide emitted, hydrogen is color-coded (Table 1.1.2). Although, there is no universal agreement over these color codes. These

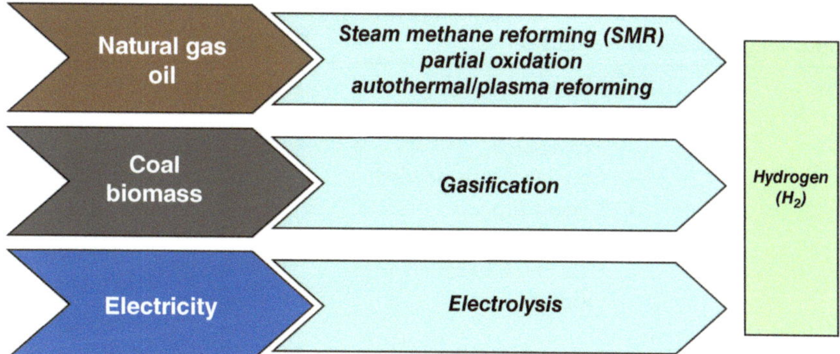

Figure 1.1.9 Industrial hydrogen production methods.

Table 1.1.2 Different types of hydrogen, their production technology, feedstocks and greenhouse gas footprint*.

Terminology	Technology	Feedstock/electricity source	GHG footprint
Green H$_2$	Gasification	Biomass	Low*
	Electrolysis	Water/wind, SPV, hydro, geothermal, tidal	Minimal
Purple/pink H$_2$		Water/nuclear	Low
Yellow H$_2$		Water/mixed-grid	Medium
Blue H$_2$	Reforming	Natural gas, coal	Low
Turquoise H$_2$	Pyrolysis	Natural gas	Solid carbon
Grey H$_2$	Reforming		Medium
Brown H$_2$	Gasification	Brown coal	High
Black H$_2$		Black coal	

*It is considered that CO_2 emitted during gasification is equal to the amount of CO_2 absorbed by the plant during its lifetime.

hydrogen production technologies are discussed briefly in the following sections.

1.1.4.1 Steam methane reforming

Industrially, most hydrogen is produced by the steam methane reforming (SMR) process. In SMR process, initially, syngas ($CO + H_2$) is produced which is followed by a water gas shift reaction to generate further quantities of hydrogen. Typically, the efficiency of the SMR process is 70–85%. In this

Table 1.1.3 Comparison of three reforming processes.

Reforming process	Advantages	Disadvantages
SR	• Mature technology with extensive industrial experience • Oxygen is not needed • Lowest process temperature • Best H_2/CO ratio for H_2 production	• Highest emissions
POX	• Desulphurization requirement is minimum • Catalyst is not required • Low methane slip	• Low H_2/CO ratio • High processing temperature • Soot formation
ATR	• Lower process temperature than POX. • Low methane slip	• Limited industrial experience • Require air or oxygen

process, around 9–12 tons of carbon dioxide is produced (depending upon the quality of feedstock) per ton of hydrogen.

1.1.4.2 Partial oxidation

Natural gas and other heavy hydrocarbons usually furnace oil, whose further treatment and utilization is difficult, are used to produce syngas using limited amount of oxygen which is known as partial oxidation or POX process. The thermal efficiency of this process is generally 60–75%. This process also releases a substantial amount of GHGs.

1.1.4.3 Autothermal reforming

Autothermal reforming (ATR) combines steam reforming (SR) and partial oxidation (POX) processes. ATR procedure creates a thermally neutral process by utilizing SR to boost hydrogen production while using the POX to generate heat. ATR is frequently carried out with less pressure than POX. This process does not require an external heat source for the reactor because POX is exothermic and ATR includes POX. However, to provide pure oxygen to the reactor, it either needs an expensive and complicated oxygen separation device, or the resulting gas is diluted with nitrogen, necessitating gas separation and purification procedures. Table 1.1.3 compares these three processes.

1.1.4.4 Coal gasification

In the coal gasification process, steam and a carefully controlled concentration of air/oxygen are used to burn coal to form syngas. This process is known for the last couple of centuries and was used to produce "town gas" or "coal gas" for municipal lighting and heating purposes. According to estimates, the gasification process is currently used to manufacture more than 30% of the methanol and around 25% of the ammonia in the globe. The typical thermal efficiency of this process is 35–50%. The hydrogen generation through the biomass gasification process is similar to the coal gasification process.

1.1.4.5 Petroleum coke gasification

Petroleum coke can also be used to produce hydrogen-rich syngas using a similar process as coal gasification. However, the syngas obtained from petroleum coke gasification is a mixture of H_2, CO, and hydrogen sulfide (H_2S) gas.

1.1.4.6 Electrolysis

Hydrogen gas can be created by splitting water with electricity in an electrolyzer. Globally about 4% of hydrogen is produced by electrolysis. The conversion efficiency of electrolysis is around 52–69% which is expected to reach up to 80% by 2030.

There are four main types of electrolyzers, that is, alkaline electrolyzer (AE), polymer electrolyte membrane electrolyzer (PEME), solid oxide electrolyzer (SOE), and anion exchange membrane electrolyzer (AEME). Due to the various types of electrolyte materials used, these electrolyzers operate slightly differently. AE is a proven and well-developed technology. PEME are commercially available, and their acceptance is increasing due to their smaller footprint. SOEs and AEME are in the developing stages. AE, PEME, and AEME operate at low temperatures (50–80°C) but SOEs operate at high temperatures (800°C). Depending on the electricity source used, electrolysis can yield absolute green hydrogen. The basic materials and state-of-the-art operating conditions of the three types of electrolyzers are shown in Table 1.1.4.

1.1.4.7 Sustainable hydrogen production methods

The current hydrogen production methods based on fossil fuels are responsible for the emission of around 830 Mt of CO_2 annually. For sustainable

Table 1.1.4 Comparison of different types of electrolyzers.

Attribute	Electrolyzer type			
	Alkaline electrolyzer	PEM electrolyzer	SOEC electrolyzer	AEM electrolyzer
Status	Well established	Established	Developing	Developing
Electrolyte	KOH/NaOH	Polymer	Y_2O_3 doped ZrO_2	Polymer
Cell temperature	60–80°C	50–80°C	700–800°C	60–80°C
Stack pressure	<30 bar	30–80 bar	–	–
Electrical efficiency (%)	63–70	56–60	55–70	74–80
System lifetime	20–30 years	10–20 years	–	–
CAPEX (USD/kW)	500–1400	1100–1800	2800–5500	–

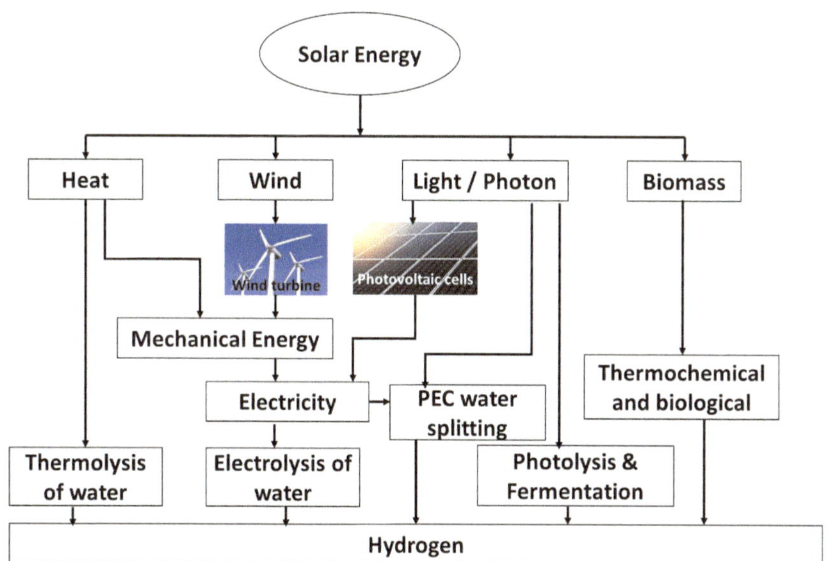

Figure 1.1.10 Sustainable hydrogen production methods.

hydrogen production, the energy and feedstocks used ought to be renewable in nature. Also, the method of hydrogen production must be cost-effective [14].

The best approach would be to produce hydrogen utilizing sustainable energy sources like solar and wind. Solar light (photon) and heat (thermal) both can be used directly or indirectly to produce hydrogen from water using various processes such as: electrolysis, thermolysis, photolysis, thermochemical, and biological conversion (Fig. 1.1.10).

Electrolytic hydrogen generation by splitting water using solar or wind-generated electricity is the only green and established technology to produce hydrogen in large quantities without any GHG emission.

Hydrogen generation using solar thermal energy could possibly be the most efficient solar path. But this technology is in the development stage presently.

Photo–electro–chemical (PEC), photolysis and fermentation are promising technologies and capable of splitting water for hydrogen production in a single step instead of two step process involved in a typical Solar Photovoltaic (SPV) system where electricity generated by SPV is used to split water using an electrolyzer. Organisms containing hydrogenase enzyme (e.g., cyanobacteria and green algae) can be used to produce hydrogen without any input or output of carbon-based molecules. All of these technologies are in the development stage.

The solar energy stored in biomass could be converted to hydrogen gas using various processes such as: gasification, pyrolysis, and fermentation. These processes could also be used to produce hydrogen from waste materials (e.g., sewage, cellulosic biomass).

All hydrogen production processes involve the consumption of water in varying quantities. The maximum theoretical water consumption would be about 9 kg of water for producing 1 kg of hydrogen through electrolysis process. However, it may be feasible to use brackish water for hydrogen production through electrolysis route. Therefore, requirement of water for hydrogen production will have to be considered while developing any project for hydrogen production.

The cost of hydrogen varies considerably depending on the process, by which it is produced, purity of hydrogen, capacity of the production facility, and transportation cost which primarily depends on the distance between the point of production and delivery apart from other factors.

1.1.5 Overview on hydrogen storage methods

Since hydrogen has a low density at room temperature, it releases little energy per unit volume for on-board uses. This is the major issue with hydrogen gas. Therefore, a proper hydrogen storage system is needed to improve its energy density. The hydrogen storage systems needed should have some important properties and selection criteria such as, high volumetric and gravimetric hydrogen capacity, operation in ambient condition, fast refueling, minimum energy loss during operation and low cost. United States Department of Energy (US DOE) has set a technical performance target (Table 1.1.5) for hydrogen storage systems for onboard light-duty vehicles. Currently, available storage options are discussed in the following sections.

1.1.5.1 Compressed storage of hydrogen

Physically, hydrogen can be stored in liquid or gas form. It can also be stored within solids (absorption) or on the surfaces of solids (adsorption). Gaseous storage of hydrogen requires high-pressure to increase the storage density. A compressor for raising pressure and pressure vessel (tanks) for storage of compressed hydrogen are needed. Hydrogen tanks for storage of compressed hydrogen at 350 bar and 700 bar pressure are used globally. Compressed gas storage is the well-established technology and available storage tanks can be classified based on their types. Table 1.1.6 provides features of each type of

Table 1.1.5 US States Department of Energy hydrogen storage target for on-board vehicles.

Storage parameter	Units	2025 target	Ultimate target
System gravimetric capacity	kWh/kg (kgH$_2$/kg system)	1.8	2.2
System volumetric capacity	kWh/L (kgH$_2$/L system)	1.3	1.7
Storage system cost	$/kWh net ($/kgH$_2$)	9	8
Fuel cost	$/gge at pump	4	4
Operating ambient temperature	°C	(−) 40/60	(−) 40/60
Operational cycle life	Cycles	1500	1500
Min delivery pressure from storage system	Bar	5	5
Max delivery pressure from storage system	Bar	12	12
System fill time during charging	Min	3–5	3–5

Table 1.1.6 Types of hydrogen storage tanks for compressed storage purpose.

Type	Material	Maximum pressure (bar)	Cost (US$/kg)	Gravimetric density (wt%)
I	Metals (steel–aluminum)	200	83	1.7
II	Metal tank with composite overwrap	200	86	2.1
III	Metal liner with full composite overwrap	700	700	4.2
IV	All composite (carbon fiber)	700	633	5.7

hydrogen storage tank. Type III and IV pressure vessels are suitable for on-board hydrogen storage in HFVs. In type IV cylinders (Fig. 1.1.11), expensive carbon-fiber composite material is used to provide strength to withstand the high pressure of the hydrogen gas. These cylinders are 70% lighter than steel cylinders, safe, and durable with a lifespan of around 20 years. However, use of carbon fiber composites increases the cost of the hydrogen storage tanks.

1.1.5.2 Hydrogen storage in liquid state

Cryogenic temperature is needed to store hydrogen in liquid state because of very low boiling point of hydrogen at atmospheric pressure (−252.8°C). Currently, the liquefaction of hydrogen is performed using the Claude

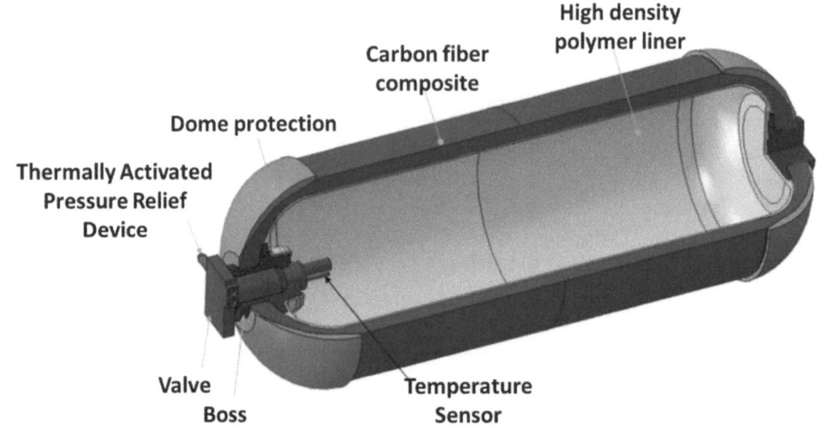

Figure 1.1.11 Cross-section view of type IV hydrogen storage tank [21].

process. This process of producing liquid hydrogen (H_2) is energy intensive. Theoretically, the minimum energy requirement for liquefaction of hydrogen depends on initial hydrogen pressure, the temperature difference between atmosphere and liquid hydrogen and lastly, the rate of ortho–para hydrogen conversion. It has been estimated that for 0.1 MPa hydrogen feed, around 3.92 kWh/kg of H_2 is consumed. Insulated cylinders with safety features have been developed to store hydrogen in liquid state. BMW developed and used their liquid storage tanks with 170 L liquid hydrogen capacity having bi-layered highly insulated tank in BMW Hydrogen 7 car. Japan has also developed a liquid hydrogen storage site in Kobe port where hydrogen is liquefied at −253°C and stored at −162°C. Liquid hydrogen is also used in spacecraft as a fuel. Loss of hydrogen gas due to boiling off is a crucial issue in liquid state hydrogen storage method. Table 1.1.7 shows the materials suitable for liquid hydrogen storage and handling.

1.1.5.3 Hydrogen storage in solid materials

Several solid-state hydrogen storage materials were proposed in recent years that can take up and release hydrogen at a specific temperature and pressure. The operational condition of such hydrogen storage materials should match with the operation conditions of the fuel cells, and it is always targeted to be operated as per the given condition by US DOE (Table 1.1.5). The solid-state hydrogen storage materials can take up hydrogen in terms of the following two mechanisms:

Table 1.1.7 Materials suitable for liquid hydrogen [24].

Material	Alloy grade	Remarks	Cost range
Aluminum alloys	2029 2219	Suitable for aerospace application	Medium–low
Titanium alloys	–	Elongation, toughness, and fracture decreases at cryogenic temperature	high
Copper alloys	–	Highly ductile and toughness at cryogenic temperature	Medium–high
Austenitic stainless steels	304	Susceptible to hydrogen embrittlement	Low
	310	Not susceptible to hydrogen embrittlement	
	316	Not susceptible to hydrogen embrittlement	
	316L	Suitable for marine environment	
	321	Highly resistant to corrosion	

(a) Physisorption: This means physical adsorption of hydrogen on the surface of the material. In this process, hydrogen molecules will be bonded by weak van der Waals forces of the material. Physisorption of hydrogen is mainly studied on materials with the higher surface area such as, for example, nanocarbons, metal organic frameworks (MOF) and polymers, etc.

Physisorption requires insulated cryo-vessels with less heat management as hydrogen remained in its molecule form and there is no enthalpy change between the loaded and empty forms of the storage. The issue is mostly with providing enough hydrogen bonding sites per volume in light carrier materials. Also, due to the lower kJ/mol range (5–8 kJ/mol H_2) of the physisorption interaction between the hydrogen molecule and the surface, working at very low temperatures may be necessary. Hence, liquid nitrogen temperature has been used to study the majority of physisorption systems (77 K).

(b) Chemisorption: This means the chemical absorption of hydrogen into the materials crystal or chemical structure. Hydrogen molecules split off into hydrogen atoms during this process, and they chemically bonded with the atoms of the host materials. Various metals, metal alloys, and chemical compounds are tested for chemisorption of hydrogen. The main advantage of storing hydrogen in chemisorbed form is the high volumetric storage density. However, this process requires a thermal management system to adsorb and release the hydrogen from the storage system. Most of the chemical storage methods are found to be either energy intensive or nonreversible in nature with few exceptions. Operating conditions (temperatures and pressures) of these materials could be in wide range depending on the physical and chemical properties of the materials. Metal hydrides such as $LaNi_5$, TiFe, etc. have shown reversibility at moderate temperature and pressure. R&D on metal hydride storage materials has been carried out extensively and has led to improvement in the hydrogen storage capacity of metal hydrides to about ~2.5 wt%. Complex hydrides (alanates, boranes, etc.) are other categories of solid storage materials which can store high amount (5–18 wt%) of hydrogen but are nonreversible in nature and requires high temperature (100–1000°C) to adsorb and desorb hydrogen.

1.1.5.4 Hydrogen storage for stationary applications

High gravimetric hydrogen storage capacity is essential for automotive and mobile applications as there is weight constrain in a moving vehicle. However, for stationary applications, the volumetric hydrogen storage capacity is the important parameter, as weight will not play any important role in a stationary energy storage system. Type I or II cylinders are generally used for stationary hydrogen storage applications.

Large underground salt caverns, domes, depleted oil, or gas cavities can be used for large scale hydrogen storage (Fig. 1.1.12). Generally, hydrogen pressure up to 50 bar can be achieved in these underground hydrogen storage methods.

1.1.6 Overview on hydrogen supply and delivery methods

An extensive hydrogen transport and distribution infrastructure must be developed for the successful implementation of hydrogen economy. Currently, hydrogen is transported and delivered using trailers in terms of gaseous compressed hydrogen and liquid hydrogen. In some cases, pipelines are also used for delivery of gaseous hydrogen. As of now, only concept studies have been developed for maritime hydrogen transport.

As far as the technical and economic competitiveness of these transport methods concerns, it will depend on the amount of hydrogen to be

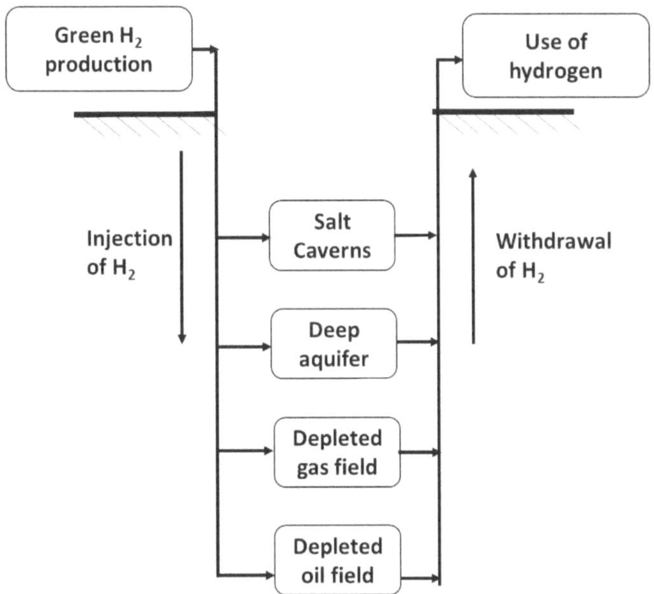

Figure 1.1.12 Underground hydrogen storage possibilities. *(From: Tarkowski R et al., RSER [22]).*

transported and delivery distances. For the transport of compressed gaseous hydrogen, the cost of compression has to be considered. Similarly, for liquid hydrogen, costs of the liquefaction plant should be taken into consideration. However, hydrogen compression is less energy intensive than liquefaction. It has been estimated that around 10–13% of the LHV energy content of hydrogen is consumed during liquefaction process. On the other hand, compression of hydrogen will be dependent on the pressure ratio of the hydrogen. Generally, current alkaline electrolyzers deliver hydrogen at 15–30 bar. Therefore, amount of energy required for compressing will be comparatively less. Currently, seamless stainless steel (SS) pressure vessels are commonly used for hydrogen transportation for short distances (<200 km) and smaller hydrogen amount (100–500 kg). Pressure vessels, like, cylinders, manifolded cylinder bullets, and tube trailers are used for this type of applications. Liquid hydrogen (H$_2$) is also transported in cylindrical super-insulated cryogenic vessels using a truck. H$_2$ transportation involves higher operating expenses (OPEX). However, the deciding factor between H$_2$ and compressed gaseous hydrogen transport will be the distance of transportation. For lower distance compressed gaseous hydrogen transport will be preferred whereas for longer distance H$_2$ transportation will be preferred.

In pipelines hydrogen can be delivered at a pressure of 10–20 bar and it has been estimated that pressures up to 100 bar can also be achieved. In the refineries and chemical industries, pipelines for supply of hydrogen have been used for more than 50 years and about 16,000 km of hydrogen pipelines are already in use. It is also envisaged that already existing NG pipelines could be successfully used for the transportation of hydrogen. However, it must be noted that the pipelines used for hydrogen transportation must be made of materials of high quality considering the physicochemical properties of hydrogen gas at high pressure. Embrittlement of pipelines, valves, manifolds, and compressor could be major factors during high-pressure hydrogen transportation and delivery. Another factor that may concern is the pressure during transportation of hydrogen in a pipeline. To ensure constant pressure throughout the pipeline, optimum design of the pipeline layout and intermittent compression could be a solution. However, it may increase the cost of hydrogen transportation. For a shorter distance and large quantity of hydrogen, pipelines would be very economic as the capital expense (CAPEX) and operational expenses (OPEX) would be less. However, with increases in distance, the CAPEX value for pipelines will increase rapidly, and the economic feasibility may depend on the amount of hydrogen. For large amounts of hydrogen, pipelines, are an economically favorable option.

Hydrogen transportation and delivery using existing NG pipelines could be a better solution. However, current condition of the existing pipelines must be evaluated in case-by-case method. As far as the mixing of hydrogen with NG streamline is concern, it has been assessed that it may be possible to add up to 30 vol% of hydrogen with NG in the existing steel pipelines, without any modifications.

Maritime hydrogen transport is new concept and is currently under consideration. In Europe and Japan many studies have been performed to validate the maritime hydrogen transport method. In these studies, maritime hydrogen vessels having various designs and different amount of hydrogen storage capacities (3600, 24,000, 50,000, and 100,000 Nm^3) have been performed. A detailed review on maritime transport studies have been done by Ustolin et al.

1.1.7 Overview on application of hydrogen

Hydrogen has been mostly used in industrial sectors as feedstocks for decades. It has been primarily used in chemical and petrochemical

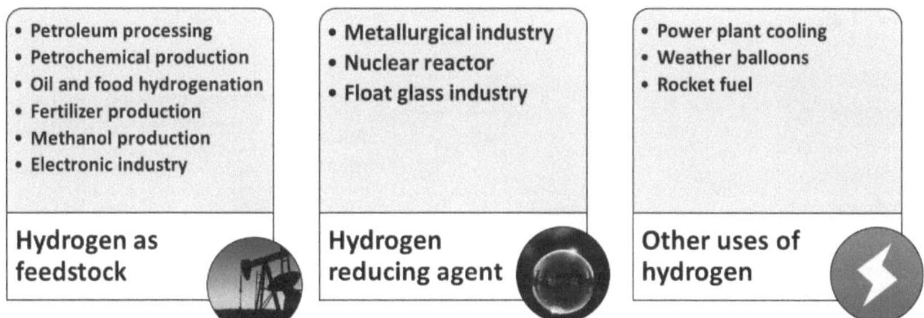

Figure 1.1.13 Use of hydrogen in different industrial sectors.

industry for the production of various chemicals and petroleum products. In petrochemical industries, hydrogen is used mainly in two processes known as hydro-cracking and hydro-treating. In hydrocracking, the larger molecules of crude oil break down to smaller molecules whereas in hydrotreating, hydrogen is used for treatment of petrochemicals, for example, sulfur removal from the oil. In the fertilizer industries, ammonia is produced using hydrogen via Haber–Bosch Process. The produced ammonia is further used for the production of various fertilizers such as urea, ammonium nitrate, etc. In addition, hydrogen is also used for as a cooling agent in thermal power plants, as a reducing agent in float glass production, as a protective gas for development of semiconductors, for welding and cutting and for hydrogenation in the food industry as well as rocket fuel. Fig. 1.1.13 shows the different industrial areas in which hydrogen is used. Ramachandran et al. [1] have reviewed the use of hydrogen in industrial sectors. Hydrogen could also be used to decarbonize the hard-to-abate industrial sectors such as the manufacturing of steel and cement, etc.

Hydrogen could be an option for energy storage, or a can be used as for complementary electricity generation in remote areas. In general, diesel generators or batteries are used for this type of application. It can especially be useful for storing energy for areas where RE resources are available. Standalone microgrid using RE electricity at the daytime, electrolyzer for hydrogen generation and later electricity generation using fuel cell could be an interesting options. Hydrogen could be stored for a long time and used whenever needed. Various demonstration projects have been installed throughout the world on RE-hydrogen system and feasibility studies were also reported.

Hydrogen is also considered as a better option for transport applications. It is believed that battery-powered EVs are best for short commutes and city

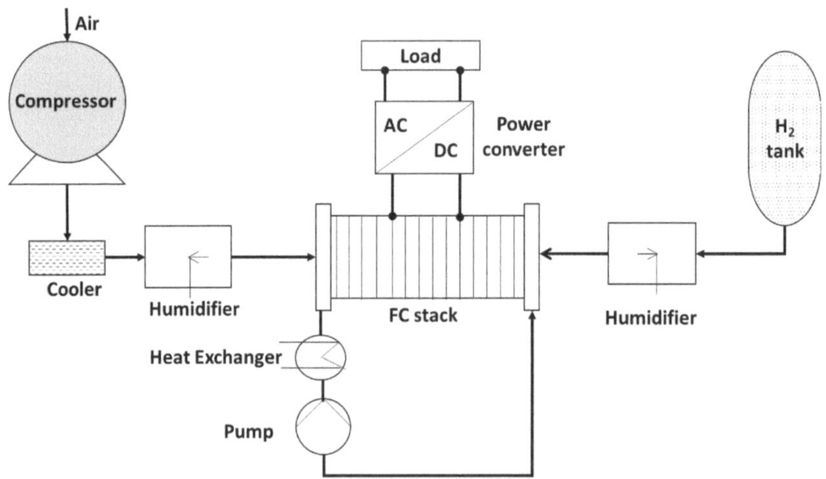

Figure 1.1.14 Components of a typical fuel cell energy generation system [23].

centers. However, lengthier trips can be difficult due to charging time. It is neither practicable, cost–effective, or ecologically friendly to use long haul trucks powered by battery. Although they do have a significant role to play, EVs are not suited for longer trips. The ability to significantly reduce CO_2 emissions from the air, the sea, and the road lies with hydrogen. With good cause, hydrogen is being considered as the better option for trucks, buses, trains, airplanes, and massive container ships. It takes only a few minutes to refuel them and installing hydrogen fuel cells might enhance the air quality as the only emissions would be water.

1.1.7.1 Role of fuel cells in hydrogen economy

A fuel cell is an electrochemical device that converts chemical energy from a fuel (usually hydrogen) and an oxidizing agent (often oxygen) into electrical energy using two redox reactions. Fuel cells, in contrast to most batteries, need a steady supply of fuel and oxygen (or air) to maintain the chemical reaction. On the other hand, a battery normally gets its chemical energy from materials that are already present there. Fuel cells can continually generate power if fuel and oxygen are available. Fig. 1.1.14 shows a typical fuel cell system for power generation application. Fuel cells come in a variety of forms, but they all share the same three essential parts: an anode, a cathode, and an electrolyte [16]. Ions (mostly proton) can move between the fuel cell's two sides through the electrolyte. At the anode, fuel (hydrogen) is subjected to oxidation reactions that generate ions (usually protons and electrons). Ions (protons) go from the anode to the cathode through the electrolyte. Whereas electrons go across an external circuit from the anode to the cathode and as a result direct current (DC) electricity is generated. At the cathode, protons, electrons, and oxygen react to produce water vapor. Depending on the fuels used, fuel cells also produce heat and very small quantities of nitrogen dioxide and other pollutants in addition to water vapor. Fuel cells typically have an energy efficiency of 40–60%, however, by utilizing waste heat in a cogeneration system, it is feasible to get efficiencies of up to 85%.

There is different type of fuel cell are available and classified depending on the type of electrolyte they use:

(a) Alkaline fuel cells (AFC)
(b) Proton exchange membrane fuel cells (PEMFC)
(c) Direct methanol fuel cells (DMFC)
(d) Molten carbonate fuel cells (MCFC)
(e) Solid oxide fuel cells (SOFC)
(f) Phosphoric acid fuel cells (PAFC)

For each type of these FCs, the electro-chemical reactions, the catalysts, the operational temperature, the fuel, and other characteristics are different (Table 1.1.8). These qualities also influence the best use cases for these cells. Each of them has unique benefits, restrictions, and possible applications. Few of these fuel cells can even operate without hydrogen as fuel.

Although it may seem evident that fuel cells may function without hydrogen, a hydrogen society without fuel cells does not make much sense in the long run given the high electricity-to-heat ratio and high total conversion efficiency of hydrogen-powered fuel cells. A fuel cell is regarded to be a better solution to fulfil the growing electricity demand in overall energy consumption. It is thought that FCs can be used in numerous applications.

PEMFCs are currently the best options for supplying electricity to fuel cell vehicles (FCVs) because they operate at low temperatures (60–90°C), have quick start-up times, are highly efficient, have good power densities, and have good weight-to-power ratios. PEMFCs also have excellent load change behavior, which means they can quickly adjust their output of electricity in response to demand. The fact that PEMFC can also function with air is an obvious advantage for their use in the transportation industry. MCFCs and SOFCs are not appropriate for use in automobiles due to their high operating temperatures and protracted start-up times. Today's test vehicles use hydrogen on par with high-end conventional vehicles. Evidently, HICE

Table 1.1.8 Different type of fuel cells and their features.

Fuel cell	Electrolyte	Temperature (°C)	Efficiency (%)	Typical electrical power (kW)	Application
Alkaline fuel cell (AFC)	KOH	60–90	45–60	>20	Submarines, spacecrafts
Phosphoric acid fuel cell (PAFC)	Immobilized liquid phosphoric acid	~200	35–40	>50	Stationary
Polymer electrolyte membrane fuel cell (PEMFC)	Proton conducting polymer	~80	40–60	<250	Vehicles, small stationary
Direct methanol fuel cell (DMFC)	Proton conducting polymer	60–130	~40	<1	Portable
Molten carbonate fuel cell (MCFC)	Immobilized mixture of sodium and potassium carbonate	650	45–60	>200	Stationary
Solid oxide fuel cell (SOFC)	Doped zirconia	700–1000	50–65	<200	Stationary

vehicles do not have a clear advantage over FCVs in terms of hydrogen consumption. However, the HICE has the option of functioning in a dual-fuel mode (diesel-hydrogen; CNG-hydrogen, etc.), which is a definite benefit with regard to the establishment of a first network of refueling stations.

Along with the PEMFC being created for vehicle propulsion, SOFC are also being studied for auxiliary power unit (APU) applications in cars because they run at extremely high temperatures and need lengthy start-up periods. In APU applications, the fuel cell can be started well in advance of an expected stop or left running for the majority of the time. The main advantages of SOFCs are their capability of tolerate hydrocarbon fuels. The SOFC's heat can be employed in the air conditioner's cooling or heating functions.

PEMFCs, MCFCs, and SOFCs are types of stationary fuel cells that can be utilized for both distributed and centralized energy generation. For cogeneration and combined-cycle power plants, the high temperatures of the exhaust gases generated by MCFCs and SOFCs allow for up to 90% system efficiency. Because of their high working temperatures, MCFCs and SOFCs have another benefit in that they can be powered directly by fossil fuels (such as NG) and do not need external reformers.

Fuel cells that may provide a few watts or 10 kW power can be adopted for portable applications. The portable fuel cells can be used for a variety of things such as charging small electronic devices like cameras, mobile phones, and laptops, etc. These FCs can be useful for traveling as well as for exploring activities such as for hikers, bikers, etc. Given that PEMFCs have issues storing hydrogen and that MCFCs and SOFCs operate at temperatures that are too high for portable devices, DMFCs are currently the best options for portable applications. Methanol can also be conveniently transported. However, DMFCs are not appropriate for mobile or fixed use due to their low efficiency (15–30%) and low power density. Further research is needed to reduce prices and enhance the peripheral systems and dependability of DMFCs, even though the technology is nearly ready for market launch. As hydrogen storage has less advantage over batteries in terms of energy density, PEMFCs are only marginally advantageous for usage in portable devices.

1.1.8 Hydrogen safety

In comparison to other fuels, hydrogen has by far the lowest ignition energy (see Table 1.1.1). Additionally, hydrogen shows the greatest spread in mixes with air in the explosion or detonation range. As a result, the creation of hydrogen and air mixes in uncontrolled conditions must be strictly avoided as there is the considerable potential of catastrophic events, mostly due to the low ignition energy and the broad detonation range. Due to these characteristics and the fact that hydrogen is 15 times lighter than air, an adjusted safety strategy is required to reap the benefits of hydrogen without subjecting people to unwarranted hazards. Due to low molecular size, ease of leakage as a gas, buoyancy, and metal embrittlement properties, hydrogen storage and use provide unique challenges that must be taken into account in order to ensure safe operation. Liquid hydrogen presents additional challenges due to its low temperature and higher energy density. However, hydrogen has no hazard classification in terms of toxicity or inherent reactivity. Prevention methods to avoid hydrogen hazards include:

(a) Purging of gas lines or storage chambers by inert gas such as N_2 before transferring hydrogen.

(b) Ignition sources should be avoided.

(c) Materials with proper codes and standards should be used for hydrogen handling.

(d) Hydrogen sensors and fire alarms must be installed.

(e) Proper ventilation is necessary.

1.1.9 International strategies on hydrogen

Due to the number of advantages of hydrogen over fossil fuels, many governments are giving priority to use of hydrogen in their policies. Global policymaking is becoming increasingly influenced by the climate change issues, as well as local air quality, supply security, and energy dependence. For these reasons, the European Commission's energy system has identified the hydrogen economy as one of its long-term priorities. In addition to Europe, the hydrogen vision is also being considered in the United States and Japan, and various national hydrogen energy initiatives have already been formed or are in the process of being developed [17]. As a result, both nationally and internationally, funding for fuel cell and hydrogen research has lately grown. It takes more than just establishing the correct payback time to put advanced, highly innovative technology, like hydrogen applications, into practice. A transition to a sustainable energy system necessitates adjustments at many societal and economic levels. To aid in the efficient and advantageous integration of hydrogen into their energy systems, many industrialized nations and regions are now developing their own hydrogen roadmaps.

Fig. 1.1.15 shows different countries and their status of hydrogen strategy. Twelve nations, including the EU, have released their national hydrogen policies to date; nine of those publications came out just last year. There is a definite acceleration in government interest, supported maybe by recent COP26 acting as a stimulant. Other 20 additional nations are presently formulating their policies, several of which aim to publish within the next few years. A few nations have had a significant influence with their hydrogen policies. Japan's early commitment sparked interest in the Asian-Pacific area, leading to the publication of own policies by South Korea and Australia shortly after. Germany was a pioneer in Europe and assisted in advancing the EU hydrogen policy. Chile has made significant progress in Latin America, and several of its neighbors are also implementing their own methods.

The policies developed by each of these nations will play a crucial role in determining how those nations might adopt hydrogen in their energy transitions. Development of hydrogen economy could take a variety of different paths depending on the use of hydrogen in different sectors and hydrogen production methods, and it might also make use of different policy instruments to promote hydrogen uptake. The priorities for hydrogen production methods and use of hydrogen in different sectors differ greatly due to various local demands in different nations.

Japan has a particular focus on the power sector to produce electricity from hydrogen. The emphasis is on decentralized small-scale hydrogen fuel cell power generation for domestic use. South Korea has set goals to use hydrogen for central power generation as well as the usage of fuel cells by decentralized systems in homes and buildings. Both Japan and South Korea are also focusing on the transportation sector, particularly the deployment of FCVs on domestic markets.

Australia has developed a strategy with an emphasis on hydrogen production and export while also considering its own use in transportation, with a focus on heavy-duty and long-haul vehicles. Australia has also target to produce green ammonia on a large-scale.

A roadmap for fuel cells and hydrogen is being created in the EU and a pan-European policy has been developed. EU has a target to develop a fully integrated hydrogen economy based on nuclear power and RE sources. Europe also aims to raise the share of RE in most of its member countries. Transportation and decarbonization of industrial sectors are the main aim for employing green hydrogen in Europe. Germany priorities on the chemical, petrochemical, and steel industries along with heavy-duty

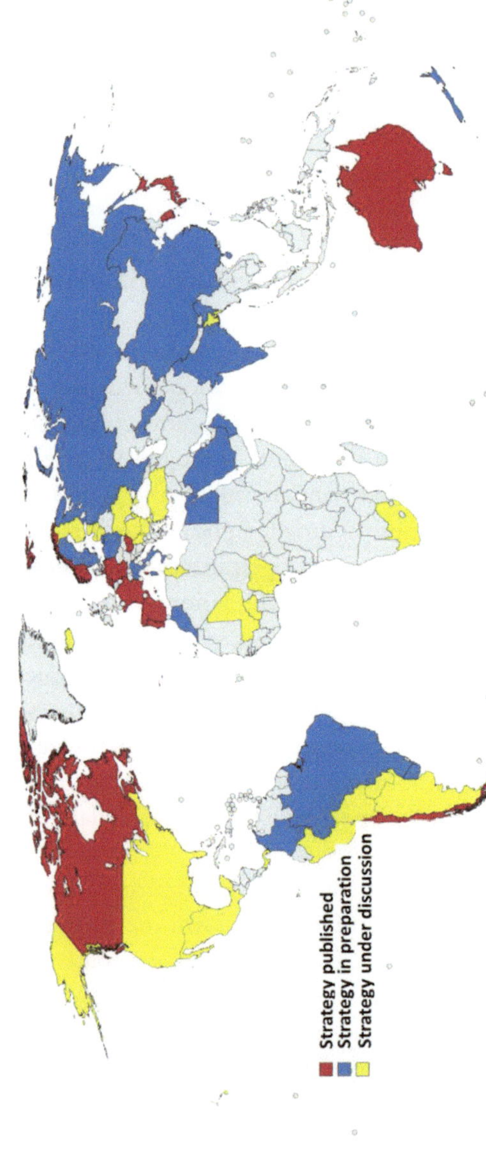

Figure 1.1.15 Different countries as per their status of hydrogen strategy. *(From: World Energy Council [18]).*

vehicles. France is concentrating on decarbonization of industrial sectors. The Netherlands is thinking of creating a hydrogen infrastructure for various sectors.

The United States and Canada have also established frameworks for hydrogen and fuel cell research and development support as well as long-term technological roadmaps. The National Hydrogen Energy Roadmap (2002) and Hydrogen Posture Plan (2004) are developed by the US DOE. India has published one of its first phases of policy and the next phases will be published soon [10].

The industrial sector and transport sector are the focus of most of the strategies. In the early stages of the market's development, HICE might be useful; later, FCV will unquestionably predominate. Most strategies says that the proportion of RE used to produce hydrogen will increase significantly after 2030. However, from a long-term perspective, every roadmap predicts a CO_2-lean production of hydrogen.

1.1.10 Summary

The world is witnessing a new era of energy transition which aims towards decarbonization of the energy sector. This energy transition is getting accelerated by several factors which include increasing energy demand, depleting fossil fuel reserves, environmental and climate change concerns, and also government regulations and actions to minimize GHG emissions to contain the global average temperature rise. It is believed that hydrogen can play an important role in this decarbonization process. The energy economy encompassing hydrogen as an energy carrier is named as "hydrogen economy" comprising four important components associated with it: hydrogen production, its storage, transportation, and utilization.

Hydrogen is not available in molecular form on the earth and the major sources from which it can be extracted are fossil fuels, water, and biomass. Currently, the most of hydrogen in the world is produced by reforming fossil fuels. Steam methane reforming, oil reforming, and gasification are the most extensively used techniques for hydrogen production from fossil fuels. However, with ever-increasing penetration of the RE based electricity from resources such as solar, wind, geothermal, etc. in the energy system at the global level, it is envisaged that green electricity can be used to produce hydrogen by splitting water through electrolysis process. As the RE resources are time and location specific, it is being increasingly believed that green hydrogen produced using green electricity will act as energy storage medium for balancing the grid as it could be used to generate electricity and heat wherever and whenever needed to meet the shortfall in demand and supply. This will not only offer a solution to the energy storage issue that is so vital for energy systems dominated by RE systems but will also contribute to the minimization of the green power curtailment by utilizing curtailed electricity for production of hydrogen that can be stored and utilized later. A hydrogen economy can play a very critical role in securing the energy supply for many nations that are heavily dependent on imported energy sources but are well endowed with RE resources.

Hydrogen should preferably be produced locally in view of the high costs of storage and transportation due to its very low gravimetric density. Nevertheless, green hydrogen can be produced economically at locations having rich RE resources and can be transported securely to consuming centers using a network of gas pipelines or other ways. Other delivery methods such as tube trailers, liquid hydrogen containers and as chemicals (NH_3, CH_3OH, etc.) can also be used depending on the amount and distances over which hydrogen needs to be delivered. For on-board vehicular application, hydrogen can be stored in composite cylinders (type-IV for cars, and type-III/IV for buses, trucks, etc.). Extensive research and development efforts are being carried out in the field of condensed hydrogen storage methods such as in carbons, zeolites, MOFs, metal hydrides, complex hydrides, and others. For two-wheeler applications, the metal hydride storage canisters/tanks option may emerge out as a better option among the others. Such canisters could be swapped to minimize hydrogen refueling time.

As of now, primary applications of hydrogen are in the industrial processes which include petroleum refineries; the production of ammonia, methanol, flat glass, vegetable oils, semiconductors; metal extraction, power plant cooling, etc. Among these industries, petroleum refineries and ammonia synthesis units are the major consumers of hydrogen, and their hydrogen requirement is met by reformation of fossil fuels presently. Therefore, the first effort for using green hydrogen could be made in the industrial processes. Other sectors such as household and transport sectors could also be decarbonized using green hydrogen. Primarily injection of hydrogen in the NG pipelines could be adopted to decarbonize applications like cooking and heating, where predominantly NG or liquefied petroleum gas are used currently. Hydrogen could further be used in propulsion of vehicles based on both IC engine as well as fuel cell technologies. Green hydrogen and fuel cell based micro-grid can primarily be used for supplying electricity to the remote areas which are without grid supplied based electricity. For meeting higher demand of electricity, extensive R&D efforts are going on to develop gas turbines that can utilize hydrogen as a fuel.

With a target of keeping the 1.5°C temperature rise within the reach by 2050 or even earlier, many nations have already developed strategies to decarbonize their economies. This is resulting in the development of a conducive policy regime by several countries that would pave the way for development of the hydrogen economy.

References

[1] R. Ramachandran, R.K. Menon, An overview of industrial uses of hydrogen, Int. J. Hydrogen Energy 23 (7) (1998) 593–598.

[2] T.C.A. Avni, A. Gupta, M, Kugelman, L, Powell, A.K., Saxena, India's energy future designing and implementing a sustainable power mix, The National Bureau of Asian Research the National Bureau of Asian Research, NBR Special Report #94, 2021.

[3] M. Ball, M. Wietschel, The Hydrogen Economy: Opportunities and Challenges, Cambridge University Press, Cambridge, UK, 2009.

[4] J. Bockris, The hydrogen economy: its history, Int. J. Hydrog. Energy 38 (2013) 2579–2588.

[5] U. Bossel, Hydrogen economy: what future? in: Current Trends and Future Developments on (Bio-) Membranes New Perspectives on Hydrogen Production, Separation, and Utilization, Elsevier, 2020, pp. 421–451.

[6] G. Crabtree, M. Dresselhaus, M. Buchanan, The hydrogen economy, Phys. Today (2004) 39–44.

[7] EIA, Annual Energy Outlook 2013 With Projection to 2040, US Energy Information Administration (EIA), 2013.

[8] M. Green, Hydrogen Safety Issues Compared to Safety Issues with Methane and Propane, Lawrence Berkeley National Laboratory, 2005.

[9] S. Hossain Md, B. Li, S. Chakraborty, R. Hossain Md, T. Rahman Md, A comparative analysis on China's energy issues and CO_2 emissions in global perspectives, Sustain. Energy 3 (1) (2015) 1–8.

[10] J. Manna, P. Jha, R. Sarkhel, C. Banerjee, A.K. Tripathi, M.R. Nouni, Opportunities for green hydrogen production in petroleum refining and ammonia synthesis industries in India, Int. J. Hydrog. Energy 46 (77) (2021) 38212–38231.

[11] G. Marbán, T. Valdés-Solís, Towards the hydrogen economy? Int. J. Hydrog. Energy 32 (2007) 1625–1637.

[12] S. Mekhilef, R. Saidur, A. Safari, Comparative study of different fuel cell technologies, Renew. Sustain. Energy Rev. 16 (2012) 981–989.

[13] National Research Council and National Academy of Engineering, The Hydrogen Economy: Opportunities, Costs, Barriers, and R&D Needs, The National Academies Press, Washington, DC, 2004.

[14] S.S. Penner, Steps toward the hydrogen economy, Energy 31 (2006) 33–43.

[15] Statista, Global primary energy consumption, 2021. Statista, (Accessed 31 January 2023).

[16] C. Stockford, N. Brandon, J. Irvine, T. Mays, I. Metcalfe, D. Book, P. Ekins, A. Kucernak, V. Molkov, R. Steinberger-Wilckens, N. Shah, P. Dodds, C. Dueso, S. Samsatli, C. Thompson, H_2FC SUPERGEN: an overview of the hydrogen and fuel cell research across the UK, Int. J. Hydrog. Energy 40 (15) (2015) 5534–5543.

[17] World Energy Council, International hydrogen strategies, Germany, 2020.

[18] World Energy Council in collaboration with EPRI and PwC, Hydrogen on the horizon: ready, almost set, go?, Working Paper on National Hydrogen Strategies, UK, 2021.

[19] IEA, Global energy review: assessing the effects of economic recoveries on global energy demand and CO_2 emissions in 2021.

[20] I. Staffell, et al., The role of hydrogen and fuel cells in the global energy system, Energy Environ. Sci. 12 (2019) 463.

[21] Fuel Cell Technologies Office, Hydrogen storage, 2017. DOE. https://www.energy.gov/sites/prod/files/2017/03/f34/fcto-h2-storage-fact-sheet.pdf (Accessed 5 January 2023).

[22] R. Tarkowski, Underground hydrogen storage: characteristics and prospects, Renew. Sustain. Energy Rev. 105 (2019) 86–94.

[23] M. Yue, H. Lambert, E. Pahon, R. Roche, S. Jemei, D. Hissel, Hydrogen energy systems: a critical review of technologies, applications, trends and challenges, Renew. Sustain. Energy Rev. 146 (2021) 111180.

[24] F. Ustolin, A. Campari, R. Taccani, An extensive review of liquid hydrogen in transportation with focus on the maritime sector, J. Mar. Sci. Eng. 10 (2022) 1222.

[25] EIA, Monthly energy consumption, http://www.eia.gov/totalenergy/data/monthly/#consumption. (Accessed 10 August 2022).

Potential market failures inhibiting the development of a green hydrogen export industry

Lee V. White[a], David Gourlay[b] and Emma Aisbett[c,]
[a]School of Regulation and Global Governance, Australian National University, Canberra, ACT, Australia
[b]Zero Carbon Energy for the Asia-Pacific Grand Challenge, Australian National University, Canberra, ACT, Australia
[c]College of Law, Australian National University, Canberra, ACT, Australia
*Authors are listed in reverse alphabetical in recognition of even distribution of contributions to the work.

1.2.1 Introduction

Hydrogen is once again emerging as a potent energy carrier and could be an important component of a transition to zero-emissions energy systems. While hydrogen can be produced in a number of ways, "green" hydrogen—that is, hydrogen produced from renewable energy sources—has the greatest ability to contribute to the goals of climate change mitigation. Green hydrogen has the potential to support the export of renewable energy from renewable-rich nations to energy-resource-constrained nations such as Japan and Germany. Although relative endowments of natural resources for green hydrogen production vary substantially across countries, the required resources are more widely distributed than fossil fuel resources. Green hydrogen thus has geopolitical advantages as an energy export/import allowing diversification and not necessarily reliant on pipeline infrastructure, and could also be a key export diversification strategy for a number of lower and middle-income countries, such as Namibia and Chile, that have limited reserves of fossil fuels for export but have abundant resources to support renewable electricity generation.

The first welfare theorem of neoclassical economics lays out a set of assumptions and shows that when these hold, the market will deliver efficient outcomes in the absence of government intervention. In the case of the hydrogen industry, for example, this would mean an efficient level of investment would arise and efficient levels of domestic and international trade would result. However, these assumptions are often not met in practice. Failures of these assumptions, so-called "market failures," prevent the market from delivering efficient outcomes. Government interventions to ameliorate these

Towards Hydrogen Infrastructure: Advances and Challenges in Preparing for the Hydrogen Economy.
DOI: https://doi.org/10.1016/B978-0-323-95553-9.00008-X

market failures can be efficiency-enhancing and hence welfare improving. Specifically here, addressing the relevant market failures will allow the emerging hydrogen export industry to deliver on its potential to contribute to climate change mitigation and diversification of global energy trade.

This chapter provides an overview of potential market failures in the green hydrogen export sector, informed by expert knowledge, to coalesce information relevant to policymakers shaping industrial policy in the green hydrogen space. It describes the potential for each of these market failures to limit industry development in the context of emerging hydrogen value chains, describes the first-best theoretically informed approach to address each market failure, and discusses emerging and best-practice green industrial policies to address these market failures for hydrogen export in practice. We draw on the list of hydrogen-exporting countries compiled by the International Energy Agency (IEA) [1] in sourcing examples.

1.2.2 Market failures

1.2.2.1 Environmental externalities

The first welfare theorem assumes that all actors in a market completely internalize the full costs and benefits of their actions—including social and environmental costs and benefits. When this is not the case, a market failure referred to as an "externality" exists. Climate change, caused by unabated greenhouse gas emissions, has been described as the biggest market failure ever seen [2]. The costs of climate change are spread across the globe but those actors responsible for greenhouse gas emissions do not pay all of these resulting costs.

Green hydrogen, with its very low life-cycle emissions footprint, is therefore likely to be undervalued in domestic and international markets, and hence underused, compared to fuels such as natural gas and oil that produce greenhouse gas emissions [3]. This extends to undervaluing green hydrogen against hydrogen produced from fossil fuels, which produce greenhouse gas emissions even when combined with carbon capture and storage (CCS); CCS itself presents ongoing challenges and often fails to deliver promised environmental benefits [4]. In scaling up hydrogen export supply chains, emissions externalities could inefficiently divert capital resources away from green hydrogen and towards fossil fuel-based hydrogen. Unpriced or mispriced greenhouse gas emissions could also lead to underinvestment in the renewable electricity generation needed to supply green hydrogen production.

1.2.2.1.1 Addressing the market failure: first-best approach

The theoretically optimal ("first-best") policy to address an externality is to price it such that the actors involved pay for the marginal social/environmental costs of their actions. This is the motivation for "carbon" or "emissions" pricing. Potential green hydrogen exporters, however, face substantial challenges in addressing emissions externalities through pricing. In addition to the usual vested interests and political economy problems that plague emissions pricing initiatives, green hydrogen exporters face jurisdictional problems: they do not have the jurisdiction to mandate emissions pricing in their export markets.

Given these challenges, it is little surprise that in existing hydrogen strategies by exporting nations, environmental externalities are rarely addressed. The exception is Chile, which in its hydrogen strategy notes a plan to establish a public-private roundtable for the purpose of discussing a path toward a carbon price [5].

The Chilean strategy explicitly notes that efforts to establish a carbon price are driven by the need to better reflect the externalities of fossil fuels used in Chile [5]. Chile's approach represents an ambition toward constrained optimal policy, where the market failure is fully addressed within its jurisdictional control. Although not directly able to affect the externality in export markets, Chile's approach has indirect benefits from an export perspective. First, it helps to grow the domestic green hydrogen industry, supporting competitiveness as will be discussed below. Second, it provides Chile with greater moral authority in international fora to push for emissions pricing to be adopted elsewhere—including in potential export markets.

1.2.2.1.2 Addressing the market failure: second-best approaches

Emissions standards represent one second-best approach to ensure that the growth of the hydrogen industry does not exacerbate climate externalities. Standards regarding maximum acceptable embodied emissions in hydrogen, particularly hydrogen called "clean," could support efforts to grow hydrogen as a fuel supporting climate change mitigation. In a scenario that lacks any carbon pricing, regulations and standards regarding counting and constraint of embodied emissions can support the development of a hydrogen industry that does not risk worsening climate outcomes. However, in absence of any policy measures to value green hydrogen over other forms of hydrogen, there is a risk that hydrogen could contribute to emissions rather than reduce them [4].

Subsidies for low-emissions technologies or products represent another second-best way of addressing climate externalities. By making low emissions options relatively more attractive to investors and consumers than higher emissions alternatives, subsidies can partially overcome the externality. However, this can be a costly approach if used as a substitute for carbon pricing [6]. As a result, subsidies are not an option for all governments. Subsidies directly aimed at export industries are even more problematic. From a fiscal perspective, the costs of subsidizing potentially large external markets can be prohibitive, and the benefits are captured in large part by foreign consumers. From a legal perspective, subsidies that target export performance are not consistent with World Trade Organization (WTO) regulations [7]. Again, we should not be surprised that subsidies for low-emissions technologies such as electrolyzers or renewable generation technologies do not explicitly appear in current hydrogen roadmaps by Australia, Canada, or Chile [5,8,9].

1.2.2.2 Asymmetric information

While unbiased, symmetric uncertainty is not a market failure, new industries are typically characterized by asymmetric, biased information. In cases of asymmetric information, one party will have more information about the product than the other, typically with the seller having more information about their product than the buyer.

For green hydrogen to be fully valued on a market (e.g., one underpinned by strong emission regulations and/or carbon pricing), it will be necessary for buyers to be able to ascertain the level of embodied emissions associated with their hydrogen purchases. While there are established methods for producers to calculate embodied emissions content, downstream market participants, including end users, may not have access to or trust in the impartiality and accuracy of this information. As a result, green hydrogen producers may not get the market premiums they otherwise could, and investment in green hydrogen supply chains will be lower than optimal.

1.2.2.2.1 Addressing the market failure

Information failures can be addressed through measures to increase transparency and the quality of information available. Emissions accounting and emissions certification systems represent one such approach. There are not yet international rules or standards in place for hydrogen trade [10,11]. Some certification systems are emerging, such as the CertifHy pilot in Europe and the TÜV SÜD standard, but uncertainty remains around certifications as these systems are still being refined [10,11]. A proliferation of competing certification systems poses the threat of excess regulatory burden for hydrogen producers and exports. Not surprisingly, all hydrogen strategies written by potential exporting nations contain an acknowledgment that it will be important to participate in and support the development of harmonized international standards and certification for green hydrogen [5,8,9]. The need for certification and standards is also recognized in international reviews of the future of hydrogen, such as by the IEA [12].

1.2.2.3 Credit constraints

Problems of asymmetric information are particularly damaging where they interact with another market failure—credit constraints. Industries characterized by a need for large, high-risk investments typically suffer inefficiently low levels of investment due to this market failure [13]. All investment involves a degree of uncertainty, but this is increased for investments in innovative new industries where there is less information to draw on [6], or where market prices do not yet exist [14]. The availability of finance is particularly important at the commercialization stage of a new product or industry to allow its deployment and diffusion [15]. Innovative projects can struggle to secure the necessary capital, or secure it at too high a cost because for innovative activities the lender does not have access to enough information to accurately assess the risk of the project [16]—in other words, the capital constraint is exacerbated by the information asymmetry between the lender and developer.

Until a green hydrogen industry is established, firms will have to make investment decisions based on imperfect foresight, or their best guess as to highly uncertain future market conditions. The IEA has identified mitigating investment risks as one of five key areas to boost momentum in green hydrogen value chains [12]. High capital costs and uncertainty about future technology developments are currently cited as barriers to electrolyzer uptake [17]. Infrastructure for storing, delivering, and exporting hydrogen over long distances is not yet established, representing a major challenge and large upfront capital outlay for the developing hydrogen industry [18]. Uncompressed hydrogen as an energy carrier has too low an energy density to store and transport economically [19] hence hydrogen must first be converted to a more energy-dense form by liquefaction, embedding in a liquid organic hydrogen carrier (LOHC), or conversion to ammonia [12,19,20], all of these conversions rely on capital investment in additional infrastructure [21], and on availability of ships.

1.2.2.3.1 *Addressing the market failure*

Where incomplete information could cause credit constraints, governments can act as a source of finance. Governments can often provide finance with a higher risk tolerance than commercial lenders, which can overcome the reluctance of a commercial lender to finance new activities [22]. Government finance can take a number of forms, such as concessional loans or the guaranteeing of debt. As the provision of finance comes with a degree of risk, governments could elect to take an equity position in green hydrogen investments in order to make a return on the investment if the venture becomes profitable. These types of support policies work by increasing the returns or reducing the risks on an investment (or both) [23].

State investment banks are a vehicle often used to overcome credit constraints and have specifically been used for new low-emissions technologies. For example, the Clean Energy Finance Corporation of Australia delivers a range of investment options for low-emissions technologies, including flexible debt and equity finance [24]. State investment banks are designed to operate like commercial lenders, but can be directed by governments to make investments that target a specific policy outcome and can take on a greater level of risk if desired due to their government backing. A record of investment from a reputable state investment bank can build trust in a technology, which can lead to a "crowding in" of private finance into the industry [15]. State investment banks could be mobilized to provide finance for early green hydrogen production projects and other projects that could establish the green hydrogen export supply chain (such as delivery and export infrastructure, and electrolyzer manufacture).

Hydrogen strategies of potential exporting nations broadly acknowledge the need to support credit and capital access for green hydrogen as an emerging innovative industry. Australia's hydrogen strategy notes the existence of the "Clean Energy Finance Corporation, the Northern Australia Infrastructure Fund, and various state initiatives [that] can provide tailored support for hydrogen-related investments," [8]. However, the strategy itself does not commit to taking any actions to expand access to green hydrogen producers to these initiatives [8]. Chile's hydrogen strategy names the launch of a 50 million USD funding round intended to "support companies and national and international consortiums to invest in scalable and replicable green hydrogen projects in Chile," with the goal of aiding growth in the local industry [5]. Canada very broadly aims to "facilitate co-funding opportunities, leveraging multiple levels of government and the private sector" [9].

1.2.2.4 Knowledge as a public good (R&D)

It is generally considered that the level of investment in research and development (R&D) is below what is needed to maximize societal welfare [25]. This is because knowledge displays a characteristic known as nonrivalry: the use of knowledge by one party does not diminish its value to others. Hence the total societal value of a piece of research is greater than what an individual firm undertaking it could gain [6,13,16,26–28]. This implies that firms' private benefits from R&D are less than the full social benefit, resulting in a lower level than that which would efficiently scale up new industries.

While the technologies to produce hydrogen are approaching commercial readiness, a significant amount of R&D still needs to occur throughout the hydrogen supply chain. Alkaline electrolyzers are currently common but not at large scales [19], while proton exchange membrane electrolyzers may scale better but are less mature and currently more expensive [17,19]. At the same time, the industry needs to take risks and innovate to learn how to produce, transport, and deploy hydrogen at scale. Liquefying hydrogen has not yet been demonstrated at scale [19], and the biggest challenge for LOHC take-up is high conversion costs requiring further research and experience to improve [12]. It is notable that ammonia provides a currently mature transportation medium. A substantial amount of basic research will be required before liquid hydrogen shipping tankers are commercially available; it was only in 2022 that the first shipment of liquid hydrogen occurred as a pilot demonstration [29].

1.2.2.4.1 Addressing the market failure

To address the knowledge as a public good, direct subsidies can be deployed to increase the level of R&D where it is needed. For green hydrogen export, subsidies can be targeted at specific processes in the supply chain that require further R&D, for example, shipping and various other processes involved with the transportation of hydrogen fuels. Direct subsidies could take the form of university funding, research grants, or industry research funding with public dissemination requirements. R&D can further be supported by government support for demonstration projects either financially or by easing coordination for the private sector.

The Australian hydrogen strategy notes Australia's existing committed R&D support for hydrogen is $146 million AUD, through the Australian Research Council, the Commonwealth Scientific and Industrial Research Organisation (CSIRO), the Australian Renewable Energy Agency, the Clean Energy Finance Corporation and the Northern Australia Infrastructure Fund, and notes as a strategic action the need to support innovation and R&D [8]. Chile plans to develop a further roadmap outlining pilots and milestones, which will "include collaboration with the awardee of the Clean Technologies Institute innovation platform with public funding of up to 193 MUSD," with planning to support the growth of green hydrogen [5]. Canada plans to establish dedicated funding for sustained R&D in hydrogen, including leveraging expertise in academia, government labs, and private sector labs [9].

1.2.2.5 Coordination failures (chicken and egg problem)

Coordination failures, or a "chicken and egg" problem, can occur in situations where the profitability of an individual investment depends on simultaneous investment decisions by other actors. This is particularly relevant for new goods or industries where new infrastructure is needed to realize the opportunity [13]. The need to scale up both supply and demand in parallel across jurisdictions adds a high degree of uncertainty to investment in infrastructure. Coordination failures can result in a suboptimal level of investment or, at worst, a sustainable industry failing to develop.

As green hydrogen export is a new industry and will require investment across the supply chain and national borders, there is a strong possibility that coordination failures could emerge. For example, a firm wishing to invest in green hydrogen production to meet export demand will need access to reliable and sufficient renewable electricity capacity, a domestic delivery network and export facilities, and will need to be sure that global supply chains are in place and importers have the required facilities. The firm may not be willing to proceed with its investment if this infrastructure is not in place—however the same is true of investors in other aspects of the supply chain. This uncertainty is increased by the fact there are no global hydrogen supply chains to build on [10]. In addition to infrastructure investment, the chicken and egg problem can also exist in the labor market—workers with the appropriate skills will need to be available as the industry grows, but workers may be unwilling to undertake the required training before the industry is established.

In the case of hydrogen, additional coordination failures are likely to occur in its transportation. Unique properties of hydrogen gas, including its low density and small molecular size, mean that existing piped storage networks and shipping tanks are not suitable for containment and transport. Domestic pipelines are estimated to be the lowest-cost approach for transporting hydrogen distances less than 1500 km [12], beyond which shipping becomes more cost-effective. Pipelines are characterized by high fixed costs, as they require large capital investments but have low (by comparison) operating costs [30]. Pipeline construction will thus only occur if an investor is confident that both sufficient demand and supply will be in place—and in the right place. For long-distance transport, hydrogen has to be converted, such as through liquefaction or the use of LOHCs [12]. To ship hydrogen, specialized infrastructure will be needed at both ends of the supply chain, including loading and receiving terminals, storage, liquefaction, and regasification facilities or conversion and reconversion plants for LOHCs [12]. All the potential hydrogen fuels (i.e., liquid hydrogen, ammonia, or LOHCs) require separate specialized infrastructures for conversation and reconversion, however, none are yet dominant. This creates uncertainties for investors as to which option is safest to rely upon and the potential for coordination failures among supply chain participants to emerge.

Coordination failures can be overcome through intra-industry collaboration. Alternatively, individual firms can seek to avoid them arising by vertically integrating. In this case, to produce and export green hydrogen a firm could operate its own renewable electricity generation, green hydrogen production, hydrogen delivery network, and export facilities and capabilities. However, this could be an inefficient outcome due to the economies of scale involved in pipelines and export infrastructure, resulting in each producer operating its own pipeline and port facilities, as has occurred in the past with liquefied natural gas (LNG) infrastructure in Australia [31]. At the same time, this option is only available for large companies and would lock smaller players out of the industry.

1.2.2.5.1 *Addressing the market failure*

Coordination failures can be addressed by government provision of institutional support for coordination among potential market participants [27]. When selecting policy instruments, a government can assume that, once the required investments are made simultaneously, all will become profitable. Therefore less financial outlay on behalf of the government should be required to overcome coordination failures than for the knowledge public good, learning by doing, or information asymmetry market failures [27].

Governments could guarantee an investor's debt to provide the certainty for it to proceed with the investment while ensuring the government only provides funds if the specific venture is unsuccessful [22,27]. These guarantees could be provided to investors across the green hydrogen export supply chain (e.g., production, transportation and shipping). This approach, however, depends on the success of individual projects and the ability of the government to pick the right projects. Another approach is for governments to provide institutional support that allows the necessary intra-industry collaboration to occur. The Chilean and Canadian hydrogen strategies both note plans to foster collaboration between sectors (such as industry, academia, and federally), though not necessarily within sectors [5,9].

Governments can also encourage coordination by facilitating the co-location of businesses across the supply chain in designated "clusters." A green hydrogen export cluster could be designated in an area favorable for both the production of green hydrogen (proximity to renewable energy resources and water) and its export (e.g., proximity to trading partners). Governments can encourage investment to occur within the cluster such as through special regulatory treatment or tax incentives. Investors within the cluster can benefit from access to shared infrastructure. The Australian government is investing $464 million in hydrogen hubs through the Clean Hydrogen Industrial Hubs program, with the intent to support cost reductions in hydrogen production [32][1].

Establishing a national strategy or "roadmap" is another approach that governments can use to give investors the confidence to go ahead with individual projects. These can create a common set of expectations amongst all businesses across the supply chain and are an explicit statement of the government's support. Roadmaps can address the coordination failures that could exist between nation-states by setting out future import demand or export potential for green hydrogen. Governments can also enter into country-to-country agreements or memoranda of understanding that either directly facilitate partnerships or provide an avenue for firms in importing and exporting countries to be working together. For example, Australia as a potential exporter has entered into partnerships with potential importers Japan and South Korea to collaborate on clean energy projects, including exploring the feasibility of hydrogen supply chains [33,34]. This included

---

[1] DCCEEW: The Australian Department of Climate Change, Energy, the Environment and Water.

collaboration between Australian and Japanese firms, with funding from both governments, to facilitate the world's first shipment of liquid hydrogen [29], noting that the hydrogen used in this demonstration was produced from unabated fossil fuels.

1.2.2.6 Network externalities

Network externalities occur when there are either positive or negative spillovers from the number of participants in a network. These are external from the perspective of a market participant (producer). Network externalities can be substantial and can lead to technological lock-in. A classic example of positive network externalities is word processing software. The benefits of using a given software depend strongly on how many of your collaborators use the same software. A classic example of negative network externalities is traffic congestion—the benefits of driving a particular network of roads are decreasing with the number of other users. The existence of network externalities, therefore, creates a value for coordination among market participants. However, network externalities can also support the emergence of a single dominant market group, again as in the case of word processing software, and can limit competition in the market.

Both positive and negative network externalities are possible for a green hydrogen industry, but the positive spillovers are likely to be the most important. The transportation and conversion phases present an example. Hydrogen can be transported in several forms (piped or compressed or liquefied H_2, ammonia, or LOHC). Each of these delivery pathways will require infrastructure with large capital costs that cannot be recovered if another pathway becomes dominant [35]. If, for example, it appears that most ports are set up with infrastructure to deal with hydrogen-transported-as-ammonia, hydrogen producers already geared for export-as-ammonia will see an advantage over producers set up for less dominant forms of the carrier. Likewise, if most producers appear to be preparing to export hydrogen-as-ammonia, it is less likely that receiving facilities will recoup value on investments for other carriers. However, as one of these has not yet emerged as dominant, the potential for network externalities contributes to coordination failures.

1.2.2.6.1 Addressing the market failure

Positive network spillovers create a value for coordination between participants in a market which could incentivize participants to create the required network (in this case, a green hydrogen export delivery pathway) without government intervention. If this network does not emerge, however, governments could have a role in facilitating coordination, such as collaboration between importing and exporting countries to support convergence on unified standards for systems of transporting hydrogen. Governments can also play a role in developing effective regulation to curb anticompetitive behaviors of any monopolies that emerge as a result of network externalities or can directly provide infrastructure that can only exist as a natural monopoly. Good quality legal frameworks can also minimize hold-up problems by enforcing contracts between network participants. Exporting countries consistently note the need for more international standardization to emerge for hydrogen trade [5,8,9].

1.2.2.7 Hold-up problem

Much of the infrastructure in the green hydrogen supply chain will be highly specific to the industry, either because it will be designed for hydrogen's unique characteristics, or because it fulfills a particular function. This infrastructure is also capital intensive requiring large, up-front investment. This creates a situation where after a firm makes a sunk investment in an asset specific to the needs of customers on the hydrogen supply chain, the bargaining power of its customers increases to the point where they (the customers) can seek to renegotiate to capture the profits of the sunk asset. This could result in too-little investment occurring or only sub-optimal investments that avoid this problem [36].

The hold-up problem is particularly acute when one or the other party invests in relationship-specific assets. Pipelines between countries are an example of a relationship-specific asset. Pipelines can be subject to hold-up problems both from customers seeking to take advantage of the sunk investment or, if the pipeline traverses a third country, from the transit country taking advantage of its position. Similar problems may occur when there are a small number of importers of a particular hydrogen product. Firms wishing to serve these customers need to invest in specific infrastructure to generate and transport this product. Due to the small number of customers, these investments become relationship-specific. Such a situation could occur, for example, if one importing country demands hydrogen embedded in a particular LOHC.

One approach that could alleviate hold-up problems is for firms to enter into long-term contracts prior to an investment being made, as this limits opportunities for ex-post renegotiation. Another approach is for firms to vertically integrate to internalize all profits and transaction costs [37]. There are, of course, limits to vertical integration in global supply chains.

1.2.2.7.1 Addressing the market failure

Government can play a role in addressing hold-up problems by ensuring the provision of quality legal institutions and frameworks to support the enforcement of such contracts, or through the establishment or participation in an export credit insurance agency. This type of institution and enforcement would not sit within a country's national hydrogen strategy and often involves international agreements and institutions. Nonetheless, hydrogen strategies can point to the need to ensure that existing institutions and agreements are consistent with the needs of the industry, and the policy approach of the government. International investment agreements, including the Energy Charter Treaty, are currently inconsistent with best-practice green industrial policy [38].

1.2.2.8 Economies of scale

1.2.2.8.1 Static internal

Internal economies of scale occur when the average cost of production is decreasing with the number of units produced by a given firm. This can occur for several reasons, such as spreading an activity's high capital costs across a larger output, the economies of increased dimensions and other technical considerations, and increased specialization [39]. Internal economies of scale typically occur when fixed (e.g., capital) costs are large relative to variable costs and capital is "lumpy." While internal economies of scale are not themselves a market failure, they can exacerbate market failures.

Green hydrogen development seems relatively unlikely to exhibit internal economies of scale. Electrolyzers are expected to be scalable and modular, with small entry costs and a relatively linear rate of returns. Likewise, renewable energy generating capacity can be added in a modular fashion. The one system component that does display economies of scale is pipelines—as the diameter of a pipeline increases, its volumetric capacity increases at a proportionally higher rate, making the costs per unit of throughput less for larger pipelines [30]. This feature, combined with the fact they can be shared by multiple producers, makes them prime candidates for shared infrastructure. Consequently, the majority of opportunities for economies of scale for green hydrogen appear to be external economies of scale, both dynamic and static.

1.2.2.8.2 Static external

External economies of scale are those factors external to a company, but within the same industry, that contribute to lower company costs when larger industry scales are achieved. For example, infrastructure and access to a trained workforce will likely be a function of the size of an industry at large, rather than driven by an individual company. A larger industry is likely to achieve external economies of scale in providing infrastructure and training. We consider external economies of scale from the perspective of hydrogen producers.

Increasing the capacity to build electrolyzers, such as investing in new factories or equipment, is capital-intensive but could result in a material reduction of the cost per unit of producing electrolyzers. This represents a potential external economy of scale for hydrogen producers, since the more producers there are in an industry, the lower the costs of capital equipment [12,19]. As mentioned previously, hydrogen pipelines display economies of scale, which means they can contribute to external economies of scale for the producers if shared. Similarly, conversion, reconversion, and liquefaction infrastructure could display economics of scale as their costs would reduce as capital costs are spread across a larger throughput. Ensuring the benefits of scale economies for these infrastructure items are shared among hydrogen producers will likely depend on regulatory intervention since the substantial economies of scale will likely drive monopsonistic or oligopolistic behavior otherwise.

1.2.2.8.3 Dynamic (learning by doing)

Since Arrow [40], it has been understood that with experience comes declines in the cost of producing a good. This "learning by doing" or "dynamic economies of scale" applies to any new industry or technology. It has been observed in other renewable energy technologies, such as solar photovoltaic (PV) and wind, which have demonstrated significant cost reductions over 20 years. Dynamic economies of scale can result in a market failure where the individual firms are not able to capture all the benefits of their learning—leading to positive spillovers for other market participants. These spillovers occur, for example, when experienced staff move to a different firm, or establish their own.

Given the current status of electrolyzer technology and application, there is every reason to predict green hydrogen production will face dynamic economies of scale. Producing hydrogen via electrolysis is currently undertaken on a small scale. As green hydrogen production is scaled up, firms will develop new techniques and improved processes that will enable green hydrogen to be produced at the lowest cost. A firm that invests in these technologies today bears not only the upfront cost of producing a unit of green hydrogen but also the cost through that experience of increasing the stock of knowledge of green hydrogen production [13]. This can create a positive spillover for future investors who benefit from this experience by having the techniques to produce a unit of green hydrogen at a lower cost. As with R&D, this implies underinvestment in green hydrogen technologies may occur, as early adopters would not be able to internalize the benefits of their investments [6,13,26].

1.2.2.8.4 Addressing the market failure of economies of scale

Economies of scale are not a market failure in and of themselves—instead, they exacerbate other market failures, particularly that of credit constraints. Thus, the best way to address this area is to address the market failure of credit constraints. That is, the financial tools described to address credit constraints (government credit and financing programs) also apply to addressing various economies of scale.

Learning by doing represents the knowledge spillovers that can arise beyond the R&D phase from the experience of production. Direct subsidies, such as grants, can also play a role where learning by doing can occur. Grants could be deployed to encourage investment in demonstration projects, such as to demonstrate the viability of green hydrogen export supply chains. Concessional loans, guarantees, or equity could be more appropriate forms of government support for processes such as the production of green hydrogen or the manufacture of electrolyzers to avoid diminishing returns. For example, in Australia, both the Australian Renewable Energy Agency and the Clean Energy Finance Corporation have funded hydrogen demonstration and pilot projects, for 1.78 billion and 300 million AUD respectively [41].

Government support for the creation of hubs and export facilities can also address external economies of scale. The Australian hydrogen strategy identifies several existing state and industry-led activities to support the formation of hydrogen hubs, and describes a plan to map existing infrastructure and with consideration hydrogen supply chain needs to shape the development of future hubs [8]. The Canadian strategy also describes the government's role in identifying opportunities for hub formation and de-risking these investments as well as noting existing export infrastructure confluences for several potential hubs [9], while the Chilean strategy notes the potential for hubs to benefit local development [5].

Table 1.2.1 Recommendations of best practice policy instrument interventions based on economic theory and summary of current policy status in exporting nations.

Market failure	Best policy instrument (from economic theory)	Status in (frontrunner) exporting nations
Environmental externalities	Carbon pricing	Political challenges remain
Information asymmetry	Certification	Politically uncontroversial but challenging due to need for international collaboration
Credit constraints	Government taking on financing risk (lower interest rates for industry)	Need to address this market failure is generally recognized and some action taken
Knowledge public good	Direct subsidy for R&D	Need to address this market failure is generally recognized and some action taken
Coordination failures	Government role in coordination (and financing of infrastructure)	Need to address this market failure is generally recognized and some action taken
Network externalities	Support for standardization of networks; regulation or provision of natural monopoly products	Need to address this market failure is generally recognized but taking action remains challenging due to need for international collaboration
Hold-up problem	Legal frameworks and institutions	Needs to be addressed external to hydrogen policy

1.2.3 Policy development and recommendations to address market failures for hydrogen exports

It is widely believed that the green hydrogen trade has the technological and economic potential to support the global net zero transition. Pervasive market failures may, however, prevent the efficient development of a green hydrogen export industry. There is, therefore, a strong argument for green industrial policy to be employed. Table 1.2.1 summarizes recommendations made in previous sections regarding best practice deployment of relevant policy instruments, based on economic theory. Table 1.2.1 further notes the policy status in exporting nations as of 2021.

In order to ensure they do not introduce "government failures" which counteract the potential benefits of addressing market failures, it is important to adhere to the three core principles of good industrial policy: embeddedness, discipline, and accountability [22]. In practice, this means that governments should develop policies to support green hydrogen export as part of an ongoing collaborative process with industry players to understand the industry's needs and to recognize when and where market failures could emerge. At the same time, governments must build into any policy instrument the appropriate (and transparent) metrics to measure success, as well as monitoring and evaluation processes to determine whether these metrics have been met. Despite the need for collaboration with industry, governments must remain in control of decision-making and have the discipline to adjust programs as needed and to withdraw support from projects deemed by these metrics to be failures [42]. Scheduled review processes can support the effective evolution of policies for energy innovation as technological systems change over time [43–45]. Transparent metrics, along with reporting requirements, will also increase the accountability of policymakers for their policies to support green hydrogen export, as will governments operating within democratic frameworks with the appropriate checks and balances built in [42]. Governments should also engage with the public as a general part of its policymaking to ensure the public's views and concerns are accounted for.

Industrial policy is largely intended to drive productivity growth but may leave it to the market to determine which technologies or supply chains to adopt [42]. In contrast, the presence of the environmental externality presented by carbon emissions justifies a green industrial policy approach. That is, there is a need to pursue a technology-specific industrial policy approach to support green hydrogen (for export or domestic use) and, additionally, to include only green hydrogen while excluding other forms with embedded carbon emissions [42]. By the same token, governments should maintain the discipline in their policymaking to withdraw all support for green hydrogen (and its export) once it is clear the industry will not develop, if alternative technologies prove superior, or once the industry is sufficiently developed that government policy is no longer needed to address market failures.

The severity and nature of the environmental externality mean that there is a time limit attached to the scale-up of innovative zero-carbon technologies to avoid catastrophic climate change [42]. This could justify more ambitious policies to support green hydrogen export, for example in the quantum of funding or length of support, than a government may have otherwise considered. While political limitations impact some of the identified market failures, many of the other options discussed in this paper could be rapidly implemented by national and regional governments of hydrogen-exporting nations.

Acknowledgment

This work was conducted as part of the ANU Grand Challenge Zero-Carbon Energy for Asia–Pacific, funded by the Australian National University.

References

[1] International Energy Agency, 2021. Global hydrogen review. https://www.iea.org/reports/global-hydrogen-review-2021.

[2] N. Stern, The Economics of Climate Change: The Stern Review, Cambridge University Press, Cambridge, 2007.

[3] J.A. Schwarzer, 2013. Industrial policy for a green economy. https://www.iisd.org/publications/report/industrial-policy-green-economy.

[4] T. Longden, et al., "Clean" hydrogen?: comparing the emissions and costs of fossil fuel versus renewable electricity based hydrogen, Appl. Energy 306 (2022) 1–14. https://doi.org/10.1016/J.APENERGY.2021.118145.

[5] Ministry of Energy Government of Chile, National green hydrogen strategy, 2020. https://energia.gob.cl/sites/default/files/national_green_hydrogen_strategy_-_chile.pdf.

[6] A.B. Jaffe, R.G. Newell, R.N. Stavins, A tale of two market failures: technology and environmental policy, Ecol. Econ. 54 (2) (2005) 164–174. https://doi.org/10.1016/j.ecolecon.2004.12.027.

[7] WTO, Anti-dumping, subsidies, safeguards: contingencies, etc, 2022. https://www.wto.org/english/thewto_e/whatis_e/tif_e/agrm8_e.htm#subsidies. (Accessed 21 September 2022).

[8] COAG Energy Council, Australia's national hydrogen strategy, 2019. https://www.industry.gov.au/sites/default/files/2019-11/australias-national-hydrogen-strategy.pdf.

[9] Hydrogen Strategy for Canada, Hydrogen strategy for Canada, 2020. https://www.nrcan.gc.ca/sites/www.nrcan.gc.ca/files/environment/hydrogen/NRCan_Hydrogen-Strategy-Canada-na-en-v3.pdf.

[10] A.V. Abad, P.E. Dodds, Green hydrogen characterisation initiatives: definitions, standards, guarantees of origin, and challenges, Energy Policy 138 (2020) 111300. https://doi.org/10.1016/j.enpol.2020.111300.

[11] L.V. White, et al., Towards emissions certification systems for international trade in hydrogen: the policy challenge of defining boundaries for emissions accounting, Energy 215 (2021) 119139. https://doi.org/10.1016/j.energy.2020.119139.

[12] IEA The Future of Hydrogen, IEA, 2019.

[13] Gillingham, K. and Sweeney, J. (2010) 'Market failure and the structure of externalities', Harnessing Renewable Energy in Electric Power Systems: Theory, Practice, Policy. Earthscan.

[14] M. Mulder, G. Zwart, (2006). Market failures and government policies in gas markets. (CPB-Memo–143). Netherlands. CPB.

[15] A. Geddes, T.S. Schmidt, B. Steffen, The multiple roles of state investment banks in low-carbon energy finance: an analysis of Australia, the UK and Germany, Energy Policy 115 (2018) 158–170. https://doi.org/10.1016/j.enpol.2018.01.009.

[16] High-Level Commission on Carbon Prices. 2017. Report of the High-Level Commission on Carbon Prices. Washington, DC: World Bank.

[17] O. Schmidt, et al., Future cost and performance of water electrolysis: an expert elicitation study, Int. J. Hydrogen Energy 42 (52) (2017) 30470–30492. https://doi.org/10.1016/j.ijhydene.2017.10.045.

[18] A.J. Chapman, T. Fraser, K. Itaoka, Hydrogen import pathway comparison framework incorporating cost and social preference: case studies from Australia to Japan, Int. J. Energy Res. 41 (14) (2017) 2374–2391. https://doi.org/10.1002/er.3807.

[19] S. Bruce, et al., National Hydrogen Roadmap: Pathways to an Economically Sustainable Hydrogen Industry in Australia, CSIRO, 2018.

[20] K. Mazloomi, C. Gomes, Hydrogen as an energy carrier: prospects and challenges, Renew. Sustain. Energy Rev. 16 (5) (2012) 3024–3033. https://doi.org/10.1016/j.rser.2012.02.028.

[21] U. Cardella, L. Decker, H. Klein, Economically viable large-scale hydrogen liquefaction, in: IOP Conference Series: Materials Science and Engineering, 171, 2017, p. 12013. https://doi.org/10.1088/1757-899X/171/1/012013.

[22] D. Rodrik, Green industrial policy, Oxf. Rev. Econ. Policy 30 (3) (2014) 469–491. https://doi.org/10.1093/oxrep/gru025.

[23] F. Polzin, et al., How do policies mobilize private finance for renewable energy?: a systematic review with an investor perspective, Appl. Energy 236 (2019) 1249–1268. https://doi.org/10.1016/J.APENERGY.2018.11.098.

[24] Clean Energy Finance Corporation (2021) About our finance. https://www.cefc.com.au/about-our-finance/. (Accessed 26 November 2021).

[25] S. Martin, J.T. Scott, The nature of innovation market failure and the design of public support for private innovation, Res. Policy 29 (4) (2000) 437–447. https://doi.org/10.1016/S0048-7333(99)00084-0.

[26] L.H. Goulder, I.W.H. Parry, Instrument choice in environmental policy, Rev. Environ. Econ. Policy 2 (2) (2008) 152–174. https://doi.org/10.1093/reep/ren005.

[27] D. Rodrik, Industrial Policy for the Twenty-First Century, Social Science Research Network, Rochester, NY, 2004.

[28] H. Schmitz, O. Johnson, T. Altenburg, Rent management: the heart of green industrial policy, IDS Work. Pap. 2013 (418) (2013) 1–26. https://doi.org/10.1111/j.2040-0209.2013.00418.x.

[29] Hydrogen Energy Supply Chain, Hydrogen energy supply chain project, 2022. https://www.hydrogenenergysupplychain.com/about-hesc/. (Accessed 21 September 2022).

[30] E.J. Omonbude, The economics of transit oil and gas pipelines: a review of the fundamentals, OPEC Energy Rev. 33 (2) (2009) 125–139. https://doi.org/10.1111/j.1753-0237.2009.00163.x.

[31] S. Reid, G. Cann, The good, the bad and the ugly: the changing face of Australia's LNG production, 2016. https://www2.deloitte.com/content/dam/Deloitte/us/Documents/energy-resources/us-er-the-changing-face-of-australias-lng-production.pdf.

[32] DCCEEW, Funding available for clean hydrogen industrial hubs, 2021. https://www.dcceew.gov.au/about/news/funding-available-for-clean-hydrogen-industrial-hubs. (Accessed 21 September 2022).

[33] A. Daisuke, Japan-Australia: building a hydrogen supply chain: the Interpreter, 2022. https://www.lowyinstitute.org/the-interpreter/japan-australia-building-hydrogen-supply-chain. (Accessed: 21 September 2022).

[34] M. Price, S. Morrisson, A. Taylor, Australia and Republic of Korea sign new deals on clean energy tech and critical minerals: media release, 2021. https://www.minister.industry.gov.au/ministers/price/media-releases/australia-and-republic-korea-sign-new-deals-clean-energy-tech-and-critical-minerals. (Accessed 21 September 2022).

[35] N. Bento, Building and interconnecting hydrogen networks: insights from the electricity and gas experience in Europe, Energy Policy 36 (8) (2008) 3019–3028. https://doi.org/10.1016/j.enpol.2008.04.007.

[36] B.E. Hermalin, M.L. Katz, Information and the hold-up problem, Rand J. Econ. 40 (3) (2009) 405–423.

[37] G. Nuryyev, H.-N.R. Chu, Effect of hold-up potential on natural gas exports, J. Econo. Develop. Stud. 2 (2014) 22.

[38] E. Aisbett, Submission to the OECD public consultation on investment treaties and climate change, 2022. https://www.oecd.org/investment/investment-policy/OECD-investment-treaties-climate-change-consultation-responses.pdf.

[39] K. Junius, Economies of scale: a survey of the empirical literature, 1997. Kiel Working Papers 813, Kiel Institute for the World Economy. https://doi.org/10.2139/SSRN.8713.

[40] K.J. Arrow, The economic implications of learning by doing, Rev. Econ. Stud. 29 (3) (1962) 155–173. https://doi.org/10.2307/2295952.

[41] Australian Government Department of Industry, Science, Energy and Resources (2021). State of hydrogen, 2021. https://www.dcceew.gov.au/energy/publications/state-of-hydrogen-2021.

[42] T. Altenburg, D Rodrik, Green industrial policy: accelerating change towards wealthy green economies, in: T. Altenburg, C. Assmann (Eds.), Green Industrial Policy: Concept, Policies, Country Experiences, German Development Institute/Deutsches Institut für Entwicklungspolitik (DIE), 2017.

[43] M.A. Dutz, D. Pilat, 'Fostering innovation for green growth: learning from policy experimentation', Making Innovation Policy Work: Learning From Experimentation, (2014) pp. 193–228. https://www.oecd-ilibrary.org/science-and-technology/making-innovation-policy-work/fostering-innovation-for-green-growth-learning-from-policy-experimentation_9789264185739-10-en.

[44] W. Lütkenhorst, et al. 'Green industrial policy: managing transformation under uncertainty, in: Deutsches Institut für Entwicklungspolitik, (2014) p. 62. http://www.die-gdi.de/uploads/media/DP_28.2014.pdf%0Ahttps://papers.ssrn.com/sol3/papers.cfm?abstract_id=2509672%0Ahttps://refubium.fu-berlin.de/handle/fub188/19217.

[45] A. Pegels, 'Lessons for successful green industrial policy, in: Green Industrial Policy in Emerging Countries, (2014) pp. 179–186. Routledge. https://doi.org/10.4324/9780203797464.

Global hydrogen economy and hydrogen strategy overview

Anuraag Nallapaneni and Soham Kshirsagar
World Resources Institute, New Delhi, India

1.3.1 Introduction

Hydrogen is the lightest and most abundant element in the universe, and it occurs naturally on earth when combined with other elements. Hydrogen, like electricity, is an energy carrier (fuel) that can be used to store, move, and deliver energy produced from other sources. It is also being used in large quantities for well over 100 years as a chemical feedstock, in fertilizer production and oil refineries. Hydrogen has significant complexity, cost, efficiency, and other safety challenges compared to the direct use of renewable electricity for decarbonizing hard-to-abate sectors. But it has been identified as a key energy intermediate to enable the decarbonization of energy systems. There is an emerging consensus that low-carbon, renewable, clean, or green hydrogen is an efficient way to address the issues related to climate change. Beyond its importance in decarbonization, hydrogen is also gaining traction as a possible pathway to increase energy independence and energy security in different countries.

With an energy content of 120 MJ/kg, hydrogen is nearly three times heavier by weight and four times lower by volume compared to petrol (45.8 MJ/kg) and diesel (45.5 MJ/kg). It is a colorless, odorless, tasteless, and nonpoisonous gas under normal conditions. It is the most abundant element in the universe, accounting for 90% of the universe by weight. However, it is not commonly found in its pure form since it readily combines with other elements. It is also the lightest element, having a density of 0.08988 g/L at standard pressure. Some of the hydrogen's comparative properties with natural gas and gasoline are listed in Table 1.3.1.

Other chemical properties that affect hydrogen as a fuel are:

- It has a high energy content per weight (nearly three times as much as gasoline), but the energy density per volume is quite low at standard temperature and pressure.

Towards Hydrogen Infrastructure: Advances and Challenges in Preparing for the Hydrogen Economy.
DOI: https://doi.org/10.1016/B978-0-323-95553-9.00002-9   59

Table 1.3.1 Fuel properties comparison [1,3].

Property	Units	Hydrogen	Methane	Gasoline
Chemical formula		H_2	CH_4	C_xH_y (x = 4–12)
Molecular weight		2.02	16.04	100–105
Density (NTP)	(kg/m^3)	0.0838	0.668	751
Viscosity	Pa·s	8.8×10^{-6}	1.1×10^{-5}	0.00037–0.00044
Normal boiling point	°C	−253	−162	27–225
Vapor specific gravity, NTP	Air = 1	0.0696	0.555	3.66
Flash point	°C	<−253	−188	−43
Flammability range in air	Vol %	4.0–75.0	5.0–15.0	1.4–7.6
Auto ignition temperature	°C	585	540	230–480
Maximum flame velocity in air	m/s	2.83	0.45	N/A

From: International Energy Agency.

- Volumetric energy density can be increased by storing the hydrogen under increased pressure or storing it at extremely low temperatures as a liquid. Hydrogen can also be adsorbed into metal hydrides.
- Hydrogen is highly flammable; it only takes a small amount of energy for ignition and combustion to occur. It also has a wide flammability range, meaning hydrogen can burn when it makes up 4–74% of the air by volume.
- Hydrogen burns with a pale blue, almost-invisible flame, making hydrogen fires difficult to see.
- The combustion of hydrogen does not produce carbon dioxide (CO_2), particulate, or sulfur emissions. It can produce nitrogen oxide (NO_X) emissions under some conditions.

Hydrogen has been tagged as a fuel of the future for decades, promising to deliver huge benefits for the global economy. With increasing capacities and decreasing costs for renewable energy green hydrogen produced from the process of electrolysis of water is quickly gaining prominence. A "hydrogen economy" is being projected as a sustainable transition to energy systems across sectors. It is expected to reduce energy consumption and curtail emissions contributing to economic growth and fulfilling energy requirements.

1.3.2 Global hydrogen consumption and fossil fuel dependence

The global consumption [1] of hydrogen is around 94 MMT as of 2021. Currently, oil refining is the most dominant consumer of hydrogen,

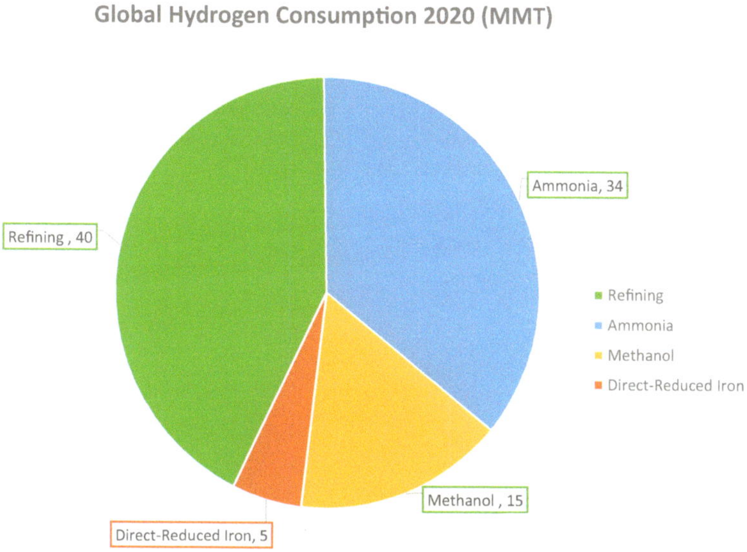

Global Hydrogen Consumption 2020 (MMT)

Figure 1.3.1 Global hydrogen sectoral consumption as of 2021. *(From: International Energy Agency).*

accounting for 40 MMT of global hydrogen consumption. The use of hydrogen in industries (as a feedstock) currently accounts for ~54 MMT, with approximately 49 MMT consumed for chemical production, which includes ammonia and methanol production, and 5 MMT for direct reduction of iron (DRI) processes for steelmaking. Of the total 94 MMT, ~70 MMT is utilized in the form of pure hydrogen for ammonia production and oil refining, with approximately ~20 MMT mixed and used with the other carbon-containing gases for methanol and steel production, with the leftover demand coming from residual gasses from industrial processes. The sectoral split of current total hydrogen consumption can be found in Fig. 1.3.1.

From a production perspective, almost all the hydrogen produced globally is derived from fossil fuels and is produced through two key pathways which are steam methane reforming (80%) and coal gasification (19%) with low carbon hydrogen with carbon capture utilization and storage (CCUS) (0.7%) and water electrolysis accounting for 0.1% of the production mix approximately.

Considering the total carbon dioxide (CO_2) footprint (kg of CO_{2e} carbon dioxide equivalent/kg of H_2) for the four production pathways used for hydrogen production, the total emissions (direct and indirect) for hydrogen production are approximately *1430 MMT of CO_{2e}*, as seen in

Table 1.3.2 Global hydrogen production emissions in 2021 [1].

Production pathway	Hydrogen production (MMT)	Average CO$_2$ footprint (MMT of CO$_{2e}$/MMT of H$_2$)	Total emissions (MMT of CO$_{2e}$)
Steam methane reforming	75.2	13	977.6
Coal gasification	17.9	25	447.5
SMR with CCUS	1	7	7
Electrolysis (RE)	0.01	1	0.01
Global hydrogen Production emissions			1432.1

From: International Energy Agency.

Table 1.3.2, which is equivalent to the combined national emissions of Canada and Australia.

It is expected that by 2050, hydrogen consumption will range between 300 and 500 MMT, with its application expanding from its existing use cases. This will make it critical for countries to produce and consume hydrogen with the lowest possible carbon footprint to meet their climate targets whilst ensuring economic development.

1.3.3 Future potential applications

With global economies gearing to meet climate aspirations and decarbonization targets, there is a rising consensus to transition current carbon-intensive fuels and feedstocks, to lower-emission alternatives and technologies, with hydrogen being one of the promising options. Given clean hydrogen's potential for emission reduction, its applications are expected to expand to sectors outside of its current use cases.

Although current applications and demand are dominated within refining, fertilizers, and industries as feedstock, projected additional demand is expected to arise from other segments, including steel, cement, heavy mobility, and the power sector. A snapshot of current and future applications can be seen in Fig. 1.3.2.

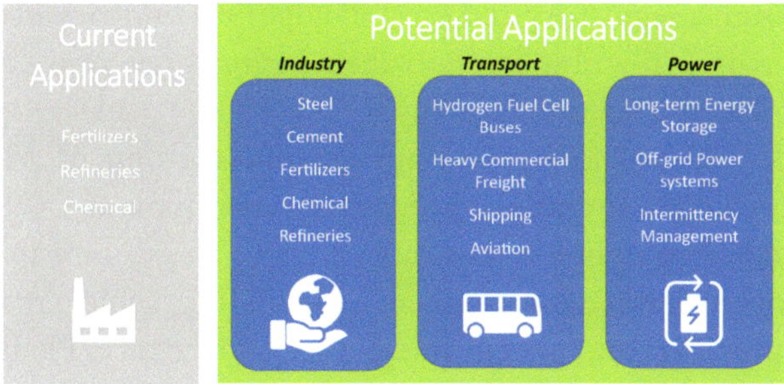

Figure 1.3.2 Hydrogen current and future application. *(From: International Energy Agency).*

1.3.4 Hydrogen commitments through global climate targets and emission reduction goals by the public and private sector

To meet emission abatement and decarbonization goals, countries are setting ambitious targets to reduce greenhouse gas (GHG) emissions. Climate concerns and increasing focus on transitioning to net zero pathways have made low-carbon hydrogen-based applications more relevant than ever. With more than 30 countries including the European Union (EU) agreeing to work together in accelerating the transition at COP26, hydrogen is expected to play an integral role in the clean energy transition globally.

Countries are setting mid and long-term emission reduction commitments to meet their nationally determined contributions (NDCs). Adoption of low-carbon hydrogen as a clean energy source, as well as a chemical feedstock, can help facilitate meeting multiple sustainable development goals (SDGs), directly and indirectly, with its impact across social, environmental, and economic spheres of sustainable development.

According to International Energy Agency (IEA) [2], hydrogen and hydrogen-based fuels are considered one of the six pillars of decarbonization along with renewable technologies, electrification of hard-to-abate sectors, CCUS, energy efficiency, and behavioral change. It is expected that the adoption of low-carbon hydrogen production technologies coupled with increasing hydrogen demand could avoid up to 60 gigatons of CO_2 emissions by 2050, whilst enabling sector coupling and adoption of other clean energy technologies.

Given low-carbon hydrogen's potential application beyond just an industrial feedstock, countries are taking necessary steps to drive cost reduction and increase clean hydrogen applications across other sectors. To enable such a transition from a regulatory and fiscal standpoint, countries are embodying their clean hydrogen ambitions into their national net zero commitments and sectoral decarbonization targets.

Alongside national and global agreements, a group of 28 companies [3] have also pledged and joined industrial consortiums at COP26 to drive the supply and demand of low-carbon hydrogen. These companies represent unique sectors including energy, oil and gas, mining, vehicle manufacturers, equipment providers and system integrators, and finance. Pledges by these 28 companies for low-carbon hydrogen adoption have been placed on both the demand and supply sides.

These companies have pledged to consume low-carbon hydrogen and create a demand of ~1.6 MMT per annum by 2030, leading to a reduction of 14 MMT of CO_2 emissions per annum, which is equivalent to taking 6 million cars off road. While similarly on the supply side, 18 MMT per annum of low-carbon hydrogen supply has been pledged by 2030, which can lead to the reduction of 190 MMT per annum of CO_2 emissions, which are the combined annual emissions of Qatar and The Czech Republic.

1.3.5 Role of hydrogen in decarbonizing global economy

Hydrogen as a potential cross-sectoral clean energy vector and chemical feedstock is witnessing unprecedented momentum as a "Swiss Army Knife" to attain decarbonization of an array of hard-to-abate sectors. Hydrogen's current and projected demand is multifaceted and interlinked in nature and can serve several end-use cases as discussed in the sections above. Depending on a country's GDP, economic strength, local resource availability, and hydrogen's current demand across sectors, there are four key drivers for hydrogen's future applications in the decades to follow:

1. Mounting energy security and food shortage concerns.
2. Meeting net zero targets and emission reduction goals.
3. Enabling cross-sectoral decarbonization, including decarbonizing existing and future hydrogen production for feedstock applications.
4. Diversification of energy systems to ensure long-term sustainability and reliability.

Hydrogen is currently produced through multiple pathways using various technologies and input feedstocks. The pathways can range from

Table 1.3.3 Carbon emissions related to hydrogen production pathways.

Technology	CO$_2$ footprint (kg of CO$_{2e}$/kg of H$_2$)
Coal Gasification (w/o CCS)	21.2 - 26
Coal Gasification (w CCS (93%))	2.6 - 6.3
Steam Methane Reforming of Natural Gas (w/o CCS)	9.5 - 13.5
Steam Methane Reforming (w CCS 93%)	1.5 - 6.2
Electrolysis using Solar Electricity	1.3 - 2.0
Electrolysis using Wind Electricity	0.6 - 1.0

Source: Towards Hydrogen Definitions based on emission intensity, International Energy Agency **CCS:** Carbon Capture & Storage, Embedded emissions from production of onshore wind turbines (12 g CO$_2$-eq/kWh) & solar PV systems (27 g CO$_2$-eq/kWh) have been considered for electrolysis-based hydrogen production, Emission values based on median of upstream, midstream and direct emissions.

fossil-based steam methane reforming (SMR) to low-emission technologies such as renewable energy-based electrolysis of water. The hydrogen "color-rainbow" involves an array of such pathways, including technologies such as SMR of natural gas, coal gasification, and electrolysis. Depending on the technology used for hydrogen production and the associated carbon footprint, color codes are assigned (as seen in Table 1.3.3) and may vary across countries.

As seen from Fig. 1.3.3, it is evident that some pathways for hydrogen production are relatively more carbon intensive than others, further highlighting some key technologies whose relevance for clean hydrogen production will be critical in coming years. While coal gasification and natural gas-based SMR technologies of hydrogen production are mature pathways contributing to the majority of current hydrogen production, renewable energy-based water electrolysis is an emerging alternative.

Globally, electrolysis of water using renewable sources is seen to be a lucrative hydrogen production pathway for the following key reasons:
- Declining costs of renewable electricity.
- Scaling of technology manufacturing (for RE and electrolyzers).

The lower carbon footprint of electrolysis-based hydrogen production technologies coupled with dropping RE prices globally makes it a promising solution for low-carbon hydrogen production. CCUS is also being considered as a pathway to reduce emissions from existing fossil-based hydrogen generating units to facilitate existing infrastructure to supply hydrogen with a reduced environmental impact.

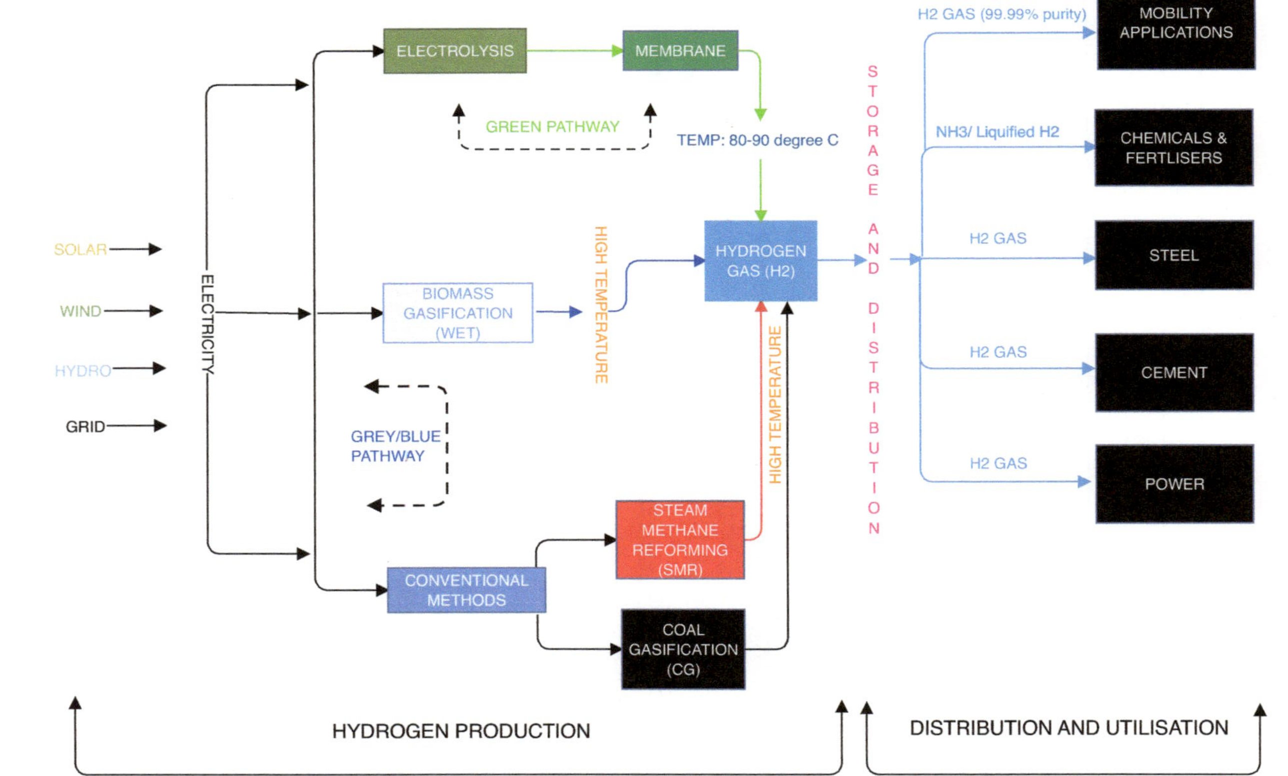

Figure 1.3.3 The hydrogen value chain. *(From: International Energy Agency).*

Biomass-based gasification and digestion technologies are also gaining traction for low-carbon hydrogen production, given their carbon-neutral nature. The current state of the technical and commercial viability of clean hydrogen production pathways coupled with other segments of the value chain including storage, distribution, and applications is at commercial nascency. Thus, it is essential that a holistic and overarching approach to developing these domestic hydrogen ecosystems and value chains is considered globally.

1.3.6 Towards a hydrogen economy

"Hydrogen economy" is a proposition of setting up an energy delivery infrastructure or development of an integrated value chain, from production to applications, based on hydrogen derived from clean energy sources. The term was first coined in the 1970s and since then has been used and modified across publications to support the idea or a vision of utilizing clean hydrogen to reduce dependence on hydrocarbons. A successful "hydrogen economy" must consist of the following pillars:

- **Integrated value chain development**: To develop a resilient hydrogen economy, it is essential that an integrated value chain development is considered starting from hydrogen production to applications. These include production technologies, storage, transport, and distribution, infrastructure (ports, refueling stations, hydrogen hubs, bunker facilities), vehicular applications, electricity/gas grid as well as regulations and legislations that govern the value chain components for hydrogen production and adoption, as seen in Fig. 1.3.3.
- **Policy focus to shift from carbon-intensive hydrogen to clean hydrogen**: Hydrogen, depending on the production technology and feedstock used, has a varying range of GHG emissions. To shift to a low carbon/clean hydrogen ecosystem, it will thus be critical that policies, legislation, and decarbonization roadmaps clearly target and prioritize this transition from an existing fossil-based hydrogen infrastructure to a market based on clean hydrogen sources.
- **Supporting financial and economic viability while driving emission reduction**: It will be pivotal to strike the balance between an economically viable transition and achieving emission reduction given the high costs of clean hydrogen. Development of appropriate public-private partnership risk-sharing frameworks, creation of demand at scale,

and ensuring financial support from governments during the initial years of the transition will determine the success of hydrogen economies.

To realize the potential of developing such hydrogen economies, countries across the globe are announcing national hydrogen strategies, with several others announcing roadmaps and other official statements in support of an array of hydrogen programs.

1.3.7 Need for national hydrogen strategies

Long-term strategies and roadmaps coupled with legislative commitments can ensure a secure, cost-effective, reliable, and climate-resilient transition to low carbon. Within the clean energy domain, it has been noted that enabling legislative mechanisms and long-term deployment strategies for successful and viable project execution can further scale investments creating a snowball effect across the sector. Such long-term strategies with short-term goals ensure that investments from both private and public stakeholders are channeled effectively in a timely manner.

These long-term strategies built on existing policies and regulatory frameworks can also help support stakeholders and international collaborations, encourage capacity building, and knowledge development as well as incentivize clean energy adoption while deterring fossil-fuel consumption. Addressing these technological, regulatory, and fiscal barriers from a deployment and infrastructure perspective through the scope of national roadmaps, policy, and legislation will increase private sector confidence and participation.

Adoption of a long-term strategy/roadmap, enabled by the combination of both supply and demand side interventions is critical to accelerate clean hydrogen uptake globally. Alongside these interventions, it is also pivotal that these strategies consider energy geopolitics, such as hydrogen's projected global tradability, bilateral partnerships, and its cross-sectoral decarbonization capabilities.

Following are five key reasons for the need for long-term hydrogen strategies with short to medium terms goal setting for establishing a resilient hydrogen economy:

- **Nascent value chain and high investment**: Considering the nascency of the hydrogen ecosystem and high technology and infrastructure CAPEX a strategic approach to build this ecosystem taking a top-to-bottom approach will be critical for clean hydrogen adoption.

- **Targeted and sector-specific demonstration projects**: To understand the technological, operational, and financial challenges of project implementation, long-term strategies can highlight the need for lighthouse/pilot projects across various sectors. These projects will be pivotal to prove technological capabilities, validate the commercial feasibility of a clean hydrogen value chain in the coming decades.
- **Cross-sectoral and value chain specific R&D (research and development)**: Nascency of the ecosystem necessities identification and allocation of critical R&D projects across the value chain (from production to utilization of clean hydrogen across sectors). Given the technical complexity of hydrogen storage and distribution and its associated costs, a clear strategic focus on specific parts of the value chain will play a critical role in enabling technology development and accelerating adoption.
- **Need for supply-demand catalyzation**: Just as any other sunrise sector, clean hydrogen deployment faces a distinctive chicken or egg conundrum from a supply-demand perspective. Enabling long-term strategies with clear policy, regulatory, and fiscal support to boost both supply and demand in a balanced and coherent manner would be key to developing a hydrogen ecosystem.
- **Geopolitics of a global hydrogen economy**: Clean hydrogen adoption is expected to take place across multiple countries. A global hydrogen economy will require an array of bilateral and multilateral partnerships across nations. Import and export of hydrogen and its derivatives will make hydrogen a globally traded commodity with strong linkages to decarbonization and national energy security.

1.3.8 Strategy announcements and legislative commitments

Hydrogen's cross-sectoral decarbonization capabilities have led to several countries announcing their national hydrogen strategies. Globally more than 30 countries have announced their official national hydrogen strategies, or roadmaps, and other official statements in support of an array of hydrogen programs.

The precedence for hydrogen strategy development was set by Japan in 2018 and further followed by Australia and South Korea. By 2020 several countries across the EU had announced their strategies, including Germany, Netherlands, and France, further supporting the development of the EU's hydrogen strategy which was launched in the same year. Several other strategies have been launched in the past two years. As of 2022, China, United States, Uruguay, and more recently Belgium have announced their national hydrogen strategies with India expected to launch its strategy soon. India has also released a green hydrogen policy earlier in 2022 as a preparatory measure to stimulate the production of green hydrogen and will soon be releasing a full-fledged national green hydrogen mission which will further aim to stimulate green hydrogen demand across existing and new applications across the domestic circuit.

Although the announcement of a long-term overarching hydrogen strategy/roadmap is one of the ways to show commitment towards the adoption of clean hydrogen, it is critical that the commitments made within these strategies are embedded and aligned within the existing or new legislations and/or acts or regulatory frameworks. Such an approach ensures that these strategies are implemented and executed effectively, whereby these strategic commitments become legislative commitments, ensuring that the strategy's short, medium- and long-term visions and goals are met.

It can also be observed that out of 16 official strategy announcements only six have enabling legislative frameworks. A key example of such a legislative-aligned framework includes Korea's Hydrogen Economy Promotion and Hydrogen Safety Management Law ("Hydrogen Law"), which has enabled the Korean strategy to be put into practice and ensure incentives are delivered as promised. It is thus imperative that the governments follow through with legislative commitments and regulatory frameworks to enable a transition to low-carbon hydrogen, turning verbal commitments into concrete actions.

1.3.9 Value chain and sectoral focus

A targeted strategic approach to improve the techno-commercial levels of existing technologies across the hydrogen value chain has garnered significant interest and acceleration poststrategy and roadmap announcements. Public and private sector investments are primarily being directed towards further reinforcing and strengthening critical segments of the value chain, whilst investments across some countries are observed to be taking the holistic value chain development approach.

It has been observed that most countries are looking at significantly ramping up their existing hydrogen production capabilities, with few focusing on R&D and investments in the storage and distribution part of the value chain. Some of the key outliers within this segment are countries like Chile and Saudi Arabia, which have a strong value chain focus on the production end as they aim to export the bulk of the clean hydrogen they would produce.

On the other side of the spectrum, countries like Japan and South Korea have a minimal focus on hydrogen production given the limited renewable energy (RE) potential and land resource constraints, thus aiming to import most of their low-carbon hydrogen. These countries are also focusing on developing and sharing their technology capabilities and expertise for the creation of a hydrogen economy.

Given hydrogen's multifaceted applications and production methods, countries are also aiming to further advance their R&D efforts toward developing technologies and hydrogen value chains. By leveraging their existing technological and geographical capabilities, nations can accelerate the techno-commercial feasibility of hydrogen projects at a much greater pace.

For example, Japan is evaluating the possibility to integrate its green hydrogen ambitions with its tidal energy potential given the country's limitations with solar resources. Similarly, countries in arid regions like Saudi Arabia and Oman should further evaluate the feasibility of green hydrogen production through the desalination of seawater as potential feedstock, whilst further pushing R&D efforts into low blue hydrogen, leveraging upon their existing capabilities within the oil and gas sector.

As opposed to the value chain focus which looks at strategies from a supply-side perspective, the sectoral focus of countries gives a clear understanding of expected demand across specific sectors and applications. Some of the key factors affecting the sectoral adoption preferences for clean hydrogen are as follows:

- RE integration.
- Cost and ease of adoption.
- Need for diversification of energy systems.
- Ensuring geopolitical dominance.
- Sectoral decarbonization targets.

For hydrogen to make a significant contribution towards a clean energy transition, it will need to be adopted in sectors where currently the applications are not present, which include sectors such as transport, building heating, steel, and power generation. Clean hydrogen's adoption as a chemical feedstock in existing applications such as refining, ammonia, and methanol production will be key to substitute existing fossil-based hydrogen demand.

Depending on the ease and cost of adoption across existing and new applications, these sectors can provide the impetus for initial demand creation, while driving cost reduction as demand scales up. The ease and low cost of adoption of transition to low-carbon hydrogen is observed to be a critical factor impacting adoption across various sectors.

For example, hydrogen adoption in industrial applications is dominantly observed given an industry's existing experience of handling and utilizing hydrogen. Apart from industrial applications, heavy mobility applications like trucks and buses seem to have an aggressive policy and investment outlook across countries, whereas maritime and aviation on the other tend have a long-term outlook, given technological and commercial nascency.

The reliability of using low-carbon hydrogen for hydrogen cars is observed to be strongly dependent on the costs of fuel cells and the availability of refueling infrastructure while for trucks it comes down to the delivered price of hydrogen. In terms of applications within buildings for heating and power generation, technological complexities and a high levelized cost of hydrogen is currently impeding factor.

1.3.10 Green hydrogen exporters and trade

Given the complexity of the hydrogen value chain and the current costs of green hydrogen production, it is observed that some countries can be bucketed into three categories, namely net exporters, net importers, and self-reliant. Alongside the availability of RE, potential for domestic demand for clean hydrogen, economic viability to export compared to domestic usage and sectoral decarbonization priorities are some of the key dynamics driving the import and export agendas of countries with national hydrogen strategies.

For example, countries like Chile, Saudi Arabia, and Oman will not necessarily have potential domestic demand for clean hydrogen but can instead cater to rising demand elsewhere, leveraging their vast RE potential. Similarly, countries like Japan and South Korea will not be able to produce most of their clean hydrogen and will bank upon exporting countries to meet their domestic demand, which will be initially directed towards mobility, given Japan's strategic policy and investment focus on hydrogen mobility applications.

Post the energy crisis triggered in the EU by Russia–Ukraine conflict, the EU is expected to import 10 MMT of renewable hydrogen by 2030, under the repower EU package. EU is thus expected to import substantial amounts of renewable hydrogen from neighboring countries through the pipeline and port infrastructure being built under the European Hydrogen Backbone (EHB) initiative. To ensure such import is met cost-effectively, bilateral partnerships with several countries across continents will play a pivotal role in ensuring a reliable supply of hydrogen into the EU and other such import dependent countries and/or regions across continents.

Some countries like Australia and Canada are looking to target both domestic and export markets and depending on domestic sectoral demand these countries are expected to determine the amount of hydrogen they would aim to export in the coming years. Factors like regional biomass availability, nuclear energy capabilities and existing natural gas infrastructure for blue hydrogen production is also observed to be determining the country's intent with the domestic hydrogen industry and its intent in terms of hydrogen's global trade.

References

[1] International Energy Agency, Global hydrogen review, 2022. https://iea.blob.core.windows.net/assets/c5bc75b1-9e4d-460d-9056-6e8e626a11c4/GlobalHydrogenReview2022.pdf.
[2] International Energy Agency, Hydrogen and fuel cells status update. https://www.iea.org/fuels-and-technologies/hydrogen.
[3] World Business Council for Sustainable Development (WBCSD), COP26 pledges. https://www.wbcsd.org/Programs/Climate-and-Energy/Energy/New-Energy-Solutions/News/28-companies-pledge-to-accelerate-use-of-decarbonized-hydrogen-at-COP26.
[4] N. Tenhumberg, K. Büker, Ecological and economic evaluation of hydrogen production by different water electrolysis technologies, Chem. Ing. Tech. 92 (10) (2020) 1586–1595. https://onlinelibrary.wiley.com/doi/full/10.1002/cite.202000090.

Hydrogen production

Recent progress and developments in photocatalytic overall water splitting

Simon Joyson Galbao[a], Sujana Chandrappa[a] and Dharmapura H. K. Murthy[a,b]
[a]Department of Chemistry, Manipal Academy of Higher Education, Manipal Institute of Technology, Manipal, Karnataka, India
[b]Center for Renewable Energy, Manipal Academy of Higher Education, Manipal Institute of Technology, Manipal, Karnataka, India

2.1.1 Introduction

Fossil fuels are the most widely used energy sources globally. However, they produce a significant amount of greenhouse gases and are not renewable [1]. Hence, sustainable energy resources must be utilized to address these challenges. Hydrogen (H_2), being environmentally benign, is predicted to replace fossil fuels. However, more than 94% of the worldwide H_2 consumed is generated by coal gasification and steam reforming of methane. Besides, these techniques require high temperatures or stringent reaction conditions and contribute to carbon footprint by producing harmful carbon-based byproducts/gases [2]. Among the other methods of H_2 generation, electrolysis has gained significant attention owing to its prospect of green H_2 generation at scale. In this approach, an electrolytic cell that produces H_2 by reducing protons from water is powered by green electricity generated by renewable resources (solar panels and wind turbines). A promising solar to hydrogen (STH) efficiency of $>10\%$ using electrolyzers is already achieved. Despite these encouraging features, installing solar panels and/or wind turbines and the associated manufacturing process demands high initial investment costs, thus making green H_2 generation from electrolyzers an expensive method [3]. As a result, green electricity generation (powering the electrolyzer) takes up over 70% of the entire cost of H_2. In this regard, photocatalytic H_2 production utilizing natural sunlight can be an alternative for a cost-effective, sustainable H_2 generation.

The discovery of photoelectrochemical splitting of water by Fujishima and Honda in 1972 set the path for significant research in photocatalysis [4].

Towards Hydrogen Infrastructure: Advances and Challenges in Preparing for the Hydrogen Economy.
DOI: https://doi.org/10.1016/B978-0-323-95553-9.00013-3

The principle of photocatalysis is derived from natural photosynthesis, where green plants facilitate thermodynamically uphill reactions (carbohydrates from CO_2 and H_2O) under sunlight [5]. Photocatalysis can be referred to as an artificial photosynthesis process, where a semiconducting photocatalyst, an artificial leaf, alters the reaction rate under light illumination. Upon absorption of light with an energy more than its band gap, electrons from the valence band (VB) of the semiconducting photocatalyst are excited to the conduction band (CB). This process consequently creates free electrons in the CB for proton reduction to generate H_2, and holes in the VB for water oxidation to form O_2 [6].

2.1.1.1 Thermodynamics of photocatalytic overall water splitting

In overall water splitting (OWS), both H_2 and O_2 are formed stoichiometrically in a thermodynamically uphill reaction (Gibbs free energy, $\Delta G° = 237$ kJ/mol). Hydrogen evolution reaction (HER) is aided by the electrons excited to CB in a photocatalyst, while the holes take part in the oxygen evolution reaction (OER).

$$H_2O \rightarrow H_2 + \tfrac{1}{2}O_2, \ \Delta G° = 237 \, kJ/mol$$

A semiconductor must satisfy two thermodynamic requirements to facilitate the OWS reaction. The semiconductor's VB maximum (VBM) must be higher than the O_2/H_2O energy level, while the CB minimum (CBM) must be lower than the H^+/H_2 energy level. The minimum band energy required for one-step photoexcitation is 1.23 eV. OWS can be carried out by two processes, that is, one-step and two-step photoexcitation. In a one-step photoexcitation system, a single photocatalyst with VBM and CBM positions straddle between the H^+/H_2 and O_2/H_2O redox potentials would realize OWS. In a two-step photoexcitation system, different photocatalysts are employed to generate H_2 and O_2 separately and independently. The hydrogen evolution photocatalyst (HEP) and oxygen evolution photocatalyst (OEP) are combined with either a solid-state electron mediator or redox mediator for the transfer of photoexcited electrons and holes. In a two-step photoexcitation system, photoexcited electrons reduce H^+ to H_2 in HEP, and holes oxidize H_2O generating O_2 in OEP [7]. The kinetics of the latter is more sluggish as there is an increase in backward electron transfer routes. Also, the need for two different types of photocatalysts with mediators introduces new material-related challenges while adding to the cost. Thus, one-step photoexcitation is preferred for OWS.

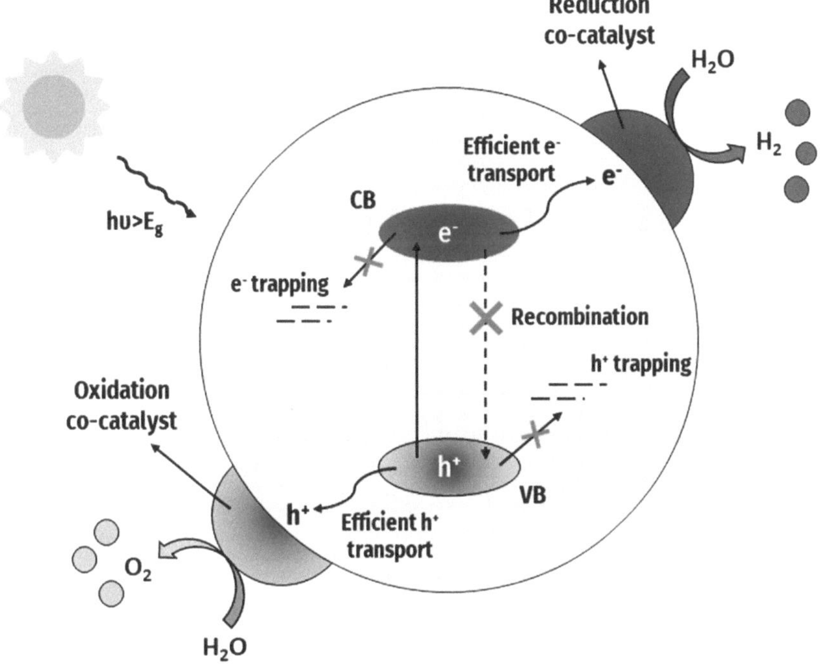

Figure 2.1.1 Mechanisms involved in overall water splitting via one-step photoexcitation.

2.1.2 Mechanism

In a typical OWS via a one-step photoexcitation system, three important steps are involved, as illustrated in Fig. 2.1.1 [8]:

  i. Absorption of light with energy equal to or greater than the band gap of the semiconducting photocatalysts.
 ii. Photogeneration of electrons and holes in CB and VB, respectively.
iii. Migration of electrons and holes from the bulk of the photocatalyst towards the surface.
 iv. Charge transfer across the photocatalyst's interface with water, aided by the cocatalysts leading to OER and HER, as described below.

$$2H_2O + 4h^+ \rightarrow 4H^+ + O_2$$

$$2H^+ + 2e^- \rightarrow H_2$$

The recombination and deep trapping (hence becoming immobile) of the photogenerated charge carriers are in kinetic competition with the

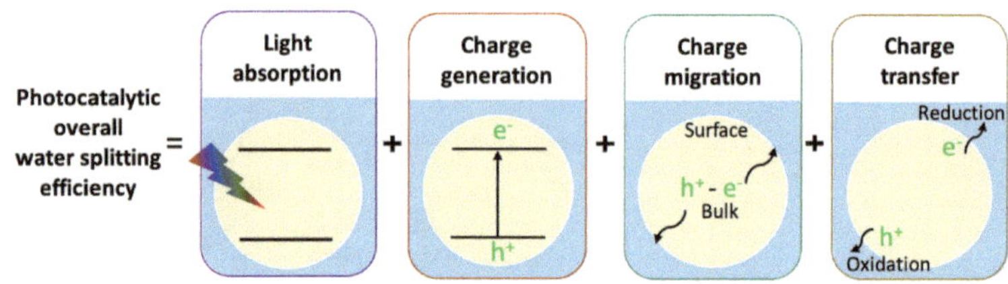

Figure 2.1.2 Factors determining the efficiency of a photocatalytic overall water splitting reaction.

electron transfer process [9]. Hence, it is necessary to effectively utilize these photoexcited electrons and holes for redox reactions before they recombine and/or get deep trapped. Therefore, increasing the charge-carrier lifetime became one of the research challenges and the rational design of photocatalysts and cocatalysts aided in addressing the same (detailed in later sections). The photocatalytic OWS efficiency collectively depends on the efficiency of light absorption, charge carrier photogeneration, migration of the photogenerated charge carriers, and finally, the charge transfer process at the surface of the photocatalyst (Fig. 2.1.2). Therefore, understanding each of these individual processes is necessary to enhance the photocatalytic efficiency of the system.

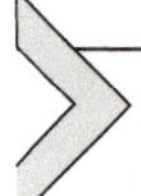

2.1.3 Measuring the photocatalytic activity and performance

For any photochemical reaction, the ratio of the reaction rate to the intensity of the light absorbed at a specific wavelength per volume and time determines the actual quantum yield [10]. However, since light transmission and scattering also occur during the light absorption of the photocatalyst, it is challenging to measure the precise number of absorbed photons. Therefore, the quantum yield in photocatalytic reactions is quantified in terms of apparent quantum yield (AQY), given by the below equation:

$$AQY = \frac{nR}{I}$$

where n denotes the number of electrons or holes utilized in the generation of one H_2 or O_2 molecule ($n = 2$ and 4 for the evolution of H_2 and O_2, respectively in one-step excitation and in Z-scheme OWS, $n = 4$ and 8), R is the number of H_2 or O_2 gas evolved at a specific time interval, and I is the number of incident photons at that time interval. At a given wavelength, a photocatalyst exhibits distinct photocatalytic activity, depending on the optical response of the photocatalyst. Consequently, AQY values change significantly with the wavelength of incident photons, allowing us to determine wavelength-dependent photocatalytic efficiency.

OWS is also measured by STH efficiency, calculated using the below equation:

$$STH = \frac{R \times \Delta G_r}{P_{sun} \times S}$$

where R denotes the rate of H_2 produced, ΔG_r is the Gibbs free energy, P_{sun} is the energy flux of sunlight (mW/cm^2), and S denotes the irradiated area of the photocatalyst [11]. The photocatalytic activity is usually quantified in terms of AQY and STH efficiency since no standard operating procedures exist to quantify and compare the efficiency of different materials [7].

The photocatalytic OWS process utilizes the two most abundant sources, that is, solar energy and water to produce H_2 and O_2. Hence, this process has immense prospects to generate cost-effective green hydrogen at scale. This chapter discusses recent advances in the OWS process and photocatalyst development toward scalable solar H_2 production.

2.1.4 Progress in OWS process and technology

Since the demonstration of the photodecomposition of water vapor on NiO-$SrTiO_3$ by Domen et al. [12], various studies featuring different types of photocatalysts have been reported. Fig. 2.1.3 highlights various classes of materials employed for photocatalytic OWS. The early research in this field was primarily focused on oxide-based perovskites and titanates. $SrTiO_3$ was the first reported photocatalyst for OWS that paved the path to a renaissance in photocatalyst design towards H_2 generation, which has recently demonstrated scalable solar hydrogen generation. In addition to oxide-based semiconductors, metal-free semiconductors [13], oxynitrides [14], and oxysulfides [15] are extensively studied for photocatalytic OWS to harvest a wider part of the solar spectrum, compared to metal oxides. Significant efforts are dedicated to enhancing the efficiency of these photocatalysts by various techniques such as tuning the morphology, enhancing charge separation efficiency [16], band gap engineering [17], and Z-scheme system [18].

The following sections discuss recent developments in OWS employing various photocatalysts and approaches to enhance efficiency.

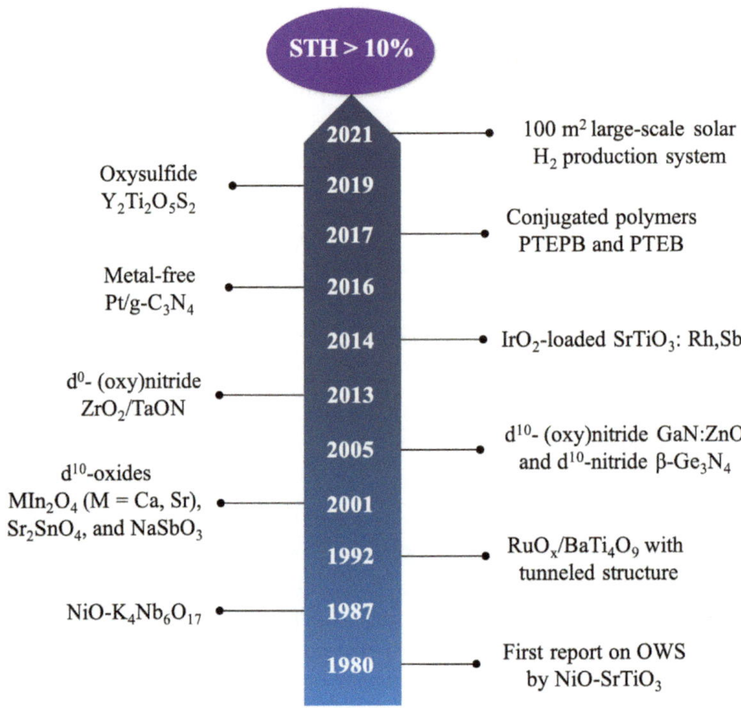

Figure 2.1.3 Timeline of various photocatalysts employed and progress in overall water splitting reaction.

2.1.5 Nitride and oxynitride-based materials

Nitrides such as tantalum nitride (Ta$_3$N$_5$) and gallium nitride (GaN) can be synthesized by nitridation of respective metal or metal oxide precursors under NH$_3$ flow [19]. Ta$_3$N$_5$ has 2.1 eV bandgap making it visible-light active. Ta 5d orbitals constitute the CB of Ta$_3$N$_5$, while VB is characterized by N 2p orbitals. Ta$_3$N$_5$ fulfils the thermodynamic requirements for OWS [20], but the sluggish kinetics associated with the rapid recombination of photoexcited electrons and holes is one of the bottlenecks for achieving improved efficiency in OWS [21]. In attempts to overcome this, Wang et al. reported OWS via one-step photoexcitation in Ta$_3$N$_5$ produced upon using KTaO$_3$ as a precursor [22]. Ta$_3$N$_5$ nanorods developed on the KTaO$_3$ edge sites on nitridation under NH$_3$ demonstrated a virtually defect-free and single-crystal nature. Ta$_3$N$_5$/KTaO$_3$ exhibited an absorption edge at around 350 nm, characteristic of KTaO$_3$ absorption, and further extended up to 600 nm due to the absorption of the pure Ta$_3$N$_5$ phase. To investigate the active sites for photocatalytic H$_2$ and O$_2$ evolution RhCl$_3$ and MnO$_x$

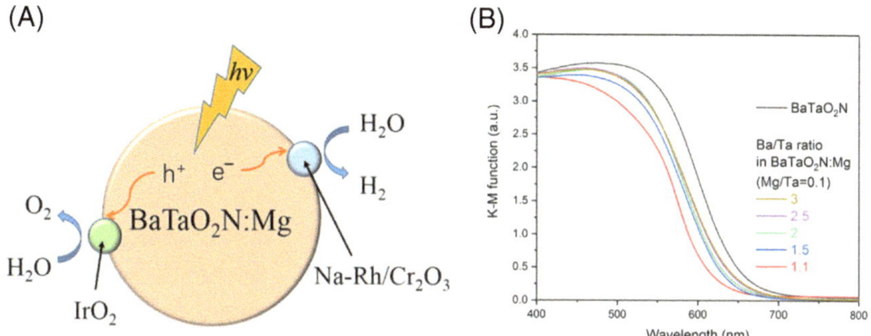

Figure 2.1.4 (A) Overall water splitting process on BaTaO$_2$N:Mg photocatalyst loaded with Cr$_2$O$_3$/Na-Rh and IrO$_2$ cocatalysts. (B) The absorption edge of BaTaO$_2$N:Mg with fixed Mg/Ta = 0.1 and varied Ba/Ta ratio [23]. *(Copyright 2022 American Chemical Society).*

were photodeposited on Ta$_3$N$_5$/KTaO$_3$. Rhodium clusters and MnO$_x$ were deposited on Ta$_3$N$_5$ nanorods which acted as reduction sites evident due to the reduction of Rh^{3+} by photoexcited electrons. An increase in nitridation time from 0.25 h to 10 h reduced the crystal quality of the Ta$_3$N$_5$ nanorods and introduced defects. Consequently, a pronounced trapping process was observed that further decreased the photocatalytic OWS efficiency. Simultaneous H$_2$ and O$_2$ were evolved via OWS under visible light ($\lambda \geq 420$ nm) in a stoichiometric ratio of 2:1 by Ta$_3$N$_5$/KTaO$_3$ loaded with Rh/Cr$_2$O$_3$. However, the STH efficiency was found to be 0.014% and the AQE was 2.2% at 320 nm, 0.22% at 420 nm, and 0.024% at 500 nm, which is orders of magnitude lower than UV-light active SrTiO$_3$:Al photocatalyst.

Oxynitrides materials offer unique advantages over nitrides and oxides. Owing to their narrow band gaps compared to similar oxides, oxynitrides are widely employed for photocatalytic applications under visible light irradiation. Li et al. synthesized Mg-doped BaTaO$_2$N by thermal nitridation in an RbCl flux [23]. The absorption edge of BaTaO$_2$N:Mg with increasing Ba/Ta ratio red-shifted close to that of pristine BaTaO$_2$N, as shown in Fig. 2.1.4B. BaTaO$_2$N:Mg was co-loaded with Cr$_2$O$_3$/Na-Rh and IrO$_2$ to produce H$_2$ and O$_2$, respectively. The maximum photocatalytic activity was observed for the material with Mg/Ta = 0.1, giving 0.4% AQY at 420 nm, while the STH efficiency measured was 0.0004%.

2.1.6 Oxysulfide-based photocatalysts

Oxysulfides are formed when sulfide ions are incorporated into the metal–oxide lattice. The sulfide ions negatively shift the VB edges of these oxides, thus narrowing the band gap and making it suitable for photocatalytic OWS. Sulfide-based photocatalysts are prone to self-oxidation, thereby hindering water oxidation. On the contrary, oxysulfides are more stable against self-oxidation due to the hybridization of S–3p and O–2p orbitals which stabilizes sulfide ions [24].

Wang et al. reported an oxysulfide photocatalyst, $Y_2Ti_2O_5S_2$, for OWS having a bandgap of 1.9 eV [15]. $Y_2Ti_2O_5S_2$ met the thermodynamic criteria for OWS at pH around 8–9. IrO_2 served as a cocatalyst for OER, while Rh/Cr_2O_3 aided HER. The photodeposition of cocatalysts in their respective reduction/oxidation sites aided in retaining their functional oxidation state, essential for efficient OWS. The effect of Rh and IrO_2 species on $Y_2Ti_2O_5S_2$ was evaluated by HER in an aqueous Na_2S–Na_2SO_3 solution and OER in an aqueous $AgNO_3$ solution, respectively. OWS was observed up to 640 nm with AQY $5.3 \pm 0.3\%$ at 420 nm and $2.3 \pm 0.1\%$ at 480 nm. The H_2 and O_2 evolution was not affected by background pressure since Cr_2O_3 shells functioned as a diffusion barrier preventing background reactions. Isotopic labelling experiments were conducted to evidence that simultaneous H_2 and O_2 gases were due to water splitting and not because self-oxidation of S^{2-} species.

2.1.7 Conjugated polymers for OWS

Conjugated polymers (CPs) have recently emerged as promising materials for photocatalytic OWS, owing to their ease of scalable production and constituting earth–abundant inexpensive elements [25]. Compared to the widely used inorganic semiconductors, the optoelectronic properties of CPs can be easily modified by varying synthetic methods and precursors [26]. However, bulk CPs are inefficient for photocatalysis due to limited light absorption, high recombination of charge carriers, low surface area [25], and thus need a significant improvement in the performance.

Wang et al. reported OWS via conjugated microporous polymer nanosheets (CMPNs) synthesized from 1,3,5-triethynylbenzene (PTEB) and 1,3,5-tris-(4-ethynylphenyl)-benzene (PTEPB). The ultrathin CMPNs synthesized reduce the charge recombination by transporting the photo-generated charge carriers to the polymer surface, enabling redox reactions. The stoichiometric ratio of H_2 and O_2 was generated with AQE of 7.65% and 10.3% at 420 nm from PTEB and PTEPB, respectively. STH efficiency reached 0.6% and 0.31% for PTEB and PTEPB, respectively, under the full solar spectrum [26]. Recently, Bai et al., reported photocatalytic OWS by a

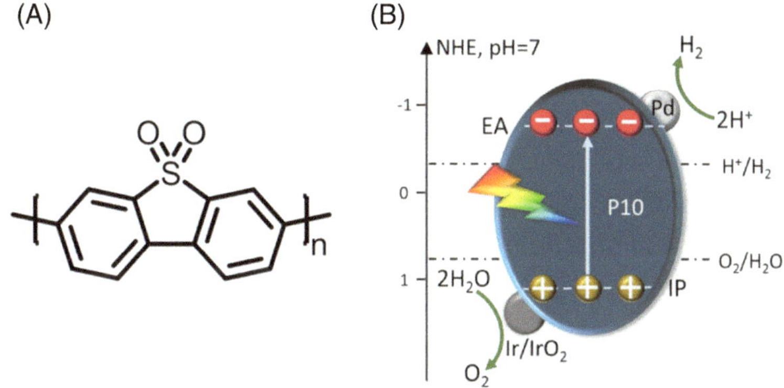

Figure 2.1.5 (A) Homopolymer of dibenzo[b,d]thiophene sulfone (P10). (B) Energy levels of P10 related to redox potentials [27]. *(Copyright 2022 Angewandte Chemie International Edition).*

homopolymer of dibenzo [b, d] thiophene sulfone (P10) (Fig. 2.1.5A) loaded with palladium and iridium oxide cocatalysts [27]. P10 contained palladium residues since it was synthesized by the Suzuki–Miyaura reaction catalyzed by Pd^0. The Pd residues in P10 acted as cocatalyst for H_2 production. Transient absorption UV/Vis spectroscopy evidenced that $P10^{(-)}$-$IrO_2^{(+)}$ state formed in the material enabled water oxidation at the IrO_2 cocatalyst. P10-Ir facilitated OWS to generate stoichiometric amounts of H_2 and O_2 for more than 60 h while yielding an AQY of 0.062% at 350 nm. The low STH efficiency of 0.0047% observed was attributed to low light absorption due to lower amounts of photocatalyst used.

2.1.8 Metal-free photocatalysts

Metal-free photocatalysts are comprised of earth–abundant materials such as carbon (C), oxygen (O), nitrogen (N), sulfur (S), and phosphorus (P). Among these, graphitic–carbon nitride (g–C_3N_4) has gained wide attention due to its tunable surface and optoelectronic properties. Specifically, g–C_3N_4 is an active material in photocatalytic research because of its optical absorption edge extended to the visible region [28]. Despite promising features, g–C_3N_4 has not yet demonstrated efficient OWS activity due to its sluggish kinetics, attributed to the lack of redox sites at the surface. The photocatalytic activity of bulk g–C_3N_4 is enhanced by tuning the optoelectronic and morphological properties and employing suitable cocatalysts [13].

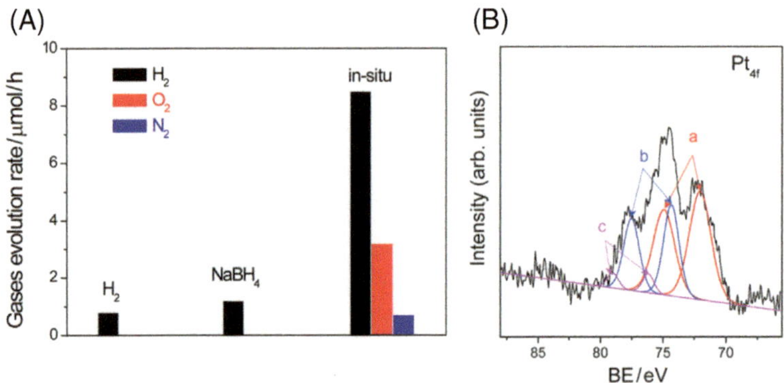

Figure 2.1.6 (A) Comparison of H_2 evolution rate and $NaBH_4$ reduced and in-situ photodeposited $Pt/g\text{-}C_3N_4$. (B) XPS spectra of Pt_{4f}. reproduced from [13] *(From: The Royal Society of Chemistry)*.

OWS by $g\text{-}C_3N_4$ without sacrificial agents was first reported by Zhang et al., wherein Pt, PtO_x, and CoO_x redox cocatalysts modified $g\text{-}C_3N_4$ simultaneously produced H_2 and O_2 [13]. Bulk $g\text{-}C_3N_4$ could not achieve an efficient OWS reaction due to densely packed graphitic layers and a lack of active redox sites. Hence $g\text{-}C_3N_4$ was modified by photodeposition of Pt on the surface to realize OWS. *In-situ* photodeposition of Pt leads to homogenous dispersion of Pt species on the surface of $g\text{-}C_3N_4$. On the contrary, photodeposition by H_2 and $NaBH_4$ reduction resulted in irregular deposition leading to no O_2 and negligible H_2 (Fig. 2.1.6A). When photodeposited material is irradiated with light of suitable wavelength, photoexcited charge carriers generated migrate to the surface of the nanosheets preventing recombination and electrons reduce Pt^{4+} adsorbed onto the surface which further aids water splitting reaction. As shown in Fig. 2.1.6B, XPS data revealed that Pt deposited was in three states, that is, Pt^0, Pt^{2+}, and Pt^{4+}. Pt^0 was responsible for H_2 evolution, whereas PtO_x promoted O_2 evolution. Additionally, it was observed that simultaneously photodeposited Pt^0 and PtO_x acted as cocatalysts for HER and OER, respectively. However, singly deposited cocatalysts exhibited poor photocatalytic activity. Cobalt was deposited to further selectively enhance the material's water oxidation and photocatalytic efficiency. Simultaneous H_2 and O_2 evolution in a stoichiometric ratio 2:1 under UV (12.2 and 6.3 µmol/h) and visible light (1.2 and 0.6 µmol/h) was greatly enhanced upon cocatalyst photodeposition, and the photocatalyst was stable for 510 h. Regardless of the AQY being 0.3% at 405 nm, which is lower than inorganic $SrTiO_3$-based photocatalysts, it is a promising candidate for OWS activity, owing to its metal-free nature, ease of synthesis and tunability.

2.1.9 ABX$_3$-type photocatalysts

Perovskites are a class of materials with a chemical formula of ABX$_3$ where A is a bigger cation in the divalent state, B is a smaller cation in the tetravalent state, and X represents an anion, typically oxygen or halogen. They are one of the extensively investigated photocatalysts/materials for photocatalytic OWS owing to their ease of synthesis, favorable energetics for water-splitting, compositional flexibility, susceptibility to doping or substituting, chemical/structural stability which endows tunable optoelectronic properties [29]. However, perovskites are fundamentally limited by their light absorption in the ultraviolet (UV) region, due to their wide band gap. In this direction, substantial efforts are dedicated to extending the optical absorption edge of these materials [30]. SrTiO$_3$ is a perovskite widely studied material for OWS with 3.2 eV band gap [19].

Takata et al. demonstrated OWS using aluminum-doped SrTiO$_3$ (SrTiO$_3$:Al) with external quantum efficiency (EQE) reaching up to 96% and internal quantum efficiency (IQE) recorded to be unity. SrTiO$_3$:Al was photodeposited with Rh and CrO$_3$ in a two-step process to simultaneously produce H$_2$ and O$_2$ in 2:1 stoichiometric ratio, respectively. Further work demonstrated stable STH efficiency exceeding 200 days of continuous operation under ambient conditions in a 100 m^2 photocatalyst modules. For cocatalysts, Rh core/Cr$_2$O$_3$ shell structure was used by reducing Rh^{3+} to Rh^0, while the consecutive reduction of Cr(VI)O$_4{}^{2-}$ by electrons led to the formation of Cr(III)$_2$O$_3$ during the photodeposition process. The high photocatalytic activity observed was due to the deposition of cocatalysts at different crystal facets, which induced an internal electric field promoting charge transfer process. The Rh/Cr$_2$O$_3$ cocatalyst was deposited on {1 0 0} crystal facets enabling electron transfer to reduce protons forming H$_2$. Furthermore, CoOOH photodeposition on {1 1 0} crystal facets facilitated hole transfer, eventually reducing the charge recombination process. Thus the photocatalyst SrTiO$_3$:Al utilized virtually all the photogenerated charge carriers for the OWS reaction. The 1% STH efficiency achieved in SrTiO$_3$:Al is close to its thermodynamic limit, due to its wide band gap, limiting light absorption of incoming sunlight primarily in UV region. Hence, narrow band gap visible-light-absorbing photocatalysts that are capable of harnessing wider part of the solar spectrum are required for further enhancing the

STH efficiency in a similar way [16]. Table 2.1.1 lists various photocatalysts, cocatalysts employed, and the activity for photocatalytic OWS reaction.

2.1.10 Role of cocatalysts in realizing efficient OWS

Cocatalysts deposited on the semiconductors' surface forms junctions/interfaces thus accelerating the reaction by promoting charge separation and transfer [34]. The cocatalysts facilitate charge injection from the photocatalyst to the adsorbed species thus reducing charge recombination [35]. Cocatalysts promoting HER and OER are called reduction cocatalysts and oxidation cocatalysts respectively (Fig. 2.1.1). The rate of OWS is determined by both HER and OER so it is important to employ dual cocatalysts to enhance the photocatalytic activity of the material. Rh, Pt, Ru, and Ir are widely used as HER cocatalysts while oxides of Fe, Co, Ir, Ru, and Ni are effective as OER cocatalysts [34].

While promoting HER and OER, the cocatalysts can also induce backward reactions, that is, hydrogen oxidation reaction (HOR) and oxygen reduction reaction (ORR). These backward reactions are thermodynamically more feasible than forward reactions. Therefore, it is crucial to suppress these reactions without affecting forward reactions. Bimetallic cocatalysts like $Rh_{2-y}Cr_yO_3$ are utilized to suppress these backward reactions [7]. Cocatalyst distribution and also placement of respective cocatalysts is majorly dependent on the method employed. The cocatalysts deposited by impregnation and adsorption techniques tend to randomly distribute on the surface of a material. Such random deposition may promote recombination of charge carriers as reduction cocatalysts may get loaded onto the hole accumulating sites or oxidation cocatalysts may deposit on electron-rich sites. Therefore, it is essential to load these cocatalysts at different crystal facets which leads to spatial charge separation and thus reduces charge recombination [7,16]. In this direction, Zang et al. revealed that the location of NiO cocatalyst deposition on $NaTaO_3$ affects the charge separation [36]. This was demonstrated by employing photoreduction and impregnation as deposition techniques which resulted in random and selective redox active sites, respectively [36]. Additionally, chemical state(s) of the elements constituting the cocatalyst vary depending on the method employed which further effect the photocatalytic efficiency [37].

Furthermore, it is evident that cocatalysts mentioned in Table 2.1.1 comprise of noble metals. Therefore, there is a crucial need for the development of nonprecious metal cocatalysts. So far particulate photocatalysts

Table 2.1.1 The overall water splitting activity of various photocatalysts via one-step photoexcitation.

Photocatalyst	Cocatalyst	Band gap	Efficiency	References
Ta_3N_5	Rh/Cr_2O_3	2 eV	AQE: 2.2% (320 nm), 0.22% (420 nm) and 0.024% (500 nm) STH: 0.014%	[22]
$Y_2Ti_2O_5S_2$	$Cr_2O_3/Rh/IrO_2$	1.9 eV	AQE: 0.36% (420 nm), 0.2% (500 nm), 0.05% (600 nm) STH: 0.007%	[15]
$g-C_3N_4$	$Pt-CoO_x$	2.80 eV	AQY: 0.3% (405 nm)	[13]
$SrTiO_3:Al$	$Rh/Cr_2O_3/CoOOH$	3.2 eV	EQE: 95.7% (350 nm), 95.95% (365 nm), 91.6% (365 nm), STH: 0.65%	[16]
$BiVO_4:In,Mo$	RuO_2	2.5 eV	AQY: 3.2% (420–800 nm)	[31]
$BiYWO_6$	RuO_2	2.71 eV	AQY: 0.17% (420 nm)	[32]
CoO	NA	2.6 eV	STH: 5%	[33]
Dibenzo [b, d] thiophene sulfone (P10)	Pd/IrO_2	NA	AQY: 0.062% (350 nm) STH: 0.0047%	[27]
PTEPB PTEB	NA	3.04 eV 3.13 eV	AQE: 10.3% and 7.6% (420 nm), STH: 0.6% and 0.31%	[26]
$BaTaO_2N:Mg$	$Cr_2O_3/Rh/IrO_2$	NA	AQY: 0.08% (420 nm), STH: 0.0004%	[23]

are employed for OWS reaction only on a laboratory scale. The scalable OWS reaction using the lab–scale particle suspension system to generate H_2 and O_2 gases has certain limitations. The overall cost of such a system is high owing to the requirement of large amount of water. Even though the depth of the reactor is 1 cm, 10 kg/m^2 volume of water is required which makes the reactor bulky. Also, if particles are suspended, they tend to settle at the bottom without continuous stirring which makes it difficult for efficient light absorption [38]. To overcome these limitations for scalable OWS reaction, the photocatalysts need to be immobilized onto a substrate forming photocatalytic reactors.

2.1.11　Scalable H$_2$ production via OWS

Based on the laboratory measurements, few efforts have been made to realize scalable photocatalytic water-splitting. The advantages and challenges of scalable solar water splitting have been explored and are discussed below.

Goto et al. developed a 1 m^2 water-splitting panel whose design is illustrated in Fig. 2.1.7A, where SrTiO$_3$:AI was immobilized by drop-casting it on a glass substrate using SiO$_2$ nanoparticles as an inorganic binder [39]. When the inner side of the panel window was hydrophilic in nature, the produced gas bubbles were released immediately from the panel, whereas the hydrophobic window allows the accumulation of gases. Thus, it is essential to use hydrophilic windows when no scavengers are used to avoid mixing of H_2 and O_2 gases which could be hazardous. After 30 min, 508 mL of H_2 was produced at 298 K and 1 atm. The STH efficiency measured was 0.4% which was less than the lab-scale experiment (0.5%), and it was attributed to the inhomogeneous photocatalyst sheet. The HER rate was constant up to 150 h, and subsequently decreased. In attempts to further increase the stability of the photocatalyst for large-scale systems, Lyu et al. reported AI-doped SrTiO$_3$ coloaded with RhCrO$_x$ cocatalyst for OWS [40]. The system showed stability up to 1300 h (Fig. 2.1.7B) with AQY of 50% at 365 nm, and STH efficiency measured was 0.3%. This work further encouraged the development of photocatalysts for long-term sunlight-driven OWS.

Recently, a 100 m^2 solar H_2 production modules incorporating 1600 reactor units was set up at the Kakioka Research Facility (University of Tokyo) [41]. The photocatalyst SrTiO$_3$:AI immobilized by silica nanoparticles was deposited on clear and frosted glass via spray method. The thickness of particulate was maintained from 4 μm to 10 μm. A trial panel reactor of an irradiation area of 9 m^2 was irradiated with the sunlight of

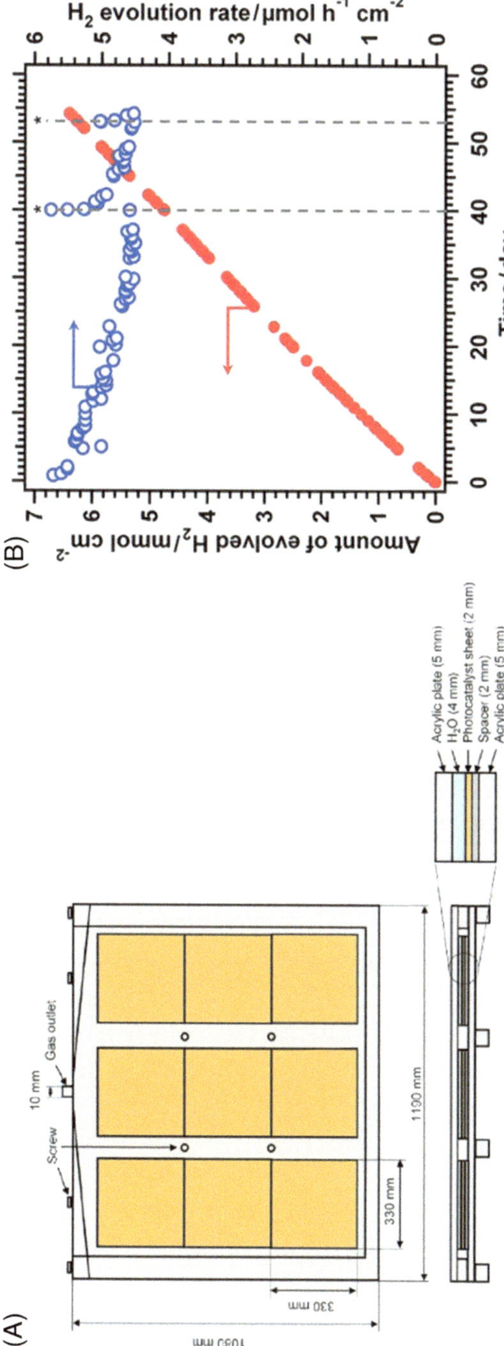

Figure 2.1.7 (A) Demonstration of 1 m² water-splitting panel [39] (Copyright 2017 Elsevier). (B) Water-splitting activity over 1300 h on SrTiO₃:Al [40]. (From: The Royal Society of Chemistry).

intensity 0.88 kW/m^2 producing H$_2$ and O$_2$ gases at a rate of 568 mL/min (11:00 am–11:30 am) equivalent to STH efficiency of 0.76%. The oxyhydrogen gas produced was separated by a gas–separating unit comprising a filter made of hollow polyimide fibers. The feed gas collected on a single day was 970 L, which contained a mixture of H$_2$ and O$_2$ at a molar ratio of 2.0 and water vapor. With an average H$_2$ purity above 94% (excluding water vapor) and H$_2$ output of 19.9 mol, the membrane separation produced 505 L of H$_2$-enriched filtrate gas that was saturated with water vapor. The H$_2$ production system was functional for a year without any hazards. Though the STH efficiency was 0.76%, less than expected, this work demonstrated the possibility of scalable sunlight-driven H$_2$ production. The low STH efficiency is attributed to the wide band gap of SrTiO$_3$ that only absorbs UV-light, which constitutes 4% of the solar spectrum. Therefore, extending the optical absorption towards visible/near-infrared region is essential to improve the STH efficiency. Further, the panels need to be developed using cost-effective photocatalytic materials, which cuts down the overall cost of the system, making solar hydrogen generation economically viable.

Recently water vapor-fed OWS reaction on SrTiO$_3$:Al coated on TiO$_x$ or TaO$_x$ nanomembranes was achieved. The reaction system was stable for 100 h and STH efficiency for CoOOH/Rh loaded SrTiO$_3$:Al was found to be 0.4%, while AQY of $54 \pm 4\%$ at 370 nm [42].

As evident from Fig. 2.1.8, the AQY of different photocatalysts is low under visible light, compared to UV light. It is necessary to extend the absorption edge up to 600 nm of the photocatalysts with 40–60% AQY for enhanced light absorption which will eventually improve the STH efficiency of the system. In addition, effective utilization of visible-light photons is also required. As shown in Fig. 2.1.9, STH efficiency in the range of 5–10% is necessary for making it economically viable [44]. This is possible by gaining insight into the optoelectronic structure and related charge carrier dynamics of the material. Hence, it is critical to design and develop photocatalysts with wide absorption edge which can be utilized for scalable water-splitting systems. Also, it is necessary to employ suitable safety measures to reduce the risk of explosive oxyhydrogen gases. Although dual cocatalysts are extensively used in OWS reactions, there is still scope for improving the morphology, composition, and size of the cocatalysts and their location sites. Accelerating material design to improve visible light absorption, good stability, defect-free nature, and high photocatalytic efficiency is necessary to commercialize scalable photocatalytic water splitting systems.

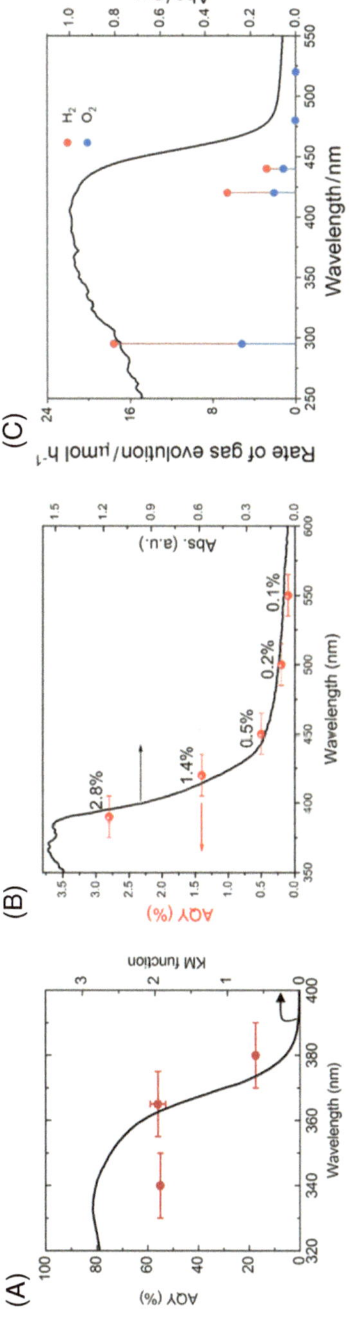

Figure 2.1.8 Wavelength-dependent (action spectra) photocatalytic activity of (A) $SrTiO_3$:Al [39]. (Copyright 2017 Elsevier). (B) 3D g-C_3N_4 [43]. (Copyright 2019 Elsevier). (C) P10-Ir [27]. (Copyright 2022 Angewandte Chemie International Edition).

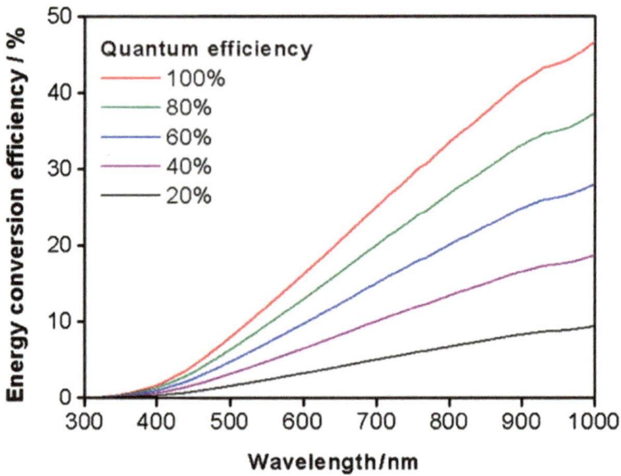

Figure 2.1.9 Relationship between the wavelength of photons available for overall water splitting and the expected solar energy conversion efficiency [45]. *(Copyright 2010 American Chemical Society).*

2.1.12 Summary and future prospects

Photocatalytic OWS is the simplest process to generate solar fuels. Recent developments demonstrated long-term stable and scalable (100 m^2) photocatalyst modules for solar H$_2$ generation. However, to make this promising technology appealing to industries for commercial application, it is necessary to further enhance the STH towards 5–10%. In this direction, the following approaches are essential to consider.

- The quantum efficiency for photocatalytic OWS reaction is almost unity in SrTiO$_3$:Al photocatalyst. However, if undoped SrTiO$_3$ is used as a photocatalyst, the efficiency reduces by orders of magnitude compared to its aluminum-doped counterpart. It means that the rational design of photocatalysts and understanding which parameter is key to realize enhanced OWS activity is critical. Similarly, elucidating the role of morphology, synthesis conditions, cocatalysts utilization, and its site of deposition, which collectively determine the OWS efficiency is also crucial.

- The development of visible-light absorbing photocatalysts with an absorption edge greater than 600 nm and an AQY exceeding 60% would aid in achieving an STH efficiency of 5%. With already demonstrated scalable solar H$_2$ generation using photocatalyst modules under ambient sunlight, it would be very promising to develop and utilize suitable visible-light harnessing photocatalyst that capture a significant part of incoming sunlight compared to oxide-based ultraviolet-light absorbing photocatalysts. In this direction, g-C_3N_4, Ta_3N_5, $BaTaO_2N$ and few conjugated polymers have shown promising OWS activity. However, these systems showed significantly smaller STH efficiency compared to the only UV-light absorbing $SrTiO_3$:AI. Hence, effective utilization of the visible-light via rationally designed photocatalysts is essential.

- The use of cocatalysts to facilitate the charge transfer process and thus avoid a backward reaction is indispensable to realize efficient OWS. However, widely utilized cocatalysts employ noble metal as a primary component making the preliminary synthesis process expensive. This imperative need necessitates the development of better cocatalyst designs that essentially utilize smaller amounts of noble metals, thereby effectively reducing the overall cost of photocatalysts. Also, the photocatalyst design strategies to selectively photodeposit these cocatalysts on the required facets of the crystallographic planes promote effective/efficient charge separation at the functional facets of the crystal.

- Though the photogeneration of charge carriers is a bulk process requiring charge carriers to migrate to the surface, the charge transfer process across the photocatalyst-reaction medium interface ultimately determines the photocatalytic efficiency. Therefore, establishing a comprehensive correlation between the dynamics of photophysical processes in the photocatalyst and photocatalytic efficiency is essential. In this regard, laser spectroscopy provides valuable information on the lifetime of charge carriers and can probe defect state formation. This will aid in understanding the charge carrier dynamics, to reveal the factors decreasing the efficiency of charge transfer process.

Future studies on developing photocatalysts for scalable OWS to produce H_2 and O_2 will aid in the transition towards a hydrogen-powered sustainable economy.

References

[1] M.A. Rosen, S. Koohi-Fayegh, The prospects for hydrogen as an energy carrier: an overview of hydrogen energy and hydrogen energy systems, Energy Ecol. Environ. 1 (1) (2016) 10–29. https://doi.org/10.1007/s40974-016-0005-z.

[2] L. García, Hydrogen production by steam reforming of natural gas and other nonrenewable feedstocks, in: Compendium of Hydrogen Energy, Elsevier, 2015, pp. 83–107 https://doi.org/10.1016/B978-1-78242-361-4.00004-2.

[3] K. Scott, Introduction to electrolysis, electrolysers and hydrogen production, in: Electrochemical Methods for Hydrogen Production, Royal Society of Chemistry, 2019, pp. 1–27 https://doi.org/10.1039/9781788016049-00001.

[4] A. Fujishima, K. Honda, Electrochemical photolysis of water at a semiconductor electrode, Nature 238 (5358) (1972) 37–38. https://doi.org/10.1038/238037a0.

[5] A. Larkum, Limitations and prospects of natural photosynthesis for bioenergy production, Curr. Opin. Biotechnol. 21 (3) (2010) 271–276. https://doi.org/10.1016/j.copbio.2010.03.004.

[6] R. Saravanan, F. Gracia, A. Stephen, Basic principles, mechanism, and challenges of photocatalysis, 2017, pp. 19–40. https://doi.org/10.1007/978-3-319-62446-4_2.

[7] S. Chen, T. Takata, K. Domen, Particulate photocatalysts for overall water splitting, Nat. Rev. Mater. 2 (10) (2017) 17050. https://doi.org/10.1038/natrevmats.2017.50.

[8] R. Li, Latest progress in hydrogen production from solar water splitting via photocatalysis, photoelectrochemical, and photovoltaic-photoelectrochemical solutions, Chin. J. Catalys. 38 (1) (2017) 5–12. https://doi.org/10.1016/S1872-2067(16)62552-4.

[9] Y. Tamaki, A. Furube, M. Murai, K. Hara, R. Katoh, M. Tachiya, Dynamics of efficient electron–hole separation in TiO_2 nanoparticles revealed by femtosecond transient absorption spectroscopy under the weak-excitation condition, Phys. Chem. Chem. Phys. 9 (12) (2007) 1453–1460. https://doi.org/10.1039/B617552J.

[10] H. Kisch, On the problem of comparing rates or apparent quantum yields in heterogeneous photocatalysis, Angew. Chem. 122 (50) (2010) 9782–9783. https://doi.org/10.1002/ange.201002653.

[11] Z. Wang, C. Li, K. Domen, Recent developments in heterogeneous photocatalysts for solar-driven overall water splitting, Chem. Soc. Rev. 48 (7) (2019) 2109–2125. https://doi.org/10.1039/C8CS00542G.

[12] K. Domen, S. Naito, M. Soma, T. Onishi, K. Tamaru, Photocatalytic decomposition of water vapour on an $NiO–SrTiO_3$ catalyst, J. Chem. Soc., Chem. Commun. (12) (1980) 543–544. https://doi.org/10.1039/C39800000543.

[13] G. Zhang, Z.-A. Lan, L. Lin, S. Lin, X. Wang, Overall water splitting by $Pt/g-C_3N_4$ photocatalysts without using sacrificial agents, Chem. Sci. 7 (5) (2016) 3062–3066. https://doi.org/10.1039/C5SC04572J.

[14] Y. Lee, et al., Zinc germanium oxynitride as a photocatalyst for overall water splitting under visible light, J. Phys. Chem. C 111 (2) (2007) 1042–1048. https://doi.org/10.1021/jp0656532.

[15] Q. Wang, et al., Oxysulfide photocatalyst for visible-light-driven overall water splitting, Nat. Mater. 18 (8) (2019) 827–832. https://doi.org/10.1038/s41563-019-0399-z.

[16] T. Takata, et al., Photocatalytic water splitting with a quantum efficiency of almost unity, Nature 581 (7809) (2020) 411–414. https://doi.org/10.1038/s41586-020-2278-9.

[17] S. Suzuki, H. Matsumoto, A. Iwase, A. Kudo, Enhanced H_2 evolution over an Ir-doped $SrTiO_3$ photocatalyst by loading of an Ir cocatalyst using visible light up to 800 nm, Chem. Commun. 54 (75) (2018) 10606–10609. https://doi.org/10.1039/C8CC05344H.

[18] Y. Kageshima, et al., Z-Scheme overall water splitting using $Zn_x Cd_{1-x}$ se particles coated with metal cyanoferrates as hydrogen evolution photocatalysts, ACS Catal. 11 (13) (2021) 8004–8014. https://doi.org/10.1021/acscatal.1c01187.

[19] S. Fang, Y.H. Hu, Recent progress in photocatalysts for overall water splitting, Int. J. Energy Res. 43 (3) (2019) 1082–1098. https://doi.org/10.1002/er.4259.

[20] D.H.K. Murthy, et al., Origin of the overall water splitting activity of $Ta_3 N_5$ revealed by ultrafast transient absorption spectroscopy, Chem. Sci. 10 (20) (2019) 5353–5362. https://doi.org/10.1039/C9SC00217K.

[21] E. Nurlaela, A. Ziani, K. Takanabe, Tantalum nitride for photocatalytic water splitting: concept and applications, Mater. Renew. Sustain. Energy 5 (4) (2016) 18. https://doi.org/10.1007/s40243-016-0083-z.

[22] Z. Wang, et al., Overall water splitting by Ta_3N_5 nanorod single crystals grown on the edges of $KTaO_3$ particles, Nat. Catal. 1 (10) (2018) 756–763. https://doi.org/10.1038/s41929-018-0134-1.

[23] H. Li, et al., One-step excitation overall water splitting over a modified Mg-doped BaTaO$_2$N photocatalyst, ACS Catal. 12 (16) (2022) 10179–10185. https://doi.org/10.1021/acscatal.2c02394.

[24] G. Zhang, X. Wang, Oxysulfide semiconductors for photocatalytic overall water splitting with visible light, Angew. Chem. Int. Ed. 58 (44) (2019) 15580–15582. https://doi.org/10.1002/anie.201909669.

[25] C. Dai, Y. Pan, B. Liu, Conjugated polymer nanomaterials for solar water splitting, Adv. Energy Mater. 10 (42) (2020) 2002474. https://doi.org/10.1002/aenm.202002474.

[26] L. Wang, et al., Conjugated microporous polymer nanosheets for overall water splitting using visible light, Adv. Mater. 29 (38) (2017) 1702428. https://doi.org/10.1002/adma.201702428.

[27] Y. Bai, et al., Photocatalytic overall water splitting under visible light enabled by a particulate conjugated polymer loaded with palladium and iridium, Angew. Chem. 134 (26) (2022). https://doi.org/10.1002/ange.202201299.

[28] M.Z. Rahman, M.G. Kibria, C.B. Mullins, Metal-free photocatalysts for hydrogen evolution, Chem. Soc. Rev. 49 (6) (2020) 1887–1931. https://doi.org/10.1039/C9CS00313D.

[29] J. Shi, L. Guo, ABO$_3$-based photocatalysts for water splitting, Prog. Nat. Sci.: Mater. Int. 22 (6) (2012) 592–615. https://doi.org/10.1016/j.pnsc.2012.12.002.

[30] V.-H. Nguyen, et al., Perovskite oxide-based photocatalysts for solar-driven hydrogen production: progress and perspectives, Solar Energy 211 (2020) 584–599. https://doi.org/10.1016/j.solener.2020.09.078.

[31] W.J. Jo, et al., Phase transition-induced band edge engineering of BiVO$_4$ to split pure water under visible light, Proceed. Nation. Acad. Sci. 112 (45) (2015) 13774–13778. https://doi.org/10.1073/pnas.1509674112.

[32] H. Liu, J. Yuan, W. Shangguan, Y. Teraoka, Visible-light-responding BiYWO$_6$ solid solution for stoichiometric photocatalytic water splitting, J. Phy. Chem. C 112 (23) (2008) 8521–8523. https://doi.org/10.1021/jp802537u.

[33] L. Liao, et al., Efficient solar water-splitting using a nanocrystalline CoO photocatalyst, Nat. Nanotechnol. 9 (1) (2014) 69–73. https://doi.org/10.1038/nnano.2013.272.

[34] J. Yang, D. Wang, H. Han, C. Li, Roles of cocatalysts in photocatalysis and photoelectrocatalysis, Acc. Chem. Res. 46 (8) (2013) 1900–1909. https://doi.org/10.1021/ar300227e.

[35] X. Tao, Y. Zhao, S. Wang, C. Li, R. Li, Recent advances and perspectives for solar-driven water splitting using particulate photocatalysts, Chem. Soc. Rev. 51 (9) (2022) 3561–3608. https://doi.org/10.1039/D1CS01182K.

[36] Q. Zhang, et al., Effect of redox cocatalysts location on photocatalytic overall water splitting over cubic NaTaO$_3$ semiconductor crystals exposed with equivalent facets, ACS Catal. 6 (4) (2016) 2182–2191. https://doi.org/10.1021/acscatal.5b02503.

[37] T. Soltani, et al., Effect of transition metal oxide cocatalyst on the photocatalytic activity of Ag loaded CaTiO$_3$ for CO$_2$ reduction with water and water splitting, Appl. Catal. B 286 (2021) 119899. https://doi.org/10.1016/j.apcatb.2021.119899.

[38] T. Hisatomi, K. Domen, Reaction systems for solar hydrogen production via water splitting with particulate semiconductor photocatalysts, Nat. Catal. 2 (5) (2019) 387–399. https://doi.org/10.1038/s41929-019-0242-6.

[39] Y. Goto, et al., A particulate photocatalyst water-splitting panel for large-scale solar hydrogen generation, Joule 2 (3) (2018) 509–520. https://doi.org/10.1016/j.joule.2017.12.009.

[40] H. Lyu, et al., An Al-doped SrTiO$_3$ photocatalyst maintaining sunlight-driven overall water splitting activity for over 1000 h of constant illumination, Chem. Sci. 10 (11) (2019) 3196–3201. https://doi.org/10.1039/C8SC05757E.

[41] H. Nishiyama, et al., Photocatalytic solar hydrogen production from water on a 100-m^2 scale, Nature 598 (7880) (2021) 304–307. https://doi.org/10.1038/s41586-021-03907-3.

[42] T. Suguro, et al., A hygroscopic nano-membrane coating achieves efficient vapor-fed photocatalytic water splitting, Nat. Commun. 13 (1) (2022) 5698. https://doi.org/10.1038/s41467-022-33439-x.

[43] X. Chen, et al., Three-dimensional porous g-C$_3$N$_4$ for highly efficient photocatalytic overall water splitting, Nano Energy 59 (2019) 644–650. https://doi.org/10.1016/j.nanoen.2019.03.010.

[44] T. Hisatomi, K. Takanabe, K. Domen, Photocatalytic water-splitting reaction from catalytic and kinetic perspectives, Catal. Lett. 145 (1) (2015) 95–108. https://doi.org/10.1007/s10562-014-1397-z.

[45] K. Maeda, K. Domen, Photocatalytic water splitting: recent progress and future challenges, J. Phys. Chem. Lett. 1 (18) (2010) 2655–2661. https://doi.org/10.1021/jz1007966.

Hydrogen storage

Current state and challenges for hydrogen storage technologies

Zainul Abdin[a,*], Chunguang Tang[b,*], Yun Liu[b] and Kylie Catchpole[a]

[a]School of Engineering, The Australian National University, Canberra, ACT, Australia
[b]Research School of Chemistry, The Australian National University, Canberra, ACT, Australia

3.1.1 Introduction

Hydrogen could be a key component of the energy transition, potentially serving as a benign and environmentally friendly fuel for transportation, heavy industry, and renewable energy-based power plants. Creating large-scale hydrogen energy value chains is necessary to ensure that hydrogen delivers on its promise. One of the most significant components of these chains is a hydrogen storage facility that connects upstream production with downstream transportation and applications. A hydrogen storage facility is also essential to address fluctuations in both production and demand. However, storing hydrogen at the GW scale (and beyond) is a significant challenge [1]. Since hydrogen is the smallest element, it is highly fugacious and has a very low density. In ambient conditions, hydrogen has a density of 0.08 kg/m^3, and it is impractical to store 1 kg in a tank of 12 m^3. In addition, the levelized costs of hydrogen storage vary according to the applications for which hydrogen is used. For example, hydrogen dispensing can be problematic due to fuel cells' high response time requirements. In addition, urea production requires flattening or refilling hydrogen storage several times a year [2]. Other applications, such as proton exchange membrane fuel cells (PEM), also require high hydrogen purity. Due to these varying constraints and the variability of energy production and demand at different timescales, different storage technologies may be necessary [3].

Hydrogen storage technologies vary in complexity depending on the components and applications involved. Fig. 3.1.1 schematically shows the components and processes involved in some common technologies, from relatively simple compressed gaseous storage to more complicated chemical storage. Currently, hydrogen is stored as a gas in pressurized tanks, which

*These authors contributed equally.

Towards Hydrogen Infrastructure: Advances and Challenges in Preparing for the Hydrogen Economy.
DOI: https://doi.org/10.1016/B978-0-323-95553-9.00012-1

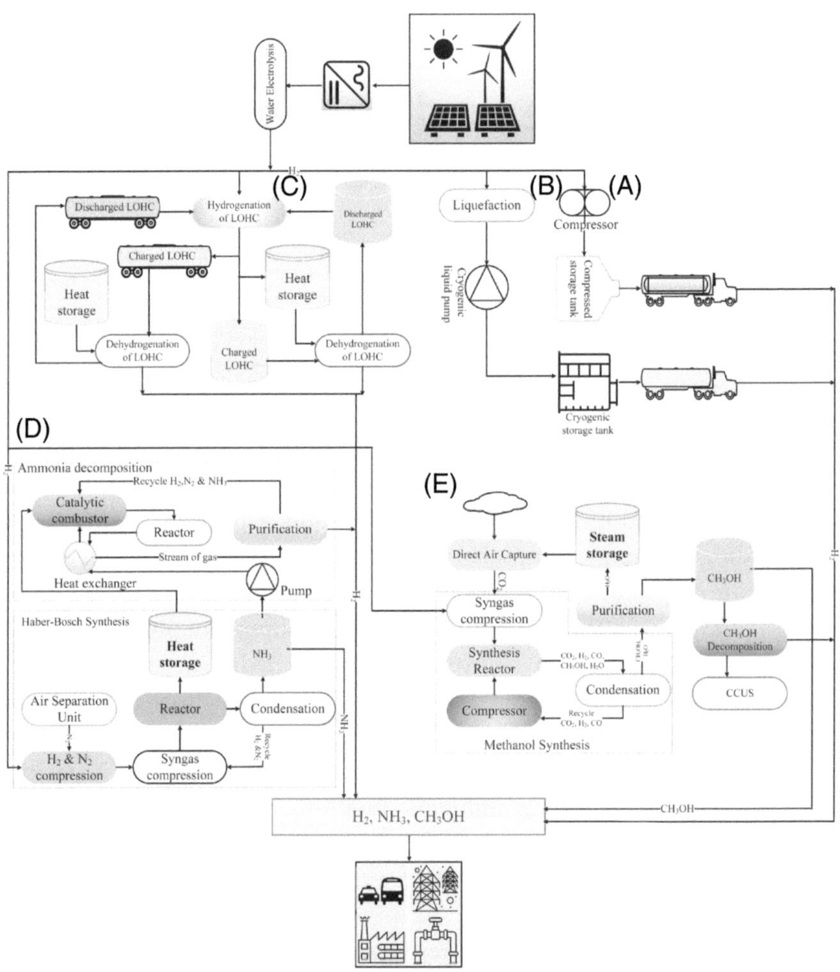

Figure 3.1.1 Schematic of different hydrogen storage systems [2]: (A) gaseous, (B) liquid, (C) LOHCs, (D) ammonia, and (E) methanol. CCUS refers to carbon capture and storage, and LOHC represents a liquid organic hydrogen carrier.

is expensive and raises safety concerns. Physical storage methods based on hydrogen densification are those in which hydrogen molecules do not react with the materials they contain. Physisorption, in which hydrogen molecules are adsorbable to materials, is also a form of physical storage. Another way to store hydrogen is to form chemical bonds with the storage medium. The types of chemical storage can be classified according to the form of the medium, such as solid hydrides, liquid organic carriers, or circular carriers. Hydrogen volumetric and gravimetric densities are compared in Fig. 3.1.2.

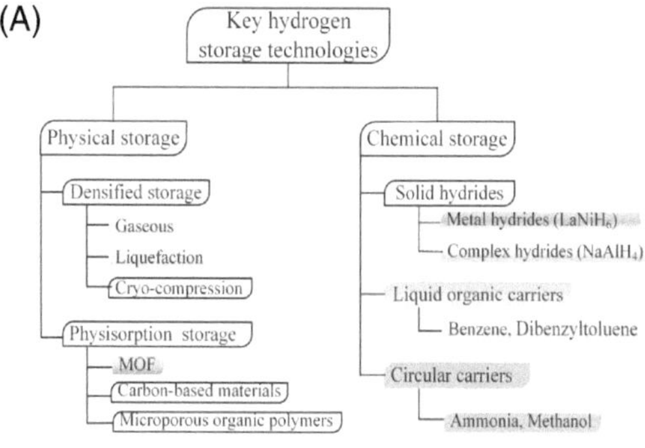

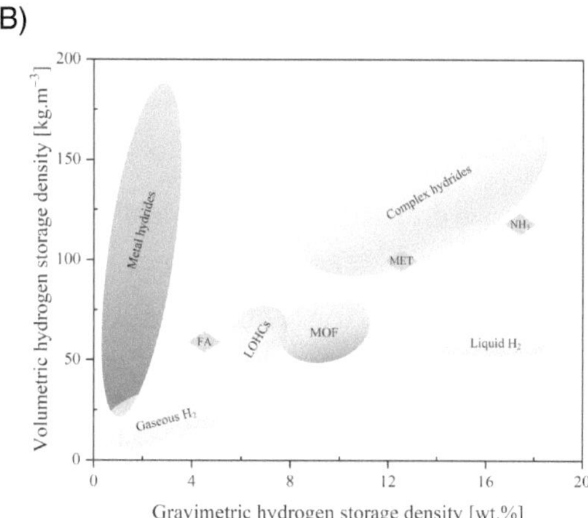

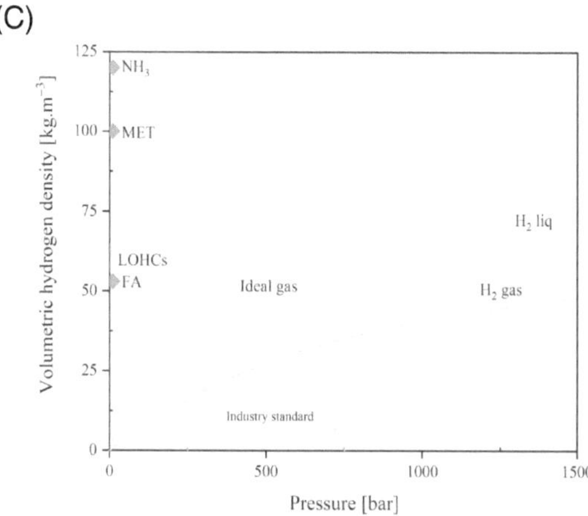

Figure 3.1.2 Energy densities of key hydrogen storage technologies [2]: (A) Technology classifications for hydrogen storage. (B) Gravimetric and volumetric densities of different hydrogen storage technologies; circular carriers: ammonia (NH_3), methanol (MET), formic acid (FA). (C) Hydrogen volumetric density as a function of pressure; the effect of system volume is not considered for liquid hydrogen as in (B).

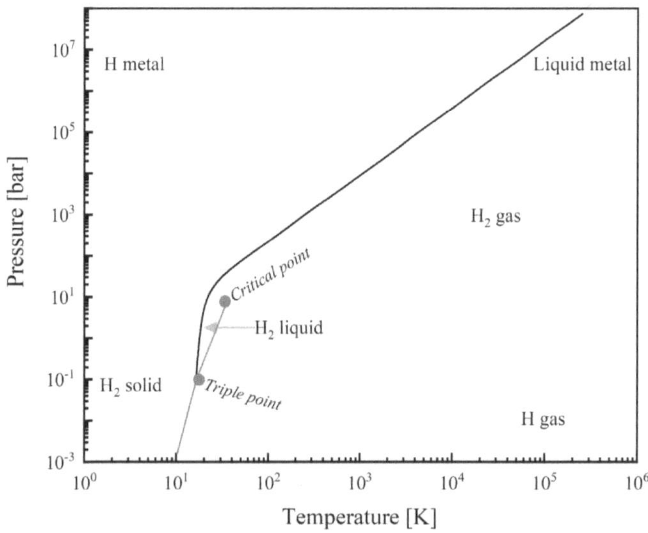

Figure 3.1.3 Primitive phase diagram for hydrogen. *(Replotted from references [4,5]).*

The volumetric and/or gravimetric hydrogen density of other methods is higher than that of traditional compression and liquefaction methods. These technologies are being developed as a result of hydrogen energy value chains. This chapter discusses the (dis)advantages and challenges of various storage technologies, as depicted in Fig. 3.1.2.

3.1.2 Physical hydrogen storage

3.1.2.1 Densified hydrogen storage

The primitive phase diagram of hydrogen in Fig. 3.1.3 illustrates that hydrogen is a liquid at ambient pressure between 20 K and 32 K and a solid below 11 K [4,5]. Hydrogen densification involves compression, liquefaction, and their combination, cryo-compression. Compression is the most common hydrogen storage method because it provides easy operation, storage, and retrieval convenience at ambient temperatures. The process of hydrogen liquefaction requires between 25% and 45% of its energy content [2]. Due to hydrogen's low boiling temperature (20 K), a sophisticated cooling system or bulky insulation is required to maintain the liquid state [6]. Additionally, hydrogen liquefaction faces several challenges, including the total volume and weight of the system, the high cost of the tank, and an exothermic ortho-para conversion (which changes hydrogen's nuclear spin from 1 to 0) [7]. Since volume and pressure are inversely related, cryo-compression at the same temperature increases gas density. When hydrogen is cooled from room temperature (300 K) to liquid nitrogen temperature (77 K), its volumetric density increases by over 3.5 times. A cryo-compression system decreases the system's volumetric and gravimetric capacity due to the additional materials required for cooling. Cryo-compression also consumes approximately 20% of the energy contained in hydrogen due to the compression process [4]. Further details on specific densification methods are provided below.

3.1.2.1.1 *Compressed gases*

Compression is commonly used to store hydrogen at high pressure in order to achieve a lower volume of hydrogen. The two key components of compression are the compressor and the storage compartments that are required to achieve storage pressure. It is generally the case that an efficient compressor requires very low compression work during compression. This work typically falls within the boundaries of adiabatic and isothermal compression; an ideal gas increases in temperature due to work done during adiabatic compression. In actual conditions, a compressor is unlikely to compress gas isothermally, so further compression can be more complex than isothermal compression, which removes heat from the system. It is, therefore, possible to utilize a multistage compressed gas process that results in compression that is as close to isothermal as possible by cooling the compressed gas after each stage [8]. As discussed below, various pressure vessels and underground spaces can be used as storage compartments.

3.1.2.1.1.1 Pressure vessels

It is common for pressure vessels to be used in industrial, automotive, commercial, and aerospace applications, ranging from tiny bottles to massive storage tanks. The optimal material for hydrogen storage is austenitic stainless or non-magnetic steel[1], aluminum alloys, and copper alloys, which are resistant to hydrogen effects at ambient temperatures. However, using high-strength steel in hydrogen storage applications must be avoided because such materials are prone to embrittlement [9].

Typically, pressure vessels have a central cylinder, two spherical domes, and a polar opening, although other shapes, such as toroids, can also be used [10]. There are four main types of pressure vessels for hydrogen storage available on the market today, based on the materials they are made from and their structural characteristics. It is common for type I vessels to be constructed with a metal cylinder typically made of carbon steel or low alloy

---

[1] Austenitic stainless steels have excellent corrosion resistance and formability.

steel. The most common, least expensive, and heavy-pressure vessels have a net volume of up to 2.5 m^3 at a pressure of 200–300 bars. They can work at pressures up to 500 bars [6,10]. In type II vessels, fiber resin composites are used primarily to wrap metal cylinders in part like a hoop wrap [10,6]. The hoop reinforcements provide the metallic liner with improved fatigue resistance and pressure tolerance due to sharing structural loads [11]. Type II vessels are approximately 30–40% lighter than type I vessels but cost about 50% more [12]. The inner metal liners used in type III vessels are composed of aluminum for gas sealing and structural integrity, and carbon fiber composites are entirely wrapped around them. There is the possibility of working with type III vessels at 450 bar pressure [10]. Although type III vessels weigh less than half that of type II vessels, their costs are nearly twice as high [12]. In type IV pressure vessels, fiber resin composites wrap the entire vessel. The structural load is carried by carbon fiber or carbon-glass composites containing polymer-like high-density polyethylene (HDPE). Type IV pressure vessels can be used to achieve a pressure range of up to 1000 bar [10]. Despite its lightness, this type of vessel is relatively expensive. Because carbon fibers are expensive, composite pressure vessels are not economically viable for large-scale applications. Metal-based containers, such as type I and II, are typically used for large-scale applications, including hydrogen refill stations and long-term energy storage [13]. Precommercial (or lab-scale) vessels are also made using composites. A first-of-its-kind pressure vessel was designed by Composites Technology Development Inc. in 2010. With a working pressure of 14 bars, type V is approximately 20% lighter than type IV [12]. However, type V has some disadvantages, including the high cost and inefficiency of storing hydrogen at such low pressures.

Storage of hydrogen at large scales can also be achieved using alternative vessels in addition to pressure vessels. Seamless tubes, multifunctional steel layered vessels (MSLVs), steel-concrete composite vessels, and wind turbine tubular towers are a few examples. A hydrogen filling station typically utilizes high-strength seamless tubes. Despite limited volumes ($\sim$0.411 m^3), they can be assembled with valves, interconnected piping, and manifolds. It is, therefore, possible to store large quantities of hydrogen in a cascade manner with valves and interconnection piping manifolds. As a result, the pipeline becomes more susceptible to leaks [14]. Seamless vessels may be able to withstand pressures as high as 800 bar; however, hydrogen embrittlement of high-strength steels may pose a threat. The vessel structures are not suited to mechanically collecting leaked hydrogen or arresting the propagation of cracks, further complicating online safety monitoring [13,15]. It is possible to resolve the problems associated with seamless vessels by utilizing MSLV. These comprise a hemispherical head with two double layers and a cylinder with three layers wound around a flat steel ribbon. There are no size restrictions on MSLVs, and the manufacturing process becomes more accessible and more efficient as the inner diameter increases. As the outer layer of MSLVs is hydrogen-compatible, they can withstand high pressures (800 bar) without hydrogen embrittlement. Furthermore, the structural design of MSLVs allows for detecting hydrogen leaks through an escape pipe, preventing any possible fires [14]. There is also the possibility of storing high-pressure hydrogen in large steel–concrete composite vessels; however, constructing these vessels is costly and complex. Due to their layered construction, composite vessels allow hydrogen to pass through holes, increasing safety, and preventing catastrophic failures [14,16]. Depending on the crossover pressure of the tower, hydrogen can be stored in the towers of wind turbines between 10 and 15 bars. For example, a 1.5 MW wind turbine of 84 m can store approximately 940 kg of hydrogen at an optimum pressure of 11 bars [14,17]. Also, fatigue damage to the towers caused by hydrogen pressure is expected to be negligible compared to the frequent stresses associated with aerodynamics. However, there is a need for further research regarding the fatigue life of the tower as well as issues such as hydrogen embrittlement and hydrogen corrosion [14,18].

It is possible to store hydrogen in metal pressure vessels that are designed to store natural gas and town gas. These vessels differ from type I vessels in many ways. The most commonly used types are gas holders, spherical pressure vessels, and pipe storage systems [3]. Under low pressure, large quantities of gas can be stored in gas holders [19]. Spherical pressure vessels are capable of storing significantly more hydrogen, around ~27 tons, than gas holders but at a lower capital and operating cost [19,20]. It is estimated that existing natural gas pipelines can store around 12 tons of hydrogen at pressures up to 100 bar [3]. Although hydrogen embrittlement remains a concern when using these pipelines, building and maintaining hydrogen pipelines is also an extremely complicated process [3,19,20]. Hydrogen pipelines are present in some industries, such as the chemical and petroleum industries. However, further research and development will be required to determine whether they are feasible [14].

3.1.2.1.1.2 Underground storage

A system that allows hydrogen to be injected, stored, and withdrawn from beneath the surface as required. Underground storage has several advantages over above-ground storage, including security, minimal interference with urban planning, minimal land requirements, and reduced costs. As a result of injection and withdrawal wells in an underground storage reservoir, which consists of a confining layer as well as injection and withdrawal wells, hydrogen is pumped into subsurface geological formations. It is possible to prevent gas leakage by enclosing formations [21,22]. The storage system uses two types of gas: cushion gas and working gas. Cushion gas remains permanently in underground storage reservoirs to prevent formation damage. Stable working gas pressure is necessary for efficient gas delivery [23]. In other words, working gas is a storage capacity that can be continuously charged and discharged [14,24]. Depleted natural gas and oil reservoirs, aquifers, salt caverns, abandoned mines, and rock caves are the five types of underground storage discussed below [2].

Depleted cap rocks in natural gas and oil reservoirs can act as confining layers because they are usually gastight. Despite depletion, residual gases can still exist in reservoirs and serve as cushion gases. Depleted reservoirs possess many of the necessary facilities and tools to convert them to hydrogen storage, making them relatively inexpensive to convert [25]. However, underground storage facilities cannot always be built in exhausted reservoirs [26]. It has not yet been proven that depleted reservoirs are capable of storing hydrogen. A number of factors contribute to this limitation, such as the depth of the reservoir, its structure, the amount of hydrogen dissolved, and the amount of hydrogen diffused into the surrounding area [27].

Aquifers are a type of underground formation composed of porous, permeable rocks able to store groundwater [28]. A dome-shaped formation is necessary to displace water sideways and downwards for gas storage. Because of their porous nature, aquifers are extremely expensive to develop [29,30]. As well as requiring additional confining layers of gas–impermeable rocks (to prevent gas leakage), they also require different tests to determine their tightness, making them more expensive than depleted reservoirs. The loss of hydrogen from aquifers may also be a concern due to mineral or microorganism reactions [29]. As a result, not much effort has been made to store hydrogen in aquifers, with the exception of some companies that store gases containing 50–60% hydrogen, such as Gaz de France [31].

The United States, Britain, and Germany already use salt caverns to store hydrogen [32]. These are formed by leaching salt from salt domes or salt beds. As salt has unique properties, salt caverns are an ideal option for storing underground gas because they are gas-tight and stable in nature [2]. Salt caverns require less cushion gas to store than other geological storage systems and require different amounts of cushion gas depending on their depth [14]. Compared with above-ground compressed storage methods, salt caverns provide enhanced safety and are significantly more economical than current and potential methods of storing hydrogen [2,33]. There are, however, several challenges associated with this system, including geological constraints, pipeline durability concerns, design constraints, concerns regarding hydrogen leakage, and legal and social concerns [2].

Although abandoned mines can be converted into high-pressure gas storage facilities, their gas tightness has not yet been proven, and testing has been difficult. As a result, abandoned mines have not been used to store gas as frequently as other geological alternatives [3]. As with conventional underground mining techniques, rock caverns are unlikely to contain high-pressure gas tightly. However, they can be sealed using a variety of methods, including natural groundwater and stainless steel linings, which serve as permeability barriers [14]. The Swedish government is currently constructing hard rock caverns for storing natural gas. They are lined with steel and can withstand a maximum pressure of 200 bar [3].

In addition to the above approaches, hydrogen can be stored underground by mixing it with other gases. Two possible methods are to blend hydrogen into natural gas pipelines and to methanate hydrogen and carbon dioxide. By doing so, natural gas is conserved. Although hydrogen can be blended, it is limited by embrittlement and leakage risks [36]. Despite the controversy regarding hydrogen percentages in a blend, many studies indicate that up to 15–20% hydrogen can be present [34,35]. By injecting hydrogen into a depleted aquifer or depleted gas reservoir, it is possible to store CO_2 underground. In the presence of a catalyst, methanation usually takes place at high temperatures. It has also been reported that methanogenic bacteria can cause the process at much lower temperatures [36,37]. This methane may be added to the natural gas pipeline in the future. However, concerns remain regarding the rate and efficiency of methane production [38].

3.1.2.1.2 Liquid hydrogen

3.1.2.1.2.1 Liquefaction

Fig. 3.1.4 illustrates the hydrogen liquefaction process as a schematic diagram consisting of compressors, heat exchangers (HX), expansion engines, and throttle valves. The liquefaction process begins with pre-compression [39]. After pre-cooling to 80 K, hydrogen is adsorbable. An impurity may be removed during this stage by freezing. Next, hydrogen is cooled below 30 K using a closed-loop cryogenic refrigeration cycle that involves continuous

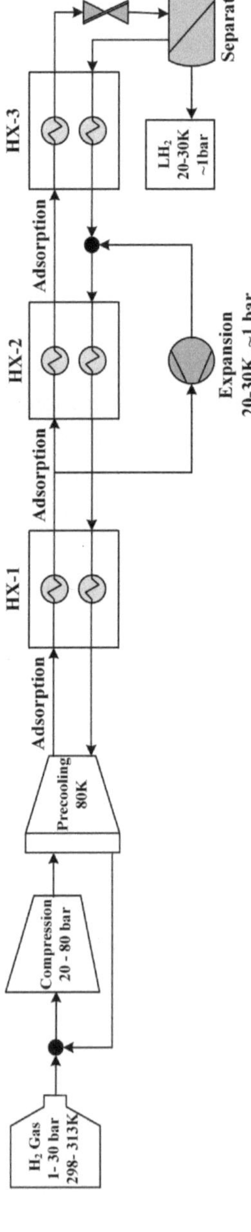

Figure 3.1.4 The schematic of the hydrogen liquefaction process. *(Replotted from reference [41]).*

or batch catalytic conversion from ortho- to para-hydrogen after adsorption [40]. A typical adiabatic expansion occurs after cooling via Joule-Thompson or turbine expansion [41].

There are three refrigeration cycles that are most commonly described in the literature for hydrogen liquefaction: the Linde-Hampson, Claude, and Brayton cycles. Linde-Hampson is an old and widely used liquefaction process; however, it is less efficient than other processes; therefore, it is primarily used on a small scale of up to 2 tons per day (TPD). As a result, the gas passes through a throttle valve before experiencing an isenthalpic Joule–Thompson (J–T) expansion and becoming liquid. After removing the liquid, the cool gas is returned to the compressor via a heat exchanger [42].

Claude cycles, on the other hand, are preferred for plants producing greater than 2 TPD. The feed gas is compressed to 50 bar, cooled to 80 K, and then converted to 47% para-hydrogen through an ortho-para hydrogen converter (O–P converter). As soon as the feed hydrogen gas has been liquefied, it is expanded at 1 bar at 20 K by the J–T valve. In industry and academia, increasing the number of heat exchangers and/or using nitrogen precooling to enhance process efficiency has been suggested [43].

As the working fluid for the Brayton cycle, nitrogen, helium, or a combination of refrigerants is used. Due to the increased efficiency of the Brayton cycle and its flexibility in fuel consumption and heat-to-power ratio, it has recently been considered for larger-scale systems. However, it is typically found only in small-scale plants [41]. The Brayton cycle uses turbines instead of Joule–Thompson throttle valves to expand. In the Claude cycle, refrigerant streams are enlarged in expanders, while a second partial stream is cooled and expanded through a throttle valve, creating a two-phase stream [44].

3.1.2.1.2.2 Cryogenic containers

In cryogenic hydrogen storage, it is imperative to minimize hydrogen boil-off losses [3]. Heat sources include ortho-para conversions, mixing and pumping energy, radiant heating, convection heating, and conduction heating. Ortho-para conversion is generally performed during the liquefaction stage of hydrogen to reduce conversion and subsequent evaporation during storage. Cryogenic containers are also designed in such a manner as to minimize the transfer of heat between the outer container wall and the liquid within [45].

In cryogenic containers, it is common to construct them with double walls and several layers of heat shielding between them, most commonly aluminum or plastic Mylar [45]. In addition, large storage containers are sometimes coated with liquid nitrogen in order to reduce heat transfer by lowering the temperature gradient within the container's outer wall. As a result of minimizing the surface-to-volume ratio, liquid hydrogen containers are typically larger and spherical. In general, cylindrical and spherical tanks have similar surface-to-volume ratios, which makes them easier to construct and more economical to operate [46].

Cryogenic storage does not prevent hydrogen evaporation. Hydrogen can be vented, allowed to build up pressure within the container, or captured and returned as part of the liquefaction process. When hydrogen is stored in a pressure vessel, pressure will gradually build up until the gas reaches its design pressure, at which it can be vented. Whenever storage times are shorter than the lock-up time or the time required for the gas pressure to reach the pressure limit, hydrogen loss does not occur [45]. Cold hydrogen gas should be removed from the liquefied hydrogen and reliquefied if the hydrogen is stored at the same location as the liquid hydrogen. As a fuel for large transportation applications, such as barges, boilers can be filled with boil-off gas [3,45].

3.1.2.1.3 Cryo-compression

In a cryo-compressed gas, hydrogen is supercritically compressed. Lawrence Livermore National Laboratory, USA, developed the concept of cryo-compressed vessels [47]. Cryo-compressed storage is a hybrid of compressed gaseous hydrogen storage and liquefied hydrogen storage. A composite pressure vessel with a metallic liner is located in the secondary insulated jacket. This jacket minimizes heat transfer from hydrogen to the environment, reducing boil-off losses from liquefied hydrogen storage while maintaining the system's energy density [10]. An insulated tank maintains a cryogenic temperature (20 K) and a high pressure (at least 30 MPa) during storage. As the tank is capable of taking high pressures, it is possible to increase the pressure in the tank significantly before hydrogen must be boiled off. Compared to liquefied hydrogen, cryo-compressed gas will be more cost-effective, as well as offer an increased degree of flexibility in terms of infrastructure since cryo-compressed tanks can be refueled with both gaseous and liquid hydrogen [48].

A cryo-compressed vessel was also tested at Argonne National Laboratory, USA, where an aluminum liner was coated with carbon fiber reinforced polymers (CFRP) in order to withstand extreme pressure. In addition, the inner layer is enclosed in a vacuum space filled with a number of highly reflective metalized plastic sheets. This vacuum space is surrounded by a stainless steel jacket similar to the Type III pressure vessel [47]. Although studies have primarily focused on the cryo-compressed vessel's storage performance, less attention has been paid to its specific structure, manufacturing process, stress distribution, and strain distribution [47–49].

BMW Group has demonstrated that cryo-compressed hydrogen gas has a greater density than liquid hydrogen after validating cryo-compressed hydrogen storage for hydrogen vehicles with high energy and long-range requirements [10].

3.1.2.2 Physisorption hydrogen storage

The physisorption process involves the adsorption of gas molecules on a surface containing van der Waals bonds and subsequent release without decomposition of those molecules. A low-temperature ($\sim-196°C$) but high-pressure preference characterizes this process because of the weak van der Waals bonds between molecular hydrogen and sorbents. The surface area of a sorbent usually determines its hydrogen physisorption capacity [50]. According to the Freundlich equation, adsorption at a given temperature increases rapidly with pressure at relatively low pressures but gradually saturates at higher pressures. The optimal pressure range for hydrogen adsorption depends on the adsorbent and application, typically 10–100 bar.

It has been widely recognized that carbon-based materials, such as activated carbon, graphite, carbon nanotubes, and carbon foams, provide excellent hydrogen physisorption properties because of their low weight, high surface area, and chemical stability [51]. Due to their high specific surface areas and tailored porosity, microporous or hyper-crosslinked polymers can also be used as energy storage materials. Synthetic routes may be utilized to optimize their interactions with hydrogen molecules [52–54]. The crystallinity and porosity of a covalent organic framework (COF) can be controlled more precisely than that of polymers [55]. Zeolites are suitable for catalysis, gas adsorption, purification, and separation due to their well-defined open pores and tunable pore sizes [56]. Additionally, zeolites are inexpensive and have been widely employed in industrial processes for decades. However, the hydrogen storage capacity (HSC) of zeolites is limited. A comprehensive experimental study has shown that zeolites have a HSC of less than 2 wt% at cryogenic temperatures. This is less than 0.3 wt% at room temperature or higher [50].

Nanoporous materials, such as metal oxide frameworks (MOFs), can be used to store hydrogen. MOFs are highly crystalline hybrid structures consisting of inorganic and organic molecules that are linked by metal clusters, ions (secondary building units), and multidentate organic ligands [57]. In order to synthesize MOFs, different organic linkers and metal nodules can be used. Due to its variable building blocks, a molecular organic compound can also be distinguished by its large surface area, high porosity, uniform pore size, and well-defined hydrogen occupancy sites. MOFs appear to be promising candidates for hydrogen physisorption in light of these characteristics [57].

3.1.3 Chemical storage

In contrast to physical hydrogen storage, where the H–H bonding of hydrogen molecules is intact, chemical storage of hydrogen entails rupturing the H-H bonds and forming chemical H–X bonding (X represents other elements), which is far stronger than the van der Waals bonds. Furthermore, chemical storage achieves higher volumetric densities than physical storage. Chemical storage can be divided into solid hydrides, liquid organic carriers, and "circular" carriers depending on the hydrogen-poor media's phase (solid, liquid, or gas). The compatibility of liquid organic and circular carriers with current fuel transportation infrastructure is a considerable advantage over solid hydrides. The following subsections discuss these chemical carriers' current states and challenges.

3.1.3.1 Solid hydrides

In general, metal hydrides and complex hydrides are two different types of solid hydrides. A wide range of substances, such as elemental metals and alloys, are included in the category of metal hydrides. These substances can dissociate hydrogen molecules at the surface and absorb hydrogen atoms into the bulk structure. Complex hydrides are metal salts with their metallic cations bonded to a complex anion with a central atom forming a covalent bond with surrounding hydrogen.

3.1.3.1.1 Metal hydrides

The first studies on hydrogen absorption into metals date back to the late 19th century [58]. However, hydrogen storage with metal hydrides has recently seen increasing research interest for their favorable volumetric and gravimetric density, safety, and energy efficiency compared to hydrogen condensation. As a result, numerous reviews of metal hydrides [59] have appeared in the literature. In addition, the US Department of Energy has created a historical metal hydride database [60] of more than 2000 entries.

Due to space limitations, we only briefly discuss the various metal hydrides and their recent developments here.

Although the majority of metallic elements can create binary metal hydrides MHx, they are usually unsuitable for hydrogen storage because of unfavorable weight capacity and/or dynamics. Among the elemental metal hydrides, MgH_2 and AlH_3 are considered the most promising because of their benefits for large-scale hydrogen storage, such as abundance, cheap price, low weight, and high potential for storing hydrogen. The key challenge for MgH_2 is the sluggish kinetics resulting from the reluctant hydrogen dissociation/association on the surface of magnesium as well as its difficult diffusion within the bulk hydride [3], which increases the temperature of dehydrogenation (over 300°C). Among the concerns for MgH_2 is its significant dehydrogenation enthalpy of 75 kJ/mol-H_2. Current promising strategies for improving the kinetics of MgH_2 (de)hydrogenation include nanostructuring of the particles, nanoconfinement, and the addition of transition metal catalytic additives [3,61]. The onset decomposition temperature of nanosized MgH_2 is 150–200°C lower than its micron counterpart [62], despite the fact that a direct connection between the dimension and dehydrogenation enthalpy of MgH_2 particles could not be established [61]. In contrast to MgH_2, AlH_3 has a very low enthalpy of dehydrogenation ($\sim$6–7.6 kJ/mol-H_2) [63] and hence can rapidly release hydrogen when temperatures are low ($\sim$100°C). However, the reverse reaction of gaseous hydrogen and aluminum metal is complicated. Therefore, the recent development of AlH_3 has focused on electrochemical and thermochemical processes for AlH_3 regeneration [3].

In practice, reversible hydrides with moderate hydrogen affinities are often based on alloys that combine strong hydriding elements, A, and weak hydriding elements, B. As a classic example, the favorable hydride formation enthalpy of compound $LaNi_5$ ($\Delta H_f = -30.9$ kJ/mol-H_2) results from the combination of strong hydriding La ($\Delta H_f = -208$ kJ/mol-H_2) and weak hydriding Ni ($\Delta H_f = -8.8$ kJ/mol-H_2). The alloys are usually intermetallic compounds with typical formulas of AB_5, AB_3, AB_2, AB, and A_2B [64]. Compared to elemental hydrides, intermetallic compound hydrides generally have more favorable kinetics, with hydrogen desorption temperatures, and pressures closer to the ambient condition (as shown in Fig. 3.1.5). However, the kinetics could vary significantly from compound to compound. In addition to the above intermetallic compounds, other intermetallic families exist, such as A_2B_7, A_6B_{23}, and A_2B_{17}. Although the family of intermetallic compound hydrides is extensive, only a few types (AB_5, AB_2, AB, and A_2B) are in practical applications, and their HSC is below 3 wt% [64,65].

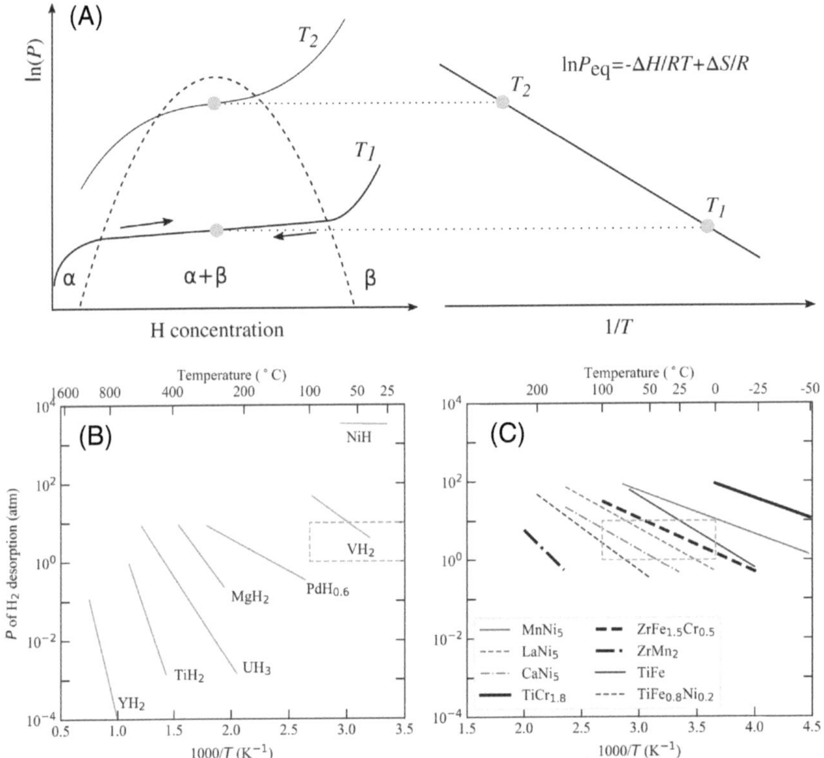

Figure 3.1.5 Effects of pressure and temperature on metal hydrides. (A) Schematic representation of isothermal pressure-composition plot of (de)hydrogenation (*left*), with the hysteresis loops neglected, and the corresponding Van't Hoff plot (*right*) obtained from the equilibrium pressure (P_{eq}). α and β represent the hydrogen-lean solid solution and hydrogen-rich hydride, respectively. (B–C) Van't Hoff plots for hydrogen desorption of elemental hydrides and intermetallic compound hydrides. The dashed boxes indicate the pressure and Htemperature range (1–10 atm, 0–100°C) selected for practical uses. Curves in (B–C) are replotted from reference [59].

As reviewed recently [65], the applications of metal hydrides include several commercial cases. For example, Griffith University in Australia has a demonstrative building designed to operate independently of the electricity grid. Its energy system includes roof-top photovoltaic panels of 376 kW, Li-ion batteries and a metal hydride reservoir from Japan Steel Works for storing 120 kg H_2 (equivalent to 2 MWh of electricity), and two 30 kW PEM fuel cells [50]. In addition, an independent solar-hydrogen system (H_2One^{TM}) was established by Toshiba Corporation. The integrated system generates hydrogen by electrolyzing water from a renewable energy source. It also uses metal hydride hydrogen storage for its 3.5 kW fuel cell and a 10 kW battery [50]. Moreover, metal hydrides are used in various fields, such as heat storage, hydrogen compressors, and nucleation reactors.

3.1.3.1.2 Complex hydrides

Complex hydrides are metal salts in which hydrogen and a core atom form a complex anion by directional covalent bonding. Typically, the core atoms belong to elements B, N, and Al, and the resulting hydrides are $(BH_4)^-$-based borohydrides, $(NH_2)^-$-based amides, and $(AlH_4)^-$-based alanates, respectively. The metal cation atoms in complex hydrides are usually alkali metals. Various derivatives of complex hydrides can be synthesized via anion substitution or the reaction of various hydrides [66]. Because of the comparatively light components used, complex hydrides have a high capacity to store hydrogen gravimetrically. However, only a few of them are reversible for dehydrogenation with the help of suitable additives or catalysts. Typical complex hydrides include $LiBH_4$, $NaAlH_4$, and $LiNH_2$.

The theoretical HSC of $LiBH_4$ is very high (18.4 wt%), but the thermodynamic and kinetic properties do not favor its partial decomposition ($LiBH_4 \rightarrow LiH + B + 1.5H_2$) with a standard enthalpy change being $\sim$67 kJ/mol-H_2. As a result, $LiBH_4$ decomposition mainly occurs above 400°C, and typically only about 50% of its theoretical capacity can be reached [67]. However, combining $LiBH_4$ with additives such as MgH_2 can reduce the decomposition enthalpy to $\sim$42 kJ/mol-H_2 by forming more stable products while increasing the reversible storage capacity to 8–10 wt% [68].

Theoretically, $NaAlH_4$ can store $\sim$7.4 wt% hydrogen, but in practice only 5.6 wt% can be reversibly produced following the reactions below:

$$3NaAlH_4 \leftrightarrow Na_3AlH_6 + 2Al + 3H_2$$

$$Na_3AlH_6 \leftrightarrow 3NaH + Al + 1.5H_2$$

The reactions have favorable thermodynamics, with the first step in equilibrium at $\sim$30°C and the second at $\sim$100°C at 1 atm. The favorable thermodynamics and adequate storage capacity make $NaAlH_4$ promising for PEM fuel cell applications. Rehydrogenation of the dehydrogenated $NaAlH_4$ is possible under moderate conditions using early transition metal chlorides [69]. For hydrogenating NaH and Al into Na_3AlH_6, only low hydrogen pressures of around 25 bar are necessary. This makes it a sensible option to use Na_3AlH_6 alone as a storage media, although this reduces the storage capacity to 1.7 wt% [70].

To avoid the release of ammonia during dehydrogenation, $LiNH_2$ must be mixed with a metal hydride, such as MgH_2 or LiH. However, this causes the loss of active storage material and is also poisonous to PEM fuel cells. Furthermore, compared with LiH, mixing MgH_2 with $LiNH_2$ results in a lower enthalpy of dehydrogenation. Therefore, integrating a $LiNH_2$-based hydrogen storage tank with high-temperature PEM fuel cells has been explored and attempted [71]. The storage system uses a mixture of $2LiNH_2\text{-}1.1MgH_2\text{-}0.1LiBH_4$ and is coupled with the intermetallic compound $ZrCoH_3$ which can release hydrogen at room temperature and hence quickly start up the system.

Complex hydrides can generally release hydrogen under mild conditions, which is a great advantage, but their hydrogen recharge is still challenging and costly. Moreover, the large-scale handling of complex hydrides is incompatible with current fuel infrastructures. These factors make practical applications of complex hydrides in a large distribution network difficult.

3.1.3.2 Liquid organic hydrogen carriers

Liquid organic molecules with conjugated π-bonds can store hydrogen via their broken π-bonds. Studies on storing hydrogen in organic molecules started more than 40 years ago, and before the term "liquid organic hydride carrier" (LOHC), which is now widely used, surfaced in the 2010s, the word "liquid organic hydride" was frequently used [72]. The potential for convenient bulk hydrogen storage and transfer using liquid organics has been established. They can be made in large quantities, work well with current infrastructure, and can be distributed and delivered continuously.

To date, a number of liquid organic molecules have been explored for hydrogen storage applications, including some typical LOHCs such as BZ (benzene), TOL (toluene), NAP (naphthalene), NEC (N-ethylcarbazole), and DBT (dibenzyltoluene). Among them, TOL and DBT are already used in industry, and the other LOHCs are either structurally significant or have special features. The selection of LOHCs is influenced by storage capacity as well as other elements, including abundance, cost, safety, thermodynamics, kinetics, and compatibility with existing infrastructure and technology. Fig. 3.1.6 compares some technical features of some typical LOHCs.

Among the aforementioned typical LOHCs, BZ has the simplest structure and is a building block for many other LOHCs. The BZ system has a theoretical HSC of ~7.2 wt%. However, BZ is highly toxic, and cyclohexane's relatively low boiling point (81°C) makes the separation of hydrogen challenging. The structures of TOL and NAP can be obtained

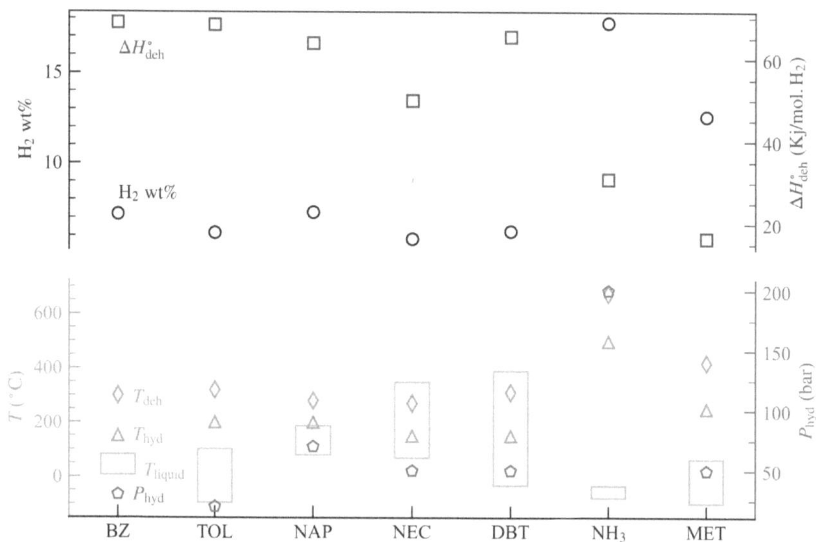

Figure 3.1.6 Properties and (de)hydrogenation parameters of some typical LOHCs and circular hydrogen carriers. (*top*) Capacity (wt%) and dehydrogenation enthalpy change. (*bottom*) Typical temperatures for (de)hydrogenation (T_{hyd} and T_{deh}) and hydrogenation pressures. The rectangles show the liquid-phase temperature range for the carriers.

from BZ by adding a methyl radical or repeating the BZ building block, which reduces the HSC to $\sim$6.2 wt% for TOL but slightly increases the capacity to $\sim$7.3% for NAP. TOL and NAP also have reduced toxicity compared to BZ. TOL is the carrier of choice for the hydrogen business of Chiyoda, Japan [2]. The demonstration plant of Chiyoda has a hydrogen production rate of 50 Nm^3/h, with the dehydrogenation rate and selectivity above 95% and 99.9%, respectively [73].

DBT is another industrial LOHC used in Germany and South Africa. The DBT system has a high HSC (6.2 wt%), low toxicity, and excellent thermal stability. The hydrogenation process of DBT is particularly attractive for industrial applications since it allows the use of hydrogen-containing gas mixtures [74]. Nevertheless, more research is needed for efficient dehydrogenation of perhydro-DBT, as incomplete dehydrogenation can substantially reduce the storage capacity after a few iterations [75].

The NEC system exhibits a significantly lower (de)hydrogenation enthalpy change than other LOHCs ($\sim$50 kJ/mol-H$_2$) and thus lower dehydrogenation temperatures due to its special composition and structure [76]. This makes it a favorable match for PEM fuel cells from a temperature perspective. However, dealkylation at relatively low temperatures was reported

[77]. Another drawback of this system is NEC's relatively high melting point, 68°C. Besides, the annual production of NEC is not more than 10,000 tons, which is a factor of concern for wide application [78].

3.1.3.3 Circular carriers

Typical circular carriers include ammonia (NH_3) and methanol (CH_3OH) from the reaction of hydrogen with N_2 and CO_2 gas, respectively. The term "circular carriers" comes from gas molecules being discharged into the air upon dehydrogenation, forming a cycle of molecules in the reaction.

Compared to the typical LOHCs, ammonia and methanol have significantly higher HSC, but their (de)hydrogenation requires more elevated temperatures despite the fact that their enthalpies of dehydrogenation are lower than those of the LOHCs.

In conjunction with hydrogen and nitrogen plants, the Haber-Bosch method is commonly employed to produce ammonia. The conventional Haber-Bosch process uses natural gas as a cheap and efficient hydrogen source for ammonia production. In contrast, a green Haber-Bosch process relies on renewable hydrogen from water electrolysis and nitrogen from air separation. The source gases make the difference between the traditional and green Haber-Bosch processes. The ammonia synthesis process prefers reduced temperatures and elevated pressures since it is exothermic and entropy-reducing. However, the synthesis is performed at high temperatures with catalysts to overcome the high energy barrier of nitrogen dissociation. Recent years have seen investigations into novel ammonia production, such as electrochemical synthesis [79].

Typically performed between 400°C and 700°C and dependent on catalysts, the decomposition of ammonia is endothermic. Although the synthesis and breakdown of ammonia are microscopically reversible, the optimal catalysts for the two processes are different [80]. While small-scale electric furnaces for ammonia decomposition are commercially available, ammonia cracking plants for massive production of pure hydrogen are still unavailable. Ammonia can also be directly used as fuel for coal-fired plants. For example, Japan's power producer Jera Co. is conducting a demonstration project to mix ammonia with coal for power generation [81].

Typically, methanol can be created using pressurized syngas, which can be obtained from various sources such as light hydrocarbons, heavy oils or carbonous solids. Green production of methanol uses water-split hydrogen and carbon dioxide recovered from the exhaust gas. In addition to its high costs, CO_2 capture consumes a great deal of energy and reduces the efficiency of the whole process. It should be remembered that the purification of synthesized methanol depends on the final consumer's needs. For instance, the methanol stream ($CH_3OH + H_2O$) might be kept in its raw state for later hydrogen production.

The majority of methanol produced today is either used directly in the transportation industry or is further processed to produce its derivatives. The production of hydrogen from steam reforming of methanol releases a large amount of CO_2 (9–12 times the weight of hydrogen), and a carbon capture unit is therefore necessary. Furthermore, as methanol dissociation is endothermic, dynamic thermal management is also required. All these factors result in additional costs and energy consumption.

It is worth briefly comparing circular carriers with LOHCs since both are in a liquid state. Compared with LOHCs, circular carriers have significantly higher HSC. However, (de)hydrogenation of circular carriers occurs at higher temperatures despite their lower (de)hydrogenation enthalpy (see Fig. 3.1.6). Besides, hydrogen separation from other gases is necessary for circular carriers but not for LOHCs. In short, both carriers have favorable and unfavorable material properties, and other relevant factors (e.g., cost) also need consideration when choosing the right approach. A recent analysis [2] estimated the major system components for ammonia, methanol, and LOHC hydrogenation plants. It indicated that ammonia and LOHCs have similar capital costs, significantly lower than methanol, assuming a daily capacity to process 500 tons of hydrogen.

3.1.4 Opportunities and challenges

A hydrogen storage system that ranges in size from kW to GW (and beyond) can reduce emissions from multiple energy sectors. In addition, hydrogen has been proven to be an effective long-term energy storage option for renewable energy systems in areas of high latitudes, such as Europe, where the intensity of renewable energy production heavily depends on the seasons. In Europe, the amount of solar energy produced in winter is approximately 60% less than in summer. As the days get shorter and colder, the amount of electricity needed increases by about 40% [82]. It is likely that batteries will be used for short-term balancing, but hydrogen can also be used for long-term storage and seasonal storage without much risk. Having hydrogen storage at GW/TW scales can benefit energy export supply chains as well. However, how hydrogen is stored will have a significant impact on the future of hydrogen as an energy vector. Hence, it is vital that mass storage techniques be developed that are not only economically viable but also environmentally benign and sustainable as well [33].

Compressed gas is the most common form of hydrogen storage. For example, Gahleitner [83] examined 48 hydrogen production plants (power-to-gas) with hydrogen storage capacities ranging from 0.2 kg to 1350 kg in a stationary setting. Approximately 88% of the hydrogen was stored in compressed gaseous form and 12% in metal hydride (MH) containers. According to Abdin [50], hydrogen storage capacities range from 0.2 kg to 450 kg, with compressed gaseous storage accounting for 74% of cases and MH storage accounting for 26%. In general, transportation relies on pressurized tanks, such as fuel-cell buses and commercial fuel-cell sedans operating at 350 bar and 700 bar, respectively [84]. On the other hand, the hydrogen market mainly uses compressed gaseous tube trailers, and there are rarely cryogenic liquid hydrogen trucks [2]. Compressed gaseous storage is dominant over other storage methods because of the mature technology and commercial availability of pressurized tanks. A downside is that the volumetric energy density is low, transportation costs are high, and compression consumes much energy.

Salt caverns are naturally occurring and can be easily developed, which makes them a cost-effective storage solution for hydrogen storage. Salt is also very corrosion-resistant, making it suitable for long-term hydrogen storage. Despite geographical limitations and other drawbacks, some countries have utilized salt caverns as energy storage methods. For example, Germany, the United Kingdom, Ireland, France, and the Netherlands have been using salt caverns [2,32]. Although cryogenic storage of large volumes of liquid hydrogen is already well established in the space industry, liquid hydrogen storage for stationary storage has not yet been demonstrated on a large scale. On the other hand, liquid hydrogen storage is highly energy-intensive ($\sim$10 kWh/kg) and capital-intensive [2].

As Mg-based hydrides are high in both gravimetric and volumetric hydrogen storage capacities (7.6 wt% and 110 g/L, respectively) and excellent in reversibility, they are thermodynamically stable, requiring a high desorption temperature of 280–300°C, with an associated enthalpy value of 75 kJ/mol-H_2. Furthermore, a lack of dissociation and recombination capabilities of H_2 molecules and H atoms on the surfaces of Mg/MgH_2 as well as the low diffusion rate of H atoms in Mg/MgH_2 matrixes, have also been shown to negatively affect the kinetic barriers for hydrogenation and dehydrogenation [85]. Nevertheless, it is also possible for FeTi alloys to be used for hydrogen storage at low temperatures due to their high capacity (1.9 wt%, >100 kg/m^3) and relatively low cost. A major technological issue, however, is initiating its activation process, which is usually carried out at high temperatures and pressures [86].

The most commonly studied metal hydride for hydrogen storage is $LaNi_5H_6$. Although $LaNi_5H_6$ contains only 1.4% hydrogen, it can store 115 kg/m^3 at room temperature and at $\sim$10 bar. Despite its slow reaction kinetics and low effective thermal conductivity, $LaNi_5$-based energy storage has already been used in off-grid buildings, such as Griffith University in Brisbane, Australia, and the Henn-na Hotel in Nagasaki [87].

The compatibility of LOHCs and circular carriers with existing liquid fuel infrastructure makes them ideal for large-scale hydrogen storage. However, LOHCs' storage capacity is typically around 6–7 wt%, and new carrier molecules with higher capacity are necessary for mobile applications, such as hydrogen vehicles. Another critical challenge for LOHCs is to find abundant, cheap, and efficient catalysts for their (de)hydrogenation. Circular carriers have significantly higher capacity than LOHCs, but they are usually used in their hydrogen-rich format for other purposes than hydrogen storage. Additionally, circular carriers are energy-intensive to decompose, and the toxic by-products are energy- and cost-intensive to recycle.

It is essential to optimize the efficiency of hydrogen storage. For example, Abdin et al. [33] have reported that circular carrier ammonia has the lowest storage efficiency at $\sim$42%, followed by methanol at $\sim$50% and compressed gas at $\sim$92%. Although LOHCs and metal hydrides are emerging technologies, they could become economically competitive with densified storage technologies, especially for long-term storage. When hydrogenation heat is used for dehydrogenation, the LOHC's storage efficiency increases from $\sim$71% to $\sim$91%. Moreover, Abdin et al. [33] have shown that salt caverns are economical for long-term storage. If salt caverns are unavailable or economically unfeasible to store hydrogen, LOHCs can serve as an alternative. Additionally, ammonia and methanol are relatively expensive to synthesize and decompose (from daily to four-monthly storage cycles) for stationary hydrogen storage. However, their compatibility with existing liquid fuel infrastructure may make them viable as energy vectors on a large scale. Furthermore, they can be used directly or as derivatives.

In order to find the most cost-effective and highly efficient storage method, many research and development activities are currently underway. Moreover, hydrogen applications can be divided into several categories, depending on the scenario in which they are being used. Selecting the most suitable storage method for each hydrogen application requires understanding its characteristics. Table 3.1.1 summarizes the merits and demerits of various hydrogen storage methods.

Table 3.1.1 Merits and demerits of different hydrogen storage approaches.

Hydrogen storage	Storage type	Merits	Demerits
Densified	Compressed	– High storage efficiency – Convenient to store – Pure hydrogen – Mature technology	– Low storage density – High storage pressure – High compression energy – Heat management needed – Expensive – Safety concern
	Liquid	– High storage density and efficiency – Pure hydrogen	– Energy and capital intensive – Boil-off – Cryogenic temperature – Heat management needed
	Cyro-compressed	– High storage density and efficiency – Reduced boil-off loss – Pure hydrogen	– Emerging stage – High pressure and low temperature – High compression/liquefaction energy
Physisorption	Porous/coordination/ network	– Highly porous – High storage capacity – High surface area – Reversible	– Emerging stage – Cryogenic temperature – Low storage capacity at ambient temperature – High storage pressure – Low operating temperature for H_2 uptake – Loss of usable H_2

(continued on next page)

Chemical	Metal hydrides	– Low working pressure and temperature – High volumetric density – Long-term stability – Reversible – Pure hydrogen	– Slow reaction kinetics – Poor thermal conductivity – Expensive – Low gravimetric density – Heat management needed – High H_2 release temperature
	Complex hydrides	– High storage density – Light weight – Low temperature and pressure for H_2 store – High stability	– Poor reversibility – High temperature required for decomposition – Expensive – Thermal management required
	Liquid organic hydrogen carriers	– Existing infrastructure – Ambient storage conditions – Long-term storage	– High enthalpy – Slow reaction kinetics – Noble metal catalyst – High energy demand at dehydrogenation – Toxic or flammable molecule – Emerging stage
	Circular carriers	– Stability – Existing infrastructure – High storage density – Ambient storage conditions	– Toxic molecule – High temperature and not mature cracking for dehydrogenation – Capital and energy-intensive – Less efficient for H_2 storage – Noxious emission during dehydrogenation if there is no carbon capture and NO_x storage

3.1.5 Conclusion

This chapter discusses various physical and chemical methods of storing hydrogen as an energy vector, as well as the opportunities and challenges associated with these methods. Even though compressed gas storage offers a near-term solution to energy storage, there is no hydrogen storage technology that is capable of meeting the ultimate energy storage objectives. In the long run, solid-state hydrogen storage materials, such as sorbents and hydrides with large surface areas, might provide an alternative to compressed gas storage if their gravimetric and volumetric capacities, kinetics, and cost can be improved to acceptable levels. As a result, cost-effective, energy-efficient methods of reclaiming spent materials are necessary to ensure their life cycle is completed efficiently and cost-effectively, particularly for solid hydrides. The industry has a long tradition of manufacturing circular carriers. However, to ensure efficient hydrogen release and management of noxious emissions, improved methods will need to be developed when using these carriers for hydrogen storage. While densified storage and circular hydrogen carriers are relatively established, the liquid organic hydrogen carrier approach is evolving. Research is currently underway to find molecules and catalysts that are more effective in terms of energy efficiency, durability, cost, safety, and other factors. Therefore, a number of different aspects must be considered when choosing the optimal hydrogen storage method. These include the cost of the hydrogen storage system, energy efficiency, environmental impact, and downstream consequences.

References

[1] S. Sharma, S.K. Ghoshal, Hydrogen the future transportation fuel: from production to applications, Renew. Sustain. Energy. Rev. 43 (2015) 1151–1158. https://doi.org/10.1016/j.rser.2014.11.093.

[2] Z. Abdin, C. Tang, Y. Liu, K. Catchpole, Large-scale stationary hydrogen storage via liquid organic hydrogen carriers, Perspective 24 (2021) 102966. https://doi.org/10.1016/j.isci.2021.102966.

[3] J. Andersson, S. Grönkvist, Large-scale storage of hydrogen, Int. J. Hydrogen Energy 44 (2019) 11901–11919. https://doi.org/10.1016/j.ijhydene.2019.03.063.

[4] A. Züttel, Materials for hydrogen storage, Mater. Today 6 (2003) 24–33. https://doi.org/10.1016/S1369-7021(03)00922-2.

[5] W.B. Leung, N.H. March, H. Motz, Primitive phase diagram for hydrogen, Phys. Lett. A 56 (1976) 425–426. https://doi.org/10.1016/0375-9601(76)90713-1.

[6] K. Hirose, M. Hirscher, Handbook of Hydrogen Storage: New Materials for Future Energy Storage, Wiley-VCH Verlag GmbH & Co. KGaA, 2010.

[7] M.S. Sadaghiani, M. Mehrpooya, Introducing and energy analysis of a novel cryogenic hydrogen liquefaction process configuration, Int. J. Hydrogen Energy 42 (2017) 6033–6050. https://doi.org/10.1016/j.ijhydene.2017.01.136.

[8] M. Stewart, Compressor fundamentals, in: Surface Production Operations, 2019, pp. 457–525. https://doi.org/10.1016/B978-0-12-809895-0.00007-7.

[9] A. Züttel, A. Remhof, A. Borgschulte, O. Friedrichs, Hydrogen: the future energy carrier, Philos. Trans. R. Soc. A Math. Phys. Eng. Sci. 368 (2010) 3329–3342. https://doi.org/10.1098/rsta.2010.0113.

[10] H. Barthelemy, M. Weber, F. Barbier, Hydrogen storage: recent improvements and industrial perspectives, Int. J. Hydrogen Energy 42 (2017) 7254–7262. https://doi.org/10.1016/j.ijhydene.2016.03.178.

[11] A. Léon, Hydrogen storage, Hydrogen Technology, Springer, Berlin Heidelberg, 2008, pp. 81–128 https://doi.org/10.1007/978-3-540-69925-5_3.

[12] R. Moradi, K.M. Groth, Hydrogen storage and delivery: review of the state of the art technologies and risk and reliability analysis, Int. J. Hydrogen Energy 44 (2019) 12254–12269. https://doi.org/10.1016/j.ijhydene.2019.03.041.

[13] J. Zheng, X. Liu, P. Xu, P. Liu, Y. Zhao, J. Yang, Development of high pressure gaseous hydrogen storage technologies, Int. J. Hydrogen Energy 37 (2012) 1048–1057. https://doi.org/10.1016/j.ijhydene.2011.02.125.

[14] A.M. Elberry, J. Thakur, A. Santasalo-Aarnio, M. Larmi, Large-scale compressed hydrogen storage as part of renewable electricity storage systems, Int. J. Hydrogen Energy 46 (2021) 15671–15690. https://doi.org/10.1016/j.ijhydene.2021.02.080.

[15] J. Zheng, L. Li, R. Chen, P. Xu, F. Kai, High pressure steel storage vessels used in hydrogen refueling station, J. Press. Vessel Technol. 130 (2008) 014503. https://doi.org/10.1115/1.2826453.

[16] M. Jawad, Y. Wang, Z. Feng, Steel–concrete composite pressure vessels for hydrogen storage at high pressures, J. Press. Vessel Technol. 142 (2020) 021202. https://doi.org/10.1115/1.4044164.

[17] R. Kottenstette, Hydrogen storage in wind turbine towers, Int. J. Hydrogen Energy 29 (2004) 1277–1288. https://doi.org/10.1016/j.ijhydene.2003.12.003.

[18] B.R.R. Bapu, J. Karthikeyan, K.V.K. Reddy, Hydrogen storage in wind turbine tower, Front. Automob. Mech. Eng. (2010) 308–312. https://doi.org/10.1109/FAME.2010.5714852.

[19] V. Tietze, S. Luhr, Near-surface bulk storage of hydrogen, in: Transition to Renewable Energy Systems, Wiley-VCH Verlag GmbH & Co. KGaA, Weinheim, Germany, 2013, pp. 659–690 https://doi.org/10.1002/9783527673872.ch32.

[20] V. Tietze, S. Luhr, D. Stolten, Bulk storage vessels for compressed and liquid hydrogen, in: Hydrogen Science and Engineering: Materials, Processes, Systems and Technology, Wiley-VCH Verlag GmbH & Co. KGaA, Weinheim, Germany, 2016, pp. 659–690 https://doi.org/10.1002/9783527674268.ch27.

[21] J.E. Clark, D.K. Bonura, R.F. Van Voorhees, An overview of injection well history in the United States of America, in: Developments in Water Science, Elsevier, 2005, pp. 3–12 https://doi.org/10.1016/S0167-5648(05)52001-X.

[22] A.M.O. Mohamed, E.K. Paleologos, Groundwater, in: Fundamentals of Geoenvironmental Engineering, Elsevier, 2018, pp. 129–159. https://doi.org/10.1016/B978-0-12-804830-6.00005-3.

[23] M. Mazarei, A. Davarpanah, A. Ebadati, B. Mirshekari, The feasibility analysis of underground gas storage during an integration of improved condensate recovery processes, J. Pet. Explor. Prod. Technol. 9 (2019) 397–408. https://doi.org/10.1007/s13202-018-0470-3.

[24] D.G. Caglayan, N. Weber, H.U. Heinrichs, J. Linßen, M. Robinius, P.A. Kukla, et al., Technical potential of salt caverns for hydrogen storage in Europe, Int. J. Hydrogen Energy 45 (2020) 6793–6805. https://doi.org/10.1016/j.ijhydene.2019.12.161.

[25] R. Tarkowski, Underground hydrogen storage: characteristics and prospects, Renew. Sustain. Energy Rev. 105 (2019) 86–94. https://doi.org/10.1016/j.rser.2019.01.051.

[26] J.G. Speight, Recovery, storage, and transportation, Natural Gas (2019) 149–186. https://doi.org/10.1016/B978-0-12-809570-6.00005-9.

[27] A. Amid, D. Mignard, M. Wilkinson, Seasonal storage of hydrogen in a depleted natural gas reservoir, Int. J. Hydrogen Energy 41 (2016) 5549–5558. https://doi.org/10.1016/j.ijhydene.2016.02.036.

[28] R.C. Bales, Hydrology, floods and droughts: overview, Encycl. Atmos. Sci. (2015) 180–184. https://doi.org/10.1016/B978-0-12-382225-3.00166-3.

[29] F. Crotogino, Larger scale hydrogen storage, Storing Energy, Elsevier, 2016, pp. 411–429 https://doi.org/10.1016/B978-0-12-803440-8.00020-8.

[30] K. Belz, F. Kuznik, K.F. Werner, T. Schmidt, W.K.L. Ruck, Thermal energy storage systems for heating and hot water in residential buildings, in: Advances in Thermal Energy Storage Systems, Elsevier, 2015, pp. 441–465 https://doi.org/10.1533/9781782420965.4.441.

[31] M. Kayfeci, A. Keçebaş, M. Bayat, Hydrogen production, in: Solar Hydrogen Production, Elsevier, 2019, pp. 45–83 https://doi.org/10.1016/B978-0-12-814853-2.00003-5.

[32] Z. Abdin, A. Zafaranloo, A. Rafiee, W. Mérida, W. Lipiński, K.R. Khalilpour, Hydrogen as an energy vector, Renew. Sustain. Energy Rev. 120 (2020) 109620. https://doi.org/10.1016/j.rser.2019.109620.

[33] Z. Abdin, K. Khalilpour, K. Catchpole, Projecting the levelized cost of large scale hydrogen storage for stationary applications, Energy Convers. Manag. 270 (2022) 116241. https://doi.org/10.1016/j.enconman.2022.116241.

[34] C.J. Quarton, S. Samsatli, Power-to-gas for injection into the gas grid: what can we learn from real-life projects, economic assessments and systems modelling? Renew. Sustain. Energy Rev. 98 (2018) 302–316. https://doi.org/10.1016/j.rser.2018.09.007.

[35] J.P. Hodges, W. Geary, S. Graham, P. Hooker, R. Goff, Injecting hydrogen into the gas network: a literature search, Policy Paper, Institution of Gas Engineers and Managers 2015.

[36] K. Stangeland, D. Kalai, H. Li, Z. Yu, CO_2 methanation: the effect of catalysts and reaction conditions, Energy Proc. 105 (2017) 2022–2027. https://doi.org/10.1016/j.egypro.2017.03.577.

[37] L. Guerra, S. Rossi, J. Rodrigues, J. Gomes, J. Puna, M.T. Santos, Methane production by a combined Sabatier reaction/water electrolysis process, J. Environ. Chem. Eng. 6 (2018) 671–676. https://doi.org/10.1016/j.jece.2017.12.066.

[38] G. Strobel, B. Hagemann, T.M. Huppertz, L. Ganzer, Underground bio-methanation: concept and potential, Renew. Sustain. Energy Rev. 123 (2020) 109747. https://doi.org/10.1016/j.rser.2020.109747.

[39] S. Krasae-in, J.H. Stang, P. Neksa, Development of large-scale hydrogen liquefaction processes from 1898 to 2009, Int. J. Hydrogen Energy 35 (2010) 4524–4533. https://doi.org/10.1016/j.ijhydene.2010.02.109.

[40] K. Ohlig, L. Decker, The latest developments and outlook for hydrogen liquefaction technology, in: AIP Conference Proceedings, 2014, pp. 1311–1317. https://doi.org/10.1063/1.4860858.

[41] S.Z. Al Ghafri, S. Munro, U. Cardella, T. Funke, W. Notardonato, J.P.M. Trusler, et al., Hydrogen liquefaction: a review of the fundamental physics, engineering practice and future opportunities, Energy Environ. Sci. 15 (2022) 2690–2731. https://doi.org/10.1039/D2EE00099G.

[42] K. Ohlig, L. Decker, Hydrogen, 4. Liquefaction, Ullmann's Encyclopedia Industrial Chemistry, Wiley-VCH Verlag GmbH & Co. KGaA, Weinheim, Germany, 2013, pp. 1–6 https://doi.org/10.1002/14356007.o13_o05.pub2.

[43] L. Yin, Y. Ju, Review on the design and optimization of hydrogen liquefaction processes, Front. Energy 14 (2020) 530–544. https://doi.org/10.1007/s11708-019-0657-4.

[44] U. Cardella, L. Decker, J. Sundberg, H. Klein, Process optimization for large-scale hydrogen liquefaction, Int. J. Hydrogen Energy 42 (2017) 12339–12354. https://doi.org/10.1016/j.ijhydene.2017.03.167.

[45] S. Schoenung. Economic Analysis of Large-Scale Hydrogen Storage for Renewable Utility Applications. Albuquerque, NM, and Livermore, CA: 2011. https://doi.org/10.2172/1029796.

[46] Z. Liu, Y. Li, G. Zhou, Study on thermal stratification in liquid hydrogen tank under different gravity levels, Int. J. Hydrogen Energy 43 (2018) 9369–9378. https://doi.org/10.1016/j.ijhydene.2018.04.001.

[47] R.K. Ahluwalia, T.Q. Hua, J.K. Peng, S. Lasher, K. McKenney, J. Sinha, et al., Technical assessment of cryo-compressed hydrogen storage tank systems for automotive applications, Int. J. Hydrogen Energy 35 (2010) 4171–4184. https://doi.org/10.1016/j.ijhydene.2010.02.074.

[48] S.M. Aceves, F. Espinosa-Loza, E. Ledesma-Orozco, T.O. Ross, A.H. Weisberg, T.C. Brunner, et al., High-density automotive hydrogen storage with cryogenic capable pressure vessels, Int. J. Hydrogen Energy 35 (2010) 1219–1226. https://doi.org/10.1016/j.ijhydene.2009.11.069.

[49] J. Moreno-Blanco, G. Petitpas, F. Espinosa-Loza, F. Elizalde-Blancas, J. Martinez-Frias, S.M. Aceves, The storage performance of automotive cryo-compressed hydrogen vessels, Int. J. Hydrogen Energy 44 (2019) 16841–16851. https://doi.org/10.1016/j.ijhydene.2019.04.189.

[50] Z. Abdin, Component models for solar hydrogen hybrid energy systems based on metal hydride energy storage. 2017. https://doi.org/10.25904/1912/1029.

[51] D.J. Durbin, C. Malardier-Jugroot, Review of hydrogen storage techniques for on board vehicle applications, Int. J. Hydrogen Energy 38 (2013) 14595–14617. https://doi.org/10.1016/j.ijhydene.2013.07.058.

[52] C.D. Wood, B. Tan, A. Trewin, H. Niu, D. Bradshaw, M.J. Rosseinsky, et al., Hydrogen storage in microporous hypercrosslinked organic polymer networks, Chem. Mater. 19 (2007) 2034–2048. https://doi.org/10.1021/cm070356a.

[53] D. Ramimoghadam, E.M. Gray, C.J. Webb, Review of polymers of intrinsic microporosity for hydrogen storage applications, Int. J. Hydrogen Energy 41 (2016) 16944–16965. https://doi.org/10.1016/j.ijhydene.2016.07.134.

[54] J. Germain, J.M.J. Fréchet, F. Svec, Hypercrosslinked polyanilines with nanoporous structure and high surface area: potential adsorbents for hydrogen storage, J. Mater. Chem. 17 (2007) 4989. https://doi.org/10.1039/b711509a.

[55] S.Y. Ding, W. Wang, Covalent organic frameworks (COFs): from design to applications, Chem. Soc. Rev. 42 (2013) 548–568. https://doi.org/10.1039/C2CS35072F.

[56] S. Liu, X. Yang, Gibbs ensemble Monte Carlo simulation of supercritical CO_2 adsorption on NaA and NaX zeolites, J. Chem. Phys. 124 (2006) 244705. https://doi.org/10.1063/1.2206594.

[57] S.P. Shet, S. Shanmuga Priya, K. Sudhakar, M. Tahir, A review on current trends in potential use of metal-organic framework for hydrogen storage, Int. J. Hydrogen Energy 46 (2021) 11782–11803. https://doi.org/10.1016/j.ijhydene.2021.01.020.

[58] XVIII, On the absorption and dialytic separation of gases by colloid septa, Philos. Trans. R. Soc. London 156 (1866) 399–439. https://doi.org/10.1098/rstl.1866.0018.

[59] G. Sandrock, A panoramic overview of hydrogen storage alloys from a gas reaction point of view, J Alloys Compd 293–295 (1999) 877–888. https://doi.org/10.1016/S0925-8388(99)00384-9.

[60] IEA/DOE/SNL HYDRIDE DATABASES. https://www.energy.gov/eere/fuelcells/databases n.d.

[61] V.A. Yartys, M.V. Lototskyy, E. Akiba, R. Albert, V.E. Antonov, J.R. Ares, et al., Magnesium based materials for hydrogen based energy storage: past, present and future, Int. J. Hydrogen Energy 44 (2019) 7809–7859. https://doi.org/10.1016/j.ijhydene.2018.12.212.

[62] J.C. Crivello, B. Dam, R.V. Denys, M. Dornheim, D.M. Grant, J. Huot, et al., Review of magnesium hydride-based materials: development and optimisation, Appl. Phys. A 122 (2016) 97. https://doi.org/10.1007/s00339-016-9602-0.

[63] J. Graetz, J.J. Reilly, V.A. Yartys, J.P. Maehlen, B.M. Bulychev, V.E. Antonov, et al., Aluminum hydride as a hydrogen and energy storage material: past, present and future, J. Alloys Compd. 509 (2011) S517–S528. https://doi.org/10.1016/j.jallcom.2010.11.115.

[64] N.A.A. Rusman, M. Dahari, A review on the current progress of metal hydrides material for solid-state hydrogen storage applications, Int. J. Hydrogen Energy 41 (2016) 12108–12126. https://doi.org/10.1016/j.ijhydene.2016.05.244.

[65] A. Baran, M. Polański, Magnesium-based materials for hydrogen storage: a scope review, Materials (Basel) 13 (2020) 3993. https://doi.org/10.3390/ma13183993.

[66] M. Paskevicius, L.H. Jepsen, P. Schouwink, R. Černý, D.B. Ravnsbæk, Y. Filinchuk, et al., Metal borohydrides and derivatives: synthesis, structure and properties, Chem. Soc. Rev. 46 (2017) 1565–1634. https://doi.org/10.1039/C6CS00705H.

[67] F. Schüth, B. Bogdanović, M. Felderhoff, Light metal hydrides and complex hydrides for hydrogen storage, Chem. Commun. (20) (2004) 2249–2258. https://doi.org/10.1039/B406522K.

[68] J.J. Vajo, S.L. Skeith, F. Mertens, Reversible storage of hydrogen in destabilized $LiBH_4$, J. Phys. Chem. B 109 (2005) 3719–3722. https://doi.org/10.1021/jp040769o.

[69] B. Bogdanović, M. Felderhoff, A. Pommerin, F. Schüth, N. Spielkamp, Advanced hydrogen-storage materials based on Sc-, Ce-, and Pr-doped $NaAlH_4$, Adv. Mater. 18 (2006) 1198–1201. https://doi.org/10.1002/adma.200501367.

[70] R. Urbanczyk, K. Peinecke, M. Felderhoff, K. Hauschild, W. Kersten, S. Peil, et al., Aluminium alloy based hydrogen storage tank operated with sodium aluminium hexahydride Na_3AlH_6, Int. J. Hydrogen Energy 39 (2014) 17118–17128. https://doi.org/10.1016/j.ijhydene.2014.08.101.

[71] M. Baricco, M. Bang, M. Fichtner, B. Hauback, M. Linder, C. Luetto, et al., SSH_2S: hydrogen storage in complex hydrides for an auxiliary power unit based on high temperature proton exchange membrane fuel cells, J. Power Sourc. 342 (2017) 853–860. https://doi.org/10.1016/j.jpowsour.2016.12.107.

[72] V. Meille, I. Pitault, Liquid organic hydrogen carriers or organic liquid hydrides: 40 years of history, Reactions 2 (2021) 94–101. https://doi.org/10.3390/reactions2020008.

[73] Y. Okada, M. Yasui, Large scale H_2 storage and transportation technology, Hyomen Kagaku 36 (2015) 577–582. https://doi.org/10.1380/jsssj.36.577.

[74] H. Jorschick, A. Bösmann, P. Preuster, P. Wasserscheid, Charging a liquid organic hydrogen carrier system with H_2/CO_2 gas mixtures, Chem. Cat. Chem. 10 (2018) 4329–4337. https://doi.org/10.1002/cctc.201800960.

[75] L. Shi, S. Qi, J. Qu, T. Che, C. Yi, B. Yang, Integration of hydrogenation and dehydrogenation based on dibenzyltoluene as liquid organic hydrogen energy carrier, Int. J. Hydrogen Energy 44 (2019) 5345–5354. https://doi.org/10.1016/j.ijhydene.2018.09.083.

[76] G.P. Pez, A.R. Scott, A.C. Cooper, H. Cheng, Hydrogen storage by reversible hydrogenation of pi-conjugated substrates. 2006. https://patents.google.com/patent/US7101530/en.

[77] C. Gleichweit, M. Amende, S. Schernich, W. Zhao, M.P.A. Lorenz, O. Höfert, et al., Dehydrogenation of dodecahydro-N-ethylcarbazole on Pt(111), Chem. Sus. Chem. 6 (2013) 974–977. https://doi.org/10.1002/cssc.201300263.

[78] N. Brückner, K. Obesser, A. Bösmann, D. Teichmann, W. Arlt, J. Dungs, et al., Evaluation of industrially applied heat-transfer fluids as liquid organic hydrogen carrier systems, Chem. Sus. Chem. 7 (2014) 229–235. https://doi.org/10.1002/cssc.201300426.

[79] I. Garagounis, V. Kyriakou, A. Skodra, E. Vasileiou, M. Stoukides, Electrochemical synthesis of ammonia in solid electrolyte cells, Front. Energy Res. 2 (2014). https://doi.org/10.3389/fenrg.2014.00001.

[80] A. Boisen, S. Dahl, J. Norskov, C. Christensen, Why the optimal ammonia synthesis catalyst is not the optimal ammonia decomposition catalyst, J. Catal. 230 (2005) 309–312. https://doi.org/10.1016/j.jcat.2004.12.013.

[81] Japan's top power producer Jera makes bet on ammonia and hydrogen. https://www.japantimes.co.jp/news/2022/03/29/business/jera-ammonia-hydrogen/ n.d.

[82] Hydrogen scaling up: a sustainable pathway for the global energy transition. n.d.

[83] G. Gahleitner, Hydrogen from renewable electricity: an international review of power-to-gas pilot plants for stationary applications, Int. J. Hydrogen Energy 38 (2013) 2039–2061. https://doi.org/10.1016/j.ijhydene.2012.12.010.

[84] Standard Reference Data n.d. https://www.nist.gov/srd/refprop.

[85] X. Ding, R. Chen, J. Zhang, W. Cao, Y. Su, J. Guo, Recent progress on enhancing the hydrogen storage properties of Mg-based materials via fabricating nanostructures: a critical review, J. Alloys Compd. 897 (2022) 163137. https://doi.org/10.1016/j.jallcom.2021.163137.

[86] A.K. Patel, D. Siemiaszko, J. Dworecka-Wójcik, M. Polański, Just shake or stir. About the simplest solution for the activation and hydrogenation of an FeTi hydrogen storage alloy, Int. J. Hydrogen Energy 47 (2022) 5361–5371. https://doi.org/10.1016/j.ijhydene.2021.11.136.

[87] Z. Abdin, C.J. Webb, E.M. Gray, One-dimensional metal-hydride tank model and simulation in Matlab–Simulink, Int. J. Hydrogen Energy 43 (2018) 5048–5067. https://doi.org/10.1016/j.ijhydene.2018.01.100.

Hydrogen storage in high entropy alloys

Abhishek Kumar[a], Nilay Krishna Mukhopadhyay[b] and Thakur Prasad Yadav[a,c]

[a]Hydrogen Energy Centre, Department of Physics, Institute of Science, Banaras Hindu University, Varanasi, Uttar Pradesh, India
[b]Department of Metallurgical Engineering, Indian Institute of Technology, Banaras Hindu University, Varanasi, Uttar Pradesh, India
[c]Department of Physics, Faculty of Science, University of Allahabad, Prayagraj, Uttar Pradesh, India

3.2.1 Introduction

Energy is a fundamental requirement of civilization and a foundational element for prosperity, growth, and reduction of poverty. Today, providing everyone with access to enough energy is a pressing challenge because of the constant demand for different energy sources. Earlier, this energy demand was met by fossil fuels like coal, petroleum, biomass, etc. Although these resources (fossil fuels) are sufficiently attractive to meet the demand for energy, they lack sufficient long-lasting storage. As a developing country, we are dependent on fossil fuels due to growing demand and fading supply. Currently, the continuous use of fossil fuels imposes massive environmental and economic costs. India is the third-largest energy consumer in the world, behind China and the United States. India is the fastest-growing energy economy and has fixed a target of 4.2% growth from 2017 to 2040. The Indian government has also spearheaded several policy reforms and initiatives for increasing the production and exploitation of domestic petroleum resources to address the priorities like energy access, energy efficiency, energy sustainability, and energy security. The production target of crude oil and natural gas for the financial year 2019–20 set by the Indian government is 35.04 million metric tons (MMT) and 34.55 billion cubic meters (BCM). From these points of view, people nowadays want new, adaptive, clean, and green energy sources to combat climate change and global warming caused by fossil fuels [1]. Because of its excellent energy range per unit mass, hydrogen is one of the most beneficial nominees among the numerous fuels available worldwide [2]. Hydrogen generation, storage, and applications are three critical components for using hydrogen as a fuel. One of the essential

Towards Hydrogen Infrastructure: Advances and Challenges in Preparing for the Hydrogen Economy.
DOI: https://doi.org/10.1016/B978-0-323-95553-9.00007-8

aspects of hydrogen as a fuel is hydrogen storage. Solid–state metal hydride is one of the most secure and effective methods of storing hydrogen [3]. The high-entropy alloys (HEAs) are new and promising fields for hydrogen storage due to an unlimited combination of alloy formation. In this chapter, we will discuss the energy sources, hydrogen storage mechanisms in HEA, some recent research going on in this field, and future aspects of the adoption of HEAs as hydrogen energy carriers.

3.2.2 Energy sources

Today's researchers are looking for a variety of energy sources by utilizing substitute technologies that would provide high reliability, maximum efficiency, and minimal pollution. There are a few fundamental sources of energy, such as:

- Nuclear fusion reaction (solar energy).
- The gravity of the earth and moon.
- Nuclear fission reaction.
- Energy in the interior of the earth.
- Energy stored in chemical bonds.

Fossil fuels provide us with a considerable amount of energy. A significant requirement of energy sources is satisfied by fossil flues. Still, fossil fuels are complex organic molecules that consist of chains of hydrogen and carbon with the general chemical formula: C_nH_{2n+2} and have a series of slow chemical reactions with oxygen in the atmosphere, causing greenhouse emissions [2]. Thus, the resource of fossil fuels is limited and it harms as well as pollutes the environment. So, there is a demand for clean, sustainable energy sources that can be environmentally friendly.

In the present times, we are constantly exploiting various energy sources. We permanently use energy sources in multiple forms of fossil fuels to fulfill our various requirements and applications. Due to this over-exploitation, humans are causing trouble for themselves as these fossil fuels may soon become extinct. At the same time, with this over-exploitation, we are continuously increasing the emission of greenhouse gases (methane, carbon dioxide, and nitrous oxide), especially carbon dioxide, which causes climate change. The National Aeronautics and Space Administration (NASA) reported a global climate change showing that the temperature of our globe increases continuously by 0.2°C per decade. This change in global temperature invites various types of disasters such as loss of ice at the poles and glaciers, increased ocean level and temperature, hurricanes, wildfires, droughts, floods, etc. As a result, this golden earth is getting worse day by day. Therefore, to avoid all these troubles in the near future, we are trying to replace these fossil fuels with a clean and sustainable energy source.

3.2.3 Hydrogen energy: sustainable energy system/advantages of hydrogen

We are currently looking for alternative energy sources that would provide high reliability, maximum efficiency, and minimal pollution. Today, fuel cells are the most environmentally friendly method of generating sustainable energy [4]. The only byproduct of using hydrogen as a fuel is water. More experts believe that hydrogen will be the proper fuel of the future because hydrogen-based fuel cells have 60% higher efficiencies. Due to its extremely low density (0.08988 g/L), hydrogen storage poses a significant problem in these three areas [2]. Liquid hydrogen has the highest energy density per unit of mass among all fossil fuels and energy carriers with a specific weight of 71 g/L. It has a great deal of potential as an eco-friendly fuel. We can produce hydrogen using a variety of resources, including biomass, nuclear power, natural gas, and renewable energy sources like solar and wind. Due to hydrogen's abundance as a renewable fuel in nature, it is the most efficient nonfossil energy source for achieving our goals. Exploring renewable energy sources that could meet future needs is highly desirable. This makes it necessary to reconsider the developments and potential of renewable hydrogen energy sources [5]. Production of energy using hydrogen is now a reality [6]. In a secure and sustainable energy system, hydrogen is presented as an energy carrier in the hydrogen-based economy of the future [7]. There will be entirely new and different methods of generating and utilizing energy. They are about to enter a brand-new era that will feature cutting-edge technology and novel fuels.

After the fossil fuel economy, hydrogen can be thought of as a synthetic fuel carrying secondary energy in the future era [8]. The energy from renewable sources can be transported over long distances from regions with abundant solar and wind resources, such as Australia or Latin America, to energy-deficient regions that are thousands of kilometers away with the help of hydrogen and hydrogen-based fuels. Hydrogen's potential can play a vital role in a clean, secure, and adorable energy future [9]. Today, since hydrogen does not cause global warming when it is produced from renewable resources, we view it as a nonpolluting energy source. Additionally, hydrogen is the only secondary energy carrier on the market that is suitable for a variety of applications. The most alluring aspect of hydrogen is that it can be produced using a variety of primary energies [10]. It has benefits for a variety of uses, including transportation, and it is portable for stationary use [11]. Furthermore, hydrogen can be used in decentralized systems without producing carbon dioxide [12]. Currently, hydrogen is already used in the chemical industry. However, if we want to use hydrogen as a source of energy, we will need technologies like fuel cells [13]. Because it can be produced from a variety of primary energies and used in a greater variety of applications, hydrogen can also serve as an energy center similar to electricity today. Hydrogen can be stored for a longer period of time than electricity; it is therefore very advantageous [14]. The various governments are committed to advancing research in the field of hydrogen energy and technology since the energy carrier promotes the stabilization of energy security and pricing as well as competition between diverse energy sources globally [15–17].

3.2.4 Hydrogen storage

The main technology for the development of hydrogen as an energy source is hydrogen storage. Of all the fuels, hydrogen has the most energy per mass. However, it has a low energy density per unit volume under ambient conditions. Therefore, it is crucial to create advanced storage technologies with the potential for greater energy density [18]. In the current onboard application scenario, hydrogen is stored as a compressed gas that typically needs high-pressure tanks, as a cryogenic liquid state (below the critical temperature 33 K), or in solid-state compounds like complex hydrides, metal hydrides, porous materials, etc. [19], which are commented regarding the technical state and the feasibility in future applications as shown in Fig. 3.2.1.

3.2.5 Types of hydrogen storage

3.2.5.1 Compression

The most typical and straightforward method of storing hydrogen is as a gas in a cylinder. However, because hydrogen has such a low density, we need to use extremely high pressure (around 300 atm or more), which is very risky and expensive. In most nations, conventional high-pressure cylinders made of reasonably inexpensive steel are regularly filled to 200 atm and tested up to 300 atm. On the other hand, applications for vehicles need about four times more hydrogen pressure. But this high pressure cannot be

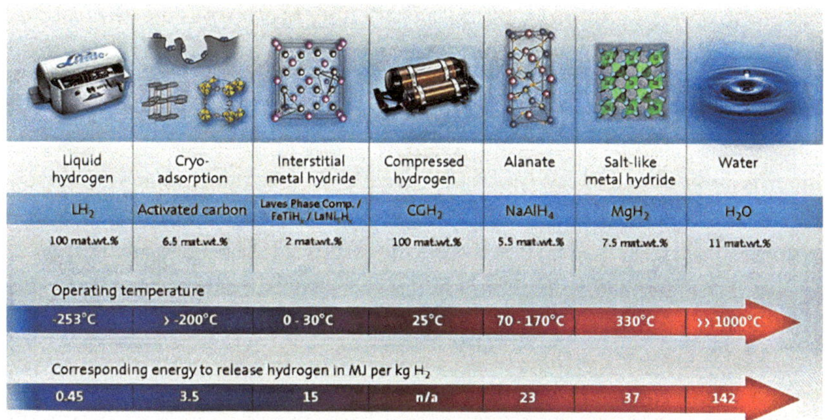

Figure 3.2.1 Hydrogen storage technologies and their respective operational conditions [87].

maintained by the cylinders that are currently on the market. Therefore, businesses set out to create cylinders that could be used in vehicles. The price of these cylinders is higher. According to a report from the Linde Company, compressed hydrogen is much more expensive than liquid hydrogen. As a result, this approach is not the best way to store hydrogen. Furthermore, the security of such pressurized cylinders is a major concern, particularly in areas and cities with dense populations.

3.2.5.2 Liquefaction

Liquid hydrogen storage is extremely expensive. A very low critical temperature (about 33.2 K) is required for hydrogen to liquefy, which is very challenging to maintain. Above this temperature, the liquid state cannot exist. Two significant problems arise during the liquefaction of hydrogen: (1) the efficiency of the liquefaction process, and (2) the boil-off of the liquid. Theoretically, 3.23 kWh/kg of work is required to liquefy one kilogram of hydrogen gas, but in practice, 15.2 kW/kg of work is required, which is almost half of hydrogen's lower heating value. Even with the best insulation technique, there will always be a loss when turning liquid hydrogen into a gas inside a cryogenic (21.2 K) vessel. An exothermic reaction that serves as a gasification heat source is the conversion of ortho- to para-hydrogen. This heat of transformation is greater than the latent heat of vaporization of common hydrogen (451.9 kJ/kg) at the standard boiling point, measuring 519 kJ/kg at 77 K and 523 kJ/kg at temperatures lower than 77 K. Since the pressure in a closed system would be extremely high (roughly 1000 MPa) at room temperature, thus we could only store liquid hydrogen.

3.2.5.3 Solid-state storage

Since hydrogen has a low molecular weight and a high molar combustion heat, it has a remarkable energy value per unit mass. However, because hydrogen has a low density (both in its liquid and gaseous states), its heating value per volume is much lower than that of conventional fuels. In solid-state storage, hydrogen is stored in the material either in molecular form (physisorption) or in atomic form (chemisorption). Solid state storage method is a practical, cost-effective, and secure method. Solid-state storage modes in metal/intermetallic hydrides were the first to be recognized as effective, secured, and economical alternatives among the various storage modes mentioned above (such as gaseous and liquification modes). Compared to other hydrogen storage methods like high-pressure and liquid hydrogen storage, these hydrides have a higher volumetric hydrogen capacity.

3.2.6 Hydrogen storage in intermetallic metal hydrides

The high-entropy alloys are a new class of intermetallic with various hydride-forming elements. Therefore, there are possibilities for hydrogen storage in HEAs. Intermetallic metal hydride is the combination of stable and unstable hydride-forming elements. The general form of intermetallic metal hydride can be defined as:

$$AB_xH_n$$

where A stands for rare and alkaline earth metals and is favorable for hydride formation, B stands for transition metal elements that form unstable metal hydrides.

The most common intermetallic types are AB, A_2B, AB_2, and AB_5 [20–22]. The first discussed AB_5 type are $LaNi_5$ type of intermetallic. In this alloy system, Ni is the nonhydride forming element, and this element work as a catalyst in the hydrogen absorption process. The amount of hydrogen abortion in metal hydride depends on the number of interstitial sites present in the hydride phase. The two criteria are most favorable for hydrogen placed in interstitial sites as shown in Fig. 3.2.2 [23,24].

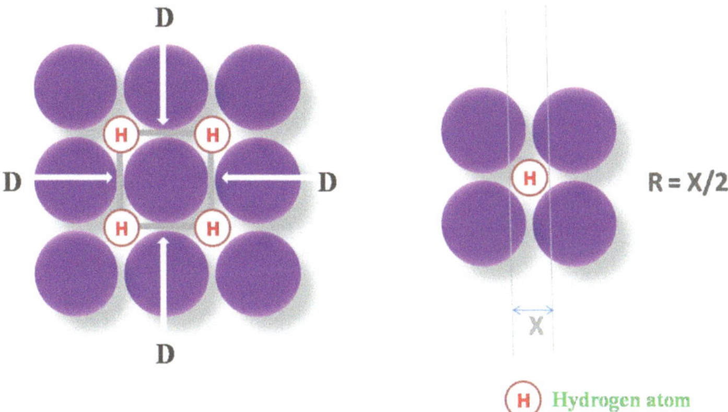

Figure 3.2.2 The distance between two hydrogen atoms at interstitial sites and the largest sphere that can fit inside the interstitial site.

1. The interatomic distance between two atoms should be at least $\sim$2.1 Å.
2. Westlake's criterion for the radius (R) of the enormous sphere on an interstitial site touching all the neighboring metallic atoms is $\sim$0.37 Å.

For designing the intermetallic for a specific application, there are many aspects [25–26]. We consider some properties like cyclic stability, thermodynamic parameters, plateau pressure at different temperatures, hydrogen charge-discharge behavior, ease of activation, easy activation, etc. [27]. The other significant effect was selecting behind the metal hydride-based HEAs storage capacity, enthalpy of dehydration, and energy density. The enthalpy varies from 30 kJ/mol to 70 kJ/mol [28,29]. Hydrogen occupies octahedral or tetrahedral sites in interstitial hydrides as shown in Fig. 3.2.3 where brown dots mark interstitial sites [30].

3.2.7 Hydrogen storage mechanism in metal

Firstly, molecular hydrogen interacts with the metal surface, then the molecular hydrogen dissociates into the atoms and diffuses into the crystalline structure. In the last step of this process, the hydrogen atom starts nucleation and triggers the hydride phase formation [31]. Due to the hydrogen atom diffusion inside the crystal lattice, the expansion of the crystal lattice occurs. The crystal lattice's expansion rate is directly proposal to the hydrogen concentration. The metals hydrogen concentration value is nearly about 2–3 A^3 per hydrogen atom [32].

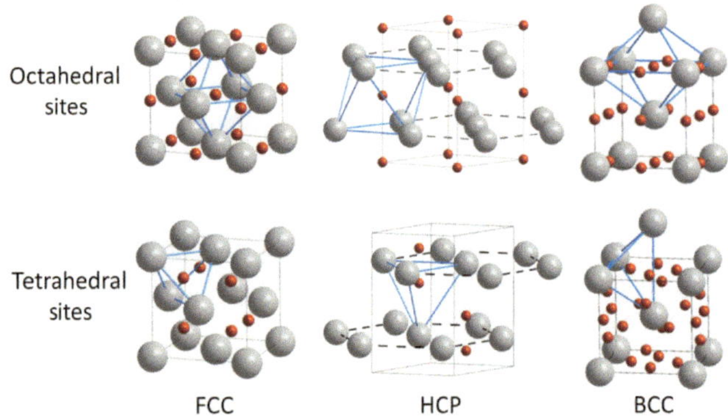

Figure 3.2.3 Possible hydrogen occupies sites: octahedral or tetrahedral sites in interstitial hydrides [30].

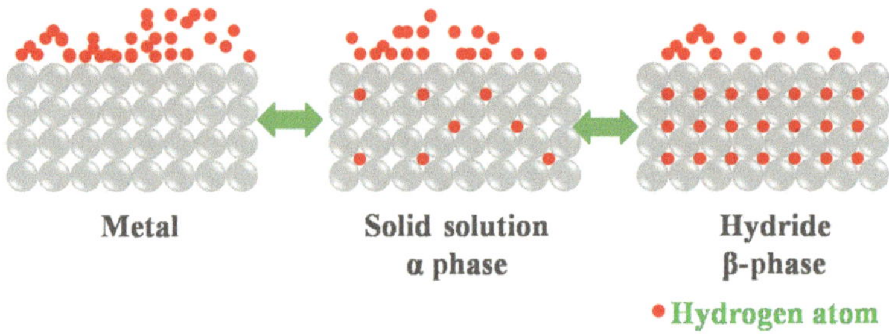

Figure 3.2.4 An illustration of the hydrogen gas absorption process in solid metal.

In Fig. 3.2.4, we can easily understand the hydrogen storage process in the interstitial site of metal. Firstly, the molecule of hydrogen is absorbed on the surface of the metal. This absorption process is done on the concept of van der Walls interaction [33]. When we provide sufficient energy, pressure, temperature, the hydrogen molecule starts dissociating into the atoms. That particular atom starts migrating into the metal's surface and diffuses into the interstitial sites of the crystalline lattice. The hydrogen atom occupies a random position inside the metal. Alpha and beta are two phases formed during this process. The randomly occupied hydrogen atoms inside metal lead to the formation of a well-known alpha phase or solid solution in the initial stage. If the hydrogen concentration is high enough as required to form an ordered array of hydrogen atoms in the interstitial voids phase transformation to beta phase takes place. The whole process can be reversible only at the required temperature and pressure. Fig. 3.2.5 illustrates a pressure-composition-temperature (PCT) diagram expressing the formation of the

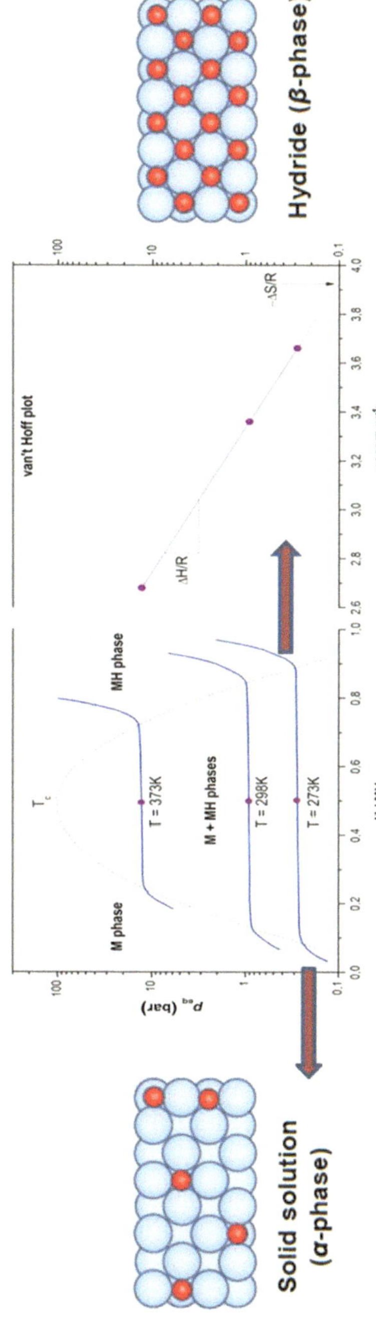

Figure 3.2.5 Illustration of a pressure-composition-temperature (PCT) diagram (*left*) for a hypothetical metal hydride (MH) and the Van't Hoff plot (*right*) from which can be plot with thermodynamic data [30].

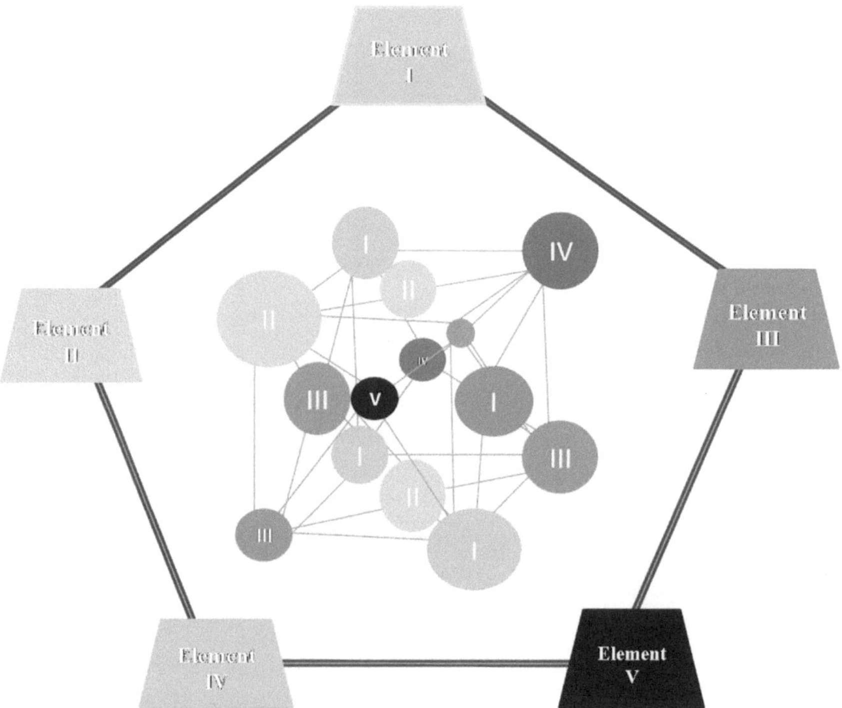

Figure 3.2.6 Diagrammatic representation of the high-entropy alloys with five major components, and a model of the structure with an element concentration range of 5–35 at%.

metal and metal hydride phases, and focusing Van't Hoff plot to calculate the entropy and enthalpy changes that occur throughout the hydrogen absorption and desorption processes.

3.2.8 Multiprincipal high entropy alloys (MPHEA)

High entropy materials (HEM) are a new class of materials that have received a lot of attention lately [34–36]. As shown in Fig. 3.2.6, HEAs contain five or more elements, each with a concentration of 5–35 atomic percentage (at %), in contrast to conventional alloys based on one principal element. The first research article on HEAs/HEMs was published by Yeh et al. [37], Cantor et al. [38], and Ranganathan [39]. The first investigation into multicomponent alloys was conducted in 1981 by Brian Cantor and his student Alain Vincent, who then presented their findings in a thesis for an undergraduate project. Here, they mixed a number of elements in an equal ratio and discovered that $Fe_{20}Cr_{20}Ni_{20}Mn_{20}Co_{20}$ composition formed a single face-centered cubic (FCC) phase. Additionally, a variety of equiatomic multicomponent alloys with six to nine elements and an FCC dendritic phase were reported. Yeh independently proposed the single-phase multiprincipal element alloy in 1995, making this idea a groundbreaking success in the study of HEAs. It is interesting to note that the high mixing entropy in multiprincipal element alloys could significantly lower the number of phases in high-order alloys, improving the material's properties [27,40].

It is inspiring history that in the late eighteenth century, the German scientist Franz Karl Achard had already studied the multicomponent equimass alloys with five to seven elements. The results of the research, however, were largely disregarded by scientists and researchers. Professor Smith presented this research finding in 1963, and it was likely the first to examine multiprincipal element alloys with five to seven elements. Multiprincipal element alloys created a completely new field of study known as HEMs/HEAs in the fields of materials science and physics as a result. Indian scientist, S. Ranganathan published a classic paper on the 'Alloyed Pleasures, Multimetallic Cocktails' in 2003, which discussed the significance of this new class of alloys [39]. Under the name HEAs/HEMs, this novel concept of equiatomic multicomponent alloys has been investigated. It has generated a great deal of interest in this field of study, which can be seen by examining the publication pattern of articles with time, as shown in Fig. 3.2.7. Before 2012, publication in the area of HEAs was not many. As a result, we can draw the conclusion that global researchers' interest in HEAs was minimal and that they were not generally aware of their significance. The developments of numerous applications and the HEAs' affordable price have drawn the attention of the scientific community, which has encouraged its development through publications [85,86]. According to Fig. 3.2.4, there was no plateau after 2013, which suggests that more articles may start to be published annually soon. In multicomponent intermetallic, the high chemical disorder in HEMs leads to high entropy of mixing, which aids in stabilizing the disordered solid solution phases with straightforward crystal structures like body-centered cubic (BCC), face-centered cubic (FCC), and hexagonal close-packed (HCP) [41]. In the case of HEAs, a wide variety of compositions, that is, 6–9 component alloys, exhibit the majority primary dendritic phase, which, after melt spinning or mechanical milling, can dissolve significant amounts of other transition metals [42].

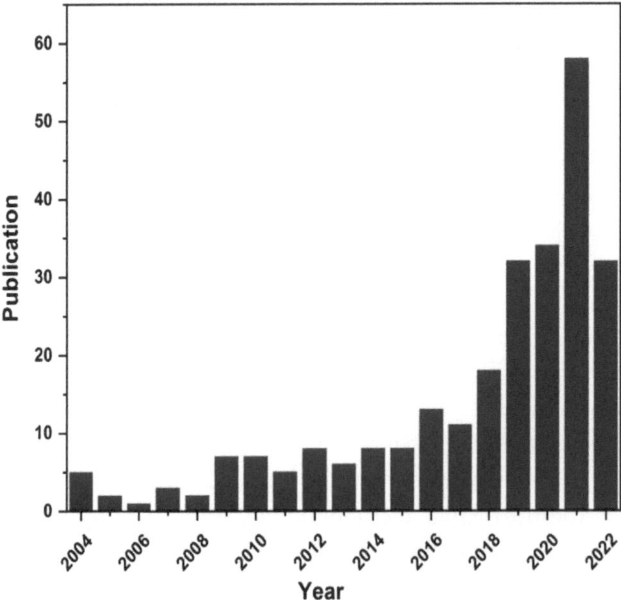

Figure 3.2.7 Number of articles published per year in 2010–2022 (until March) with "high-entropy alloy" and refine by "hydrogen storage" in the title, keywords, and abstract. *(From: Web of Science).*

Unexpectedly, the alloy primarily consists of a single primary phase, and finally, the total number of phases is consistently lower than the maximum number of phases permitted under equilibrium conditions and even lower than the maximum number permitted under nonequilibrium conditions according to the Gibbs phase rule. HEMs, therefore, possess a number of distinctive and desirable properties, such as high-temperature strength, adhesive wear properties, thermodynamic stability, high hardness and strength, distinctive electrical and magnetic properties, considerable strengthening and homogeneous deformation, tensile properties including high ultimate tensile strength and reasonable ductility, corrosion behavior, etc. [43–47]. HEMs are exotic to the scientific community because these properties are rarely seen in conventional alloys. These HEMs properties, however, also depend on alloy design and manufacturing processes [48,83,84]. The process of phase choice and microstructure evolution of HEMs can be significantly influenced by the synthesis route [49]. Therefore, various strengthening mechanisms, such as grain boundary and precipitation strengthening, have been introduced to improve the properties of the HEMs in order to develop technologically essential materials for various applications, such as environments, particularly in the nuclear industry, damage-resistant materials, tool materials, turbines, aerospace industries, hydrogen storage, etc. The number of interstices sites is influenced by the materials' microstructure and structure, which is very helpful for efficient hydrogen storage [50]. There have only been a few studies [51], despite the continued high interest in developing HEMs with high strength for energy-saving applications, such as in transportation and energy. The alloys along with the Laves phase [36] provide new opportunities for studying materials for hydrogen storage. Some reported Hydrogen Storage based high-entropy alloys, and C14 Laves phase-based HEAs, whose hydrogen capacity is higher than 1 wt% are given in Table 3.2.1. Designing a new equiatomic multicomponent HEM is therefore highly motivating; especially for the scientific community to understand various phase and microstructure evolution phenomena since the fundamental mechanism underlying phase evolution and HEM stability is still a subject of active research.

3.2.9 The concept and thermodynamic parameters of high-entropy alloys

The conceptual thermodynamic prediction support that if the structure try to minimums Gibbs free energy (G) under isobaric and isothermal condition, then it follow the relation given as:

$$G = H - TS \qquad (3.2.1)$$

We can see that the entropy(S) and enthalpy (H) of the system are directly correlated to predicting the equilibrium state at the predicted temperature. There is no condition to select equimolar elements in HEAs, but it should be between 5 and 35 at %. Research work on this topic is entirely limitless. The Hume–Rothery rule can identify mutual solubility in an alloy system, which depends on crystal structure, valance electron concentration (VEC), atomic size differences, and electronegativity. In standard alloys, the phase formation criteria we use are atomic size difference, enthalpy of mixing, and configurational entropy. In the class of HEA, we get less number of phases expected from a prediction by the phase rules. The lower number of phases in multicomponent HEA is due to low diffusivities and high configurational entropy. A statistical thermodynamic equation can calculate the configurational entropy of a system.

$$\Delta S_{conf} = k \, ln \, w \qquad (3.2.2)$$

Table 3.2.1 Some reported HS (Hydrogen Storage) based high-entropy alloys whose hydrogen capacity is higher than 1 wt%.

S.No.	Alloy Composition	Synthesis method	Alloy System	Storage Capacity	References
1.	MgAlTiFeNi	MA	BCC And TiH_2	1.0 wt%	[89]
2.	TiZrHfMoNb	Arc Melting	BCC	1.18 wt%	[74]
3.	TiZrNbTa	Arc Melting	BCC	1.40 wt%	[77]
4.	TiZrNbHfTa	Arc Melting	BCC	2.00 H/M	[78]
5.	TiVZrNbHf	Arc Melting	BCC	2.70 wt%	[62]
6.	TiZrNbMoV	LENS	BCC	2.30 wt%	[71]
7.	$Ti_{0.30}V_{0.25}Zr_{0.10}Nb_{0.25}Ta_{0.10}$	Arc Melting	BCC	2.5 wt%	[90]
8.	CoFeMnTiVZr	Arc Melting	C14 Laves Phase	1.91 wt%	[91]
9.	ZrTiVCrFeNi	LENS	C14 Laves Phase	1.81 wt%	[68]
10.	$Ti_{0.325}V_{0.275}Zr_{0.125}Nb_{0.275}$	Arc Melting	BCC	2.5 wt%	[92]
11.	$Mg_{0.10}Ti_{0.30}V_{0.25}Zr_{0.10}Nb_{0.25}$	MA	BCC	2.7 wt%	[93]
12.	$V_{35}Ti_{30}Cr_{25}Fe_5Mn_5$	Arc Melting	BCC	3.51 wt%	[82]
13.	$Ti_{0.95}Zr_{0.07}Mn_{1.15}Cr_{0.7}V_{0.15}$	Induction levitation melting	C14 Laves phase	1.83 wt%	[64]
14.	Ti–Zr–V–Cr–Ni	Arc Melting	C14 Laves phase	1.52 wt%	[27,88]

where w stands for the number of way of energy shared by the different particles and k stand for Boltzmann's constant. For calculating the configurational entropy in n element with contain mole fraction equation express as:

$$\Delta S_{\text{conf}} = -R \sum_{i=1}^{n} x_i \, ln \, x_i \tag{3.2.3}$$

Yeh et al. [52] also express the formula for HEA or solid solution as:

$$\Delta S_{\text{conf}} = -k \, ln \, w = -R \left(\frac{1}{n} ln \frac{1}{n} + \frac{1}{n} ln \frac{1}{n} \pm - - - - \frac{1}{n} ln \frac{1}{n} \right) \tag{3.2.4}$$

where R stands for ideal gas constant. Yeh et al. [52] express that if $\Delta S_{\text{conf}} \geq$ 1.5R, then that type of alloy will be designated as high entropy alloy. When the ΔS_{conf} range between 1R and 1.5R represents medium entropy alloy, $\Delta S_{\text{conf}} \leq R$ represents low entropy alloy, also known as the traditional alloy. The properties of the microstructure of HEA are affected by four primary effects, known as the four core effects. The four core effects namely defined as the high entropy effect, cocktail effect, sluggish diffusion, and severe lattice distortion effect. The high entropy effect makes the microstructure much simpler than predicted by enhancing the solid solution formation. The addition of light elements for HEAs formation in equiatomic or nonequiatomic ratio, decreases the density of alloy but increases the strength of alloy and the hardness of the alloy due to strong cohesive bonding between elements. The strength of HEAs is also affected by interelemental reaction and lattice distortion. The sluggish diffusion effect expresses that the phase transformation and diffusion in HEA are much slower than in conventional alloys. The HEA synthesized by mainly different sizes of the element leads to distortion in the lattice. The larger atom displaced or pushed the small neighbors' atom, and the strain energy combined with lattice distortion increases the overall free energy of HEA. This effect affects the physical properties of HEAs, like solid solution strengthening. Due to dislocation movement and lattice distortion, this effect also affects the thermoelectric properties of materials. The term lattice distortion δ can be expressed in the following manner.

$$\delta_r = 100 \sqrt{\sum_{i=1}^{n} C_i \left(1 - \frac{r_i}{\bar{r}} \right)^2} \, \bar{r} = \sum_{i=1}^{n} C_i r_i \tag{3.2.5}$$

where $\bar{r}$ is the molar average atomic radius, C_i is an atomic fraction of ith element.

The mixing enthalpy (ΔH_{mix}) is $-15 \sim 5$ kJ/mol, and the calculation formula of ΔH_{mix} is [53]:

$$\Delta H_{mix} = \sum_{\substack{i=1 \\ i \neq j}}^{n} 4 H_{AB} C_i C_j \tag{3.2.6}$$

where C_j is the mole fraction of the jth element, and $I \neq j$; H_{AB} is the enthalpy of mixing between the A element and the B element.

$\Delta \chi$ is expressed as [54]:

$$\Delta \chi = 100 \sqrt{\sum_{i=1}^{n} C_i \left(1 - \frac{\chi_i}{\bar{\chi}}\right)^2} \tag{3.2.7}$$

χ_i is the electronegativity of the ith element; $\bar{\chi}$ is the molar average electronegativity.

Yang et al. [55] believed that $T_m . \Delta S_{\text{mix}}$ was the driving force for solid solutions formation, and $|\Delta H_{mix}|$ was the resistance of solid solutions formation, and thus proposed a new parameter Ω to reflect the common effect of ΔS_{mix} and ΔH_{mix} expressed as [55]:

$$\Omega = \frac{\Delta T_m \Delta S_{\text{mix}}}{|\Delta H_{\text{mix}}|} \tag{3.2.8}$$

where $T_m = \sum_{i=1}^{n} C_i \, (T_m)$

Wang et al. [56] presented a parameter Υ that could reflect the difference between the maximum and minimum atomic radius, and through statistics found that $\Upsilon < 1.175$ was a necessary condition to form a solid solution. Υ is expressed as:

$$\Upsilon = \left(1 - \sqrt{\frac{(r_s + \bar{r})^2 - \bar{r}^2}{(r_s + \bar{r})^2}}\right) \bigg/ \left(1 - \sqrt{\frac{(r_l + \bar{r})^2 - \bar{r}^2}{(r_l + \bar{r})^2}}\right) \tag{3.2.9}$$

where r_s and r_l are the alloy's largest and smallest atomic radii, respectively. HEAs composition is complex, and there is no unified theoretical system for composition design. The semi-empirical criterion proposed by scholars

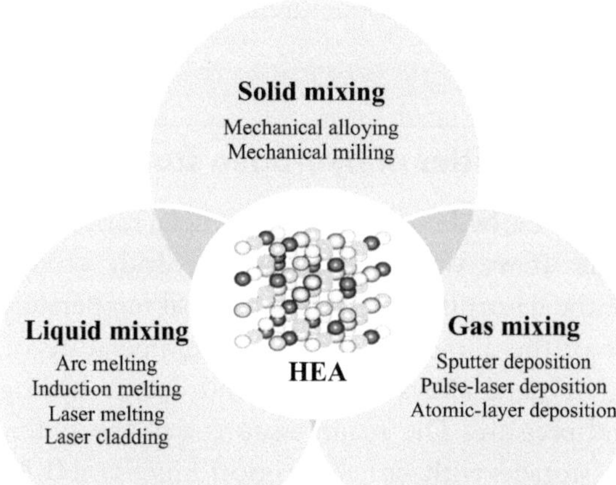

Figure 3.2.8 Scheme indicating the concepts of high-entropy alloy synthesis methods.

is only applicable to part of HEAs, so the composition design theory of HEAs needs further improvement.

3.2.10 Synthesis techniques of HEMs

The HEMs can be synthesized by a variety of processing route involving solid, liquid, and gaseous states like vacuum arc melting, induction melting, mechanical alloying (MA), sputtering, spark plasma sintering (SPS), etc. as shown in Fig. 3.2.8. The melting and casting route gives the material in the shape of ribbons, bars, and rods. Mechanical alloying followed by sintering is the most promising and highly usable solid–state processing route for homogeneous nanocrystalline materials. When we talk about thin–film and thick layers, sputtering cladding and plasma nitriding methods are used as surface modification techniques. The most popular techniques used in high entropy materials synthesis are vacuum arc melting, melt spinning, mechanical alloying followed by sintering and vacuum induction melting; all processing routes are similar for nonequiatomic and equiatomic–based high entropy materials in the solid–state processing route. MA [57] is a process using a high-energy ball mill to ensure for homogeneity of alloy firstly this technique was developed by Benjamin [58] and his co-workers for the synthesis Ni-base superalloy. Fecht et al. [59] first reported that the materials can give rise to amorphous or nanocrystalline materials due to through continuous deformation, fracture, and particle welding during high-energy ball milling.

3.2.11 Possibilities of hydrogen storage in HEMs

In metal hydride, hydrogen is stored in metal forming a covalent bond with metal atoms. If we want to use metal hydride as a fuel carrier, we must make sure the desorption is quick at normal temperature and pressure. However, most of the alloys require a high temperature of around 100–200°C. Hence, we need alloys to absorb and desorb hydrogen at normal temperature and pressure. The major issue is that the system should have good hydrogen storage with fast desorption kinetics. HEA is one of the emerging fields to design an alloy system for these required properties. With the help of theoretical prediction and elemental designing, we can develop the suitable alloys for this purpose.

Nowadays, we have two concepts to design HEAs for hydrogen storage: (1) the phase-oriented design and (2) the elemental property-orientated design. In the case of phase-orientated alloy design, we select the phase (BCC or Laves phases), which has a high tendency to store maximum hydrogen. And if we choose an elemental property-based system, we choose those elements which are hydride forming elements—the hydride forming element help to maximize absorption kinetics. And the nonhydride forming elements do a favor for desorption. Both the hydride and nonhydride forming elements play an essential role in hydrogen storage performance. Due to this, the researcher also investigated enhancing the hydrogen storage capacity in HEAs.

In most cases, Ti–V-based alloy (BCC solid solution) have been selected [20] to enhance hydrogen storage. For vanadium, the diffusion rate is very high, but it is reduced in presence of titanium. Still, the Ti–V-based alloy shows the maximum hydrogen storage capacity. To improve desorption kinetics, Cr addition in this Ti–V alloy exhibits promising result for hydrogen storage performance. The maximum hydrogen absorption is up to 4.37 wt% under ambient condition is reported for composition Ti_2CrV [60], but the main problem with this is that they do not exhibit good desorption kinetics, thus requiring activation at high temperatures. Hence, we should design the microstructure in HEAs that can be favorable for hydrogen storage applications. The storage capacity also depends on interstitial sites available for hydrogen atoms in HEA, which depends on structural factors. The intermetallic phase-based alloys open a new area of research on hydrogen storage materials. Alloys whose compositions are like Zr/Ti–V and Zr/Ti–V–X (X = Fe, Mn, Co, Cr, or Ni) are most favorable for high hydrogen capacity and fast absorption and desorption kinetics at a normal temperature [35]. Therefore, designing the HEA using hydride-forming elements and single-phase orientated can give rise to interesting results in the future. Because in HEAs, we can explore the materials' world for hydrogen storage by adding elements with required favorable parameters.

The first investigation into HEAs for hydrogen storage was done by Kao et al. [61]. They studied the hydrogen storage properties of multiprincipal component $CoFeMnTi_xV_yZr_z$ alloys and reported 0.03–1.80 wt% hydrogen storage. Thereafter, some literature claimed higher hydrogen storage. Sahlberg et al. [62] reported 2.7 wt% hydrogen storage in TiZrHfNbV HEAs. This alloy system has only a single BCC phase. Recently, Liu et al. [54] reported 3.51 wt% hydrogen storage in $V_{35}Ti_{30}Cr_{25}Fe_5Mn_5$ HEA, which is also a BCC system. The maximum reported hydrogen storage in Laves phases is found to be only 1.91 wt% [63], and the BCC phase maximum reported hydrogen storage is 3.51 wt%. So, we can say that BCC has good hydrogen storage properties than the Laves phase but Laves phase has better absorption and desorption kinetics than the other phases. So, in this context, it has an advantage over the BCC phase. HEAs give new and broad areas for research on the Laves phases for hydrogen storage applications. Zhou et al. [64] reported a study on low-vanadium TiZrMnCrV-based alloys for high-density hydrogen storage. In recent research, all the tests of hydrogen storage have done only C14 laves phase due to its maximum interstitial sites to capture hydrogen in their voids. Recently, Chen et al. [65] focused on the stability of phase during absorption and desorption of Hydrogen in TiZrFeMnCrV multicomponent HEAs. This alloy system belonging to the C14 type Laves phase family was synthesized by Arc melting and proceeded for hydrogen storage after mechanical milling. The maximum reported hydrogen absorption of this alloy for the first cycle is 1.80 wt%, and for the second cycle, it is about 1.76 wt%. They report hydrogen storage capacity slightly fluctuated between 1.76 wt% and 1.73 wt% approx. each cycle. Previously, Edalati et al. [51] also reported TiZrCrMnFeNi HEA, which belongs to C14 laves phase. The maximum hydrogen absorption reported was 1.7 wt%. As the research on the development of HEAs for hydrogen storage is going at a faster rate, soon we will get more interesting data on the study of hydrogen storage behavior of HEAs.

3.2.12 Hydrogen storage in BCC HEAs

Most of the literature hardly reported hydrogen storage performance in multiphase and intermetallic compounds [51,61,66–69] some literature also reported single-phase HEAs for hydrogen absorption [63,70–78]. However, a couple of literatures [73,74,79] reported that hydrogen absorption properties correlate with fundamental properties of HEAs like chemical composition, lattice constant, lattice distortion, and valence electron concentration (VEC). But no single concept could be adopted for predictinghydrogen storageproperties HEAs. Some systems possess different chemical compositions that can play an important role in hydrogen absorption in multicomponent HEA. And yet it is not clearly understood that individual elements or a mixture of multielements are responsible for the overall properties of HEAs [41]. Basically, HEAs for hydrogen storage applications are designed by phase-oriented or elemental properties. As per the literature survey, we can conclude that BCC phase has a better tendency to store a maximum % of hydrogen compared to any intermetallic phases. However, hydrogen absorption and desorption kinetics are superior in intermetallic phases compared to the BCC phases. In this section,we will discuss some salient results pertaining to the hydrogen storage performance of BCC HEAs.

Kunce et al. [71] reported the first study on hydrogen storage performance in a single-phase TiVZrNbMo HEA in equimolar composition. This HEA was synthesized using laser-engineered net shaping (LENS). The author reported 0.6 wt% hydrogen storage at 50°C under 85 bar equilibrium pressure. This HEA shows low hydrogen storage capacity but shows fast absorption kinetics. Lack of proper activation before hydrogenation and the degree of surface oxidation can be the reason for poor hydrogen storage capacity. The LENS technique was used to synthesize this HEA. This technique method is used for synthesizing material with a porous microstructure. The author also concluded that the lattice distortion did not favor the hydrogen storage capacity, but nowadays, most research articles claim that lattice distortion is one of the favorable aspects of hydrogen storage capacity. In the same way, Sahlberg et al. [62] reported that the TiVZrNbHf HEA was synthesized using arc melting. The reported phase of the system is a pure single BCC phase. This system contains all the hydride-forming elements that can be one of the reasons behind the remarkable hydrogen storage capacity of about 2.7 wt% hydrogen storage. This system has high lattice distortion due to a large atomic size mismatch ($\delta = 6.8\%$), which is favorable for high wt% hydrogen uptake. This reported hydrogen storage capacity is more noticeable than other reported maximum hydrogen capacities of $\sim$2 H/M [70]. In this HEA, the authors stated that the reason behind the superior hydrogen storage is that the hydrogen atoms occupy both the tetrahedral and octahedral interstitial sites in the BCC phase. Zlotea et al. [78] reported single-phase TiZrNbHfTa HEA for which uptake is 1.7 wt% hydrogen at 300°C. It shows the two plateau pressures in the PCI curve on the reported data. In this analysis, the dihydride phase accrues at 23 bar equilibrium pressure, and the mono hydride phase accrues at low equilibrium pressure. The authors also reported that this HEA went through phase transformation during hydrogenation. The BCC structure transformed into the FCC structure. With the help of TDS measurement, the said hydrogen desorption activation energy is found 80.2 kJ/mol.

In a different study, Zhang et al. [77] focused on the hydrogen performance effect on the chemical composition analysis before and after heat treatment. They used X-ray photoelectron spectroscopy before and after activation for determining the chemical composition of the surface of TiZrNbTa HEA. The surface oxide transforms into hydroxides and other complex hydrides. This property allows in favor of faster hydrogen absorption kinetics. But this HEA does not have a good storage capacity. The hydrogen storage capacity of this HEA is about 1.7 wt%. Zepon et al. [76] reported the BCC phase-based $MgZrTiFe_{0.5}Co_{0.5}Ni_{0.5}$ HEA, which was synthesized using high-energy ball milling in an inert condition. The maximum capacity of this HEA is about 1.2 wt% at 300°C. The resulting hydride phase contains both the BCC and FCC phases. The authors also reported phase transformation during the hydrogen cycle. Shen et al. [75] noticed phase transformation in TiZrNbMoHf HEA after the first hydrogenation from BCC to FCC hydride phase. The phase transformation is also reversible when they perform hydrogenation cycling. The hydrogen capacity up to 1.2 wt% was determined by thermo gravimetric analysis's mass loss. In further investigation into this system same research group studied the addition of Mo concentration in this base alloy. The 10 at % addition of Mo in the base HEA had increased the hydrogen storage capacity up to 1.5 wt%.

In the investigation of the lattice distortion effect in hydrogen sorption performance, Nygård et al. [73] have investigated a series of $TiVZr_{(z)}Nb_{(1-z)}$ and $TiVZr_{(1+z)}Nb$ multi principal entropy alloy (MPEAs). In these MPEAs, the authors tuned the lattice distortion with the help of varying Zr. The hydrogen storage capacity of these MPEAs lies between 1.8 H/M and 2.0 H/M. Interestingly, they found out that alloys with less Zr-content can withstand high temperatures (up to 1000°C) for hydrogen desorption, recovering the initial BCC structure. Meanwhile, the alloys with Zr–content $\leq$12.5 at % decomposed in multiple BCC phases at high temperatures. The same research group also studied various MPEAs concerning VEC [73]. They observed that the temperature for desorption kinetics increased with the higher value of VEC. That explores the relationship between VEC with hydride stability. The value of VEC higher than 5.0 is found good for the hydrogen absorption capacity. These properties have also been seen in classical BCC alloys [41]. Recently, Sleiman et al. [80] focused on the microstructure properties in $TiHfZrNb_{1-x}V_{1+x}$ HEA for x = 0, 0.1, 0.2, 0.4, and 0.6. In this article, the author reported a pure single BCC phase at x = 0. In this study, the author substitute's niobium with vanadium and the result is that the BCC phase is replaced by the hcp and FCC phases. The storage capacity in this HEA is found to be $\sim$2 wt%.)

3.2.13 Hydrogen storage in HEAs at room temperature

Most of the HEAs require activation at high temperatures to perform hydrogen storage. Some reported alloys absorb hydrogen at high temperatures (573 K), such as TiVZrNbHf [80], MgTiVCrFe [72], and TiZrNbHfTa [78]. It is also reported that some other HEAs such as TiZrNbMoV [71], TiZrHfScMo [81], TiVZrNb [73], and TiVCrNb [79], can absorb hydrogen at room temperature ($\sim$30 K). We require the HEA to store hydrogen at room temperature and prefer a good hydrogen absorption/desorption cycle without phase change during application. That type of alloy is designed by the limiting value of VEC, perfect element selection, which is thermodynamic stable for good reversibility of hydrogen absorption and

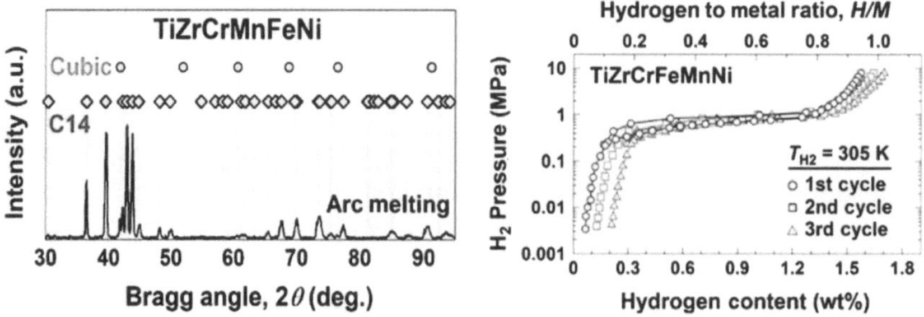

Figure 3.2.9 Reversible room temperature hydrogen storage with fast kinetics in high-entropy alloy [51].

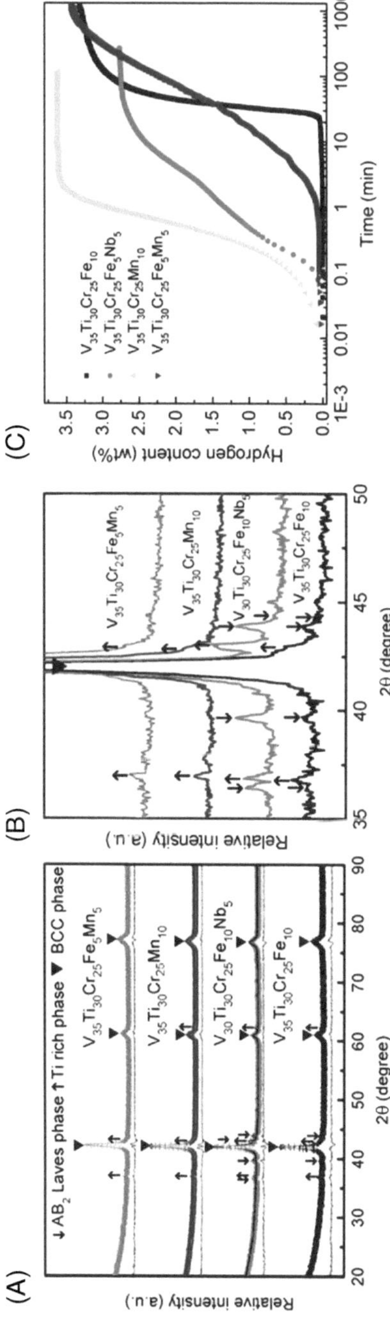

Figure 3.2.10 XRD pattern of the alloy sample with 2Θ range (A) 20–90, (B) 35–50 and activation kinetics at room temperature under 2000 kPa of hydrogen pressure [54].

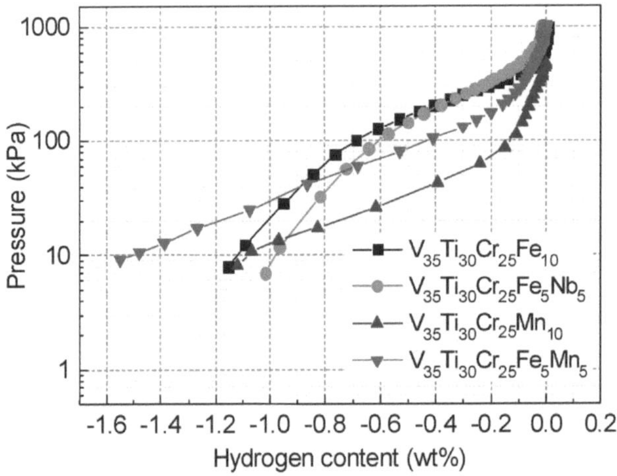

Figure 3.2.11 Pressure–composition–temperature desorption curves of the alloy samples after 1st cycle hydrogen absorption at 25°C.

desorption. Nygard et al. [73] predicted that if any HEA system has a VEC set of about 6.4, that system will easily perform dehydrogenation at room temperature. Edalati et al. [51] reported a hexanary AB$_2$ (A = hydride forming element, B = nonhydride forming element) type HEA that contains TiZrCrMnFeNiequiatomic component synthesized by Arc melting contain C14 Laves phase. This alloy system follows the desired VEC values related to hydrogen absorption and desorption at room temperature. This system's reported hydrogen absorption and desorption capacity is 1.7 wt% with fast kinetics at 303 K. Interestingly, this HEA did not require any activation at high temperatures. It could easily perform absorption and desorption in the presence of 3.9 Mpa hydrogen pressure and also at room temperature.

Following this approach, the highest hydrogen storage capacity was reported in Ti–V–Zr based four alloys such as: (1) V$_{35}$Ti$_{30}$Cr$_{25}$Fe$_{10}$, (2) V$_{35}$Ti$_{30}$Cr$_{25}$Mn$_{10}$, (3) V$_{30}$Ti$_{30}$Cr$_{25}$Fe$_{10}$Nb$_5$, and (4) V$_{35}$Ti$_{30}$Cr$_{25}$Fe$_5$Mn$_5$ high entropy alloys [82]. They all are synthesized by arc melting technique under an argon atmosphere (Fig. 3.2.9). XRD (X-ray diffraction) reports of this system conclude that they all belong to the BCC phases. The XRD pattern has shown in Fig. 3.2.10. Among these four BCC HEA systems, V$_{35}$Ti$_{30}$Cr$_{26}$Fe$_5$Mn$_5$ alloy revealed the highest reversible capacity, about 3.65 wt%.

These HEAs perform hydrogen absorption under 2000 kPa hydrogen pressure at room temperature; shown in Fig. 3.2.10C. We can see that the order of hydrogen absorption of these HEAs is as: V$_{35}$Ti$_{30}$Cr$_{25}$Mn$_{10}$ >

$V_{30}Ti_{30}Cr_{25}Fe_5Mn_5 > V_{35}Ti_{30}Cr_{25}Fe_{10} > V_{30}Ti_{30}Cr_{25}Fe_{10}Nb_5$. Generally, hydrogen storage capacity increases with increased cell volume. But in these HEAs, hydrogen desorption pressure decreases with increased cell volume as: $V_{35}Ti_{30}Cr_{25}Fe_{10} > V_{35}Ti_{30}Cr_{25}Fe_5Mn_5 > V_{35}Ti_{30}Cr_{25}Mn_{10}$. The PCT desorption curve for this HEA is shown in Fig. 3.2.11.

3.2.14 Summary

In this chapter, we have outlined the benefits of hydrogen energy for potential use in the future. The most efficient and secured storage method for hydrogen is solid–state storage. We have discussed how metallic alloys can store hydrogen. The chapter covers the main reported sources of synthesis like induction melting, arc melting, mechanical alloying, etc. Recent research suggested that the strongest hydride formers, or efficient hydride-forming HEAs, would be among the best solid-state hydrogen storage materials. It is possible that the higher hydrogen storage capacity can be attained by a high entropy material possessing a distorted lattice due to the strain effect, which further encourages hydrogen to occupy both tetrahedral and octahedral sites. Therefore, we should explore the new class of high entropy alloys for designing efficient hydrogen storage materials.

Acknowledgment

The authors gratefully acknowledge the discussion with all their collaborators, who have immensely contributed to the collaborative research work, a part of which has been discussed in this chapter.

References

[1] D. Pukazhselvan, R.S.S. Saravanan, T.P. Yadav, Towards sustainable green energy development and insights on few scientific problems leading to less carbon economy, Adv. Sci. 1 (2012) 302–318. https://doi.org/10.1166/rase.2012.1019.

[2] O.N. Srivastava, T.P. Yadav, R.R. Shahi, S.K. Pandey, M.A. Shaz, A. Bhatnagar, Hydrogen energy in India: storage to application, Proc. Indian Nation. Sci. Acad. 81 (2015) 915–937. https://doi.org/10.16943/ptinsa/2015/v81i4/48303.

[3] T.P. Yadav, R.R. Shahi, O.N. Srivastava, Synthesis, characterization and hydrogen storage behaviour of AB_2 ($ZrFe_2$, $Zr(Fe_{0.75}V_{0.25})_2$, $Zr(Fe_{0.5}V_{0.5})_2$ type materials, Int. J. Hydrogen Energy 37 (2012) 3689–3696. https://doi.org/10.1016/j.ijhydene.2011.04.210.

[4] A. Midilli, I. Dincer, Key strategies of hydrogen energy systems for sustainability, Int. J. Hydrogen Energy 32 (2007) 511–524. https://doi.org/10.1016/j.ijhydene.2006.06.050.

[5] A. Pareek, R. Dom, J. Gupta, J. Chandran, V. Adepu, P.H. Borse, Insights into renewable hydrogen energy: recent advances and prospects, Mater. Sci. Energy Technol. 3 (2020) 319–327. https://doi.org/10.1016/j.mset.2019.12.002.

[6] A. Midilli, M. Ay, I. Dincer, M.A. Rosen, On hydrogen and hydrogen energy strategies. II: future projections affecting global stability and unrest, Renew. Sustain. Energy Rev. 9 (2005) 273–287. https://doi.org/10.1016/j.rser.2004.05.002.

[7] I. Dincer, C Acar, Smart energy solutions with hydrogen options, Int. J. Hydrogen Energy 43 (2018) 8579–8599. https://doi.org/10.1016/j.ijhydene.2018.03.120.

[8] G. Badea, R.A. Felseghi, I. Aschilean, S.M. Raboaca, T.M Soimosan, The role of hydrogen as a future solution to energetic and environmental problems for residential buildings, in: AIP Conference Proceedings, 1917, 2017.

[9] N. Afgan, A Veziroglu, Sustainable resilience of hydrogen energy system, Int. J. Hydrogen Energy 37 (2012) 5461–5467. https://doi.org/10.1016/j.ijhydene.2011.04.201.

[10] I. Dincer, Green paths for hydrogen production, Int. J. Hydrogen Energy 37 (2012) 1954–1971. https://doi.org/10.1016/j.ijhydene.2011.03.173.

[11] H. Jeon, S. Kim, K. Yoon, Fuel cell application for investigating the quality of electricity from ship hybrid power sources, J. Mar. Sci. Eng. 7 (1-22) (2019) 241. https://doi.org/10.3390/jmse7080241.

[12] M. İnci, Ö. Türksoy, Review of fuel cells to grid interface: configurations, technical challenges and trends, J. Cleaner Prod. 213 (2019) 1353–1370. https://doi.org/10.1016/j.jclepro.2018.12.281.

[13] O.Z. Sharaf, M.F. Orhan, An overview of fuel cell technology: fundamentals and applications, Renew. Sustain. Energy Rev. 32 (2014) 810–853. https://doi.org/10.1016/j.rser.2014.01.012.

[14] C. Pötzinger, M. Preißinger, D Brüggemann, Influence of hydrogen-based storage systems on self-consumption and self-sufficiency of residential photovoltaic systems, Energies 8 (2015) 8887–8907. https://doi.org/10.3390/en8088887.

[15] A. Midilli, M. Ay, I. Dincer, M.A. Rosen, On hydrogen and hydrogen energy strategies: I: current status and needs, Renew. Sustain. Energy Rev. 9 (2005) 255–271. https://doi.org/10.1016/j.rser.2004.05.003.

[16] M. Momirlan, T.N. Veziroglu, The properties of hydrogen as fuel tomorrow in sustainable energy system for a cleaner planet, Int. J. Hydrogen Energy 30 (2005) 795–802. https://doi.org/10.1016/j.ijhydene.2004.10.011.

[17] T.N. Veziroglu, S Sahin, 21st century's energy: hydrogen energy system, Energy Convers. Manag. 49 (2008) 1820–1831. https://doi.org/10.1016/j.enconman.2007.08.015.

[18] S.K. Verma, A. Bhatnagar, V. Shukla, P.K. Soni, A.P. Pandey, T.P. Yadav, …, O.N. Srivastava, Multiple improvements of hydrogen sorption and their mechanism for MgH_2 catalyzed through $TiH_2@$ Gr, Int. J. Hydrogen Energy 45 (2020) 19516–19530. https://doi.org/10.1016/j.ijhydene.2020.05.031.

[19] S.K. Pandey, A. Bhatnagar, S.S. Mishra, T.P. Yadav, M.A. Shaz, …, O.N. Srivastava, Curious catalytic characteristics of Al–Cu–Fe quasicrystal for de/rehydrogenation of MgH_2, J. Phys. Chem. C 121 (2017) 24936–24944. https://doi.org/10.1021/acs.jpcc.7b07336.

[20] S.K. Pandey, J. Singh, T.P. Yadav, O.N. Srivastava, Material tailoring of Ti–V–Fe–Zr for improving sorption behaviour employing off stoichiometric concentration of native element Ti, Mater. Focus 3 (2014) 28–35. https://doi.org/10.1166/mat.2014.1131.

[21] R.R. Shahi, T.P. Yadav, M.A. Shaz, O.N Srivastva, Effect of processing parameter on hydrogen storage characteristics of as quenched Ti45Zr38Ni17 quasicrystalline alloys, Int. J. Hydrogen Energy 36 (2011) 592–599. https://doi.org/10.1016/j.ijhydene.2010.10.031.

[22] T.P. Yadav, R.M. Yadav, D.P. Singh, Mechanical milling: a top down approach for the synthesis of nanomaterials and nanocomposites, J. Nanosci. Nanotechnol. 2 (2012) 22–48. https://doi.org/10.5923/j.nn.20120203.01.

[23] A.C. Switendick, Band structure calculations for metal hydrogen systems, Zeitschrift für PhysikalischeChemie 117 (1979) 89–112. https://doi.org/10.1524/zpch.1979.117.117.089.

[24] D.G. Westlake, A geometric model for the stoichiometry and interstitial site occupancy in hydrides (deuterides) of LaNi$_5$, LaNi$_4$Al and LaNi$_4$Mn, J. Less-Common. Metals 91 (1983) 275–292. https://doi.org/10.1016/0022-5088(83)90322-3.

[25] D.K. Rai, T.P. Yadav, V.S. Subrahmanyam, O.N. Srivastava, Structural and Mössbauer spectroscopic investigation of Fe substituted Ti–Ni shape memory alloys, J. Alloys Compd. 482 (2009) 28–32. https://doi.org/10.1016/j.jallcom.2009.03.164.

[26] T.P. Yadav, N.K. Mukhopadhyay, R.S. Tiwari, O.N. Srivastava, On the evolution of quasicrystalline and crystalline phases in rapidly quenched Al-Co–Cu–Ni alloy, Mater. Sci. Eng. R. Rep. 449 (2007) 1052–1056. https://doi.org/10.1016/j.msea.2005.12.098.

[27] T.P. Yadav, A. Kumar, S.K. Verma, N.K. Mukhopadhyay, High-entropy alloys for solid hydrogen storage: potentials and prospects, Trans Indian Natl. Acad. Eng. 7 (2022) 147–156. https://doi.org/10.1007/s41403-021-00316-w.

[28] R.R. Shahi, A. Bhatanagar, S.K. Pandey, V. Shukla, T.P. Yadav, M.A. Shaz, …, O.N. Srivastava, MgH$_2$–ZrFe$_2$H$_x$ nanocomposites for improved hydrogen storage characteristics of MgH$_2$, Int. J. Hydrogen Energy 40 (2015) 11506–11513. https://doi.org/10.1016/j.ijhydene.2015.03.162.

[29] R.R. Shahi, T.P. Yadav, M.A. Shaz, & O.N. Srivastava. Synthesis characterization and hydrogenation behaviour of as quenched Ti41.5+XZr41.5-XNi17(x= 0, 3.5, 11.5 and 13.5) nano quasicrystalline ribbons. J. Phys. Conf. Ser. 809 (2017) 012011. https://doi.org/10.1088/1742-6596/809/1/012011.

[30] K.T. Møller, T.R. Jensen, E. Akiba, H. Li, Hydrogen: a sustainable energy carrier, Progr. Nat. Sci.: Mater. Int. 27 (2017) 34–40. https://doi.org/10.1016/j.pnsc.2016.12.014.

[31] Q. Lai, M. Paskevicius, D.A. Sheppard, C.E. Buckley, A.W. Thornton, M.R. Hill, Q. Gu, J. Mao, …, K.F. Aguey-Zinsou, hydrogen storage materials for mobile and stationary applications: current state of the art, Chem. Sus. Chem. 8 (2015) 2789–2825. https://doi.org/10.1002/cssc.201500231.

[32] A. Züttel, A. Remhof, A. Borgschulte, O. Friedrichs, Hydrogen: the future energy carrier, Philos. Trans. A Math Phys. Eng. Sci. 368 (2010) 3329–3342. https://doi.org/10.1098/rsta.2010.0113.

[33] J.M Banuelos, Refractory High-Entropy Alloys for Hydrogen Storage (doctoral thesis), Institut de Chimie et des Matériaux de Paris-Est, CNRS, 2020.

[34] S.S. Mishra, S. Mukhopadhyay, T.P. Yadav, N.K. Mukhopadhyay, O.N Srivastava, Synthesis and characterization of hexanary Ti–Zr–V–Cr–Ni–Fe high-entropy Laves phase, J. Mater. Res. 34 (2019) 807–818. https://doi.org/10.1557/jmr.2018.502.

[35] S.S. Mishra, T.P. Yadav, O.N. Srivastava, N.K. Mukhopadhyay, K Biswas, Formation and stability of C14 type Laves phase in multi component high-entropy alloys, J. Alloys Compd. 832 (2020) 153764. https://doi.org/10.1016/j.jallcom.2020.153764.

[36] T.P. Yadav, S. Mukhopadhyay, S.S. Mishra, N.K. Mukhopadhya, O.N. Srivastava, Synthesis of a single phase of high-entropy Laves intermetallics in the Ti–Zr–V–Cr–Ni equiatomic alloy, Philos. Mag. Lett. 97 (2017) 494–503. https://doi.org/10.1080/09500839.2017.1418539.

[37] J.W. Yeh, S.K. Chen, J.Y. Gan, S.J. Lin, T.S. Chin, T.T. Shun, C.H. Tsau, S.Y. Chou, Formation of simple crystal structures in Cu-Co-Ni-Cr-Al-Fe-Ti-V alloys with multiprincipal metallic elements, Metall. Mater. Trans. A 35 (2004) 2533–2536. https://doi.org/10.1007/s11661-006-0234-4.

[38] B. Cantor, I.T.H Chang, P. Knight, A.J.B. Vincent, Microstructural development in equiatomic multicomponent alloys, Mater. Sci. Eng. A Struct. Mater. 375-377 (2004) 213–218. https://doi.org/10.1016/j.msea.2003.10.257.

[39] S. Ranganathan, Alloyed pleasures: multimetallic cocktails, Curr. Sci. 85 (2003) 1404–1406.

[40] M.H. Tsai, J.W. Yeh, High-entropy alloys: a critical review, Mater. Res. Lett. (2) (2014) 107–123. https://doi.org/10.1080/21663831.2014.912690.

[41] D.B. Miracle, O.N Senkov, A critical review of high entropy alloys and related concepts, Acta Mater. 122 (2017) 448–511. https://doi.org/10.1016/j.actamat.2016.08.081.

[42] Y.F. Ye, Q. Wang, J. Lu., C.T. Liu, Y Yang, High-entropy alloy: challenges and prospects, Mater. Today 19 (2016) 349–362. https://doi.org/10.1016/j.mattod.2015.11.026.

[43] I. Basu, J.Th.M.D. Hosson, Strengthening mechanisms in high entropy alloys: fundamental issues, Scr. Mater. 187 (2020) 148–156. https://doi.org/10.1016/j.scriptamat.2020.06.019.

[44] K. Biswas, J.W. Yeh, P.P. Bhattacharjeec, J.T.M. DeHosson, High entropy alloys: key issues under passionate debate, Scr. Mater. 188 (2020) 54–58. https://doi.org/10.1016/j.scriptamat.2020.07.010.

[45] E.P. George, W.A. Curtin, C.C Tasan, High entropy alloys: a focused review of mechanical properties and deformation mechanisms, Acta Mater. 188 (2020) 435–474. https://doi.org/10.1016/j.actamat.2019.12.015.

[46] P. Sathiyamoorthi, H.S. Kim, High-entropy alloys with heterogeneous microstructure: processing and mechanical properties, Prog. Mater. Sci. 123 (2020) 100709. https://doi.org/10.1016/j.pmatsci.2020.100709.

[47] X. Yan, Y. Zhang, Functional properties and promising applications of high entropy alloys, Scr. Mater. 187 (2020) 188–193. https://doi.org/10.1016/j.scriptamat.2020.06.017.

[48] M. Vaidya, G.M. Muralikrishna, B.S. Murty, High-entropy alloys by mechanical alloying: a review, J. Mater. Res. 34 (2019) 664–686. https://doi.org/10.1557/jmr.2019.37.

[49] A.S. Sharma, S. Yadav, K. Biswas, B. Basu, High-entropy alloys and metallic nanocomposites: processing challenges, microstructure development and property enhancement, Mater. Sci. Eng. R: Rep. 131 (2018) 1–42. https://doi.org/10.1016/j.mser.2018.04.003.

[50] M. Dornheim, N. Eigen., G. Barkhordarian, T. Klassen, R. Bormann, Tailoring hydrogen storage materials towards application, Adv. Eng. Mater. 8 (2006) 377–385. https://doi.org/10.1002/adem.200600018.

[51] P. Edalati, R. Floriano, A. Mohammadi, Y. Li, G. Zepon, H.W. Li, K Edalati, Reversible room temperature hydrogen storage in high-entropy alloy TiZrCrMnFeNi, Scr. Mater. 178 (2020) 387–390. https://doi.org/10.1016/j.scriptamat.2019.12.009.

[52] J.W. Yeh, Alloy design strategies and future trends in high-entropy alloys, JOM 65 (2013) 1759–1771. https://doi.org/10.1007/s11837-013-0761-6.

[53] Y. Zhang, Y.J. Zhou, J.P. Lin, G.L. Chen, P.K. Liaw, Solid-solution phase formation rules for multi-component alloys, Adv. Eng. Mater. 10 (2008) 534–538. https://doi.org/10.1002/adem.200700240.

[54] B. Liu, J.F. Wu, Y.W. Cui, Q.Q. Zhu, G.R. Xiao, S.Q. Wu, G.H. Cao, Z. Ren, Structural evolution and superconductivity tuned by valence electron concentration in the Nb-Mo-Re-Ru-Rh high-entropy alloys, J. Mater. Sci. 85 (2021) 11–17. https://doi.org/10.1016/j.jmst.2021.02.002.

[55] X. Yang, Y. Zhang, Prediction of high-entropy stabilized solid-solution in multi-component alloys, Mater. Chem. Phys. 132 (2012) 233–238. https://doi.org/10.1016/j.matchemphys.2011.11.021.

[56] Z.J. Wang, Y.H. Huang, Y. Yang, J. Wang, C.T. Liu, Atomic-size effect and solid solubility of multicomponent alloys, Scr. Mater. 94 (2015) 28–31. https://doi.org/10.1016/j.scriptamat.2014.09.010.

[57] N.K. Mukhopadhyay, T.P. Yadav, O.N. Srivastava, An investigation on the transformation of the icosahedral phase in the Al-Fe-Cu system during mechanical milling and subsequent annealing, Philos. Mag. A 82 (2002) 2979–2993. https://doi.org/10.1080/01418610208239629.

[58] J.S. Benjamin, Dispersion strengthened superalloys by mechanical alloying, Metall. Trans. 1 (1970) 2943–2951. https://doi.org/10.1007/BF03037835.

[59] H.J. Fecht, G. Han, Z. Fu, W.L. Johnson, Metastable phase formation in the Zr Al binary system induced by mechanical alloying, J. Appl. Phys. 67 (1990) 1744–1748. https://doi.org/10.1063/1.345624.

[60] A. Kumar, K. Shashikala, S. Banerjee, J. Nuwad, P. Das, C.G.S Pillai, Effect of cycling on hydrogen storage properties of Ti_2CrV alloy, Int. J. Hydrogen Energy 37 (2012) 3677–3682. https://doi.org/10.1016/j.ijhydene.2011.04.135.

[61] Y.H. Kao, S.K. Chen, J.H. Sheu, J.T. Lin, W.E. Lin, J.W. Yeh, S.J. Lin, T.H. Liou, C.W. Wang, Hydrogen storage properties of multi-principal-component $CoFeMnTi_xV_yZr_z$ alloys, Int. J. Hydrogen Energy 35 (2010) 9046–9059. https://doi.org/10.1016/j.ijhydene.2010.06.012.

[62] M. Sahlberg, D. Karlsson, C. Zlotea, U. Jansson, Superior hydrogen storage in high entropy alloys, Sci. Rep. 6 (2016) 36770. https://doi.org/10.1038/srep36770.

[63] J. Montero, C. Zlotea, G. Ek, J.C. Crivello, L. Laversenne, M. Sahlberg, TiVZrNb multi-principal-element alloy: synthesis optimization, structural, and hydrogen sorption properties, Molecules 24 (2019) 2799. https://doi.org/10.3390/molecules24152799.

[64] P. Zhou, Z. Cao, X. Xiao, L. Zhan, S. Li, Z. Li, L. Jiang, L. Chen, Development of Ti-Zr-Mn-Cr-V based alloys for high-density hydrogen storage, J. Alloys Compd. 875 (2021) 160035. https://doi.org/10.1016/j.jallcom.2021.160035.

[65] J. Chen, Z. Li, H. Huang, Y. Lv, B. Liu, Y. Li, Y. Wu, J. Yuan, Y. Wang, Superior cycle life of TiZrFeMnCrV high entropy alloy for hydrogen storage, Scr. Mater. 212 (2022) 114548. https://doi.org/10.1016/j.scriptamat.2022.114548.

[66] S.K. Chen, P.H. Lee, H. Lee, H.T. Su, Hydrogen storage of $C14-Cr_uFe_vMn_wTi_xV_yZr_z$ alloys, Mater. Chem. Phys. 210 (2018) 336–347. https://doi.org/10.1016/j.matchemphys.2017.08.008.

[67] I. Kunce, M. M. Polański, T. Czujko, Microstructures and hydrogen storage properties of La-Ni-Fe-V-Mn alloys, Int. J. Hydrogen Energy 42 (2017) 27154–27164. https://doi.org/10.1016/j.ijhydene.2017.09.039.

[68] I. Kunce, M. Polanski, J Bystrzycki, Structure and hydrogen storage properties of a high entropy ZrTiVCrFeNi alloy synthesized using laser engineered net shaping (LENS), Int. J. Hydrogen Energy 38 (2013) 12180–12189. https://doi.org/10.1016/j.ijhydene.2013.05.071.

[69] V. Zadorozhnyy, B. Sarac, E. Berdonosova, T. Karazehir, A. Lassnig, C. Gammer, M. Zadorozhnyy, S. Ketov, S. Klyamkin, J. Eckert, Evaluation of hydrogen storage performance of ZrTiVNiCrFe in electrochemical and gas-solid reactions, Int. J. Hydrogen Energy 45 (2020) 5347–5355. https://doi.org/10.1016/j.ijhydene.2019.06.157.

[70] D. Karlsson, G. Ek, J. Cedervall, C. Zlotea, K.T. Moller, T.C. Hansen, …, M. Sahlberg, Structure and hydrogenation properties of a HfNbTiVZr high-entropy alloy, Inorg. Chem. 57 (2018) 2103–2110. https://doi.org/10.1021/acs.inorgchem.7b03004.

[71] I. Kunce, M. Polanski, J. Bystrzycki, Microstructure and hydrogen storage properties of a TiZrNbMoV high entropy alloy synthesized using laser engineered net shaping (LENS), Int. J. Hydrogen Energy 39 (2014) 9904–9910. https://doi.org/10.1016/j.ijhydene.2014.02.067.

[72] M.O. Marco, Y. Li, H.W. Li, K. Edalati, R. Floriano, Mechanical synthesis and hydrogen storage characterization of MgVCr and MgVTiCrFe high-entropy alloy, Adv. Eng. Matr. 22 (2020) 1901079. https://doi.org/10.1002/adem.201901079.

[73] M.M. Nygård, G. Ek, D. Karlsson, M. Sahlberg, M.H. Sørby, B.C. Hauback, Hydrogen storage in high-entropy alloys with varying degree of local lattice strain, Int. J. Hydrogen Energy 44 (2019) 29140–29149. https://doi.org/10.1016/j.ijhydene.2019.03.223.

[74] H. Shen, J. Hu, P. Li, G. Huang, J. Zhang, J. Zhang, Y. Mao, H. Xiao, X. Zhou, X. Zu, X. Long, S. Peng, Compositional dependence of hydrogenation performance of Ti-Zr-Hf-Mo-Nb high-entropy alloys for hydrogen/tritium storage, J. Mater. Sci 55 (2020) 116–125. https://doi.org/10.1016/j.jmst.2019.08.060.

[75] H. Shen, J. Zhang, J. Hu, J. Zhang, Y. Mao, H. Xiao, X. Zhou, X. Zu, A novel TiZrHfMoNb high-entropy alloy for solar thermal energy storage, Nanomaterials 9 (2019) 248. https://doi.org/10.3390/nano9020248.

[76] G. Zepon, D.R. Leiva, R.B. Strozi, A. Bedoch, S.J.A. Figueroa, T.T. Ishikawa, W.J Botta, Hydrogen-induced phase transition of $MgZrTiFe_{0.5}Co_{0.5}Ni_{0.5}$ high entropy alloy, Int. J. Hydrogen Energy 43 (2018) 1702–1708. https://doi.org/10.1016/j.ijhydene.2017.11.106.

[77] C. Zhang, Y. Wu, L. You, X. Cao, Z. Lu, X. Song, Investigation on the activation mechanism of hydrogen absorption in TiZrNbTa high entropy alloy, J. Alloys Compd. 781 (2019) 613–620. https://doi.org/10.1016/j.jallcom.2018.12.120.

[78] C. Zlotea, M.A. Sow, G. Ek, J.P. Couzinié, L. Perrière, I. Guillot, J. Bourgon, K.T Møller, T.R. Jensen, E. Akiba, M Sahlberg, Hydrogen sorption in TiZrNbHfTa high entropy alloy, J. Alloys Compd. 775 (2019) 667–674. https://doi.org/10.1016/j.jallcom.2018.10.108.

[79] M.M. Nygård, G. Ek, D. Karlsson, M.H. Sørby, M. Sahlberg, B.C Hauback, Counting electrons: a new approach to tailor the hydrogen sorption properties of high-entropy alloys, Acta Mater. 175 (2019) 121–129. https://doi.org/10.1016/j.actamat.2019.06.002.

[80] S. Sleiman, J. Huot, Microstructure and first hydrogenation properties of $TiHfZrNb_{1-x}V_{1+x}$ Alloy for x = 0, 0.1, 0.2, 0.4, 0.6 and 1, Molecules 27 (2022) 1054. https://doi.org/10.3390/molecules27031054.

[81] J. Hu, H. Shen, M. Jiang, H. Gong, H. Xiao, Z. Liu, G. Sun, X. Zu, A DFT study of hydrogen storage in high-entropy alloy TiZrHfScMo, Nanomaterials 9 (2019) 461. https://doi.org/10.3390/nano9030461.

[82] J. Liu, J. Xu, S. Sleiman, X. Chen, S. Zhu, H. Cheng, J Huot., Microstructure and hydrogen storage properties of Ti–V–Cr based BCC-type high entropy alloys, Int. J. Hydrogen Energy 46 (2021) 28709–28718. https://doi.org/10.1016/j.ijhydene.2021.06.137.

[83] S.S.M. Pauzi, W. Darham, R. Ramli, M. Harun, M.K. Talari, Effect of Zr addition on microstructure and properties of $FeCrNiMnCoZr_x$ and $Al_{0.5}FeCrNiMnCoZr_x$ high entropy alloys, Trans. Indian Inst. Met. 66 (2013) 305–308. https://doi.org/10.1007/s12666-013-0264-8.

[84] T.P. Yadav, R.S. Tiwari, O.N. Srivastava, Effect of Cu substitution in Al–Co–Ni decagonal quasicrystals, J. NonCryst. Solids 334 (2004) 39–43. https://doi.org/10.1016/j.jnoncrysol.2003.11.009.

[85] J.W. Yeh, S.K. Chen, S.J. Lin, J.Y. Gan, T.S. Chin, T.T. Shun, C.H. Tsau, S.Y. Chang, Nanostructured high-entropy alloys with multiple principal elements: novel alloy design concepts and outcomes, Adv. Eng. Mater. 6 (2004) 299303. https://doi.org/10.1002/adem.200300567.

[86] Y. Zou, S. Maiti, W. Steurer, R Spolenak, Size-dependent plasticity in an $Nb_{25}Mo_{25}Ta_{25}W_{25}$ refractory high-entropy alloy, Acta Mater. 65 (2014) 85–97. https://doi.org/10.1016/j.actamat.2013.11.049.

[87] R. Helmolt, U. Eberle, Fuel cell vehicles: status 2007, J. Power Source 165 (2007) 833–843. https://doi.org/10.1016/j.jpowsour.2006.12.073.

[88] A. Kumar, T.P. Yadav, N.K. Mukhopadhyay, Notable hydrogen storage in Ti–Zr–V–Cr–Ni high entropy alloy, Int. J. Hydrogen Energy 47 (2022) 22893–22900. https://doi.org/10.1016/j.ijhydene.2022.05.107.

[89] K.R. Cardoso, V. Roche, A.M. Jorge Jr, F.J. Antiqueira, G. Zepon, Y. Champion, Hydrogen storage in MgAlTiFeNi high entropy alloy, J. Alloys Compd. 858 (2021). https://doi.org/10.1016/j.jallcom.2020.158357.

[90] J. Montero, G. Ek, L. Laversenne, V. Nassif, G. Zepon, M. Sahlberg, C. Zlotea, Hydrogen storage properties of the refractory Ti–V–Zr–Nb–Ta multi-principal element alloy, J. Alloys Compd. 835 (2020). https://doi.org/10.1016/j.jallcom.2020.155376.

[91] B. Sarac, V. Zadorozhnyy, E. Berdonosovac, Y.P. Ivanovde, S. Klyamkin, S. Gumrukcu, A.S. Sarac, Artem Korol, D. Semenov, M. Zadorozhnyy, A. Sharma, A.L. Greer, J. Eckert, Hydrogen storage performance of the multi-principal-component

CoFeMnTiVZr alloy in electrochemical and gas–solid reactions, RSC Adv. 10 (2020). https://doi.org/10.1039/D0RA04089D.

[92] J. Montero, C. Zlotea, G. Ek, J.C. Crivello, L. Laversenne, M. Sahlberg, TiVZrNb Multi-Principal-Element Alloy: Synthesis Optimization, Structural, and Hydrogen Sorption Properties, Molecules 24 (2019). https://doi.org/10.3390/molecules24152799.

[93] J. Montero, G. Ek, M. Sahlberg, C. Zlotea, Improving the hydrogen cycling properties by Mg addition in Ti–V–Zr–Nb refractory high entropy alloy, Scripta Materialia 194 (2021). https://doi.org/10.1016/j.scriptamat.2020.113699.

Further reading

A. Kumar, T.P. Yadav, M.A. Shaz, N.K. Mukhopadhyay, Hydrogen Storage Performance of C14 Type $Ti_{0.24}V_{0.17}Zr_{0.17}Mn_{0.17}Co_{0.17}Fe_{0.08}$ High Entropy Intermetallics, Trans. Indian Natl. Acad. Eng. (2023). https://doi.org/10.1007/s41403-023-00390-2.

Hydrogen storage technology

Peng Lv[a,b]
[a]Engineering Research Center of Nuclear Technology Application, Ministry of Education, East China University of Technology, Nanchang, China
[b]School of Water Resources and Environmental Engineering, East China University of Technology, Nanchang, China

3.3.1 Compressed hydrogen storage

Compressed hydrogen storage technology is one of the most commonly used hydrogen storage methods in our daily life. In factories and scientific laboratories, compressed hydrogen storage technology has been widely used to supply hydrogen to instruments [1]. Recently, compressed hydrogen storage technology has also been used in Toyota MIRAI and Hyundai NEXO hydrogen fuel cell vehicles [2]. The fundamental principle of compressed hydrogen storage technology is to compress hydrogen in a hydrogen tank by increasing the pressure at a certain temperature and volume. Generally speaking, the pressure of common compressed hydrogen storage cylinders used in factories and scientific laboratories is around 15–20 MPa. This technology has significant advantages, such as mature technology, convenient operation, fast hydrogen charging and discharging, and low cost. However, the disadvantages of this technology are also obvious, such as low volumetric hydrogen storage density, high hydrogen compression energy consumption, and safety risks [3]. An important way to solve the above disadvantages is to develop new types of compressed hydrogen storage tanks.

Over the past few decades, researchers have developed many types of compressed hydrogen storage tanks [2–10]. The schematic representation of the four types of hydrogen storage tanks is shown in Fig. 3.3.1. It is very clear that the compressed hydrogen storage tanks mainly include type I (all-metal hydrogen storage tank), type II (thick metal liner, hoop wrapped with glass fiber reinforced polymer [GFRP]), type III (metal liner, fully wrapped with carbon fiber reinforced polymer [CFRP]) and type IV (plastic liner, fully wrapped with CFRP) [4,6,9,11].

Towards Hydrogen Infrastructure: Advances and Challenges in Preparing for the Hydrogen Economy.
DOI: https://doi.org/10.1016/B978-0-323-95553-9.00001-7

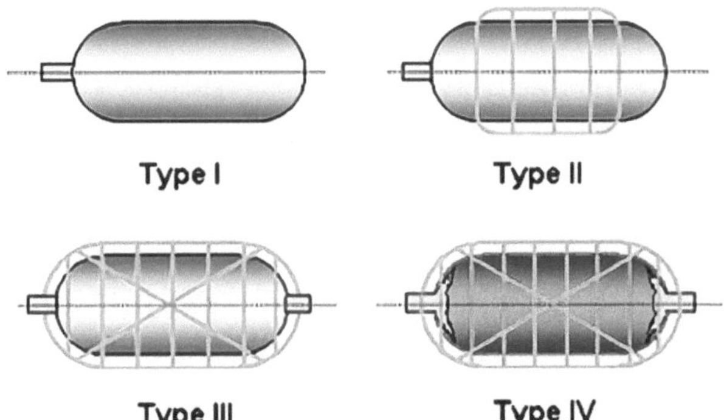

Figure 3.3.1 Schematic representation of the four types of hydrogen storage tanks [4].

3.3.1.1　Type I (all-metal hydrogen storage tank)

Type I hydrogen storage tank is made up entirely of metal. The kind of hydrogen storage tank has the characteristics of large weight, low cost, and simple manufacturing process [4,12]. In the earliest days, the compressed hydrogen was usually stored in an all-metal hydrogen storage tank. Limited by the material of the all-metal hydrogen storage tank, the maximum hydrogen storage pressure of the early all-metal hydrogen tank was around 15–30 MPa and the hydrogen storage density per unit mass was about 1.0 wt% [6,8]. When selecting the metal material for the all-metal hydrogen storage tank, the problem of hydrogen embrittlement of the material must be considered to prevent hydrogen dissolved in the material from causing stress concentration, which will lead to cracks inside the material [2,4,6,10]. Therefore, all-metal hydrogen storage tanks are generally made up of metal materials that have a certain resistance to hydrogen embrittlement, such as stainless steel, copper, aluminum, and titanium [2,4,12].

3.3.1.2　Type II (thick metal liner, hoop wrapped with GFRP)

Type II hydrogen storage tank is composed of a thick metal liner and an outer hoop-wrapped GFRP layer, which can significantly improve the pressure resistance and hydrogen storage density per unit mass[2,4,13]. Compared to type I hydrogen storage tank, type II hydrogen storage tank reduces the weight by 30–40% and increases the cost by about 50% [13]. The metal liner is generally composed of metal materials such as aluminum and titanium with anti-hydrogen embrittlement properties. The pressure resistance of this type of hydrogen storage tank is generally 30 MPa [11]. We can adjust the pressure resistance of type II hydrogen storage tank by adjusting the material of the metal liner and the type of external GFRP layer.

3.3.1.3 Type III (metal liner, fully wrapped with CFRP)

Type III hydrogen storage tank is composed of a metal liner and an outer fully wrapped CFRP layer [2,13]. Compared with type II hydrogen storage tank, type III hydrogen storage tank has a thinner metal liner and is fully wrapped by CFRP. The pressure resistance of this type of hydrogen storage tank is generally 35–70 MPa and it has been widely used in China [11].

3.3.1.4 Type IV (plastic liner, fully wrapped with CFRP)

Type IV hydrogen storage tank has a plastic liner and an outer fully wrapped CFRP layer [2,13]. Due to the use of non-metallic materials for the tank, this type of hydrogen storage tank has a significant reduction in weight, exhibits superior lightweight characteristics, and is suitable for use in mobile applications [2,6]. At the same time, the hydrogen storage density per unit mass is also significantly improved. Therefore, this type of hydrogen tank (70 MPa) has been used in Toyota MIRAI and Hyundai NEXO hydrogen fuel cell vehicles [2,14]. The biggest difference between type IV hydrogen storage tank and type III hydrogen storage tank lies in the choice of liner materials. Type IV hydrogen storage tank adopts the plastic liner instead of the metal liner, which greatly reduces the system weight. The plastic liner is generally made of high-density polyethylene (HDPE), which has advantages such as good chemical stability, excellent heat resistance and cold resistance, and high rigidity and toughness [2,11,15,16]. The pressure resistance of this type of hydrogen tank is generally 70 MPa. The 100 MPa hydrogen tank is also under development. In terms of type IV hydrogen storage tank, Quantum made a positive contribution. This company has developed more than twenty models of type IV hydrogen storage tanks up to 2021 [17].

3.3.2 Liquid hydrogen storage

The liquid hydrogen storage technology refers to the compression of hydrogen, cryogenically cooled to 21 K, into liquid hydrogen and stored in an adiabatic vacuum container [6,18]. The advantage of this method is good volumetric hydrogen densities and high hydrogen purity. The density of liquid hydrogen at ambient pressure reaches 70 kg/m^3, which is about 845 times the density of hydrogen under standard conditions [5,14]. Liquid hydrogen storage is more suitable for long-distance transportation than compressed hydrogen storage. Due to the extremely low boiling point (−252.78°C and ambient pressure) of liquid hydrogen, the temperature difference between the environment, and the inside of the container is extremely large [18]. Therefore, the requirements for thermal insulation of liquid hydrogen containers are very high [3,19].

3.3.2.1 Hydrogen liquefaction technology

In Fig. 3.3.2, there are three main types of hydrogen liquefaction technologies including Linde–Hampson cycle (or Joule–Thomson cycle), Claude cycle, and helium Brayton cycle [20–29]. The Linde–Hampson cycle is the earliest gas liquefaction method used in the industry. Its device is simple in structure and stable in the operation process. Therefore, the Linde–Hampson is considered the simplest gas liquefaction process [20,21,25]. Fig. 3.3.2A shows the Linde–Hampson cycle with liquid nitrogen pre-cooling. In this cycle, the compressed hydrogen is pre-cooled by liquid nitrogen and then cooled in a heat exchanger. Subsequently, the compressed and cooled hydrogen enters the throttle valve for isenthalpic Joule–Thomson expansion. A portion of the compressed hydrogen becomes liquid hydrogen and another portion is recycled back to the heat exchanger [21,29,30]. The Claude cycle was invented by Georges Claude in 1902. At present, the Claude cycle is widely used by large-scale hydrogen liquefaction plants. Fig. 3.3.2B and C shows the Claude cycle without/with liquid nitrogen pre-cooling. Compared with the Linde–Hampson cycle, the Claude cycle combines isenthalpic Joule–Thomson expansion and expander. In addition, the expander can produce a lower temperature than isenthalpic Joule–Thomson expansion. Therefore, the Claude cycle shows a higher work efficiency than the Linde–Hampson cycle[20–22,24,29]. Helium Brayton cycle is very complex and mainly includes helium Brayton cycle (d), helium Brayton cycle with pre-cooling using liquid nitrogen (e), and 2-stages helium Brayton cycle (f) in Fig. 3.3.2 [21,30]. It must be pointed out that helium is usually used as a refrigerant rather than a liquefier in the helium Brayton cycle [21,31].

3.3.2.2 Container construction and materials of liquid hydrogen

Due to the low boiling point (21 K) and latent heat of liquid hydrogen, the evaporation of liquid hydrogen is an important problem that must be considered [18]. The evaporation of liquid hydrogen is closely related to the

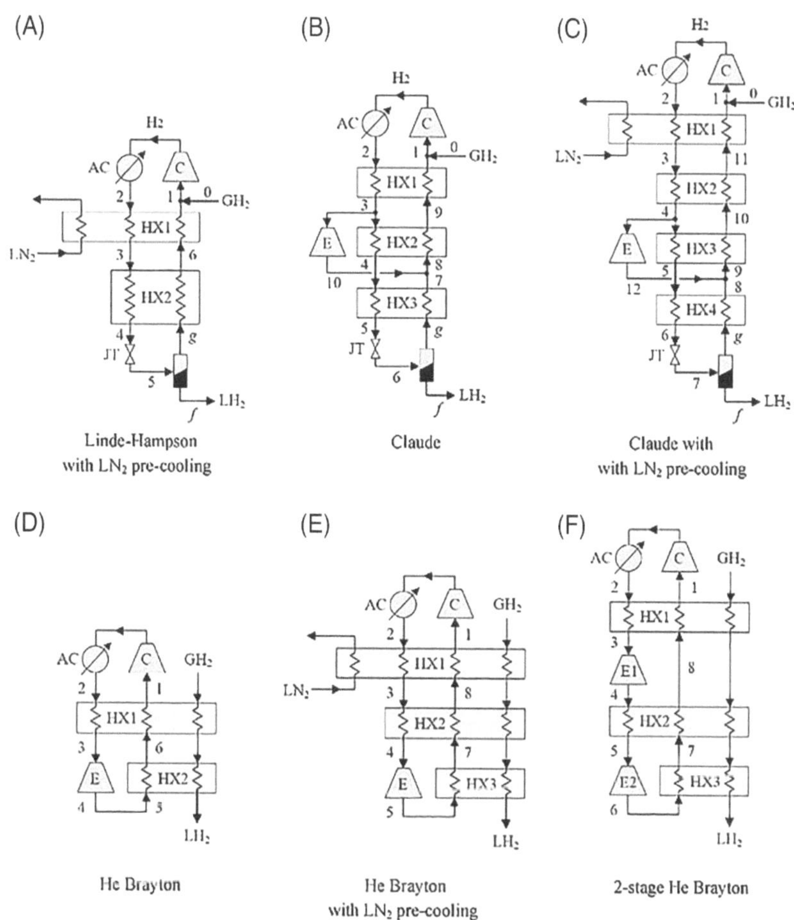

Figure 3.3.2 Linde–Hampson cycle with liquid nitrogen pre-cooling (a), Claude cycle (b), Claude cycle with liquid nitrogen pre-cooling, standard helium Brayton cycle (d), helium Brayton cycle with liquid nitrogen pre-cooling (e) and 2-stages helium Brayton cycle (f). Legend (C, compressor; AC, after cooler; E, expander; HX, heat exchanger; LN_2, liquid nitrogen; GH_2, gas hydrogen; LH_2, liquid hydrogen) [30].

thermal insulation performance of the liquid hydrogen storage container. Therefore, researchers have conducted a lot of research on the thermal insulation properties of liquid hydrogen storage containers, including the structure of the container, and the choice of materials [19,21,29,32–34].

The thermal insulation technology of liquid hydrogen storage container mainly includes four types: conventional external thermal insulation, vacuum thermal insulation, vacuum powder thermal insulation, and vacuum multilayer thermal insulation [19,21,29,34]. The thermal insulation layer of conventional external thermal insulation is generally composed of materials with low density and low thermal conductivity, such as pearl sand, foam glass, rigid polyurethane foam, balsa wood, etc. [18,19,21]. Vacuum thermal insulation refers to pumping a high vacuum in the middle of the vacuum thermal insulation interlayer to reduce gas convection heat transfer and gas heat conduction. This is because lowering the gas pressure can reduce gas convection heat transfer and gas heat conduction [19,34]. Vacuum powder insulation refers to filling powder materials with small grain size (such as perlite, expanded SiO_2, etc.) between the vacuum thermal insulation interlayer, and maintaining a vacuum of $10^{-2}- 10^{-3}$ Pa [19,29]. Vacuum multi-layer insulation is composed of multiple layers of high-reflectivity metal foils and low-thermal conductivity spacers. And the vacuum degree of the vacuum thermal insulation interlayer must be around 10^{-4} Pa [18,19,29].

When choosing the material of the liquid hydrogen storage container, we must consider the influence of the ultra-low temperature, hydrogen embrittlement, and permeability on the container [21]. At present, we mainly use stainless steel and aluminum alloys to prepare the large-scare liquid hydrogen storage container. For mobile applications, researchers are developing a new type of liquid hydrogen storage container that is composed of lightweight GFRP/CFRP and metallic inner [18].

3.3.3 Solid-state hydrogen storage

In recent years, solid-state hydrogen storage technology has attracted extensive attention as a new hydrogen storage method. It is considered to be one of the most promising hydrogen storage technologies for long-term and mobile hydrogen storage application because of its advantages of good safety and high volumetric hydrogen storage density [6,7,26,35–37]. Solid-state hydrogen storage technology requires some specific materials as hydrogen storage media. These specific materials are called hydrogen storage materials. For this reason, we must find and develop high-performance hydrogen storage materials, which has become a top priority. There are many kinds of hydrogen storage materials, which can be mainly divided into physical adsorption hydrogen storage materials and chemical hydrogen storage materials. Physical adsorption hydrogen storage materials mainly include carbon nanomaterials, metal organic frameworks (MOFs), covalent organic frameworks (COFs), etc. Chemical hydrogen storage materials mainly include hydrogen storage alloys and complex hydrides.

3.3.3.1 Physical adsorption hydrogen storage materials

In case of physical adsorption, hydrogen storage materials store hydrogen through the weak interaction between hydrogen molecules and the surface of hydrogen storage materials.

Carbon materials have a long history. It mainly covers activated carbon, carbon fiber, carbon nanotube, graphene, and so on. This type of material has a large specific surface area and a large number of pores, which can adsorb a certain amount of hydrogen at low temperature [38–43]. For example, Zhang et al. [43]. prepared porous carbon materials with largest specific surface area of 1703 m^2/g from biomass pyrolysis vapors. The prepared porous carbon materials showed the hydrogen adsorption capacity of 1.53 wt% at 77 K at atmospheric pressure. Sawant et al. [42] studied boron-doped carbon nanotubes and found that the maximum hydrogen adsorption capacity was 0.497 wt% at 273 K under 1.6 MPa pressure.

Metal organic frameworks (MOFs) have also been found very promising hydrogen storage material due to their huge specific surface area, uniform pores, and easily tunable crystal structure. In the past two decades, researchers have prepared a variety of MOFs materials by adjusting the crystal structure of MOFs and tested their adsorption properties for hydrogen [39,44–50]. Ren et al. [47]. synthesized Zr-fumarate MOFs at 240°C which has shown 1.38 wt% hydrogen adsorption capacity at 77 K and 1 bar. Butova et al. [45]. designed a new simple, fast, scalable technique to synthesize MOF-801, which adsorbed around 1.1 wt% hydrogen under ambient conditions. On the other hand COFs are made up of with light elements without any metal element and show lighter skeleton density [51–56]. COFs also have huge specific surface area. Furukawa and Yaghi [35] found that the hydrogen adsorption capacity of COF-102 and COF-103 were higher than 7.0 wt% in 77 K and 9 MPa. Xia et al. [51]. reported the different hydrogen adsorption capacities of a series of COF-320–X (X=OH, Cl, NO_2, NH_2, CH_3, and CN). COF-320–NH_2 showed the highest hydrogen adsorption capacity of 0.766 wt% among all COF-320–X materials.

3.3.3.2 Chemical hydrogen storage materials

Chemical hydrogen storage materials can be divided into hydrogen storage alloys and complex hydrides. Hydrogen storage alloys play an important role in solid-state hydrogen storage technology. In this part, we mainly introduce hydrogen storage alloys for hydrogen storage application. During the process of hydrogenation of hydrogen storage alloys, hydrogen is first catalytically decomposed into hydrogen atoms on the surface of hydrogen storage alloy. Then the hydrogen atoms diffuse into in the gap of the unit cell of the hydrogen storage alloy to form metal hydrides [29]. It must be pointed out that the reaction process is reversible, thereby realizing the absorption and desorption of hydrogen. Generally speaking, good hydrogen storage alloy requires the following strong points including high hydrogen storage capacity, fast hydrogen absorption/desorption rate, low hydrogenation enthalpy, moderate hydrogen release platform, easy activation, rich raw materials, low cost and so on [18,29]. At present, the developed practical hydrogen storage alloys (intermetallics) mainly include: AB_5, AB, AB_2, and A_2B type hydrogen storage alloys.

3.3.3.2.1 *AB$_5$ type hydrogen storage alloys*

The AB_5 type hydrogen storage alloy is represented by $LaNi_5$, which is the most in-depth researched hydrogen storage alloy. $LaNi_5$ hydrogen storage alloy has $CaCu_5$ type crystal structure. Its space group is P6/mmm. Its theoretical hydrogen storage capacity is around 1.38 wt%. $LaNi_5$ hydrogen storage alloy shows good activation and anti-poisoning properties. But its shortcomings are also very obvious, mainly including high cost, poor cycle property, and low hydrogen storage capacity during the hydrogen absorption and desorption process [29,57–61].

In order to solve the shortcomings of $LaNi_5$ hydrogen storage alloy, some researchers have carried out a lot of work on it. Compared to traditional arc melting preparation process, some researchers investigated the effect of different new preparation processes (ball milling [61–64], rapid solidification [65,66], and surface modification [67–69], etc.) on the hydrogen storage properties of $LaNi_5$ hydrogen storage alloy. Related studies proved that these new preparation processes are helpful for the improvement of the hydrogen storage properties of $LaNi_5$ hydrogen storage alloy. Aoyagi et al. [63]. found that the initial hydrogen absorption rate of $LaNi_5$ hydrogen storage alloy could be improved after ball milling. This is because ball milling could reduce the particle size effectively and create new reactive surfaces. Yao et al. [66]. thought that rapid solidification was very helpful to promote the formation of homogeneous composition of $LaNi_5$. Kuang et al. [68]. used CuCl to modify the surface of $LaNi_5$ and found that the surface modification was a good way to improve the electrochemical activity of $LaNi_5$. In addition, some researchers found another way to improve the hydrogen storage properties of $LaNi_5$ by adding or substituting certain elements (Al [70], Co [71], Fe [72], Zn [73], Mn [74], Cu [75], Sn [58], Ga

[76], In [60], Ge [77], etc.). For example, Liu et al. [70]. prepared $LaNi_{5-x}Al_x$ ($x = 0$, 0.25, 0.5, 0.75, and 1.0) tritium-storage alloys and found that the increase of Al content resulted in a reduction of the absorption/desorption plateau pressures and the hydrogen storage capacity of $LaNi_{5-x}Al_x$ alloys. Pandey et al. [72]. found that the substitution of Ni by Fe has a positive effect on the hydrogen storage capacity of $LaNi_5$ hydrogen storage alloy. Among all compositions, La $(Ni_{0.8}Fe_{0.2})_5$ presented the highest hydrogen storage capacity (around 2.2 wt%).

3.3.3.2.2 AB type hydrogen storage alloys

TiFe hydrogen storage alloy is a typical representative of AB type hydrogen storage alloy. TiFe hydrogen storage alloy has CsCl type crystal structure. TiFe hydrogen storage alloy also has excellent hydrogen storage properties and received extensive attention. Its theoretical hydrogen storage capacity is around 1.86 wt%, which is much higher than that of the traditional AB_5 type (such as $LaNi_5$) hydrogen storage alloy [78,79]. In addition, Ti and Fe elements are rich in resources and cheap, making TiFe hydrogen storage alloy very suitable for large-scale industrial production and application. However, the traditional TiFe hydrogen storage alloy has difficulties in activation (many hydrogen absorption and desorption cycles are required at high temperature and hydrogen pressure) [80–82]. Its resistance to impurity gas poisoning is also very poor. If the hydrogen gas mixes O_2 content greater than 0.01% or CO content greater than 0.03%, the hydrogen storage capacity of TiFe hydrogen storage alloy will be sharply attenuated [29].

At present, the mechanical deformation and element substitution (addition) are two main ways to improve the hydrogen storage properties of TiFe hydrogen storage alloy. Some researchers studied the effect of mechanical deformation methods (such as ball milling [82,83], cold rolling [84], groove rolling [81], and high-pressure torsion [85], etc.) on the hydrogen storage properties of TiFe hydrogen storage alloy. Although the activation properties and the resistance to impurity gas poisoning properties of the mechanically deformed TiFe alloy were improved to a certain extent, but their hydrogen storage capacities may be reduced. For example, Emami et al. [82]. found that ball milling could obviously reduce the grain size and crystallite size of TiFe hydrogen storage alloy and lower the difficulty of activation process. But the hydrogen storage capacity of TiFe hydrogen storage alloy after ball milling was only 1.3–1.5 wt%. Similarly, Vega et al. [84]. used cold rolling to prepare TiFe ingot. After cold rolling, the crystallite size of TiFe hydrogen storage alloy was reduced and hydrogen absorption rate was high. However, the hydrogen storage capacity was only 1.4 wt%, which was lower than 1.86 wt%. For element substitution (addition) method, some researchers studied the element (such as B [86], S [87], Mg [88], Al [89], Fe [90], Cr [91], Cu [92], Mn [93], V [79], Ni [94], Co [95], Y [88], Zr [96], La [97], Pr [98], Sm [99], etc.) substitution (addition) on the hydrogen storage properties of TiFe hydrogen storage alloy. After element substitution (addition), some TiFe-based hydrogen storage alloys did not need annealing treatment and could absorb the hydrogen directly without activation at room temperature and low hydrogen pressure. But element substitution (addition) still may lead to a reduction in the hydrogen storage capacity of TiFe hydrogen storage alloy. For example, Jain et al. [94]. studied the effect of Zr, Ni, and Zr_7Ni_{10} on hydrogen storage properties of TiFe hydrogen storage alloy. They found that all alloys with additives can absorb the hydrogen directly without any activation under 2 MPa H_2 at 40°C. But the hydrogen storage capacity is lower than pure TiFe hydrogen storage alloy. Yang et al. [88]. reported the effect of the addition of Cr, Mn and Y on the microstructure and hydrogen storage properties of TiFe hydrogen storage alloy. They found that all alloys could absorb hydrogen under 3 MPa H_2 pressure and below 30°C. But their hydrogen storage capacities were only around 1.5 wt%. Therefore, how to improve the activation property while maintaining the hydrogen storage capacity of TiFe hydrogen storage alloy is an urgent problem to be solved in the future.

3.3.3.2.3 AB_2 type hydrogen storage alloys

AB_2 type hydrogen storage alloys mainly include Ti-based hydrogen storage alloys ($TiMn_2$ [100–104], $TiCr_2$ [105–107], etc.) and Zr-based hydrogen storage alloys ($ZrCr_2$ [108,109], $ZrMn_2$ [110–112], ZrV_2 [113], etc.). AB_2 type hydrogen storage alloys have three kinds of crystal structures: C14 ($MgZn_2$ type, hexagonal system), C15 ($MgCu_2$ type, cubic system), and C36 ($MgNi_2$ type, hexagonal system). But only C14 and C15 can be used for hydrogen storage materials [29].

For AB_2 type hydrogen storage alloys, their metal hydrides formed in hydrogenation process usually very stable. This problem seriously hinders their practical application [114]. In order to improve the hydrogen storage properties of AB_2 type hydrogen storage alloys, the researchers have carried out a lot of research about the element addition or substitution to replace part of A or B [102,103,106,108,109,112–114]. Huang et al. [102]. studied the effect of V substitution for Mn on the hydrogen storage properties of $TiMn_{2-x}V_x$ alloys. Their hydrogen storage capacity increased with increasing V content. Bulyk et al. [108]. prepared the $Zr_{1-x}Ti_xCr_2$ alloys by replacing Zr with Ti to investigate the effect of Ti on the hydrogen-induced phase-structure transformations. Wu et al. [113] reported the effect of adding Ni on the $Zr(V_{1-x}Ni_x)_2$ intermetallic compounds. They found that the increase of Ni content resulted in the decrease of hydrogen absorption capacities and the increase of equilibrium pressure.

3.3.3.2.4 A_2B type hydrogen storage alloys

There are two typical representatives of A_2B type hydrogen storage alloys. One is Mg_2Ni hydrogen storage alloy, the other one is Ti_2Ni hydrogen storage alloy. Mg_2Ni hydrogen storage alloy has attracted much attention due to its high theoretical hydrogen storage capacity (3.6 wt%), abundant resources and low price. However, Mg_2Ni hydrogen storage alloy also has many disadvantages, such as difficult activation, slow hydrogen absorption and desorption, high hydrogen absorption and desorption temperature, poor reaction kinetics, and so on [115–118]. Especially, the formed Mg_2NiH_4 hydrides during the hydrogenation of Mg_2Ni hydrogen storage alloy is very stable. Therefore, the practical application of Mg_2Ni hydrogen storage alloy is hindered [115,119,120].

Now there are two main ways to improve the hydrogen storage properties of Mg_2Ni hydrogen storage alloy. One way is to study the preparation process of Mg_2Ni alloy, and the other is to study the addition or substitution of one or more elements to Mg_2Ni. For the preparation process, the researchers have developed many methods including ball milling [121–123], hydriding combustion synthesis [124], chemical synthetic method [125,126], induction field activated combustion synthesis [127], high-pressure torsion [128], forging [129], laser sintering method [130], surface coating [131], and melt spinning [132–134]. For instance, Kumar et al. [126]. reported the positive effect of the hydrogen storage properties of Mg_2Ni that was prepared by using polyol reduction chemical synthetic method. Si et al. [130]. used the laser sintering method to prepare Mg_2Ni hydrogen storage alloy. The obtained Mg_2Ni alloy could be activated easily at 573 K. For the addition or substitution of one or more elements, many elements (Al [119,135], Ti [119,136,137], Zr [119], B [138], C [138], Si [138], V [136], La [139], Pd [140], Y [141,142], Nd [118,143], Zn [143], Cu [117,142], Mn [135,144], and Cr [145], etc.) have been studied to improve the hydrogen storage properties of Mg_2Ni . For example, Xie et al. [118]. found the substitution of Nd for Mg could enhance the hydrogen storage properties of Mg_2Ni obviously. In addition, the $Mg_{1.9}Nd_{0.1}Ni$ alloy showed the best hydrogen storage properties including the improved kinetics, storage capacity, and reversibility. Song et al. [141]. studied the microstructure and the hydrogen storage properties of $Mg_{67}Ni_{33-x}Y_x$ alloys by doping Y element. The related results showed that the prepared $Mg_{67}Ni_{32}Y$ alloy had good hydrogen storage capacity. It could absorb around 96% of the maximum hydrogen storage capacity (3.79 wt %) within only 8 minutes. In addition to this, some researchers also prepared Mg_2Ni composites. For example, Hou et al. [122]. reported the improvement hydrogen storage properties of Mg_2Ni with multi-walled carbon nanotubes (MWCNTs). Zhao et al. [146]. investigated the hydrogen storage properties of $Mg_2Ni + 10$ wt% NbN composite and found the excellent hydrogenation/dehydrogenation kinetics properties of this composite. The improvement of the hydrogen storage properties of Ti_2Ni alloy also focused on the innovation of the preparation methods and the addition or substitution of elements, which will not be described in detail here [147–153].

3.3.3.2.5 Complex hydrides

Complex hydrides are mainly composed of metal cations (such as Li, Mg, Na, etc.) and hydrogen-containing coordination anions (such as AlH_4^-, NH_2^-, BH_4^-, etc.)[37,154–156]. This type of hydrogen storage materials has high hydrogen storage density. For example, the hydrogen storage density of $LiBH_4$ and $Mg\ (BH_4)_2$ is as high as 18.5 and 14.8 wt% [156]. Although these materials have a very high hydrogen storage density, they are difficult to reversibly absorb and release hydrogen. In 1997, Bogdanovic′ et al. [157]. used Ti-based compounds as catalysts and successfully made $NaAlH_4$ release hydrogen under mild conditions for the first time. This discovery ignited the enthusiasm for research on the hydrogen storage properties of complex hydrides. Up to now, the researchers have synthesized and studied numerous complex hydrides including $LiAlH_4$, $KAlH_4$, $CaAlH_4$, $LiNH_2$, $NaNH_2$, KNH_2, $Mg\ (NH_2)_2$, $Ca\ (NH_2)_2$, $LiBH_4$, $NaBH_4$, KBH_4, $Mg\ (BH_4)_2$, $Ca\ (BH_4)_2$, $Al\ (BH_4)_2$, $Y\ (BH_4)_2$, etc. But these complex hydrides are basically at the stage of laboratory research and are still far from practical applications. In the future, our main work is to realize the reversible absorption and desorption of hydrogen from complex hydrides under mild conditions.

References

[1] L.E. Klebanoff, Hydrogen Storage Technology: Materials and Applications, CRC Press, 2016.
[2] M.R. Usman, Hydrogen storage methods: review and current status, Renew. Sustain. Energy Rev. 167 (2022) 112743.

[3] C. Tarhan, M.A. Çil, A study on hydrogen, the clean energy of the future: hydrogen storage methods, J. Energy Storage 40 (2021) 102676.

[4] H. Barthélémy, Hydrogen storage: industrial prospectives, Int. J. Hydrogen Energy 37 (22) (2012) 17364.

[5] O. Faye, J. Szpunar, U. Eduok, A critical review on the current technologies for the generation, storage, and transportation of hydrogen, Int. J. Hydrogen Energy 47 (29) (2022) 13771.

[6] I.A. Hassan, H.S. Ramadan, M.A. Saleh, D. Hissel, Hydrogen storage technologies for stationary and mobile applications: review, analysis and perspectives, Renew. Sustain. Energy Rev. 149 (2021) 111311.

[7] Y. Ma, X.R. Wang, T. Li, J. Zhang, J. Gao, Z.Y. Sun, Hydrogen and ethanol: production, storage, and transportation, Int. J. Hydrogen Energy 46 (54) (2021) 27330.

[8] B.C. Tashie-Lewis, S.G. Nnabuife, Hydrogen production, distribution, storage and power conversion in a hydrogen economy: a technology review, Chem. Eng. J. Adv. 8 (2021) 100172.

[9] L. Zhao, Q. Zhao, J. Zhang, S. Zhang, G. He, M. Zhang, T. Su, X. Liang, C. Huang, W. Yan, Review on studies of the emptying process of compressed hydrogen tanks, Int. J. Hydrogen Energy 46 (43) (2021) 22554.

[10] J. Zheng, X. Liu, P. Xu, P. Liu, Y. Zhao, J. Yang, Development of high pressure gaseous hydrogen storage technologies, Int. J. Hydrogen Energy 37 (1) (2012) 1048.

[11] Z. Wang, Y. Wang, S. Afshan, J. Hjalmarsson, A review of metallic tanks for H_2 storage with a view to application in future green shipping, Int. J. Hydrogen Energy 46 (9) (2021) 6151.

[12] M. Ayvaz, S.İ. Ayvaz, İ. Aydin, A novel method for determining effects of fire damage on the safety of the Type I pressure hydrogen storage tanks, Int. J. Hydrogen Energy 43 (44) (2018) 20271.

[13] R. Moradi, K.M. Groth, Hydrogen storage and delivery: review of the state of the art technologies and risk and reliability analysis, Int. J. Hydrogen Energy 44 (23) (2019) 12254.

[14] M. Li, Y. Bai, C. Zhang, Y. Song, S. Jiang, D. Grouset, M. Zhang, Review on the research of hydrogen storage system fast refueling in fuel cell vehicle, Int. J. Hydrogen Energy 44 (21) (2019) 10677.

[15] Y. Sun, H. Lv, W. Zhou, C. Zhang, Research on hydrogen permeability of polyamide 6 as the liner material for type IV hydrogen storage tank, Int. J. Hydrogen Energy 45 (46) (2020) 24980.

[16] D. Wang, B. Liao, C. Hao, A. Wen, J. Zheng, P. Jiang, C. Gu, P. Xu, Q. Huang, Acoustic emission characteristics of used 70 MPa type IV hydrogen storage tanks during hydrostatic burst tests, Int. J. Hydrogen Energy 46 (23) (2021) 12605.

[17] Quantum Fuel Systems Scalable hydrogen fuel systems and infrastructure. https://www.qtww.com/product/hydrogen/.

[18] C. Wu, Y. Li, Y. Li, Hydrogen Storage and Transportation, Chemical Industry Press, Beijing, 2021.

[19] W. Peschka, Thermal insulation, storage and transportation of liquid hydrogen, in: W. Peschka (Ed.), Liquid Hydrogen: Fuel of the Future, Springer, Vienna, 1992. https://doi.org/10.1007/978-3-7091-9126-2_4.

[20] M. Aasadnia, M. Mehrpooya, Large-scale liquid hydrogen production methods and approaches: a review, Appl. Energy 212 (2018) 57.

[21] M. Aziz, Liquid hydrogen: a review on liquefaction, storage, transportation, and safety, Energies 14 (18) (2021).

[22] U. Cardella, L. Decker, H. Klein, Roadmap to economically viable hydrogen liquefaction, Int. J. Hydrogen Energy 42 (19) (2017) 13329.

[23] D. Kang, Y. Sungho, B.-K. Kim, Review of the liquid hydrogen storage tank and insulation system for the high-power locomotive, Energies 15 (12) (2022).

[24] S. Krasae-in, J.H. Stang, P. Neksa, Development of large-scale hydrogen liquefaction processes from 1898 to 2009, Int. J. Hydrogen Energy 35 (10) (2010) 4524.

[25] T.K. Nandi, S. Sarangi, Performance and optimization of hydrogen liquefaction cycles, Int. J. Hydrogen Energy 18 (2) (1993) 131.

[26] S. Niaz, T. Manzoor, A.H. Pandith, Hydrogen storage: materials, methods and perspectives, Renew. Sustain. Energy Rev. 50 (2015) 457.

[27] R.J. Thomas, P. Ghosh, K. Chowdhury, Exergy analysis of helium liquefaction systems based on modified Claude cycle with two-expanders, Cryogenics 51 (6) (2011) 287.

[28] A.T. Wijayanta, T. Oda, C.W. Purnomo, T. Kashiwagi, M. Aziz, Liquid hydrogen, methylcyclohexane, and ammonia as potential hydrogen storage: comparison review, Int. J. Hydrogen Energy 44 (29) (2019) 15026.

[29] X. Li, Hydrogen and Hydrogen Energy, China Machine Press, Beijing, 2012.

[30] H.-M. Chang, K.N. Ryu, J.H. Baik, Thermodynamic design of hydrogen liquefaction systems with helium or neon Brayton refrigerator, Cryogenics 91 (2018) 68.

[31] J. Kumar, S.R. Nair, R.S. Menon, M. Goyal, N.A. Ansari, A. Chakravarty, V. Joemon, Helium refrigeration system for hydrogen liquefaction applications, IOP Conf. Ser.: Mater. Sci. Eng. 171 (2017) 012029.

[32] J.E. Fesmire, Layered composite thermal insulation system for nonvacuum cryogenic applications, Cryogenics 74 (2016) 154.

[33] J.E. Fesmire, W.L. Johnson, Cylindrical cryogenic calorimeter testing of six types of multilayer insulation systems, Cryogenics 89 (2018) 58.

[34] H.M. Tariel, J.C. Boissin, M.P. Segel, Thermal insulation for liquid hydrogen space tankage. In Advances in Cryogenic Engineering, Springer Boston, MA, 1967, p 274.

[35] H. Furukawa, O.M. Yaghi, Storage of hydrogen, methane, and carbon dioxide in highly porous covalent organic frameworks for clean energy applications, J. Am. Chem. Soc. 131 (25) (2009) 8875.

[36] L.M. Kustov, A.L. Tarasov, J. Sung, D.Y. Godovsky, Hydrogen storage materials, Mendeleev Commun. 24 (1) (2014) 1.

[37] S.-i. Orimo, Y. Nakamori, J.R. Eliseo, A. Züttel, C.M. Jensen, Complex hydrides for hydrogen storage, Chem. Rev. 107 (10) (2007) 4111.

[38] M. Doğan, P. Sabaz, Z. Bicil, B. Koçer Kizilduman, Y. Turhan, Activated carbon synthesis from tangerine peel and its use in hydrogen storage, J. Energy Inst. 93 (6) (2020) 2176.

[39] K.K. Gangu, S. Maddila, S.B. Mukkamala, S.B. Jonnalagadda, Characteristics of MOF, MWCNT and graphene containing materials for hydrogen storage: a review, J. Energy Chem. 30 (2019) 132.

[40] J. Juan-Juan, J.P. Marco-Lozar, F. Suárez-García, D. Cazorla-Amorós, A. Linares-Solano, A comparison of hydrogen storage in activated carbons and a metal: organic framework (MOF-5), Carbon 48 (10) (2010) 2906.

[41] E. Mosquera-Vargas, R. Tamayo, M. Morel, M. Roble, D.E. Díaz-Droguett, Hydrogen storage in purified multi-walled carbon nanotubes: gas hydrogenation cycles effect on the adsorption kinetics and their performance, Heliyon 7 (12) (2021) e08494.

[42] S.V. Sawant, M.D. Yadav, S. Banerjee, A.W. Patwardhan, J.B. Joshi, K. Dasgupta, Hydrogen storage in boron-doped carbon nanotubes: Effect of dopant concentration, Int. J. Hydrogen Energy 46 (79) (2021) 39297.

[43] H. Zhang, Y. Zhu, Q. Liu, X. Li, Preparation of porous carbon materials from biomass pyrolysis vapors for hydrogen storage, Appl. Energy 306 (2022) 118131.

[44] A. Ahmed, D.J. Siegel, Predicting hydrogen storage in MOFs via machine learning, Patterns 2 (7) (2021) 100291.

[45] V.V. Butova, I.A. Pankin, O.A. Burachevskaya, K.S. Vetlitsyna-Novikova, A.V. Soldatov, New fast synthesis of MOF-801 for water and hydrogen storage: modulator effect and recycling options, Inorganica Chim. Acta 514 (2021) 120025.

[46] H.W. Langmi, J. Ren, B. North, M. Mathe, D. Bessarabov, Hydrogen storage in metal-organic frameworks: a review, Electrochim. Acta 128 (2014) 368.

[47] J. Ren, N.M. Musyoka, H.W. Langmi, B.C. North, M. Mathe, X. Kang, S. Liao, Hydrogen storage in Zr-fumarate MOF, Int. J. Hydrogen Energy 40 (33) (2015) 10542.

[48] Y. Wang, Z. Lan, X. Huang, H. Liu, J. Guo, Study on catalytic effect and mechanism of MOF (MOF = ZIF-8, ZIF-67, MOF-74) on hydrogen storage properties of magnesium, Int. J. Hydrogen Energy 44 (54) (2019) 28863.

[49] S.J. Yang, H. Jung, T. Kim, J.H. Im, C.R. Park, Effects of structural modifications on the hydrogen storage capacity of MOF-5, Int. J. Hydrogen Energy 37 (7) (2012) 5777.

[50] S. Yu, G. Jing, S. Li, Z. Li, X. Ju, Tuning the hydrogen storage properties of MOF-650: a combined DFT and GCMC simulations study, Int. J. Hydrogen Energy 45 (11) (2020) 6757.

[51] L. Xia, F. Wang, Q. Liu, Effects of substituents on the H_2 storage properties of COF-320, Mater. Lett. 162 (2016) 9.

[52] S.S. Han, H. Furukawa, O.M. Yaghi, W.A. Goddard, Covalent organic frameworks as exceptional hydrogen storage materials, J. Am. Chem. Soc. 130 (35) (2008) 11580.

[53] N. Huang, P. Wang, D. Jiang, Covalent organic frameworks: a materials platform for structural and functional designs, Nat. Rev. Mater. 1 (10) (2016) 16068.

[54] X.-D. Li, J.-H. Guo, H. Zhang, X.-L. Cheng, X.-Y. Liu, Design of 3D 1,3,5,7-tetraphenyladamantane-based covalent organic frameworks as hydrogen storage materials, RSC Adv. 4 (47) (2014) 24526.

[55] W. Shangguan, H. Zhao, J.-Q. Dai, J. Cai, C. Yan, First-principles study of hydrogen storage of sc-modified semiconductor covalent organic framework-1, ACS Omega 6 (34) (2021) 21985.

[56] E. Tylianakis, E. Klontzas, G.E. Froudakis, Multi-scale theoretical investigation of hydrogen storage in covalent organic frameworks, Nanoscale 3 (3) (2011) 856.

[57] K. Asano, Y. Hashimoto, T. Iida, M. Kondo, Y. Iijima, Hydrogen diffusion by substitution of aluminum and cobalt for a part of nickel in $LaNi_5$-H alloy, J. Alloys Compd. 395 (1) (2005) 201.

[58] E.M. Borzone, A. Baruj, M.V. Blanco, G.O. Meyer, Dynamic measurements of hydrogen reaction with $LaNi_{5-x}Sn_x$ alloys, Int. J. Hydrogen Energy 38 (18) (2013) 7335.

[59] J.-K. Chang, D.-N.S. Shong, W.-T. Tsai, Effect of Ni content on the electrochemical characteristics of the $LaNi_5$-based hydrogen storage alloys, Mater. Chem. Phys. 83 (2) (2004) 361.

[60] H. Drulis, A. Hackemer, L. Folcik, K. Giza, H. Bala, Ł. Gondek, H. Figiel, Thermodynamic and electrochemical hydrogenation properties of $LaNi_{5-x}In_x$ alloys, Int. J. Hydrogen Energy 37 (21) (2012) 15850.

[61] B. Joseph, B. Schiavo, Effects of ball-milling on the hydrogen sorption properties of $LaNi_5$, J. Alloys Compd. 480 (2) (2009) 912.

[62] R.A. Andrievski, B.P. Tarasov, I.I. Korobov, N.G. Mozgina, S.P. Shilkin, Hydrogen absorption and electrocatalytic properties of ultrafine $LaNi_5$ powders, Int. J. Hydrogen Energy 21 (11) (1996) 949.

[63] H. Aoyagi, K. Aoki, T. Masumoto, Effect of ball milling on hydrogen absorption properties of FeTi, Mg_2Ni and $LaNi_5$, J. Alloys Compd. 231 (1) (1995) 804.

[64] L. Zaluski, A. Zaluska, P. Tessier, J.O. Ström-Olsen, R. Schulz, Catalytic effect of Pd on hydrogen absorption in mechanically alloyed Mg_2Ni, $LaNi_5$ and FeTi, J. Alloys Compd. 217 (2) (1995) 295.

[65] D. Chen, L. Chen, K. Yang, M. Tong, R. Long, W. Sun, Y. Li, Electrochemical performance of Sn-substituted, $LaNi_5$-based rapidly solidified alloys, J. Alloys Compd. 724 (1999) 293–295.

[66] Q. Yao, Y. Tang, H. Zhou, J. Deng, Z. Wang, S. Pan, G. Rao, Q. Zhu, Effect of rapid solidification treatment on structure and electrochemical performance of low-Co AB_5-type hydrogen storage alloy, J. Rare Earths 32 (6) (2014) 526.

[67] C. Deng, P. Shi, S. Zhang, Effect of surface modification on the electrochemical performances of $LaNi_5$ hydrogen storage alloy in Ni/MH batteries, Mater. Chem. Phys. 98 (2) (2006) 514.

[68] G. Kuang, Y. Li, F. Ren, M. Hu, L. Lei, The effect of surface modification of $LaNi_5$ hydrogen storage alloy with CuCl on its electrochemical performances, J. Alloys Compd. 605 (2014) 51.

[69] H. Yu, Y. Wu, S. Chen, Z. Xie, Y. Wu, N. Cheng, X. Yang, W. Lin, L. Xie, X Li, Pd-modified $LaNi_5$ nanoparticles for efficient hydrogen storage in a carbazole type liquid organic hydrogen carrier, Appl. Catal. B: Environ. 317 (2022) 121720.

[70] G. Liu, D. Chen, Y. Wang, K. Yang, Experimental and computational investigations of $LaNi_{5-x}Al_x$ (x = 0, 0.25, 0.5, 0.75 and 1.0) tritium-storage alloys, J. Mater. Sci. Technol. 34 (9) (2018) 1699.

[71] K. Asano, Y. Yamazaki, Y. Iijima, Hydriding and dehydriding processes of $LaNi_{5-x}Co_x$ (x=0–2) alloys under hydrogen pressure of 1–5 MPa, Intermetallics 11 (9) (2003) 911.

[72] S.K. Pandey, A. Srivastava, O.N. Srivastava, Improvement in hydrogen storage capacity in $LaNi_5$ through substitution of Ni by Fe, Int. J. Hydrogen Energy 32 (13) (2007) 2461.

[73] B. Rożdżyńska-Kiełbik, W. Iwasieczko, H. Drulis, V.V. Pavlyuk, H. Bala, Hydrogenation equilibria characteristics of $LaNi_{5-x}Zn_x$ intermetallics, J. Alloys Compd. 298 (1) (2000) 237.

[74] X. Chen, J. Xu, W. Zhang, S. Zhu, N. Zhang, D. Ke, J. Liu, K. Yan, H. Cheng, Effect of Mn on the long-term cycling performance of AB_5-type hydrogen storage alloy, Int. J. Hydrogen Energy 46 (42) (2021) 21973.

[75] M. Spodaryk, N. Gasilova, A. Züttel, Hydrogen storage and electrochemical properties of $LaNi_{5-x}Cu_x$ hydride-forming alloys, J. Alloys Compd. 775 (2019) 175.

[76] A. Drašner, Ž.Ž. Blažina, Hydrogen sorption properties of the $LaNi_{5-x}Ga_x$ alloys, J. Alloys Compd. 359 (1) (2003) 180.

[77] C. Witham, R.C. Bowman Jr, B. Fultz, Gas-phase H_2 absorption and microstructural properties of $LaNi_{5-x}Ge_x$ alloys, J. Alloys Compd. 574 (1997) 253–254.

[78] T. Ha, S.-I. Lee, J. Hong, Y.-S. Lee, D.-I. Kim, J.-Y. Suh, Y.W. Cho, B. Hwang, J. Lee, J.-H. Shim, Hydrogen storage behavior and microstructural feature of a $TiFe–ZrCr_2$ alloy, J. Alloys Compd. 853 (2021) 157099.

[79] J.Y. Jung, Y.-S. Lee, J.-Y. Suh, J.-Y. Huh, Y.W. Cho, Tailoring the equilibrium hydrogen pressure of TiFe via vanadium substitution, J. Alloys Compd. 854 (2021) 157263.

[80] E.M. Dematteis, D.M. Dreistadt, G. Capurso, J. Jepsen, F. Cuevas, M. Latroche, Fundamental hydrogen storage properties of TiFe-alloy with partial substitution of Fe by Ti and Mn, J. Alloys Compd. 874 (2021) 159925.

[81] K. Edalati, J. Matsuda, A. Yanagida, E. Akiba, Z. Horita, Activation of TiFe for hydrogen storage by plastic deformation using groove rolling and high-pressure torsion: similarities and differences, Int. J. Hydrogen Energy 39 (28) (2014) 15589.

[82] H. Emami, K. Edalati, J. Matsuda, E. Akiba, Z. Horita, Hydrogen storage performance of TiFe after processing by ball milling, Acta Materialia 88 (2015) 190.

[83] V.Y. Zadorozhnyy, G.S. Milovzorov, S.N. Klyamkin, M.Y. Zadorozhnyy, D.V. Strugova, M.V. Gorshenkov, S.D. Kaloshkin, Preparation and hydrogen storage properties of nanocrystalline TiFe synthesized by mechanical alloying, Prog. Nat. Sci.: Mater. Int. 27 (1) (2017) 149.

[84] L.E.R. Vega, D.R. Leiva, R.M. Leal Neto, W.B. Silva, R.A. Silva, T.T. Ishikawa, C.S. Kiminami, W.J. Botta, Mechanical activation of TiFe for hydrogen storage by cold rolling under inert atmosphere, Int. J. Hydrogen Energy 43 (5) (2018) 2913.

[85] K. Edalati, J. Matsuda, H. Iwaoka, S. Toh, E. Akiba, Z. Horita, High-pressure torsion of TiFe intermetallics for activation of hydrogen storage at room temperature with heterogeneous nanostructure, Int. J. Hydrogen Energy 38 (11) (2013) 4622.

[86] S.M. Lee, T.P. Perng, Effects of boron and carbon on the hydrogenation properties of TiFe and Ti1.1Fe, Int. J. Hydrogen Energy 25 (9) (2000) 831.

[87] V.Y. Zadorozhnyy, S.N. Klyamkin, M. Yu. Zadorozhnyy, M.V. Gorshenkov, S.D Kaloshkin, Mechanical alloying of nanocrystalline intermetallic compound TiFe doped with sulfur and magnesium, J. Alloys Compd. 615 (2014) S569.

[88] T. Yang, P. Wang, C. Xia, N. Liu, C. Liang, F. Yin, Q. Li, Effect of chromium, manganese and yttrium on microstructure and hydrogen storage properties of TiFe-based alloy, Int. J. Hydrogen Energy 45 (21) (2020) 12071.

[89] V.Y. Zadorozhnyy, S.N. Klyamkin, M.Y. Zadorozhnyy, O.V. Bermesheva, S.D. Kaloshkin, Mechanical alloying of nanocrystalline intermetallic compound TiFe doped by aluminum and chromium, J. Alloys Compd. 586 (2014) S56.

[90] K.B. Park, J.O. Fadonougbo, C.-S. Park, J.-H. Lee, T.-W. Na, H.-S. Kang, W.-S. Ko, H.-K. Park, Effect of Fe substitution on first hydrogenation kinetics of TiFe-based hydrogen storage alloys after air exposure, Int. J. Hydrogen Energy 46 (60) (2021) 30780.

[91] J.Y. Jung, S.-I. Lee, M. Faisal, H. Kim, Y.-S. Lee, J.-Y. Suh, J.-H. Shim, J.-Y. Huh, Y.W. Cho, Effect of Cr addition on room temperature hydrogenation of TiFe alloys, Int. J. Hydrogen Energy 46 (37) (2021) 19478.

[92] E.M. Dematteis, F. Cuevas, M. Latroche, Hydrogen storage properties of Mn and Cu for Fe substitution in TiFe0.9 intermetallic compound, J. Alloys Compd. 851 (2021) 156075.

[93] H. Shang, Y. Zhang, Y. Li, Y. Qi, S. Guo, D. Zhao, Effects of adding over-stoichiometrical Ti and substituting Fe with Mn partly on structure and hydrogen storage performances of TiFe alloy, Renew. Energy 135 (2019) 1481.

[94] P. Jain, C. Gosselin, J. Huot, Effect of Zr, Ni and Zr_7Ni_{10} alloy on hydrogen storage characteristics of TiFe alloy, Int. J. Hydrogen Energy 40 (47) (2015) 16921.

[95] V. Kumar, P. Kumar, K. Takahashi, P. Sharma, Hydrogen adsorption studies of TiFe surfaces via 3-d transition metal substitution, Int. J. Hydrogen Energy 47 (36) (2022) 16156.

[96] A.K. Patel, B. Tougas, P. Sharma, J. Huot, Effect of cooling rate on the microstructure and hydrogen storage properties of TiFe with 4 wt% Zr as an additive, J. Mater. Res. Technol. 8 (6) (2019) 5623.

[97] T. Zhai, Z. Wei, Z. Yuan, Z. Han, D. Feng, H. Wang, Y. Zhang, Influences of La addition on the hydrogen storage performances of TiFe-base alloy, J. Phys. Chem. Solids 157 (2021) 110176.

[98] H. Shang, Y. Zhang, Y. Li, J. Gao, W. Zhang, X. Wei, Z. Yuan, L. Ju, Effect of Pr content on activation capability and hydrogen storage performances of TiFe alloy, J. Alloys Compd. 890 (2022) 161785.

[99] Y. Zhang, H. Shang, J. Gao, W. Zhang, X. Wei, Z. Yuan, Effect of Sm content on activation capability and hydrogen storage performances of TiFe alloy, Int. J. Hydrogen Energy 46 (48) (2021) 24517.

[100] J.L. Bobet, B. Chevalier, B. Darriet, Crystallographic and hydrogen sorption properties of $TiMn_2$ based alloys, Intermetallics 8 (4) (2000) 359.

[101] J.L. Bobet, B. Darriet, Relationship between hydrogen sorption properties and crystallography for $TiMn_2$ based alloys, Int. J. Hydrogen Energy 25 (8) (2000) 767.

[102] T. Huang, Z. Wu, G. Sun, N. Xu, Microstructure and hydrogen storage characteristics of $TiMn_{2-x}V_X$ alloys, Intermetallics 15 (4) (2007) 593.

[103] B.-H. Liu, D.-M. Kim, K.-Y. Lee, J.-Y. Lee, Hydrogen storage properties of $TiMn_2$-based alloys, J. Alloys Compd. 240 (1) (1996) 214.

[104] S. Semboshi, N. Masahashi, S. Hanada, Effect of composition on hydrogen absorbing properties in binary $TiMn_2$ based alloys, J. Alloys Compd. 352 (1) (2003) 210.

[105] M. Coduri, S. Mauri, C.A. Biffi, A. Tuissi, A new method for simple quantification of Laves phases and precipitates in $TiCr_2$ alloys, Intermetallics 109 (2019) 110.

[106] D.S. dos Santos, M. Bououdina, D. Fruchart, Structural and thermodynamic properties of the pseudo-binary $TiCr_{2-x}V_x$ compounds with $0.0 \leq x \leq 1.2$, J. Alloys Compd. 340 (1) (2002) 101.

[107] R. Sato, I. Tajima, T. Nakagawa, H. Uchida, Kinetics of hydrogen absorption by $TiCr_2$ and Ti–Cr based alloys at low temperatures, J. Alloys Compd. 580 (2013) S21.

[108] I.I. Bulyk, Y.B. Basaraba, Y.O. Dovhyj, Influence of Ti on the hydrogen-induced phase-structure transformations in the $ZrCr_2$ intermetallic compound, Intermetallics 14 (7) (2006) 735.

[109] A. Drašner, Ž. Blažina, The influence of Si and Ge on the hydrogen sorption properties of the intermetallic compound $ZrCr_2$, J. Alloys Compd. 199 (1) (1993) 101.

[110] L. Dai, S. Wang, L. Wang, Y. Yu, G.-j. Shao, Preparation of $ZrMn_2$ hydrogen storage alloy by electro-deoxidation in molten calcium chloride, Trans. Nonferrous Met. Soc. China 24 (9) (2014) 2883.

[111] H. Fujii, M. Saga, T. Okamoto, Magnetic, crystallographic and hydrogen absorption properties of YMn_2 and $ZrMn_2$ hydrides, J. Less Common Met. 130 (1987) 25.

[112] H. Fujii, V.K. Sinha, F. Pourarian, W.E. Wallace, Effect of hydrogen absorption on the magnetic properties of $ZrMn_{2-x}Fe_x$ ternaries with a C14 structure, J. Less Common Met. 85 (1982) 43.

[113] T. Wu, X. Xue, T. Zhang, R. Hu, H. Kou, J. Li, Role of Ni addition on hydrogen storage characteristics of ZrV_2 Laves phase compounds, Int. J. Hydrogen Energy 41 (24) (2016) 10391.

[114] W. Jiang, C. He, X. Yang, X. Xiao, L. Ouyang, M. Zhu, Influence of element substitution on structural stability and hydrogen storage performance: a theoretical and experimental study on $TiCr_{2-x}Mn_x$ alloy, Renew. Energy 197 (2022) 564.

[115] I.C. Atias-Adrian, F.A. Deorsola, G.A. Ortigoza-Villalba, B. DeBenedetti, M. Baricco, Development of nanostructured Mg_2Ni alloys for hydrogen storage applications, Int. J. Hydrogen Energy 36 (13) (2011) 7897.

[116] Baroutaji, A., Arjunan, A., Ramadan, M., Alaswad, A., Achour, H., Abdelkareem, M. A., Olabi, A.-G. Nanocrystalline Mg_2Ni for hydrogen storage, in: Olabi, A.-G. (Ed.), Encyclopedia of Smart Materials, Elsevier: Oxford, 2022, https://doi.org/10.1016/B978-0-12-815732-9.00061-9.

[117] D. Vyas, P. Jain, J. Khan, V. Kulshrestha, A. Jain, I.P. Jain, Effect of Cu catalyst on the hydrogenation and thermodynamic properties of Mg2Ni, Int. J. Hydrogen Energy 37 (4) (2012) 3755.

[118] D.H. Xie, P. Li, C.X. Zeng, J.W. Sun, X.H. Qu, Effect of substitution of Nd for Mg on the hydrogen storage properties of Mg_2Ni alloy, J. Alloys Compd. 478 (1) (2009) 96.

[119] M. Anik, Improvement of the electrochemical hydrogen storage performance of Mg_2Ni by the partial replacements of Mg by Al, Ti and Zr, J. of Alloys Compd. 486 (1) (2009) 109.

[120] T. Hongo, K. Edalati, M. Arita, J. Matsuda, E. Akiba, Z. Horita, Significance of grain boundaries and stacking faults on hydrogen storage properties of Mg_2Ni intermetallics processed by high-pressure torsion, Acta Materialia 92 (2015) 46.

[121] M. Anik, Electrochemical hydrogen storage capacities of Mg_2Ni and MgNi alloys synthesized by mechanical alloying, J. Alloys Compd. 491 (1) (2010) 565.

[122] X. Hou, R. Hu, T. Zhang, H. Kou, J. Li, Hydrogenation behavior of high-energy ball milled amorphous Mg_2Ni catalyzed by multi-walled carbon nanotubes, Int. J. Hydrogen Energy 38 (36) (2013) 16168.

[123] H. Kou, X. Hou, T. Zhang, R. Hu, J. Li, X. Xue, On the amorphization behavior and hydrogenation performance of high-energy ball-milled Mg_2Ni alloys, Mater. Charact. 80 (2013) 21.

[124] X. Liu, Y. Zhu, L. Li, Structure and hydrogenation properties of nanocrystalline Mg_2Ni prepared by hydriding combustion synthesis and mechanical milling, J. Alloys Compd. 455 (1) (2008) 197.

[125] Y. Ikeda, T. Ohmori, Study on chemical synthetic method to prepare Mg_2Ni hydrogen absorbing alloy, Int. J. Hydrogen Energy 34 (13) (2009) 5439.

[126] L.H. Kumar, B. Viswanathan, S.S. Murthy, Hydrogen absorption by Mg_2Ni prepared by polyol reduction, J. Alloys Compounds 461 (1) (2008) 72.

[127] Y. Kodera, N. Yamasaki, T. Yamamoto, T. Kawasaki, M. Ohyanagi, Z.A. Munir, Hydrogen storage Mg_2Ni alloy produced by induction field activated combustion synthesis, J. Alloys Compd. 446-447 (2007) 138.

[128] Á. Révész, M. Gajdics, E. Schafler, M. Calizzi, L. Pasquini, Dehydrogenation-hydrogenation characteristics of nanocrystalline Mg_2Ni powders compacted by high-pressure torsion, J. Alloys Compd. 702 (2017) 84.

[129] N. Skryabina, V. Aptukov, P. de Rango, D. Fruchart, Effect of temperature on fast forging process of Mg-Ni samples for fast formation of Mg_2Ni for hydrogen storage, Int. J. Hydrogen Energy 45 (4) (2020) 3008.

[130] T.Z. Si, D.M. Liu, Q.A. Zhang, Microstructure and hydrogen storage properties of the laser sintered Mg_2Ni alloy, Int. J. Hydrogen Energy 32 (18) (2007) 4912.

[131] T. Yougen, S. Jianbin, X.U. Yijun, J. Jinzhi, Effects of chemical coating with Ni on electrochemical properties of Mg_2Ni hydrogen storage alloys, Rare Met. 26 (6) (2007) 611.

[132] Z. Yanghuan, Z. Guofang, Y. Tai, H. Zhonghui, G. Shihai, Q. Yan, Z. Dongliang, Electrochemical and hydrogen absorption/desorption properties of nanocrystalline Mg_2Ni-type alloys prepared by melt spinning, Rare Met. Mater. Eng. 41 (12) (2012) 2069.

[133] Y. Zhang, C. Li, Y. Cai, F. Hu, Z. Liu, S. Guo, Highly improved electrochemical hydrogen storage performances of the Nd–Cu–added Mg_2Ni-type alloys by melt spinning, J. Alloys Compd. 584 (2014) 81.

[134] Y.-h. Zhang, F. Hu, Z.-g. Li, K. Lü, S.-h. Guo, X.-l. Wang, Hydrogen storage characteristics of nanocrystalline and amorphous Mg_2Ni-type alloys prepared by melt spinning, J. Alloys Compd. 509 (2) (2011) 294.

[135] H.-c. Zhong, J.-b. Xu, C.-h. Jiang, X.-j. Lu, Microstructure and remarkably improved hydrogen storage properties of Mg_2Ni alloys doped with metal elements of Al, Mn and Ti, Trans. Nonferrous Met. Soc. China 28 (12) (2018) 2470.

[136] E. Grigorova, M. Khristov, M. Khrussanova, P. Peshev, Addition of 3d-metals with formation of nanocomposites as a way to improve the hydrogenation characteristics of Mg_2Ni, J. Alloys Compd. 414 (1) (2006) 298.

[137] C.K. Lin, C.K. Wang, P.Y. Lee, H.C. Lin, K.M. Lin, The effect of Ti additions on the hydrogen absorption properties of mechanically alloyed Mg_2Ni powders, Mater. Sci. Eng.: A 449-451 (2007) 1102.

[138] H. Ding, S. Zhang, Y. Zhang, S. Huang, P. Wang, H. Tian, Effects of nonmetal element (B, C and Si) additives in Mg_2Ni hydrogen storage alloy: a first-principles study, Int. J. Hydrogen Energy 37 (8) (2012) 6700.

[139] X. Hou, R. Hu, T. Zhang, H. Kou, W. Song, J. Li, Microstructure and electrochemical hydrogenation/dehydrogenation performance of melt-spun La-doped Mg_2Ni alloys, Mater. Charact. 106 (2015) 163.

[140] X. Shan, J.H. Payer, J.S Wainright, Increased performance of hydrogen storage by Pd-treated $LaNi_{4.7}Al_{0.3}$, $CaNi_5$ and Mg_2Ni, J. Alloys Compd. 426 (1) (2006) 400.

[141] W. Song, J. Li, T. Zhang, H. Kou, X. Xue, Microstructure and tailoring hydrogenation performance of Y-doped Mg_2Ni alloys, J. Power Sources 245 (2014) 808.

[142] H. Sun, X. Gao, F. Zhao, Y. Li, Y. Zhang, H. Ren, Interactions of Y and Cu on Mg_2Ni type hydrogen storage alloys: a study based on experiments and density functional theory calculation, Int. J. Hydrogen Energy 45 (53) (2020) 28974.

[143] W.-j. Song, J.-s. Li, T.-b. Zhang, X.-j. Hou, H.-c. Kou, X.-y. Xue, R. Hu, Microstructure and hydrogenation kinetics of Mg_2Ni-based alloys with addition of Nd, Zn and Ti, Trans. Nonferrous Met. Soc. China 23 (12) (2013) 3677.

[144] Z. Zhang, J. Jin, H. Zhang, X. Qi, Y. Bian, H. Zhao, First-principles calculation of hydrogen adsorption and diffusion on Mn-doped Mg_2Ni (010) surfaces, Appl. Surf. Sci. 425 (2017) 148.

[145] D. Vyas, P. Jain, G. Agarwal, A. Jain, I.P. Jain, Hydrogen storage properties of Mg_2Ni affected by Cr catalyst, Int. J. Hydrogen Energy 37 (21) (2012) 16013.

[146] X. Zhao, S. Han, Y. Zhu, X. Chen, D. Ke, Z. Wang, T. Liu, Y. Ma, Investigation on hydrogenation performance of Mg_2Ni+10wt.% NbN composite, J. Solid State Chem. 221 (2015) 441.

[147] M. Anik, B. Baksan, T.Ö. Orbay, N. Küçükdeveci, A.B. Aybar, R.C. Özden, H. Gaşan, N. Koç, Hydrogen storage characteristics of Ti_2Ni alloy synthesized by the electro-deoxidation technique, Intermetallics 46 (2014) 51.

[148] M. Balcerzak, J. Jakubowicz, T. Kachlicki, M. Jurczyk, Hydrogenation properties of nanostructured Ti_2Ni-based alloys and nanocomposites, J. Power Sources 280 (2015) 435.

[149] B. Hosni, X. Li, C. Khaldi, O. ElKedim, J. Lamloumi, Structure and electrochemical hydrogen storage properties of Ti_2Ni alloy synthesized by ball milling, J. Alloys Compd. 615 (2014) 119.

[150] J.-j. Li, J.-f. Zhou, X.-y. Zhao, M. Yang, L.-q. Ma, X.-d. Shen, Electrochemical hydrogen storage properties of non-equilibrium $Ti_{2-x}Mg_xNi$ alloys, Trans. Nonferrous Met. Soc. China 25 (11) (2015) 3729.

[151] B. Luan, N. Cui, H.K. Liu, H.J. Zhao, S.X. Dou, Effect of cobalt addition on the performance of titanium-based hydrogen-storage electrodes, J. Power Sources 55 (2) (1995) 197.

[152] B. Luan, N. Cui, H. Zhao, S. Zhong, H.K. Liu, S.X. Dou, Studies on the performance of $Ti_2Ni_{1-x}Al_x$ hydrogen storage alloy electrodes, J. Alloys Compd. 233 (1) (1996) 225.

[153] X. Zhao, L. Ma, Y. Ding, X. Shen, Structural evolution and electrochemical hydrogenation behavior of Ti_2Ni alloy, Intermetallics 18 (5) (2010) 1086.

[154] I. Bürger, L. Komogowski, M. Linder, Advanced reactor concept for complex hydrides: hydrogen absorption from room temperature, Int. J. Hydrogen Energy 39 (13) (2014) 7030.

[155] I.P. Jain, P. Jain, A. Jain, Novel hydrogen storage materials: a review of lightweight complex hydrides, J. Alloys Compd. 503 (2) (2010) 303.

[156] M.B. Ley, L.H. Jepsen, Y.-S. Lee, Y.W. Cho, J.M. Bellosta von Colbe, M. Dornheim, M. Rokni, J.O. Jensen, M. Sloth, Y Filinchuk, Complex hydrides for hydrogen storage: new perspectives, Mater. Today 17 (3) (2014) 122.

[157] B. Bogdanović, M. Schwickardi, Ti-doped alkali metal aluminium hydrides as potential novel reversible hydrogen storage materials[1], J. Alloys Compd. 253–254 (1997) 1–9.

Hydrogen transportation and distribution

Arash Aghakhani[a,b], Nawshad Haque[b], Cesare Saccani[a], Marco Pellegrini[a] and Alessandro Guzzini[a]
[a]Department of Industrial Engineering, University of Bologna, Bologna, Italy
[b]Commonwealth Scientific and Industrial Research Organization (CSIRO energy), Melbourne, Australia

4.1.1 Hydrogen delivery infrastructures

For the proper operation of large socio-technical systems such as hydrogen production, appropriate infrastructure is required [1]. Hydrogen delivery is one of the main contributors to hydrogen fuel economy, energy consumption, and emissions. A proper hydrogen delivery pathway can increase the market penetration of this technology. This chapter focuses on the delivery infrastructures to bring hydrogen fuel from production facilities to the end users. Considering the location of production facilities hydrogen production can be categorized as centralized and decentralized, which are related to H_2 production in large-scale plants called off-site, and on the spot, called on-site production, respectively [2]. In the case of centralized hydrogen production appropriate hydrogen delivery infrastructure including transmission and distribution is an integral part of the widespread use of hydrogen. Transmission is related to the hydrogen delivery from the production sites to the city gates or the areas of consumption, and distribution is associated with the delivery from the city gates to the fueling stations or end users [3].

The specific cost of centralized hydrogen production is lower than the decentralized specific production cost due to the economy of scale [2]. Even though on-site hydrogen production is costly, in districts with low hydrogen demand and a far distance between production plants and end users or located on islands, decentralized hydrogen is preferred because of less delivery and distribution cost. In the case of remote users, the preferred option for hydrogen generation is on-site natural gas steam reforming or electrolysis [4].

An important parameter to determine the transport infrastructure scale is the amount of hydrogen demand. With the expansion of the hydrogen market and increase in hydrogen fuel acceptance, the development of

Towards Hydrogen Infrastructure: Advances and Challenges in Preparing for the Hydrogen Economy.
DOI: https://doi.org/10.1016/B978-0-323-95553-9.00003-0

transport infrastructures is predictable. If green hydrogen becomes desirable, longer delivery distances might be required [5] because the suitable locations with the potential for renewable energy production are usually located at far distances from city gates.

There are three primary pathways for hydrogen delivery from production plants to the end users: (1) tube-trailer or pressurized tank delivery, (2) pipeline delivery, and (3) cryogenic tankers or liquid delivery [5,6]. These pathways could be used either for transmission or distribution of hydrogen, which results in the opportunity for pathways combined for transporting hydrogen to the end users and hydrogen refueling stations (HRS) [5]. An example of pathways combination assessment is provided by Elgowainy et al. [5] which the following combinations were considered: pipeline and liquid delivery pathway, pipeline and tube-trailer delivery pathway. Other pathways such as barge or rail transportation are possible, but they are almost the same as trucks with less geographic availability [7]. Alterations in temperature and pressure can change the hydrogen form from gaseous to liquid or vice versa which leads to transportation means differences. For the sake of economic aspects in terms of hydrogen storage and delivery, it is desired to reduce the large hydrogen gas volume. Hydrogen gas has very low volumetric energy density under standard conditions, that is, standard temperature and atmospheric pressure, approximately more than three orders of magnitude less than gasoline; under this condition volume of 1 kg of hydrogen is 11 m^3 [8,9].

Hydrogen can be delivered in gaseous form, liquid form, or by means of carrier chemicals. Gaseous delivery can be performed via truck (tube trailer) or pipeline, the liquid form can be delivered via tank truck and the delivery of carriers can be done by means of tank trucks or pipeline [7].

The choice of the delivery pathway depends mostly on the regional sources, geographical characteristics, forecasted demand, district population, consuming behaviors, population density, size of HRS, and market penetration of hydrogen-consuming units such as fuel cell vehicles [1,3]. In addition, industries' decision in terms of the type of fuel is a turning point in the hydrogen pathway options. In districts with high market penetration, high population density, and large HRS pipeline is the most favorable pathway while in low-demand districts with low population and population density, low market penetration and small HRS compressed gas trucks could be the main pathway. For districts with a large population with low population density and small HRS, a liquid hydrogen truck is the desired pathway for hydrogen delivery [10].

Currently, the most used hydrogen delivery pathway to stations is trucks, while delivery by means of the pipeline is rarer [11]. It is predicted that, in 2050, approximately all stationary hydrogen consumers, such as industries, households, and services, will be connected to regional pipeline networks and liquefied hydrogen will be used to supply HRS demand [4]. An example of accelerating decarbonization via hydrogen infrastructure is European Hydrogen Backbone (EHB), which aims to accelerate decarbonization in Europe based on existing and new pipelines. According to the EHB report, a 53,000 km pipeline is required by 2040, 60% of repurposed natural gas pipelines, and 40% of new pipeline extension, which could cost approximately €80–143 billion [12].

If the produced hydrogen is used as fuel in a hydrogen vehicle (HV), the final destination of hydrogen delivery will be HRS. Each delivery mode has a different type of HRS based on the hydrogen form: (1) HRS relied on compressed gas, which further compression from the station delivery pressure to the required hydrogen vehicle storage pressure is needed, (2) HRS relied on liquid hydrogen delivery [9]. Thus, the delivery pathway should consist of several major steps and components or elements such as liquefiers, pipelines, trucks, terminals, and HRS [5]. In this case, the system components included in the delivery pathway regarding the hydrogen form and type of pathway could be as follow [9]:

- Gas pipeline: Compression and storage at H_2 plant, gas pipelines, HRS, compressor, high-pressure storage, and dispensers.
- Compressed gas trucks: Compression and storage at H_2 plant, compressed gas trucks, HRS, compressor, high-pressure storage, and dispensers.
- Liquid H_2 trucks: Liquefaction and storage at the H_2 plant, liquid H_2 trucks, HRS, liquid hydrogen storage, liquid hydrogen pump, high-pressure storage, and dispensers.

Compression and storage at the H_2 plant are required for compressed gas delivery by pipeline or truck, while in liquid hydrogen delivery, liquefaction, storage at the central plant, and liquid hydrogen storage tanks are needed.

It can be seen from the system components that storage is an integral part of hydrogen delivery which is used to overcome the imbalances between demand and supply. Storage is provided at HRS and at the production plant to meet the time variations in hydrogen demand and to provide a reliable and acceptable hydrogen supply [9]. Hydrogen storage is a relatively expensive component of hydrogen delivery infrastructure [7]. Storage characteristics such as size and pressure depend on the variation in demand and production. Considering the temporal variation in demand there could be two types of storage for hydrogen delivery: short-term and long-term. Short-term is related to hourly variations in demand and long-term covers seasonal variations in demand and outages in hydrogen production plant schedule [5]. Studies based on economic optimization of gaseous hydrogen pipeline pathways summarized that for long-term storage, in the presence of suitable geologic storage, compressed gas storage in geologic formation is desirable, otherwise liquid storage can be used; for short-term storage, if the system comprises HRS, the desired storage is low-pressure ($\sim$170 bar) compressed gas storage at the HRS [5].

It is possible to mix various pathways and storage types to evaluate the economic and environmental aspects. Examples of combination could be as follow as represented by Elgowainy et al [5]:

- Liquid delivery pathway with liquid long-term storage: In this combination liquid H_2 terminal is co-located with the hydrogen production plant, and storage tanks are used for transmission and delivery to HRS.

- Mixed-mode pipeline and liquid delivery pathway with long-term geologic storage: In this combination, produced hydrogen at the central plant is stored in a geological storage system, and transmission is done through the pipeline to the city gate in which the liquid hydrogen terminal is located, at the liquid hydrogen terminal tankers are filled and the distribution is done by the tankers.

- Mixed-mode pipeline and liquid delivery pathway with liquid long-term storage: This combination is similar to the previous one except for long-term storage which is replaced by a compressor and storage tank.

- Mixed-mode pipeline and tube-trailer delivery pathway with long-term geologic storage: In this combination, produced hydrogen at the central plant is compressed and stored for long-term in geologic storage, the pipeline is used for transmission to the city gate, gaseous hydrogen terminal is located at the city gate which tube-trailer deliver hydrogen as distribution section.

- Mixed-mode pipeline and tube-trailer delivery pathway with liquid long-term storage: This combination is similar to the previous one, the difference is the first stage after hydrogen production and before transmission via pipeline, this stage is allocated to liquefaction and liquid storage.

- Tube-trailer delivery pathway with long-term geologic storage: In this combination geologic storage is used for long-term storage, the gaseous hydrogen terminal is located before transmission and the tube-trailer delivers hydrogen from the terminal to HRS, in this pathway distribution and transmission are performed by means of tube-trailer.

- Tube-trailer delivery pathway with long-term liquid storage: This combination is similar to the previous combination, except for the storage type, which is liquid storage, thus, the liquefaction stage is needed.
- Pipeline delivery pathway with long-term geologic storage: Both transmission and distribution are performed by means of pipeline, compression is required, and compressed hydrogen is stored in geologic storage.
- Pipeline delivery pathway with long-term liquid storage: This combination is similar to the previous combination, except for the storage type, which is liquid storage, thus, a liquefaction stage is needed.

4.1.1.1 Tube-trailers

To increase the total performance efficiency of hydrogen delivery it is required to reduce its volume, which is done by means of compressor. Tube trailers can be used to deliver high-pressure gas. The hydrogen delivery system can include a stationary compressor at the central plant, in which the tube trailers can be filled under their specified pressure. The structure of tube trailers includes several tubes which are designed to resist the desired pressure [5]. The tube trailer is shipped by means of a truck cab (tractor) to the HRS [9]. 12–20 long steel cylinders mounted on a truck trailer bed are used in commercial tube trailers which should be regulated by means of the relevant government sector [9] and for safe delivery, tubes are protected inside a protective frame [8]. In the United States, the maximum limit of gas pressure on the truck is 160 atm [9]. The medium amount of compressed gaseous hydrogen is delivered by truck in tubes with a pressure range of 200–500 bars or in gas cylinders. As an example, the capacity of a tube trailer with steel cylinders is around 25,000 L of compressed hydrogen at 200 bars, which is equal to 420 kg of hydrogen [8]. However, the delivered capacity is less than the nominal one since it is not possible to completely discharge the tube trailer. If the end users of hydrogen are vehicles, considering the hydrogen pressure delivered by tube-trailer and the required pressure for vehicles it might be needed to have an extra compression phase at the HRS [9]. The loading procedure of tubes is relatively a time-consuming part of this pathway. It could be mostly in the range of 6–10 hours which varies based on the characteristics of the tubes and compressors such as the temperature limit of the tube and compressor capacity [5]. The required number of trailers depends on the demand; to supply the demand properly and overcome the long loading time in the HRS case the number of tube trailers can be defined as the number of truck cabs plus the number of HRS, which the tube trailer can be left at the station, thus, drivers and trucks should not wait for the procedure of loading and unloading [9].

The capacity of this pathway can increase by utilizing materials with higher resistance, for example, lighter tank materials such as composite materials for gas tubes and gas cylinders, which are capable of working at higher pressure [8]. Carbon fiber and glass fiber are examples of composite materials which can affect the efficiency of the pathway. Tubes made of these kinds of materials can deliver hydrogen at higher pressure and capacity because of their low weight and higher pressure resistance [5]. The capacity of a tube trailer can increase to around 39,600 L of hydrogen by using superlight cylinder materials made of carbon fiber composite with high-density polyethylene liners [8]. This capacity could be improved to around 1000 kg by cooling the compressed gas [5]. Weisberg et al. [10] suggested delivering hydrogen in glass fiber vessels by truck and tube trailers, carrying hydrogen at approximately 70 MPa and 200 K to increase the efficiency of the hydrogen delivery and reduce the cost. They stated that by cooling down hydrogen to 200 K it will be 35% more compact, furthermore, at this temperature glass fibers are almost 50% stronger which increases the capacity of the pathway without using carbon fiber composite which is more costly. In addition, faster refueling without overtemperatures and overpressures can be done by cooling down the hydrogen [10].

Tube trailers have several advantages, in terms of infrastructure requirement it is the simplest pathway, in addition, compression costs at refueling stations and hydrogen loss are low. Despite the advantage, tube trailers have several drawbacks, the high manufacturing cost of composite overwrapped pressure vessels which could be 70% of the total cost, limitation of the tank dimension and maximum pressure capacity, and low storage capacity of the tube trailer [3].

4.1.1.2 Pipelines

A pipeline pathway could be the best choice for hydrogen delivery when large amounts of hydrogen are required to be delivered to several stationary consumers [4]. It can be considered a low–cost pathway for long-term hydrogen delivery and large volumes of hydrogen [7]. The pipeline pathway can be considered as a network with two performances: (1) supplying energy to the consumers regarding the demand by connecting the production sites to the distribution network and end users. (2) Storage performance, which can store considerable volumes of hydrogen which depends on the capacity of the pipeline and the difference between supply and demand [4,13]. The current pipeline networks are not proportional to the hydrogen demand. Around 2600 km of hydrogen pipelines are currently under operation in the United States which are located in districts with high demand such as chemical plants and petroleum refineries [14]. Fig. 4.1.1 includes information about the length of the existing pipelines for hydrogen delivery in the United States, Europe, and the rest of the world by location based on the study in 2016 by the Pacific Northwest National Laboratory [15], the world total length of the hydrogen pipelines is estimated around 4550 km, while 2600 km is associated with the United States, 1600 km is operating in Europe.

According to Tzimas et al. [4] study based on three scenarios in Europe, by 2050, approximately 35,000 km of high-pressure transmission pipelines, 400,000 km of medium pressure sub-transmission pipelines and 1–4 km distribution pipeline might be needed, which could cost between 700 and 2200 thousand million euros and the main expenses will be required for the distribution network development. Similar to tube trailers, lightweight materials can be used to reduce the pathway cost. An alternative to the low-alloy steel which is used in hydrogen delivery could be glass fibre reinforced polymer (GFRP) which provides a better lifecycle performance and lower capital cost in comparison to steel pipelines [16].

An option to reduce the pipeline delivery cost is blending hydrogen with natural gas (NG) and using the existing natural gas infrastructure for hydrogen delivery. The estimated length of the required pipeline for hydrogen delivery without blending with natural gas in the United States is approximately hundreds of thousands of kilometers which led policymakers and researchers to investigate the use of existing natural gas pipeline networks to transmit and distribute hydrogen [3]. There could be two main reasons behind the blending of hydrogen with natural gas: (1) using the existing natural gas network for hydrogen delivery, (2) improving the renewable content of natural gas [17]. A challenging issue in blending hydrogen with natural gas and delivering with existing pipelines is the reaction of hydrogen with pipeline and other components (i.e., compressors, valves, and instruments) material which could result in detrimental consequences. Various blending limitation has been recommended by researchers. Faye et al. [8] stated that less than 5–15% volumetric mixture of hydrogen with natural gas results in minor issues, which are correlated with the composition of

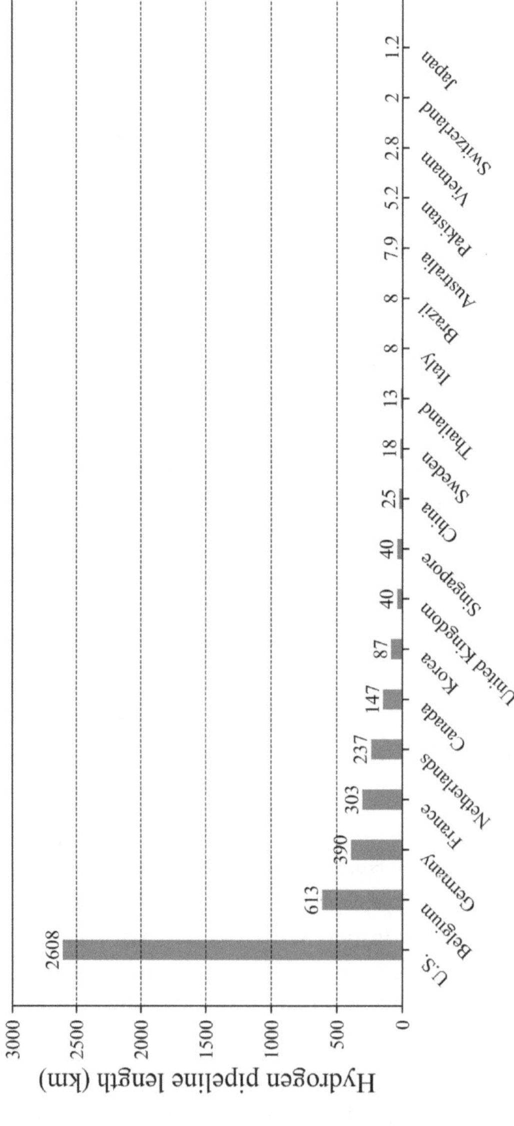

Figure 4.1.1 Existing hydrogen pipelines length, 2016.

natural gas and the characteristics of the existing infrastructure. This amount is limited to 17% by Gondal et al. [13]. It should be noted that high amounts of blended hydrogen lead to the necessity of household appliance conversion and enhanced compression capacity for industrial users along the delivery pathway [8]. Pellegrini et al. [18] proposed up to 10% blending to the Italian natural gas network. The proposed percentage by the European Commission for the cap of hydrogen blending EU-wide by 2030 is 5% [19]. Ogbe et al. [17] proposed a stochastic model to optimize the quality/concentration of hydrogen in the natural gas pipeline and minimize the capital and operating cost by optimizing the location of power-to-gas (PtG) units [17].

In the pipeline design procedure, the capacity of the network is higher than the average flow rate based on the studies on demand and demand variation, to overcome the temporal variation of flow and also provide the possibility of network expansion [10]. Pipelines can be designed underground and aboveground. In the case of aboveground, pipelines should be protected by paint against atmospheric corrosion [20].

4.1.1.2.1 Specifications and characteristics
4.1.1.2.1.1 Advantages
Considering the emission of burning fuel in tube trailers and cryogenic tankers pathways and the lengthy lifetime of large-scale pipeline infrastructures the latter is considered the most environmentally friendly choice of hydrogen delivery [8]. In addition, since pipelines are mostly buried underground, and there is less probability of incidents or accidents and human exposure, this pathway is safer and more dependable compared to other delivery modes.

4.1.1.2.2 Challenges
Despite the advantages of pipeline infrastructure for hydrogen delivery, several significant challenges should be overcome to make this pathway more reliable. The most important issue is related to the high capital expenditure relative to hydrogen demand initially, which goes back to the cost of installation including civil work, pipes, and equipment cost. In addition, losses of hydrogen gas along the delivery pipeline are higher than the other fuels. Thus, it is required to properly design the network's components to prevent gas leakage from seals, gaskets, and valves to eliminate safety hazards [8], either in the blending mode with natural gas or in pipelines specifically allocated to hydrogen delivery. Hydrogen has a very low density which is approximately 1/8th of natural gas, which results in the necessity of hydrogen

compression to the pressure of around 10–20 bars to improve the delivery rate [8]. Standard reciprocating compressors are not completely satisfying for hydrogen compression and standard centrifugal ones are not sufficiently efficient because of the small size of the hydrogen molecule [7].

The medium in which the pipeline should be installed plays an important role in the performance of the infrastructure. Carbon steel pipelines are susceptible to the hydrogen embrittlement (HE) phenomenon. Pipeline history, including the frequency and intensity of pressure variation, can be used to predict embrittlement [13]. Cohesive forces in steel structure and deformation mechanisms of steel could be changed through hydrogen, which under specific conditions such as specific loading results in ductility and toughness reduction of steel. In addition, there could be cyclic variation in pressure due to demand alteration, in which cracks growth might be sensitive under fatigue loads [7]. Therefore, the pipeline should be capable of resisting acidic soil environment and hydrogen inside the pipeline. Furthermore, materials that are used in the pipeline should be capable of bearing hydrogen pressure, operating temperature, other loadings such as backfill soil, and be able to resist freezing. In urban areas that are highly populated and there is a dense network of infrastructures it is not possible to construct a new pipeline network for hydrogen, which is a relatively new and unknown fuel for energy customers. For the pipelines which are installed underground, there would be a probability of damage due to lightning strikes and ground fault conditions, thus, electrical continuity to the surface should be avoided [20].

In the case of blending hydrogen with natural gas, since the existing networks are quite old in some cases, it might not be possible to use them as the infrastructure for hydrogen delivery. The old pipelines were probably designed and installed based on the requirements that are not suitable for the condition of the hydrogen mixture, for example, allowable defect sizes, inspection techniques, carbon steel weld, and grad of carbon steel. Moreover, the pipeline network has likely been affected by minor or major damages during its service time, such as deterioration and wall thinning due to corrosion, third-party damage, and soil movement. In addition, during the pipeline lifespan, there might be replacement and repair in the maintenance process. In the case of blending hydrogen with natural gas in plastic pipeline networks, since polymer materials have high porosity and hydrogen has a higher diffusion coefficient in comparison to natural gas, as it is the smallest element in the periodic table, they are not appropriate infrastructure choice for mixing hydrogen with natural gas [8]. Therefore, substantial uncertainty in pipeline conditions is inevitable, which makes it challenging to combine hydrogen with existing natural gas networks [7]. Therefore, using the existing natural gas pipelines requires performing intensive testing of pipes, welds, and joints to control the suitability for hydrogen blending [13].

4.1.1.3 Cryogenic tankers

In this mode of delivery, gaseous hydrogen is liquefied by decreasing temperature below $-253°C$ through the liquefaction process, then it can be stored in large tanks at the liquefaction plant. In comparison to the compressed gas in containers, the result is denser hydrogen, and this pathway is capable of delivering significantly more hydrogen [8]. This method could be suitable for moderate to high demands, above 500 kg/day, and medium to long distances [3,10]. There are three main steps in hydrogen delivery by cryogenic tankers pathway: (1) liquefaction, (2) storage, and (3) delivery by cryogenic tankers to the end users [3].

Cryogenic liquid truck's capacity is around 65 m^3, and a nominal capacity of 4600 kg hydrogen [5], which is almost more than 10 times the quantity of hydrogen that can be delivered using the tube trailer pathway. It reduces the number of trucks, drivers, and trips that are required to supply the demand and reduces fuel consumption for truck transport. A cryogenic liquid hydrogen truck includes a truck cab and a large single liquid hydrogen tank which is mounted on a trailer. Unlike the compressed gas tube trailer which was possible to leave the trailer at the HRS, it is not possible to leave the liquid hydrogen trailers, so the assumptions for the number of trucks to optimize the performance in the tube trailer are not applicable here, and the number of liquid hydrogen tank trailers is always equal to the number of truck cabs [10].

4.1.1.3.1 Advantages and disadvantages
There are several advantages and disadvantages to the cryogenic tanker pathway. Capital cost and required energy for liquefaction are considerably higher than the energy and cost of compression [10], energy loss during the liquefying process is significant, around 40% of energy is lost, in addition, there would be loss through evaporation, boil-off of liquid hydrogen and unloading procedure at HRS, which could be around 6% of the total capacity of the truck. Therefore, it is advised to minimize the number of HRS that each truck should serve on each trip to prevent huge losses. Furthermore, it is highly likely that liquefied hydrogen is not used for stationary applications, which roots in the high and constant demand that makes the delivery

of liquified hydrogen impractical by using trucks, since it requires large cryogenic storage facilities and also a huge number of trucks is needed to supply the demand without any interruption [4]. Tanker filling time in comparison to gaseous hydrogen is faster, approximately 2–3 hours which results in less waiting time for drivers and trucks [5,8]. To reduce heat transfer and reduce boil-off, the ratio of volume to the surface of the tank should be higher. When the diameter increases the volume increases faster than the surface area, thus spherical and cylindrical tanks are suitable for this pathway, however, cylindrical tanks are cheaper and easier to construct in comparison to the spherical ones with the same volume to surface ratio. The evaporated hydrogen can be vented or stays in the tank until the tank design pressure, then it must be vented. The time at which the gas pressure reaches the design pressure is called lock-up time. The evaporated hydrogen can be used in the process in which gaseous hydrogen is needed. If the storage time which is required to use gaseous hydrogen is shorter than the lock-up time, hydrogen loss due to boil-off would be zero, because gaseous hydrogen can be used before venting into the atmosphere. However, this amount of hydrogen does not have significant safety issues since it diffuses rapidly in the air. If the liquid hydrogen is stored at the same site of liquefaction, it is possible to pull out the hydrogen gas from the storage vessel and re-liquify it [21].

4.1.1.4 Under development pathways

In addition to the mentioned pathways, there are several pathways that are not mature and are under development, such as chemical reactions or physisorption with a material [6]. Liquid organic hydrogen carriers (LOHCs) can be considered long-term storage and can be used for long-distance hydrogen delivery. In this pathway, organic molecules are hydrogenated by loading liquid hydrogen molecules into organic molecules, and they will be dehydrogenated and hydrogen molecules will be unloaded to consume. The benefit of carrying hydrogen by organic molecules is related to the condition in which LOHCs are transported, they can be delivered under ambient conditions and using existing crude oil infrastructures [22]. An example of LOHCs could be toluene-methylcyclohexane, in which Carvalho et al. [22] performed an economic analysis and found that the break-even price of hydrogen loading and release would be around 1.9 \$/kg of hydrogen. In another study, Brigljević et al. [23] performed a technological evaluation, cost estimation, and market performance analysis of a large-scale dehydrogenation system with a capacity of 1000 m^3/h hydrogen and no CO_2 emission. Four LOHCs were assessed in this study, Eutectic biphenyl/diphenylmethane mixture (EUT), 2-(N-Methylbenzyl)pyridine (MBP), N-phenylcarbazole (NPC), and N-ethylcarbazole (NEC), and also a case using all four LOHCs in a temperature-cascade process. They found that the main costs are related to LOHCs price and shipping costs, and the total capital investment for this size of plant could be around 24–44 million USD. By the evaluation of the market performance, they found that for the hydrogen which could be sold at 3.5 USD per kilogram of hydrogen the purchase prices of LOHC chemicals must be 5.44, 4.74, 4.01, and 4.12 USD/kg for EUT, MBP, NPC, and NEC respectively. Another hydrogen carrier could be ammonia which is evaluated by Malenkov et al. [24]. They concluded that utilizing ammonia as a hydrogen carrier in industrial quantities could be logical when using sea or trail delivery for distances over 2000–3000 km. The capacity of ammonia to store hydrogen is 123 kg hydrogen in one m^3 of liquid ammonia at 10 bars, around 17.7% (wt) [25]. An ammonia cracking plant is needed to extract hydrogen from ammonia [26]. Methanol is another hydrogen carrier, but the problem with methanol is its carbon dioxide emission when it is used and also the direct carbon capture from air to make methanol which considerably increases the cost.

Another option is bonding hydrogen chemically to metal and alloys, which is called metal hydride. Hydrides can store hydrogen in the range of around 2–6% of the weight of hydrides, but since they have high volumetric storage density, they could be a suitable choice for hydrogen delivery. The advantage of this pathway is associated with the procedure pressure, in which hydrogen can bond to some metals at or below atmospheric pressure and can be released at higher pressures after heating (thermolysis), higher pressures could be achieved by applying higher temperature. The heating breaks the bond between hydrogen and metal. Hydrogen can bond to the metal up to around 90% of metal capacity under low pressure, the rest of the capacity can be achieved by increasing the pressure. It should be noted that the reaction of hydrogen with metal is an exothermic reaction in which the released heat should be removed to prevent hydride from heating up. Hydrogen extraction of up to 90% could be achieved simply by applying heat, while the remaining 10% is related to the strong bond of hydrogen and metal which is not easy to break [21]. In addition to thermolysis which is an endothermic reaction, hydrogen can be released from hydride via hydrolysis (reaction with water), which is an exothermic reaction. An example of metal hydride for hydrolysis is sodium borohydride ($NaBH_4$). The most suitable metals for the large-scale storage of hydrogen are magnesium hydride (MgH_2) and aluminium hydride (AlH_3). Magnesium has a high theoretical capacity of hydrogen storage, approximately 7.6% (wt) and it is cheap and widely available. The problem with magnesium hydride is the strong bond that needs higher energy for dehydrogenation. The advantage of aluminium hydride is its weak bond which the dehydrogenation is simpler in comparison to magnesium hydride, but the problem of aluminium hydride roots from the reaction of aluminium and hydrogen gas which can be done under extreme pressures [25].

4.1.2 Environmental impact

An important issue regarding the market penetration and acceptance of technology is its environmental impact. Frank et al. [27] performed a life-cycle analysis of greenhouse gas emissions from hydrogen delivery for different pathways and different distances, the result is shown in Fig. 4.1.2. These results include the emission due to liquefaction, pipeline, terminal for tube trailer, trucking, and HRS. It can be seen that the pipeline delivery mode has the lowest amount of greenhouse gas (GHG) emission for 1 kg hydrogen delivery. In addition, the Tube trailer is the most sensitive pathway to distance since the variation in the GHG emission by increasing the distance is more significant.

Among the three main pathways, the energy consumption of pipelines is considerably lower than the other modes of delivery. The only energy type which is used in the pipeline pathway is electricity while in tube-trailer and cryogenic tankers pathways diesel fuel is used in addition to electricity as a truck fuel. Electricity is used for gas compression and liquefaction. Tube-trailer and cryogenic tanker trucks consume considerable diesel fuel by increasing the delivery distance because of their limited capacity. In comparison to compressed gas trucks, pipeline pressure is lower and lower energy is required to overcome the friction losses along the pipeline if the network is sized adequately. In addition, in the pipeline delivery, the energy loss variation at a wide range of flow, 2–100 ton/day, is almost negligible, and the majority of energy consumption is related to hydrogen gas compression at the pipeline inlet [10]. The energy that is required for the hydrogen gas compression is considerable in tube trailer and pipeline pathways, it could be approximately 40% of the delivered hydrogen energy content, lower heating value (LHV). The total energy which is used in the cryogenic tankers' pathway is higher than the tube-trailer delivery mode, mainly due to the amount of energy that is used for liquefaction which is higher than the energy required for hydrogen gas compression [5,10]. However, the approximated value was different in Weisberg et al. [10] study. They stated that in the pipeline pathway the energy used for hydrogen gas compression to around 1000 psi equals 2 to 3% of the energy content of

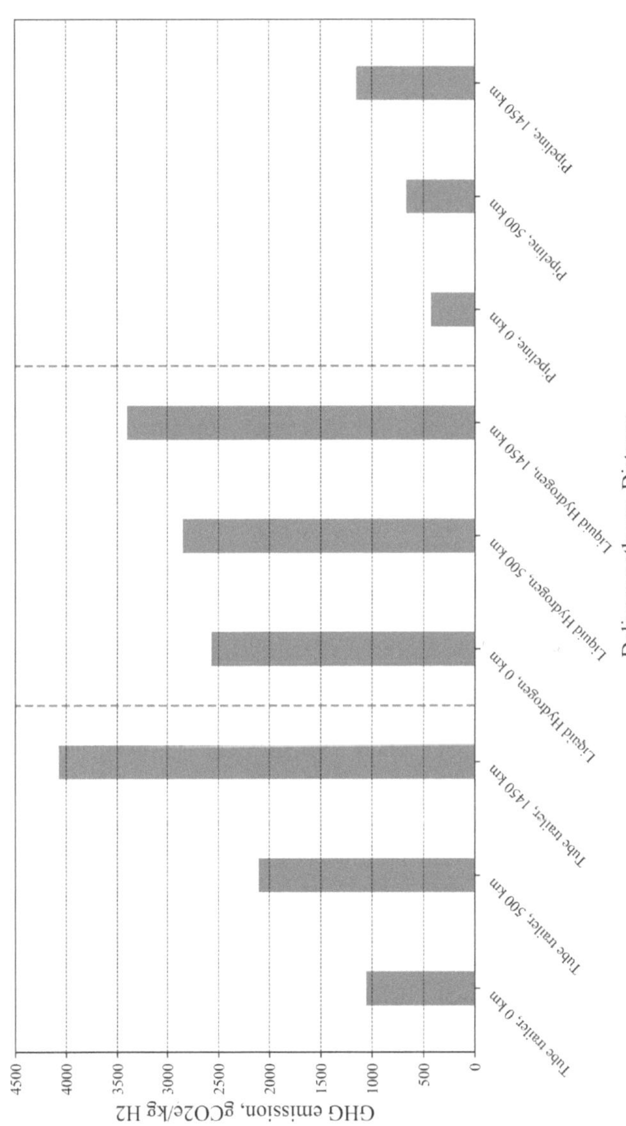

Figure 4.1.2 Life-cycle greenhouse gas emissions for delivery by tube trailer for gaseous H_2, by tanker for liquid H_2, or by pipelines in various delivery distances [27].

hydrogen which is much less than 40% which is evaluated by Elgowainy et al. [5], in addition, they reported the liquefaction energy consumption as around 33% of the hydrogen energy content. These differences could be due to the scale of the experimental test, the efficiency of the technologies for compression and liquefaction, and conditions such as temperature and pressure, in addition, improper calibration could lead to incorrect results. According to the report provided by the U.S. Department of Energy [28], the theoretical energy to isothermally compress hydrogen from 20 bar to 350 bar, which is around 5000 psi, is 1.05 kWh/kg H_2 while to reach 700 bar, around 10,000 psi, the theoretical energy is 1.36 kWh/kg H_2. Considering the LHV of hydrogen which is 33.33 kWh/kg the theoretical energy to isothermally compress hydrogen would be approximately 3–4% of Hydrogen energy content. The theoretical energy to liquefy hydrogen from the standard condition is reported 3.9 kWh/kg $H_{2(l)}$, which is around 10% of the hydrogen energy content. The practical required energy is much higher than the theoretical one, the practical hydrogen compression of produced hydrogen on-site could be approximately 5–20% of hydrogen LHV, and for liquefaction it could be 30–40% of LHV.

4.1.3 Cost of hydrogen transportation and distribution

Even though hydrogen fuel has a less environmental impact in comparison to fossil fuels, but it is not sufficient to penetrate the market. An important issue in market penetration is the price of fuel, thus, the oil price in comparison to hydrogen plays an important role for the acceptance of hydrogen technology. Hydrogen delivery can be considered one of the main contributors to the overall hydrogen fuel economy. Zero-emission technologies become more competitive considering the world's concern about climate change and carbon price. But to stay in the competition it is required to reduce the costs. Hydrogen infrastructure has a high initial investment which takes a long period to reach the break-even point due to uncertainty in demand and reliability by society. Thus, investors could be sensitive to investment risk in hydrogen technology. Consumers are an integral part of an infrastructure success because if there is not any demand for hydrogen there would not be any investment interest by companies in it [1]. An increase in hydrogen demand leads to an increase in production and cost reduction, and the cost reduction could increase the demand, this loop can continue until reaching a stable condition. However, Supplying the increased demand needs larger infrastructures for delivery.

4.1.3.1 Pipeline

The Pipeline pathway seems to be the most economical mode for hydrogen delivery for large quantities and long distances [29]. The cost of hydrogen pipeline infrastructure varies considerably, this cost includes initial investment, operational costs, and maintenance costs. However, the maintenance and operational costs of pipeline networks in comparison to the initial investment are relatively low [8]. The pipeline pathway costs are associated with the geography of the terrain in which the network should be installed, the diameter of the pipeline, the equipment along the network, the labour and material cost in each district, the choice of material, and the operating condition [4]. The total cost per unit length of a pipeline at a large pipe diameter network is sensitive to the pipe diameter (while at a smaller diameter pipelines diameter is less effective, because materials costs are a small portion of the total cost in small-dimeter networks), installation and rights of way costs are the main contributors to the overall cost [10]. Lee et al. [30], compared the cost of hydrogen delivery in gaseous and liquid forms in Seoul, Korea, based on a techno-economic analysis. They found that the gaseous form has a better performance in terms of economics, however, cost changes due to variation in the capacity of the station and the market penetration, especially when these factors are small.

The cost target for 1 kg hydrogen delivery by the year 2017 was less than 1 dollar [7], which is not achieved. Pawel et al. [7] indicated that the pipeline cost for hydrogen delivery could cost \$2 M/mile or more for larger diameters. They estimated that to reach the goal of less than 1 kg hydrogen delivery it is required to reduce the pipeline cost to around \$0.5 M/mile. Among the initial costs, materials and labour for the network construction are the main expenses, around 70% of the total pipeline cost, which is almost constant over time. The remaining 30% of costs are associated with regulatory fees, right-of-way purchase, engineering, surveying, equipment, management, and administration [7]. The costs related to the remaining 30% cost cannot be influenced by research and development (R&D), thus, the main focus could be on material and labour to achieve the mentioned cost. Pawel et al. [7] stated that for cost reduction of around 50% or more per mile, it is required to reduce the costs of labour and materials by approximately 80% or more, which is almost unlikely to be achieved. However, polymer pipelines might be a choice for the material with improvement in their characteristics such as preventing hydrogen loss due to the high porosity of polymers. An alternative for steel pipelines is fibre-reinforced polymer (FRP) pipelines which can be installed in extended lengths, depending

Table 4.1.1 Cost analysis of pipeline delivery [11], total cost includes production, terminal, delivery, and retail.

Hydrogen production rate (kg/day)	Number of HRS	Pipeline pressure (bar)	Total cost ($/kg)	Delivery cost ($/kg)	Delivery cost $/(kg.km)
400	8	40	21.6	7.79	0.13
400	8	1000	17.1	7.83	0.13
950	16	40	10.7	1.75	0.03
950	16	1000	7.1	1.75	0.03
1500	24	40	8.5	0.69	0.01
1500	24	1000	5.3	0.69	0.01

on the diameter, which reduces the number of joints/welds per unit length of the pipeline, which reduces costs related to handling, fit-up, and joining by labourer and inspection costs, such as leak-checking and non-destructive tests. In addition, FRP pipelines require less corrosion protection processes such as painting for buried steel and cathodic protection [7]. These advantages in cost reduction suggest the development of FRP pipeline infrastructure as an area for research and development.

Based on the study by Fraunhofer Institute (Germany) [19] the increase in the blending percentage of hydrogen to the natural gas leads to the end-user price increase, especially for industrial customers, they estimated by blending 5%, 10%, and 20% of hydrogen to natural gas the price increase for the end-users in European Union could be 1.3%, 9.9%, and 23.8%, respectively. This increase at higher percentages is more significant.

To reduce the cost of hydrogen delivery, Penev et al. [11] proposed a system which is called the HyLine system, this system would produce hydrogen at urban industrial or commercial power plants. In this system, hydrogen would be compressed to 15,000 psi and stored at the production sites, then delivered through a high-pressure pipeline to HRS. They considered 6 scenarios for pipeline delivery. The total length of pipeline was approximately 61.15 km in all scenarios. The difference between scenarios is related to the pipeline pressure, hydrogen production rate, and number of HRS. The cost assessment result is provided in Table 4.1.1.

Transmission costs are associated with the transportation distance and the hydrogen demand quantity while distribution stage costs are related to demand volume and the size of the targeted district. However, these costs are highly dependent upon the geography of the delivery route and infrastructure complexity [2]. For remote districts, small cities, and rural communities where the demand for hydrogen is low and at the early market of hydrogen fuel, it is not logical economically to deliver hydrogen via pipeline to the end-users. For large power plants, around 1000 tons/day, the pipeline pathway is the most cost-effective choice considering the high demand which could cost around 2.73 $/kg of hydrogen delivery [8,6]. This cost is based on the pipeline pressure of 1000 psi, transmission length of 100 km, 80 HRS, 2 trunk pipelines each 112.6 km for distribution, and average demand of 80,000 kg/day. Assuming the delivery to the furthest consumer, the total length of the pipeline would be 212.6 km, thus, the hydrogen delivery cost would be approximately 0.01 $/(kg.km).

4.1.3.2 Tube trailer

The main parameters to determine the hydrogen delivery cost utilizing the tube-trailer pathway are the costs of truck cabs, tubes, and tube trailers, driver cost, fuel consumption for delivery and its price, and operations and maintenance [10]. In other words, the cost of hydrogen delivery by tube-trailer pathway can be formulated as a function of transported capacity and distance, which includes other costs such as compression, storage, and road transportation, this term is called the Levelized cost of transporting hydrogen (LCOTH). LCOTH should be minimized by optimization of the capacity of each compressed gas truck. LCOTH can be reduced by increasing the transport capacity and increased by increasing the truck trip distance. LCOTH has a range of 3.02 $/kg to 0.5 $/kg [31] which depends on the amount of delivery and distance. For the daily demand of 50 ton per day, the cost for 20, 75, 250, 350, and 400 km distance delivery were estimated around 0.006, 0.004, 0.003, 0.004, and 0.004 $/(kg.km), respectively. According to the analysis performed by Weisberg et al. [10], in 2019, based on information extracted from the U.S. Department of Energy the cost of hydrogen delivery by tube-trailer pathway could be less than 1 $/kg H_2. Faye et al. [8] revealed that the hydrogen delivery cost by tube-trailer in a small-scale power plant would be around $2.86 cost per kg of hydrogen, considering 100 km round-trip distance, the cost would be 0.03 $/(kg.km). Moreno-Blanco et al. [32] compared the transport cost of hydrogen gas compressed to 875 bar (high pressure) and at the temperature of 200 K with a 350-bar trailer. For this purpose, the high-pressure hydrogen was considered to be delivered in a thermally insulated trailer and dispensing was performed directly from the trailer, it could result in the elimination of the station compressor, cascade, and refrigerator, which lead to a reduction in complexity and cost of the system. They concluded that the total delivery cost at 875 bars can be 0.58 $/kg H_2 less than 350 bar, 2.96 and 2.38 $/kg H_2

Table 4.1.2 Tube trailer cost based on Lahnaoui et al. [33] study, 2021.

	CGH					LOHC	LH
Storage pressure (bar)	180	250	350	500	540	1	1
Net truck capacity (kg)	350	668	885	1100	1230	1500	3600
Tube trailer cost (k$)	455	621	815	1250	1416	67	2049

Table 4.1.3 Cost analysis of liquid trucks delivery [11], total cost includes production, liquefaction, terminal, delivery, and retail. Total distance is considered 2*61.15 km for roundtrip.

Hydrogen production rate (kg/day)	Number of HRS	Total cost ($/kg)	Delivery cost ($/kg)	Delivery cost $/(kg.km)
400	8	16.4	0.22	0.002
950	16	13.3	0.22	0.002
1500	24	12.7	0.22	0.002

for 350 bar and 875 bar, respectively. Considering 100 km distance from production site and city gate and assuming 50 km for distribution, the total roundtrip would be 300 km which results in around 0.008–0.01 $/(kg.km). Lahnaoui et al. [33] reported tube trailer costs for different forms of hydrogen, compressed gas (CGH), LOHC, and liquid hydrogen (LH) at different pressures as it is shown in Table 4.1.2, it should be noted that the reported values for tube trailer cost are related to the trailer cost, not the hydrogen delivery cost.

4.1.3.3 Cryogenic liquid trucks

Even though the liquid hydrogen tank trailers cost more than tube trailers due to liquefaction process cost, the total hydrogen delivery cost could be lower considering that the cost of delivery, including truck fuel, and driver labour cost, per unit of hydrogen can be lower. The high cost of liquefaction is associated with two factors, high energy consumption for liquefaction, and the high capital cost of equipment for liquefaction [7]. Hydrogen delivery via this pathway and dispensing at the fuel station could cost approximately 1.40–2.42 $/kg of hydrogen [8]. Thermal insulation is an effective issue to reduce the cost of cryogenic tankers pathway. It is possible to reduce the costs by reducing the losses due to evaporation, and boil–off of liquid hydrogen. Penev et al. [11] considered 3 scenarios for liquid hydrogen delivery at three different production rate and the number of HRS. The results are provided in Table 4.1.3.

4.1.3.4 Comparison of pathways

One of the most important parameters in hydrogen delivery is the amount of hydrogen which can be delivered via each pathway. Fig. 4.1.3 shows the mass of hydrogen which can be delivered in 1 m^3 of materials. Various types of delivery are considered such as 100% hydrogen in the pipeline, 5% and 15% blending with natural gas, tube trailer, cryogenic liquid tanker, and hydrogen carrier materials. This figure is plotted based on the information provided in the appendix. It can be seen that magnesium hydride and methanol can deliver the highest amount of hydrogen in the same volume of delivery compared to other modes of delivery, 110.20 and 99 kg, respectively, which shows their potential to reduce the cost of delivery.

Several parameters are effective in the most affordable choice of hydrogen delivery pathway among the main three mentioned pathways, such as the geography of the delivery route and market characteristics including city population and size, population density, and end-user type (e.g., size and number of HRS, and fuel cell vehicles market penetration). Weisberg et al. [10] stated that the pipeline pathway is suitable for dense areas with high hydrogen demand, the compressed gas truck pathway is suitable for small stations with very low demand, and liquid delivery is suitable for long distances and moderate hydrogen demand.

Tzimas et al. [4], in 2007, by an economic assessment expressed that the difference between the cost of hydrogen delivery to a vehicle as the end-user when hydrogen is produced in a large-scale steam reforming plant, then liquefied and delivered by truck to a HRS which is located at a distance of 100 km, or it is produced in small scale on-site infrastructure by means of steam reforming technology and compressed to 700 bar is not significant [4].

Demir et al. [6] compared economical aspect of three scenarios for hydrogen delivery, (1) transmission via pipeline to the city gate, distribution by liquid tanker, (2) using geological storage, transmission, and distribution by hydrogen tube trailers, and (3) using geological storage, transmission, and distribution via pipeline. They indicated that the lowest Levelized cost of delivery is obtained by the third scenario, 2.73 \$/kg H$_2$, and the highest cost was related to the first scenario, 8.02 \$/kg H$_2$. The Levelized cost is the lifetime costs divided by energy production.

The storage cost of liquid hydrogen is significantly less than high-pressure hydrogen. It is possible to increase the amount of stored hydrogen by liquefaction, therefore the reliability of the system to supply the variation of the demand can be achieved with a relatively lower cost in comparison to compressed hydrogen. However, it should be noted that the liquefaction

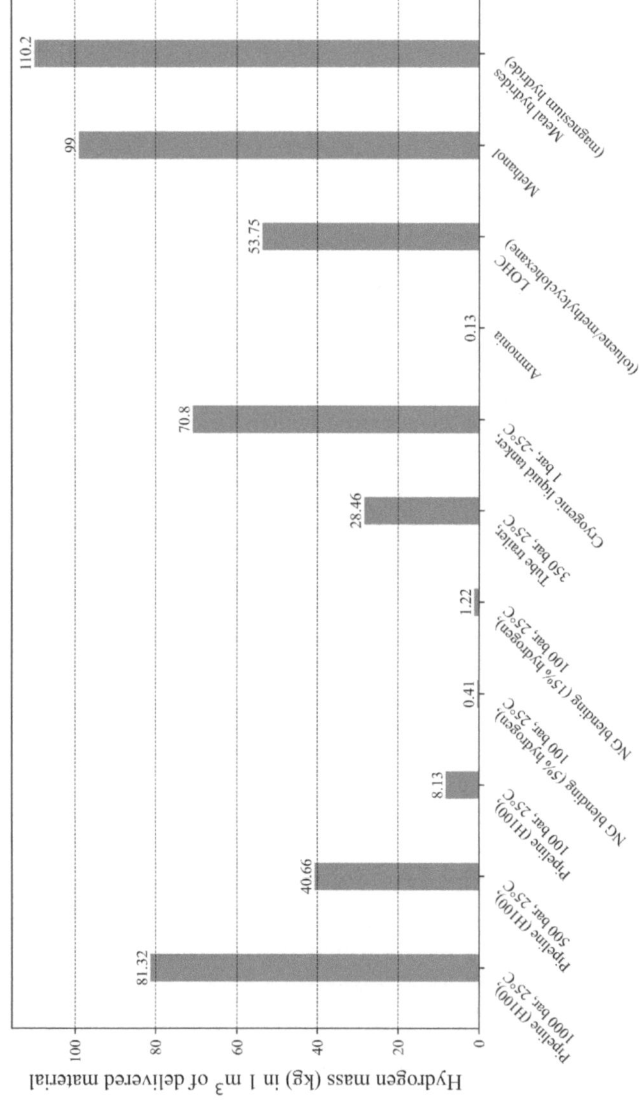

Figure 4.1.3 Mass of delivered hydrogen in 1 m³ of delivered material. H100 refers to 100% hydrogen.

Table 4.1.4 Hydrogen delivery cost comparison via different pathways in case "A," numbers are extracted from diagrams in the European Union report [26]. All deliveries are considered via shipping except pipeline.

Case A, total cost of delivery and preparation processes for delivery and reverse processes	H_2 pipeline		Compressed H_2		Liquefied H_2		LOHC		Ammonia	
	Hi	Lo	Hi	Lo	Hi	Lo	Hi	Lo	Hi	Lo
	0.85	0.73	1.04	0.86	1.27	0.89	2.25	1.12	3.11	1.53
Contribution of transport and storage	90%	95%	67%	81%	30%	36%	12%	24%	7%	14%

Table 4.1.5 Hydrogen delivery cost comparison via different pathways in case "B," numbers are extracted from diagrams in the European Union report [26].

Case B, total cost of delivery and preparation processes for delivery and reverse processes	Compressed H_2		Liquefied H_2		LOHC		Ammonia	
	Hi	Lo	Hi	Lo	Hi	Lo	Hi	Lo
	5.21	4.42	3.73	2.67	5.17	3.94	6.83	4.84
Contribution of transport and storage	72%	83%	58%	58%	14%	19%	37%	51%

cost is a crucial factor to define the lowest cost pathway and liquid hydrogen is preferred when it is required to store a large amount of hydrogen [10].

In the report published by the European Union [26], in 2021, two hydrogen production rates and delivery distance scenarios, named case "A" and case "B," were economically assessed considering various pathways for each scenario which the results are provided in Tables 4.1.4 and 4.1.5. Values reported in this table are hydrogen delivery costs in USD per kg of hydrogen, including transport, storage, compression, liquefaction, process of bonding in a carrier, and the reversing procedure to have gaseous hydrogen at the desired pressure and purity. In case "A," which is an industrial use of hydrogen, the production rate was considered 1 Mt hydrogen per year and the delivery distance of 2500 km. In case "B," the production rate was considered 100 kt hydrogen per year and the delivery distance of 3000 km, including 2500 km shipping, and 500 km train and truck. In addition, two electricity cost scenarios were considered in this report, low price (Lo) and high price (Hi). In Lo, the production cost at the production site and consumption site were considered 11.83 \$/MWh and 59.15 \$/MWh, respectively. In Hi, the production cost at the production site and consumption site were considered 59.15 \$/MWh and 153.79 \$/MWh, respectively. Waste heat was considered in case "A" at a cost of 23.66 \$/MWh.

It can be seen that in the case "A" the maximum cost is associated with ammonia carrier (Hi), while it has the minimum contribution of transport

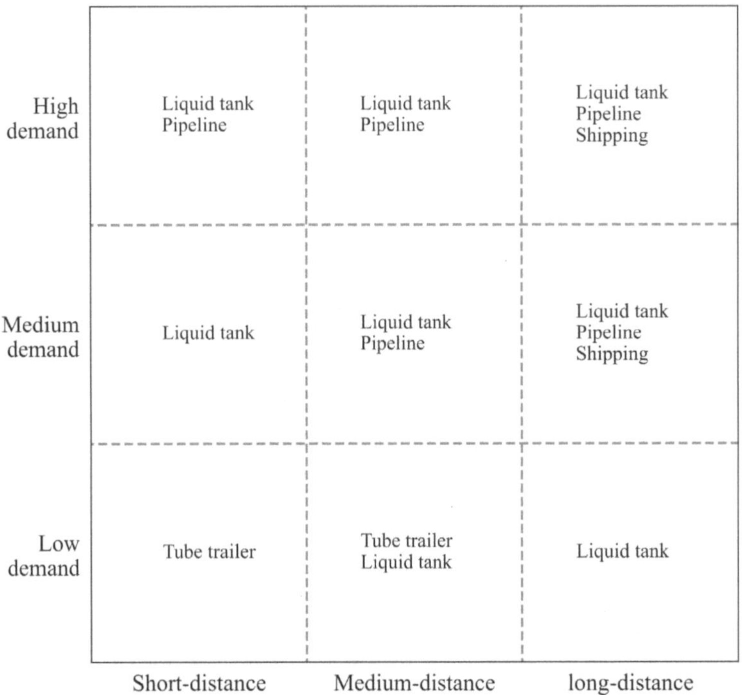

Figure 4.1.4 Pathway choice regarding distance and demand.

and storage. The maximum contribution of transport and storage in the case "A" is related to the H_2 pipeline (Lo) while it has the lowest total cost. In the case "B," the maximum cost is related to ammonia carrier (Hi), the same as the case "A," and the minimum cost is associated with the liquefied H_2 (Lo). The minimum and the maximum contribution of transport and storage in case "B" are related to LOHC (Hi) and LOHC (Lo), respectively. These numbers show the importance of processes required for hydrogen preparation for delivery and the reverse processes to have gaseous hydrogen at the desired pressure and purity.

Fig. 4.1.4 depicts the suitable choice of pathway regarding demand and distance. This figure is summarized based on the abovementioned studies. To choose the best mode of hydrogen delivery for each project a techno-economic assessment is necessary.

4.1.3.5 HRS

The size of the HRS can affect the total cost of hydrogen delivery, the greater the station size the lower the station cost per unit of hydrogen, therefore, the less the total cost of hydrogen delivery. Liquid storage cost is less than gas storage, in addition, the cost of compressors for gas compression is higher than the cost of liquid hydrogen pumping [10]. The components which are usually required in HRS are: hydrogen production unit, hydrogen compressor, hydrogen storage, purification unit, hydrogen gas booster, cooling unit, safety equipment, and dispenser. However, variation in components is associated with the delivery and production method. The cost of HRS is related to the components which are used in the station. The most contributor to the cost of HRS is the compressor [34]. The cost of HRS in 2018 ranged between €1.2 and €2 million, and the expected cost for 2023 is €0.6 to €1.6 million [34].

4.1.4 Safety in hydrogen transportation and distribution

Considering the higher diffusion of hydrogen in comparison to natural gas, around five times, due to the small size of the hydrogen molecule, there would be a higher risk of leakage of hydrogen from valves, seals, and gaskets, which could lead to safety hazards. The annual loss of hydrogen as a result of leakage could be around 0.0005–0.001% of the total volume of delivered hydrogen. Among the pipelines' materials, cast iron and fibrous cement pipelines have a higher leakage risk. Currently, polyethylene pipelines are mostly used for the natural gas network [13], however, there is a problem of high porosity for this material. Gondal et al. [13] stated that although leakage of hydrogen is much higher than natural gas, due to its high energetic content the leakage is negligibly small. However, considering the flammability range and small amount of energy that is required for ignition the leakage could lead to problems [8]. Deformation mechanisms and cohesive forces of steel pipelines can change significantly due to the presence of hydrogen, which under specific loading reduces ductility and toughness [7]. That might cause the pipeline gets damaged and uncontrollable release of hydrogen happens which is potentially hazardous for humans, other species, and surroundings properties [29]. Mouli-Castillo et al. [35] assessed the risk of hydrogen release upstream and downstream of the domestic gas meter in a proposed H100 network (pipelines include 100% hydrogen, and there is no blending with natural gas). They found that the risk of release for the proposed H100 network is less than the existing natural gas network by a factor of 88%. Lee et al. [30], compared the safety of hydrogen delivery in gaseous and liquid form in Seoul, Korea, based on a quantitative risk analysis. They found that liquid form has a better performance in terms of safety, however, risk alters due to variation in the capacity of the station and the market penetration, especially when these factors are small.

Several hazardous conditions could be considered during the lifetime of hydrogen pipelines [20]:

- use of unsuitable materials and equipment under all operating conditions,
- hydrogen embrittlement,
- external corrosion due to inappropriate cathodic protection,
- damage by third parties,
- pipeline improper operation and maintenance of the pipeline,
- leaks at valve packing, gaskets, and other components,
- pipeline over pressurization,
- unsuitable inerting procedure,
- the radiation of a vent fire or a flare,
- uncommon applied loads due to natural happenings such as landslides, floods, earthquakes, or non-natural, for example, the crossing of roads, railways, and
- other structures impact, such as high-power electrical grids, electrical railways.

To mitigate the risk of these hazardous conditions it is needed to consider several aspects of design, construction, and maintenance procedures [20]:

- pipeline thickness increment,
- pipelines route alteration based on safety issues,
- reducing pipeline operating pressure,
- using proper pipeline materials,
- deepen the pipeline,
- physical protections, such as concrete coating,
- control of third-party interference,
- valves and other components isolation,
- regular maintenance,
- performing non-destructive tests on welds,
- pipeline inerting,
- pipeline marking, and
- monitoring and performing mass balance for leak detection.

To increase the public acceptance of hydrogen it is required to increase safety. To achieve this target, standards and codes are needed for the construction and performance of the hydrogen delivery infrastructure. It is possible to use odorants that help the identification of gas leakage and reduce the risk of leakage. In a case study for gas escape detection of hydrogen and natural gas performed by Mouli-Castillo et al. [36] it is found that the untrained test participants can recognize the escaping gas which is odorized with New Blend (78% tert-Butylthiol, 22% Dimethyl Sulfide), and Standby Odorant 2 (34% Odorant NB, 64% Hexane) when the concentration is 1% of the air.

4.1.5 Policies

If fossil fuels are used for hydrogen delivery in a large-scale hydrogen delivery chain the amount of greenhouse gas emission would be significant, thus, certification and labeling of renewable hydrogen should also be considered for the full delivery chain rather than only hydrogen production [26]. Cost-effective, environmentally friendly, and easy operating hydrogen delivery infrastructures have importance in case of the investment risks of stakeholders [6], thus, proper policies should be applied for stakeholders' confidence. Demand is an integral part of defining public policies since it affects the economy of technology. The transition from fossil fuels to hydrogen fuel is not straightforward and public support is required to achieve a mature technology, especially in R&D. Hydrogen technology is relatively new and rapid progress is expected in the early stages, thus, hydrogen end-users such as hydrogen car users might wait for cheaper and better technologies, for example, better, and more affordable hydrogen cars and more available HRS. The absence of infrastructure is an important problem in hydrogen delivery. Investors in hydrogen infrastructures could be infrastructure promoters such as oil companies whose currently main investment is in the conventional energy network, and they might want to make benefit from them as much as possible. Investment is unlikely by private investors in an infrastructure that has high initial costs and the demand amount is unclear [1]. Carbon taxes and trading schemes which are related to controlling carbon emissions could influence the choice of the delivery pathway [10].

Blending hydrogen into the natural gas network will increase costs for end users, by approximately 16% for households and 43% for industrial users at the blending level of 20 vol-%, which shows the pricing policies' impact [19].

4.1.6 Hydrogen transportation model

To design a hydrogen transportation model different kinds of information are needed to fully address all contributors to the delivery cost. This information includes temporal and spatial demand variation, demand growth over time considering social, economic, technological, and policy alteration, city characteristics (such as city radius, number of HRS, and their distribution), population, and population density. The model should satisfy the consumers' needs. Assuming the hydrogen cars as end-users the following factors should be taken into account to make hydrogen consumption convenient for consumers: (1) minimizing the average distance that consumers must travel to reach a HRS, (2) the distance which the trucks should travel from the city gate to the HRS or the length of the pipeline if pipeline pathway is used as the delivery mode for distribution, and (3) the distribution of demand considering the distribution of HRS in the city [10]. However, other parameters such as hydrogen price and capacity of the hydrogen production site should be taken into account, since the optimization of the total cost of hydrogen production and delivery is the final goal, not only minimizing the delivery cost.

Hydrogen Delivery Scenario Analysis Model (HDSAM) is a model which was designed to analyse different scenarios of hydrogen delivery. This model was developed under the hydrogen analysis (H_2A) project, by the U.S. Department of Energy. The hydrogen delivery in this model is considered from the central production to vehicles. This model includes each delivery system component with related performance characteristics and cost. In addition to the mentioned parameters, this model included a scenario model which made it possible to assess the system performance under various market supply and demand conditions. Inputs to define the scenarios included pathway or pathways combination, number of HRS and their distribution, number of hydrogen vehicles, urban population, and expected revenue [7,37].

In another study, Gim et al. [2] built a cost-effective central hydrogen production using a transportation model in Korea. They used a window-based software developed by Scharge, which is called LINGO. By utilizing LINGO they determined the optimal hydrogen delivery volumes for supply and demand sites. They formulated the hydrogen transportation problem in the form of a general transportation model, as provided in Eq. (4.1.1), in which hydrogen plants and HRS are assumed as the hydrogen sources are hydrogen destinations, respectively. To solve the problem, z value (which is the total transportation cost) should be minimized.

$$z = \sum_{i=1}^{m} \sum_{j=1}^{n} c_{ij} x_{ij} \tag{4.1.1}$$

$$\sum_{j=1}^{n} x_{ij} = a_i, \quad i = 1, 2, \ldots, m \tag{4.1.2}$$

$$\sum_{i=1}^{m} x_{ij} = b_j, \quad j = 1, 2, \ldots, n \tag{4.1.3}$$

$$x_{ij} \geq 0, \quad \text{all } i \text{ and } j \tag{4.1.4}$$

where

x_{ij} = delivery volume from hydrogen plant i to end-user j,
c_{ij} = unit delivery cost from hydrogen plant i to end-user j,
m = number of hydrogen plants,
n = number of location of end-users,
a_i = hydrogen supply from plant i, and
b_j = hydrogen demand at end-user j.

Eq. (4.1.2) is used to define the hydrogen supply from each production site and the result of Eq. (4.1.3) should be equal to hydrogen demand at the destination. They assumed that the total supply is equal to the total demand, which means the model is balanced.

As it can be seen demand prediction is an important challenge in Gim's model. The diffusion model is a method to predict the demand for new products which are not mature and have not had enough market penetration. The diffusion model that is used by Gim et al. [2] is Lawrence–Lawton's diffusion model for the estimation of the number of fuel cell vehicles and the amount of hydrogen. The general form of Lawrence–Lawton's diffusion model is provided in Eq. (4.1.5):

$$S_t = \frac{1 + P}{1 + \left(\frac{1}{P}\right)e^{-RV}} - P \qquad (4.1.5)$$

where

St = accumulated rate time,
P = initial market parameter,
R = diffusion rate parameter, and
V = current year + degree of maturity base year.

The initial market parameter is associated with the market share of the product in the base year, which is very low for the new products and is considered around 0.1–1.0%. The diffusion rate parameter depends on the product characteristics and is considered the measurement unit of market diffusion. Maturity is related to the duration in which a product exists in the market. The higher the duration in the market the higher value of maturity. Products with subsidies and incentives have lower maturity value or are considered new products. Based on Eq. (4.1.5) and the survey, Gim et al. [2] estimated the demand in Korea as 12,000 tons, 330,000 tons, and 3,545,000 tons in 2020, 2030, and 2040, respectively. Based on the information about the location of destinations and hydrogen production sites, and using the diffusion model for the demand forecast, Gim et al. [2] estimated the delivery volumes from hydrogen plant i to end-user j in 2040 in various districts in Korea to minimize the cost of delivery.

Hydrogen demand for future scenarios should include market penetration, population growth, and hydrogen vehicle and appliance technologies development. In addition, increasing the number of hydrogen vehicles in public transport, such as hydrogen–powered buses, increase the number of people who use hydrogen for daily transport. However, these considerations could be only projections, because unpredicted circumstances might occur such as COVID-19 that can reduce the use of public transport or decrease the travel distance of private cars due to lockdown.

Gim et al. [2] model can be modified by considering other parameters such as hydrogen price, and carbon tax. Since more than one hydrogen production site is considered and the price of hydrogen varies depending on the method and technology to produce hydrogen, authority pricing policies of each district, the source of energy for hydrogen production, gas compression, and liquefaction, therefore, it is important to consider the price of hydrogen in the Gim's formula. For example, hydrogen production via PEM electrolyser is more expensive in comparison to alkaline electrolyser [33]. Hydrogen production cost also includes the consumed energy for hydrogen production and any other steps such as compression and liquefaction, thus, the energy price difference is already considered implicitly by the hydrogen production cost. Hydrogen generation cost information can be collected based on surveys, price inquiries, and empirical equations. For instance, Lahnaoui et al. [33] used the following equation which is based on the National Renewable Energy Laboratory (NREL) model for hydrogen production to define the hydrogen production cost.

$$
LCOPH = \begin{cases} \dfrac{55P_e + 1.6}{100}\left(174 - 13.11\ln\left(Tp_d\right)\right) & \text{for } Tp_d \in [1, 10] \text{ ton per day} \\[2mm] \dfrac{55P_e + 1.6}{100}\left(67 - 1.74\ln\left(Tp_d\right)\right) & \text{for } Tp_d \in [10, 200] \text{ ton per day} \end{cases}
$$

where

$LCOPH =$ levelized cost of hydrogen production,

$P_e =$ electricity price, and

$Tp_d =$ plant capacity.

For future scenarios, hydrogen cost can be forecasted by performing a sensitivity analysis of the parameters that define the hydrogen price and their probable future development. However, unpredictable situations could occur which make a considerable difference between predicted and actual prices such as war and energy crises.

In addition to hydrogen price alteration in each hydrogen production site, it is likely that the amount of carbon dioxide emission would be different due to the different sources of energy and technology to produce hydrogen. Furthermore, carbon dioxide emission of different modes of hydrogen delivery should be considered in the total cost. Since the general equation is assumed to be used within various districts the carbon tax might be different because this term is related to the authority's policies of each district. It is assumed that the carbon tax is calculated based on the location of the hydrogen production site. There might be a hydrogen production plant with huge emissions within a short distance from the HRS which the total cost of hydrogen production, carbon tax, and delivery becomes higher in comparison to a hydrogen production plant located at a longer distance using the energy source of renewable energies.

The proposed model to cover the hydrogen production cost and the carbon tax is shown in Eq. (4.1.6), by minimizing T the optimized delivery cost can be achieved. After optimization, it is possible to remove the hydrogen production cost and the carbon tax related to hydrogen production to define the total cost of hydrogen delivery in the whole network. It should be noted that in this equation instead of volume of hydrogen the mass of hydrogen is used, thus, this equation can be used for various pathways with different hydrogen forms, for example, gaseous hydrogen, liquid hydrogen, or carriers in which the hydrogen content should be considered in this equation.

$$T = \sum_{i=1}^{m} \sum_{j=1}^{n} (c_{dij} + c_{ctpij} + c_{ctdij} + c_{Hpij}) M_{ij} \tag{4.1.6}$$

$$\sum_{j=1}^{n} M_{ij} = A_i, \quad i = 1, 2, \ldots, m \tag{4.1.7}$$

$$\sum_{i=1}^{m} M_{ij} = B_j, \quad j = 1, 2, \ldots, n \tag{4.1.8}$$

$$M_{ij} \geq 0, \quad \text{all } i \text{ and } j \tag{4.1.9}$$

where

$T =$ total cost of hydrogen production, carbon tax, and delivery,

$M_{ij} =$ delivery mass from hydrogen plant i to end–user j,

$c_{dij} =$ unit delivery cost from hydrogen plant i to end–user j,

$c_{ctpij} =$ unit carbon tax for hydrogen produced at plant i which is delivered to end–user j,

$c_{ctdij} =$ unit carbon tax for hydrogen delivery from plant i to end–user j,

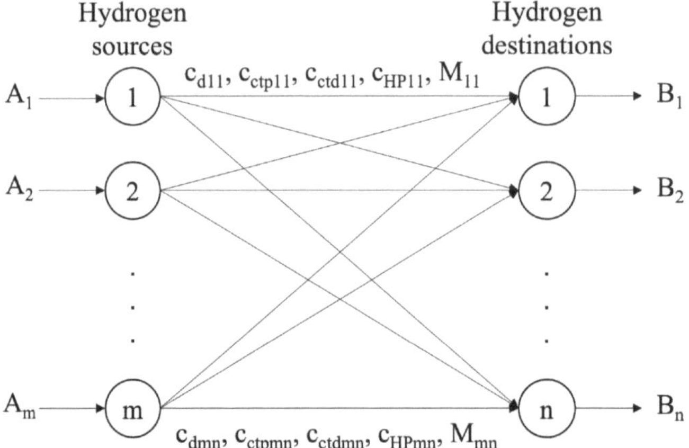

Figure 4.1.5 network representation of the total hydrogen production, delivery, and carbon tax model.

c_{Hpij} = unit hydrogen production cost at hydrogen plant i which is delivered to end-user j,

m = number of hydrogen plants,

n = number of location of end-users,

A_i = mass of hydrogen supply from plant i, and

B_j = mass of hydrogen demand at end-user j.

After optimizing the total hydrogen production, delivery, and carbon tax cost, the hydrogen amounts which are supplied from each hydrogen generation plant to each end user are defined and can be used in Eq. (4.1.10) to calculate the whole delivery cost.

$$D = \sum_{i=1}^{m} \sum_{j=1}^{n} (c_{dij} + c_{ctdij}) M_{ij} \qquad (4.1.10)$$

where

D = total cost of hydrogen delivery and carbon tax associated with the delivery.

The network of the hydrogen production sources and destinations to define the total cost of hydrogen production, hydrogen delivery, and carbon tax can be depicted in Fig. 4.1.5.

The capacity of storage should be considered as the limiting value for both hydrogen plants and HRS (hydrogen destination in this model). However, for the system development, it is possible to use this model to define the increase in storage size by assuming a higher limit for the storage and rerun the model. In addition, by using this model it is possible to decide about the location of the new hydrogen generation plant and HRS. Furthermore, this model can be applied to the NG pipeline considering various possible percentages of hydrogen. This percentage could be related to the age of different parts of the network, and other parameters which are mentioned in the previous sections. Hydrogen should be separated from the mixture of NG and hydrogen to use in HV, thus, a scenario for combining pathways could be combining hydrogen with NG, delivering by a transmission line, then separating hydrogen for distribution, in this way appliances in the domestic sector do not need adaptability to hydrogen, but other infrastructures for distribution of hydrogen and hydrogen separation from NG are needed. It is possible to perform sensitivity analysis on Eq. (4.1.6) regarding various parameters such as storage size, the capacity of tankers, pipeline capacity, and other parameters to find the optimum values.

A probable scenario for hydrogen delivery could be using a hydrogen hub. Including transmission to hydrogen hub and then delivery by other pathways as distribution part considering the districts demand to reduce the cost of delivery. As a result, the cost of maintenance of the storage step could be less in comparison to maintenance in several hydrogen generation plant storage. A hydrogen hub can be used in the case of imported hydrogen or when the production is higher than the storage capacity of the hydrogen production plant. It is possible to evaluate the impact of hydrogen hub construction on the total delivery cost by using the model proposed in this section.

4.1.7 Conclusion

Hydrogen delivery is one of the main steps in hydrogen supply. There are three main infrastructures for hydrogen delivery, tube trailers, pipelines, and cryogenic tankers. Among these pathways, pipelines are used to supply high-demand districts, while tube trailers and medium cryogenic tankers are mainly used for smaller customers. Several challenges are mentioned in this chapter that should be addressed to reduce the cost and environmental impact of hydrogen delivery. These challenges are related to cost, required hydrogen pressure, hydrogen losses, storage, the reaction between hydrogen and delivery infrastructure materials, geography of the delivery route, maintenance, and monitoring. The pipeline delivery mode has the lowest GHG emission among the three main pathways, while the lower emission between the tube trailer and the cryogenic tank is defined based on the distance of delivery. The cost of delivery is associated with various parameters such as distance and density of users in a district. Thus, for each project, it is required to perform a techno-economic analysis to find the most suitable pathway. The pipeline delivery mode is likely to be the safest and most economical way of hydrogen distribution considering the lower likelihood of accidents and exposure to humans. A transport model is proposed in this chapter to minimize the cost of the delivery, which includes the cost of hydrogen production, carbon tax for hydrogen production, and carbon tax for hydrogen delivery.

Appendix

Mass of delivered hydrogen in 1 m^3 of delivered materials.

Density of gaseous hydrogen is calculated based on the ideal gas equation. Cryogenic liquid tanker information is extracted from [38].

	Pressure (bar)	Temperature (°C)	Hydrogen content-volume	Hydrogen content-mass	Density (kg/m^3)	Mass of delivered hydrogen (kg) in 1 m^3 of the whole
Pipeline (H100, 100% hydrogen)-1000 bar	1000	25	100%	100.00%	81.32	81.32
Pipeline (H100, 100% hydrogen)-500 bar	500	25	100%	100.00%	40.66	40.66
Pipeline (H100, 100% hydrogen)-100 bar	100	25	100%	100.00%	8.13	8.13
Pipeline-NG blending (5% hydrogen)	100	25	5%	0.51%	79.89	0.41
Pipeline-NG blending (15% hydrogen)	100	25	15%	1.69%	72.34	1.22
Tube trailer	350	25	100%	100.00%	28.46	28.46
Cryogenic liquid tanker	1	-252	100%	100.00%	70.8	70.80
Ammonia	ambient	ambient	–	17.65%	0.73	0.13
LOHC (toluene/ methylcyclohexane)	ambient	ambient	–	6.20%	867	53.75
Methanol	ambient	ambient	–	12.50%	792	99.00
Metal hydrides (magnesium hydride)	ambient	ambient	–	7.60%	1450	110.20

Hydrogen mass content in NG blending 5% volumetric.

	Hydrogen	Natural gas
Percentage (volumetric)	5	95
Density (kg/m^3)	8.13	83.672
Volume in 1 m^3 mix (m^3)	0.05	0.95
Mass in 1 m^3 mix (kg)	0.40	79.49
Mixed density (kg/m^3)	79.89	
Mass content	0.51%	

Hydrogen mass content in NG blending 15% volumetric.

	Hydrogen	Natural gas
Percentage (volumetric)	15	85
Density (kg/m^3)	8.13	83.672
Volume in 1 m^3 mix (m^3)	0.15	0.85
Mass in 1 m^3 mix (kg)	1.22	71.12
Mixed density (kg/m^3)	72.34	
Mass content	1.69%	

References

[1] N. Bento, Investing in the hydrogen delivery infrastructure: methodology for a public policy, Energy Stud. Rev. 17 (2) (2010).

[2] B. Gim, K. Boo, S. Cho, A transportation model approach for constructing the cost effective central hydrogen supply system in Korea, Int. J. Hydrogen Energy 37 (2) (2012) 1162–1172.

[3] R. Moradi, A.K. Groth, Hydrogen storage and delivery: review of the state of the art technologies and risk and reliability analysis, Int. J. Hydrogen Energy 44 (23) (2019) 12254–12269.

[4] E. Tzimas, P. Castello, A.S. Peteves, The evolution of size and cost of a hydrogen delivery infrastructure in Europe in the medium and long term, Int. J. Hydrogen Energy 32 (10-11) (2007) 1369–1380.

[5] A. Elgowainy, M. Mintz and M. Gardiner, Distribution networking, in: *Handbook of Hydrogen Energy*, 2014, 935–956.

[6] M. Demir, I. Dincer, Cost assessment and evaluation of various hydrogen delivery scenarios, Int. J. Hydrogen Energy 43 (22) (2018) 10420–10430.

[7] S.P.M. Gardiner, Hydrogen delivery: infrastructure, challenges, and materials needs …, in: In Effects of Hydrogen on Materials: Proceedings of the 2008 International Hydrogen Conference, 1, Jackson Lake Lodge, Grand Teton National Park, Wyoming, USA, ASM International, 2008, p. 316.

[8] O. Faye, J. Szpunar, U. Eduok, A critical review on the current technologies for the generation, storage, and transportation of hydrogen, Int. J. Hydrogen Energy 47 (29) (2022) 13771–13802.

[9] C. Yang, J. Ogden, Determining the lowest-cost hydrogen delivery mode, Int. J. Hydrogen Energy 32 (2) (2007) 268–286.

[10] A.H. Weisberg, S.M. Aceves, F. Espinosa-Loza, E. Ledesma-Orozco, A.B. Myers., Delivery of cold hydrogen in glass fiber composite pressure vessels, Int. J. Hydrogen Energy 34 (24) (2009) 9773–9780.

[11] M. Penev, J. Zuboy, C. Hunter, Economic analysis of a high-pressure urban pipeline concept (HyLine) for delivering hydrogen to retail fueling stations, Transport. Res. Part D: Transport Environ. 77 (2019) 92–105.

[12] E. H. Backbone, Estimated investment & cost, European hydrogen backbone, https://ehb.eu/page/estimated-investment-cost. (Accessed 19 September 2022).

[13] I. Gondal, M. Sahir, Prospects of natural gas pipeline infrastructure in hydrogen transportation, Int. J. Energy Res. 36 (15) (2012) 1338–1345.

[14] Office of Energy Efficiency & Renewable Energy, Hydrogen pipelines, United States hydrogen and fuel cell technologies office, https://www.energy.gov/eere/fuelcells/hydrogen-pipelines. (Accessed 29 August 2022).

[15] Pacific Northwest National Laboratory, Hydrogen tools, Pacific Northwest National Laboratory, 2016. https://h2tools.org/hyarc/hydrogen-data/hydrogen-pipelines. (Accessed 29 August 2022).

[16] D.B. Smith, B.J. Frame, L.M. Anovitz, A.C. Makselon, Feasibility of using glass-fiber-reinforced polymer pipelines for hydrogen delivery, in: Pressure Vessels and Piping Conference, 50435, 2016, V06BT06A036.

[17] E. Ogbe, U. Mukherjee, M. Fowler, A. Almansoori, A. Elkamel., Integrated design and operation optimization of hydrogen commingled with natural gas in pipeline networks, Industr. Eng. Chem. Res. 59 (4) (2019) 1584–1595.

[18] M. Pellegrini, A. Guzzini, C. Saccani, A preliminary assessment of the potential of low percentage green hydrogen blending in the Italian natural gas network, Energies 13 (2020) 5570.

[19] J. Bard, N. Gerhardt, P. Selzam, M. Beil, D. M. Wiemer and M. Buddensiek, The limitations of hydrogen blending in the European gas grid: a study on the use, limitations and cost of hydrogen blending in the European gas grid at the transport and distribution level, 2022. https://www.iee.fraunhofer.de/content/dam/iee/energiesystemtechnik/en/documents/Studies-Reports/FINAL_FraunhoferIEE_ShortStudy_H2_Blending_EU_ECF_Jan22.pdf. (Accessed 29 August 2022).

[20] European Industrial Gases Association, Hydrogen transportation pipelines, IGC doc 121/04/e globally harmonised document, 2004. https://h2tools.org/sites/default/files/Doc121_04%20H2TransportationPipelines.pdf. (Accessed 29 August 2022).

[21] W. Amos, Costs of Storing and Transporting Hydrogen, National Renewable Energy Laboratory, Golden, CO, 1999.

[22] M.C. Carvalho, J. Marques, H.A. Matos, J.F. Granjo, Long-distance hydrogen delivery using a toluene-based liquid organic hydrogen carrier system, Comput. Aided Chem. Eng. 50 (2021) 1647–1652.

[23] B. Brigljević, M. Byun, H. Lim, Design, economic evaluation, and market uncertainty analysis of LOHC-based, CO_2 free, hydrogen delivery systems, Appl. Energy 274 (2020) 115314.

[24] A.S. Malenkov, V.Y. Naumov, S.I. Shabalova, D.M. Kharlamova, Economic feasibility assessment of using ammonia for hydrogen transportation, SMART Automatics and Energy, Springer, 2022, pp. 89–98.

[25] J. Andersson, S. Grönkvist, Large-scale storage of hydrogen, Int. J. Hydrogen Energy 44 (23) (2019) 11901–11919.

[26] European Union, Assessment of hydrogen delivery options- science for policy briefs, 2021, JRC124206, https://joint-research-centre.ec.europa.eu/system/files/2021-06/jrc124206_assessment_of_hydrogen_delivery_options.pdf. (Accessed 26 August 2022).

[27] E.D. Frank, A. Elgowainy, K. Reddi, A. Bafana, Life-cycle analysis of greenhouse gas emissions from hydrogen delivery: a cost-guided analysis, Int. J. Hydrogen Energy 46 (43) (2021) 22670–22683.

[28] U.S. Department of Energy, M. Gardiner, Energy requirements for hydrogen gas compression and liquefaction as related to vehicle storage needs, 2009. https://www.hydrogen.energy.gov/pdfs/9013_energy_requirements_for_hydrogen_gas_compression.pdf. (Accessed 31 August 2022).

[29] A. Witkowski, A. Rusin, M. Majkut, K. Stolecka., Comprehensive analysis of hydrogen compression and pipeline transportation from thermodynamics and safety aspects, Energy 141 (2017) 2508–2518.

[30] Y. Lee, U. Lee, K. Kim, A comparative techno-economic and quantitative risk analysis of hydrogen delivery infrastructure options, Int. J. Hydrogen Energy 46 (27) (2021) 14857–14870.

[31] A. Lahnaoui, C. Wulf, H. Heinrichs, D. Dalmazzone, Optimizing hydrogen transportation system for mobility via compressed hydrogen trucks, Int. J. Hydrogen Energy 44 (35) (2019) 19302–19312.

[32] J. Moreno-Blanco, G. Camacho, F. Valladares, S.M. Aceves., The cold high-pressure approach to hydrogen delivery, Int. J. Hydrogen Energy 45 (51) (2020) 27369–27380.

[33] A. Lahnaoui, C. Wulf, D. Dalmazzone, Optimization of hydrogen cost and transport technology in France and Germany for various production and demand scenarios, Energies 14 (744) (2021).

[34] D. Apostolou, G. Xydis, A literature review on hydrogen refuelling stations and infrastructure: current status and future prospects, Renew. Sustain. Energy Rev. 113 (2019) 109292.

[35] J. Mouli-Castillo, S.R. Haszeldine, K. Kinsella, M. Wheeldon, A. McIntosh, A quantitative risk assessment of a domestic property connected to a hydrogen distribution network, Int. J. Hydrogen Energy 46 (29) (2021) 16217–16231.

[36] J. Mouli-Castillo, G. Orr, J. Thomas, N. Hardy, M. Crowther, M. Wheeldon, A. McIntosh, A comparative study of odorants for gas escape detection of natural gas and hydrogen, Int. J. Hydrogen Energy 46 (27) (2021) 14881–14893.

[37] M. Mintz, J. Gillette, A. Elgowainy, M. Paster, M. Ringer, D. Brown, A.J. Li, Hydrogen delivery scenario analysis model for hydrogen distribution options, Transport. Res. Record 1 (1983) 114–120.

[38] A. Züttel, Materials for hydrogen storage, Mater. Today 6 (9) (2003) 24–33.

Commercially available resources for physical hydrogen storage and distribution

Pranjali Sharma, Akash Kumar Burolia, Ananya Mandal and Swati Neogi

Chemical Engineering Department, Indian Institute of Technology Kharagpur, Kharagpur, West Bengal, India

4.2.1 Introduction

Undoubtedly, the two most frequently mentioned important challenges of the 21st century are the increasing global energy demand and the rise in carbon emissions causing global warming. Therefore, carbon-free power must be generated to meet the growing need for energy while safeguarding the environment [1]. Renewable energy is an alternative and the current interest of every country to end the reliance on depleted fossil fuels. Therefore, a strategic road map has been developed by many countries to approach clean, accessible, affordable, sustainable, and reliable energy [2–4]. There are four major stages as given in Fig. 4.2.1 for adopting a given energy resource as a fuel, to fulfill the fuel requirements the road map comprises (1) production → (2) storage → (3) distribution → (4) utilization. All stages are interdependent. The required production rate, purity index, and cost of hydrogen depend on its utilization [5]. Distribution and storage are the bottlenecks to connect the commercial and industrial sectors. Though many well matured and developed renewable resources are available in the global market, including hydro, wind, geothermal, solar, bioenergy, and tidal energy. These technologies are geographically dependent and inaccessible in many locations. The alternative is hydrogen fuel, which has been gaining more attention in the last decade due to its nongeographical dependency and zero emissions. Many nations view direct hydrogen and hydrogen-powered fuel cells as crucial energy alternatives for constructing sustainable energy infrastructure as reliable power resources for stationary electricity and heat generation in the transportation, industrial, and residential sectors [6].

Towards Hydrogen Infrastructure: Advances and Challenges in Preparing for the Hydrogen Economy.
DOI: https://doi.org/10.1016/B978-0-323-95553-9.00010-8

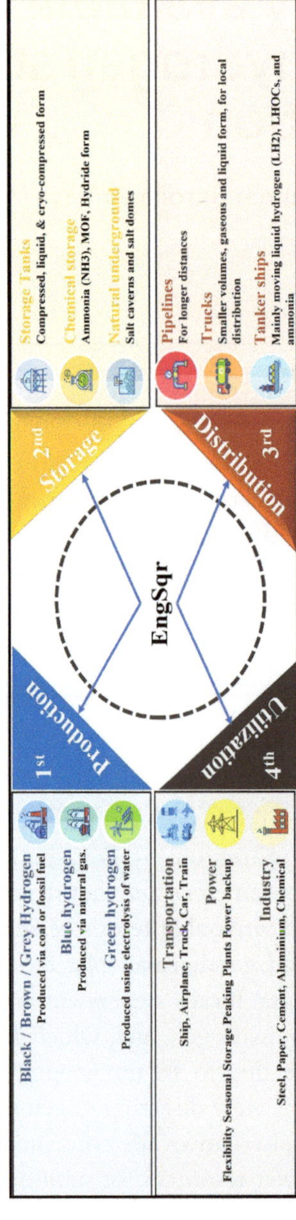

Figure 4.2.1   Energy square for adopting hydrogen as a fuel.

4.2.1.1 Hydrogen fuel past, present, and future

Hydrogen is endowed as a clean energy vector for its applications in various industries. It is found in nature as complex molecules, most notably in water and hydrocarbons. Numerous methods, including water electrolysis, coal/biomass gasification, natural gas reforming, photoelectrolysis, high-temperature water splitting, and biological processes produce hydrogen from nuclear energy, fossil fuels, and renewable resources [5]. Presently, hydrogen at the industrial scale is globally produced from coal gasification, natural-gas reforming, or water electrolysis. The most common hydrogen production process in the chemical and petrochemical sector is the steam reforming of natural gas. Currently, it is the most affordable production technique and releases the least CO_2 emissions out of all production methods involving fossil fuels.

The next most cost-effective technique for producing hydrogen is coal gasification because natural gas prices are expected to increase soon. However, even coal deposits have an expiration date and are not limitless. Hydrogen produced by electrolysis is of high purity and is produced using readily accessible renewable water. Our goal for green hydrogen production can be achieved if the electricity used to power the electrolyzer is likewise derived from renewable sources. Biomass competes with other fuels for thermal and electrical power generation, and biomass gasification for hydrogen production is predicted to be the future hydrogen source [2].

Presently, hydrogen is used in petroleum refining, ammonia production, chemical hydrogenation reactions, and during the refining of metals like lead, nickel, tungsten, copper, zinc, and molybdenum [7]. Its use can be extended for future applications to replace fossil fuels as a green and abundantly available fuel. It is more beneficial than gasoline as an automotive fuel, its energy value is 141.9 MJ/kg, which is three times more gravimetric energy dense than gasoline, which is 47.4 MJ/kg [6].

The storage and distribution of hydrogen in its energy square, are major challenges due to its higher flammability range, reactivity, and lower volumetric storage density [8]. These are the probable reasons for any nation to resist becoming a complete hydrogen economy. Recognizing proper storage and distribution solutions of hydrogen for economical utilization is the major bottleneck for adopting hydrogen as a future energy fuel.

A master plan for the choice of hydrogen storage and distribution systems to replace any form of energy with hydrogen fuel as per the end-user requirements should be created. Various pre-existing methods can be chosen to produce hydrogen in either gaseous or liquid forms per the utilization requirements. The fuel can directly be stored, transported, and distributed through different channels depending on the required physical state. The tailorability of the hydrogen production process and the diverse class of hydrogen storage/distribution solutions aid in conserving the latent heat of phase change or the specific heat capacity for heating/cooling the hydrogen fuel for different applications. The detailed study of the utilization sites and the corresponding requirements for hydrogen fuel further aid in choosing appropriate storage and distribution techniques for better efficiency and lower operational cost.

4.2.1.2 Challenges with hydrogen fuel

There are many advantageous characteristics of hydrogen such as high energy content, zero harmful emission, and abundant availability, making it a potential future fuel. However, some limitations constantly challenge its utilization in the automotive industry [8,9].

4.2.1.2.1 Molecular size

The molecules of hydrogen are very small compared to other fuel or nonfuel gases. Therefore, in a polymeric material, the permeation of gas is more severe due to the availability of free volume in its structural composition and it is influenced by (1) type of polymer, (2) morphology, (3) test environment, and (4) additives [10]. Therefore, the maximum allowable hydrogen permeation rate based on ISO 15869 in the polymeric liner of type IV vessel is 2.0 Ncc/h/L and 2.8 Ncc/h/L of water capacity at 350 bar and 700 bar respectively.

4.2.1.2.2 Diffusivity

The diffusion coefficient of hydrogen gas in air is approximately four times higher (6.1×10^{-5} m^2/sec at NTP condition) than natural gas (1.6×10^{-5} m^2/sec). Therefore, in partially confined space, hydrogen can accumulate underneath a roof/compartment, leading to a disastrous situation [11]. Diffusion can create two significant issues, that is, the formation of the flammable gas mixture in air and blistering inside the material.

Blistering can cause structural failure due to the degradation of material strength. It occurs when atomic hydrogen is absorbed and diffused in the material due to heterogeneity. Inside these voids, atomic hydrogen combines with another atomic hydrogen to form molecular hydrogen [12]. The molecular hydrogen does not diffuse further and generates enormous pressure on the material which could be sufficient to rupture the material.

4.2.1.2.3 Reactivity

The hydrogen in atomic form is highly reactive; it combines with most elements to form hydrides, for example, NaH. Hydrogen embrittlement (HE) is generally encountered due to the reactiveness of hydrogen atoms with metal, in which brittle hydride is formed as a consequence of the diffusion or dissolution of atomic hydrogen in the material. As per AMPP (The Association for Materials Protection and Performance), hydrogen embrittlement is responsible for the loss of material toughness and ductility because of the excessive presence of reactive hydrogen atoms. Hydrogen embrittlement creates severe issues in the metallic liner of a type III hydrogen pressure vessel or metallic pipeline carrying pressurized hydrogen. Hydrogen embrittlement is classified into two types (1) external HE: caused by external stress application on the structure and (2) internal HE: caused by the dilution of hydrogen during the processing of the material [13].

4.2.1.2.4 Density

Hydrogen is approximately seven times lighter than natural gas (CH_4: 0.65 kg/m^3 vs H_2: 0.09 kg/m^3 at 273 K) [11]. However, on the other hand, its lower density creates challenges for mobile applications due to the low volumetric storage energy density of 0.01 MJ/L at ambient conditions as shown in Fig. 4.2.2 [6]. To improve the volumetric storage density, the gas must be compressed up to 1000 bar.

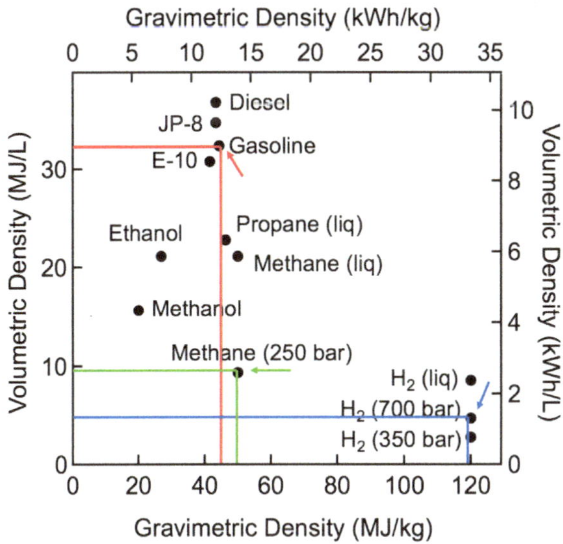

Figure 4.2.2 Energy storage of various fuels in a pressure vessel [6].

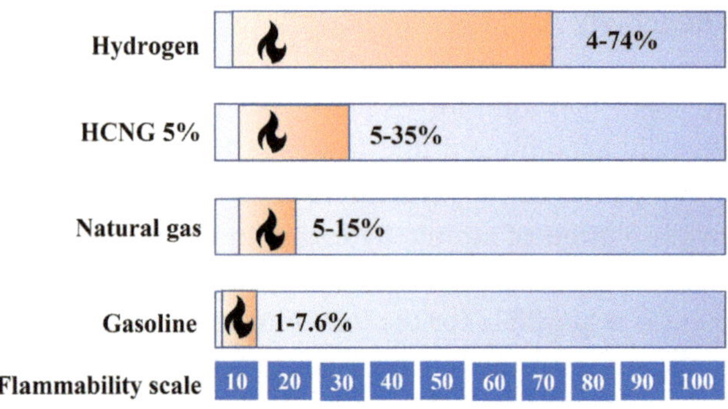

Figure 4.2.3 Flammability range for various gaseous fuels.

4.2.1.2.5 *Flammability*

Hydrogen is a highly flammable gas that produces a pale blue flame if burned with 100% purity. It has a significantly higher flammability range of 4–74 vol% in the air compared to other fuel gas, as given in Fig. 4.2.3 [14].

Therefore, some international organizations have established well standards for its safe utilization such as ISO 15869, HySafe, ISO 19881, and SAE J2579.

4.2.2 Physical hydrogen storage

Hydrogen is stored in three physical forms: (1) compressed hydrogen, that is, gaseous phase, (2) cryo hydrogen, which is a liquid phase, and (3) cryo-compressed hydrogen, which is double phase storage, i.e., gas–liquid phase [15]. The volumetric energy storage density at ambient conditions is much less than gasoline. Therefore, storing hydrogen at ambient conditions (NTP) using underground hydrogen storage is preferable as space occupancy is not a significant issue in this case. For stationary and portable applications with limited storage volume, high gravimetric capacity is required for which hydrogen is stored in compressed form. Cold/cryo hydrogen has higher volumetric density, i.e., when hydrogen is stored and transported at a cryogenic temperature of −253°C and 10 bar pressure. Cryo-compressed hydrogen (CcH_2) storage contains high volumetric storage density, but it requires a sophisticated thermal management system to maintain cryogenic temperature (−253°C) and high compression energy to pressurize the liquid-gas mixture up to 350 bar [16]. In this section, these physical hydrogen storage solutions are discussed in detail. The density of the gas is increased by pressurizing the gas from ambient to high pressure or reducing the gas temperature to low-temperature conditions, or a combination of both.

4.2.2.1 Underground gas storage

Hydrogen is a 100% clean energy source that emits no harmful byproducts and gives higher energy output than present fuel alternatives. It can also be used to generate electricity from renewable sources using fuel cells. However, the demand for electric power is constantly fluctuating; therefore, excess hydrogen needs to be stored such that it is readily available for its sudden demand or can be set aside without too much processing.

Underground hydrogen storage can be used for this purpose as it can be formed at different geological structures based on the type of storage, location, subsurface conditions, and economic considerations. There are four types of structures where hydrogen can be stored: salt caverns, depleted oil and gas fields, hard rock caverns, and deep aquifers [17].

Till 2010, there were 642 underground hydrogen storages exploited worldwide among which 476 were depleted hydrocarbon deposits, 82 aquifers, and 76 salt caverns [18]. Most of them are situated in North America, including 399 in the United States and 50 in Canada. Europe was second with 130 storage spaces, followed by the Commonwealth of Independent States (CIS) countries (50), Asia and Oceania (12), and one facility in South America and one in Argentina [19].

This method has ambiguities such as finding suitable locations and capacity, in situ reactions, unexpected fluids movement or hydrodynamics, excessive production of water during drainage of hydrogen, leakage causing contamination, and other environmental problems.

4.2.2.2 Compressed gas storage

Compressed hydrogen (CH_2) gas is a well-matured and widely accepted fuel in automotive industries because it does not require any secondary operation, such as a complex thermal management unit for liquid hydrogen to utilize it. However, compressed hydrogen gas still has a significant downside; it has a low volumetric density at normal temperature and pressure [9]. Therefore, improving the volumetric storage density requires storing at a very high pressure of up to 1000 bar. This can be achieved by storing the gas in vessels that can withstand high pressure easily, such as advanced composite pressure vessels.

There are five types of pressure vessels differentiated based on the material of construction, as shown in Fig. 4.2.4 [15].

4.2.2.2.1 Performance

The weight and load-bearing capacity are the two major parameters to analyze the performance of pressure vessels for compressed hydrogen gas storage. Both parameters are majorly dependent on the material properties, such as type of composite and liner as given in Fig. 4.2.5. Type IV pressure vessels manufactured by carbon composites provide the highest weight performance due to the light-weighed polymeric liner and complete load sharing by composite as shown in Fig. 4.2.5A, while glass fiber type IV pressure vessels are less expensive than carbon fiber vessels but have higher vessel weight as depicted in Fig. 4.2.5B. Though cheaper, glass fibers undergo strength degradation under environmental conditions such as chemical or moisture exposure [20].

4.2.2.2.2 Safety

The safety concerns for hydrogen gas storage include fatal accidents due to the failure of cylinders storing gas at a high compression pressure of up to 1000 bar, wide flammability range (4–74%), high reactivity, and diffusivity. There are many factors that can cause vessel failure due to external and internal damage, as given in Fig. 4.2.6 [21]. Failure caused due to the aforementioned factors can be prevented by taking safety precautions such as a thermally activated pressure relief device to restrict over-pressurization, dome geometry optimization for higher impact resistance, and safety factor based on ISO 15869 standards for the design and manufacture of hydrogen pressure vessels as given in Table 4.2.1.

4.2.2.3 Cold gas storage at low temperatures and normal pressure

The density of the gas increases with a decrease in the storing temperature. The density of the gas increases minutely with a decrease in the temperature at normal pressure as seen in Fig. 4.2.7A. Places requiring stationary storage with the ability to create low temperatures at low energies can use cold storage [22]. The normal pressure is one atmosphere which is equivalent to 1.01325 bar. At a storing pressure of less than 7 bar, not much change in the density of hydrogen gas is observed on reducing the temperature up to −155°C. The density of the gas is 0.1269 kg/m^3 at 25°C at atmospheric

Type	Vessel geometry	Description	Material of construction	Operating pressure	History	Technology maturity
I		Metallic vessel	Stainless-Steel, Aluminium, Iron.	Upto 300 bar	1880 for military use	+++
II	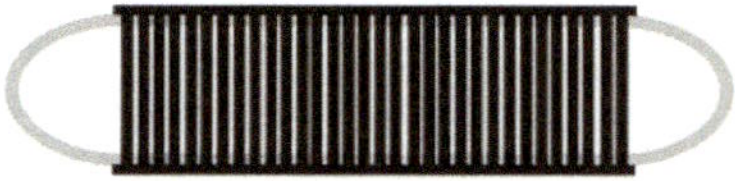	Metallic vessel/liner + Composite hoop wrapped	**Liner material:** Stainless-Steel, Aluminium, Iron **Composite material:** Glass, Carbon, Kevlar.	No limit		+
III	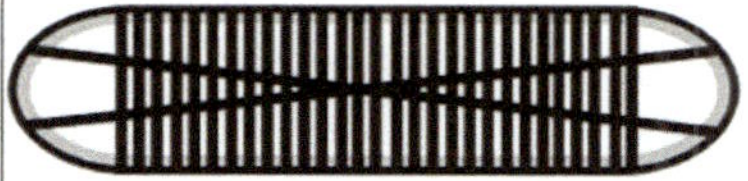	Metallic liner + Composite over wrapped	**Liner material:** Stainless-Steel, Aluminium, Iron **Composite material:** Glass, Carbon, Kevlar.	Upto 500 bar		++
IV	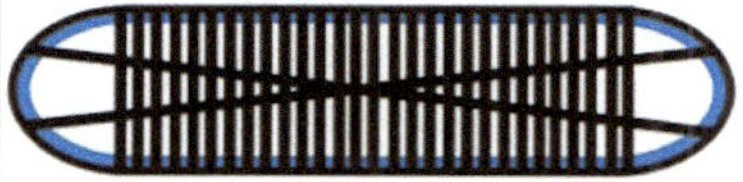	Polymeric liner + Composite over wrapped	**Liner material:** Stainless-Steel, Aluminium, Iron **Composite material:** Glass, Carbon, Kevlar.	Upto 1000 bar	2001, 1st prototype demonstration	++
V	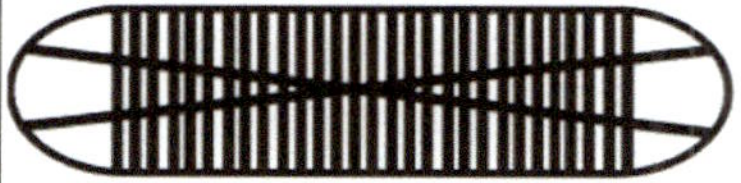	Linerless, Fully composite vessel	**Liner material:** No liner **Composite material:** Carbon, other material may be used.	Upto 13.3 bar	2010 Developed by Composite Technology Development Inc	-

Figure 4.2.4 Types of pressure vessels and their current status.

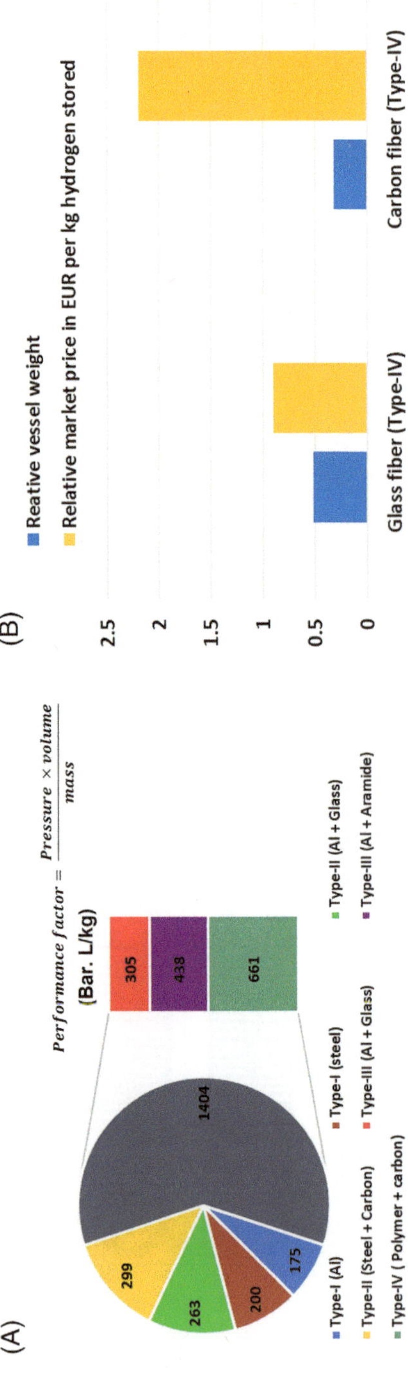

Figure 4.2.5 Comparison of pressure vessel types based on their performance and cost (A) Performance analysis of various types of pressure vessels [20] and (B) comparison of vessel performance based on the type of material used [58].

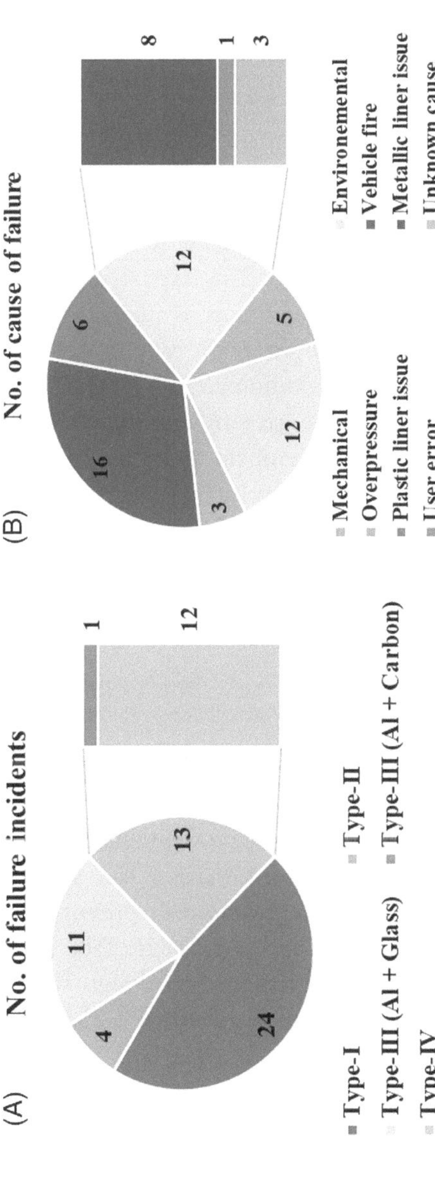

Figure 4.2.6 Number of pressure vessel failure incidents [21] (A) Comparison of types of pressure vessels and failure and (B) cause of failure.

Table 4.2.1 Nominal safety factor for hydrogen storage vessel.

Vessel types/construction	Type I	Type II	Type III	Type IV
All metal	2.25			
Glass		2.4	3.4	3.5
Aramid		2.25	2.9	3.0
Carbon fiber, P <350 bar		2.25	2.25	2.25
Carbon fiber, P ≥350 bar		2.0	2.0	2.0

pressure. However, it increases to a value of 7.09 kg/m^3 and 13.85 kg/m^3 at a temperature of $-155°C$ when stored at 35 and 70 bar, respectively. Such pressures can easily be attained in Type-1 pressure vessels [23]. For stationary storage in large confinements as illustrated in Fig. 4.2.7B, this increase in the density value improves the system storage capacity significantly. The system involves the combined use of liquid nitrogen or oxygen as the coolant fluid to reduce the gas temperature to around $-155°C$ in the described capacity. The boil-off temperature of liquid nitrogen and oxygen is $-195.8°C$ and $-183°C$, respectively. Hydrogen gas can be enclosed in a conductive enclosure surrounded by liquid nitrogen or oxygen for cold hydrogen storage, maintaining the possible gas temperature at around 35–70 bar pressure for better storage performance. Higher amounts of hydrogen gas can be stored in the proposed assembly without adding extra cooling energy.

4.2.2.4 Cold-compressed gas storage

Cryo-compressed gas storage is a two-phase system involving saturated liquid and gaseous phase hydrogen. It provides the highest volumetric energy storage, eliminates boil-off issues in routine use, and improves thermal endurance with comparative cost and no over-pressurization during filling compared to other physical forms, as given in Table 4.2.2. As the name suggests, cryo-compressed means the vessel design should be capable of maintaining a cryogenic temperature of up to $-253°C$ and withstanding a high pressure of up to 350 bar. In 2011, BMW presented a hydrogen fuel vehicle prototype equipped with type III cryo-compressed hydrogen storage vessels [24]. The system consists of a type III pressure vessel which was enveloped by multilayer insulation surrounded by a vacuum jacket with a metallic casing as shown in Fig. 4.2.8A.

The significant advantage of the cryo-compressed storage technique is the flexibility in fueling because the vessel is designed to operate at cryo

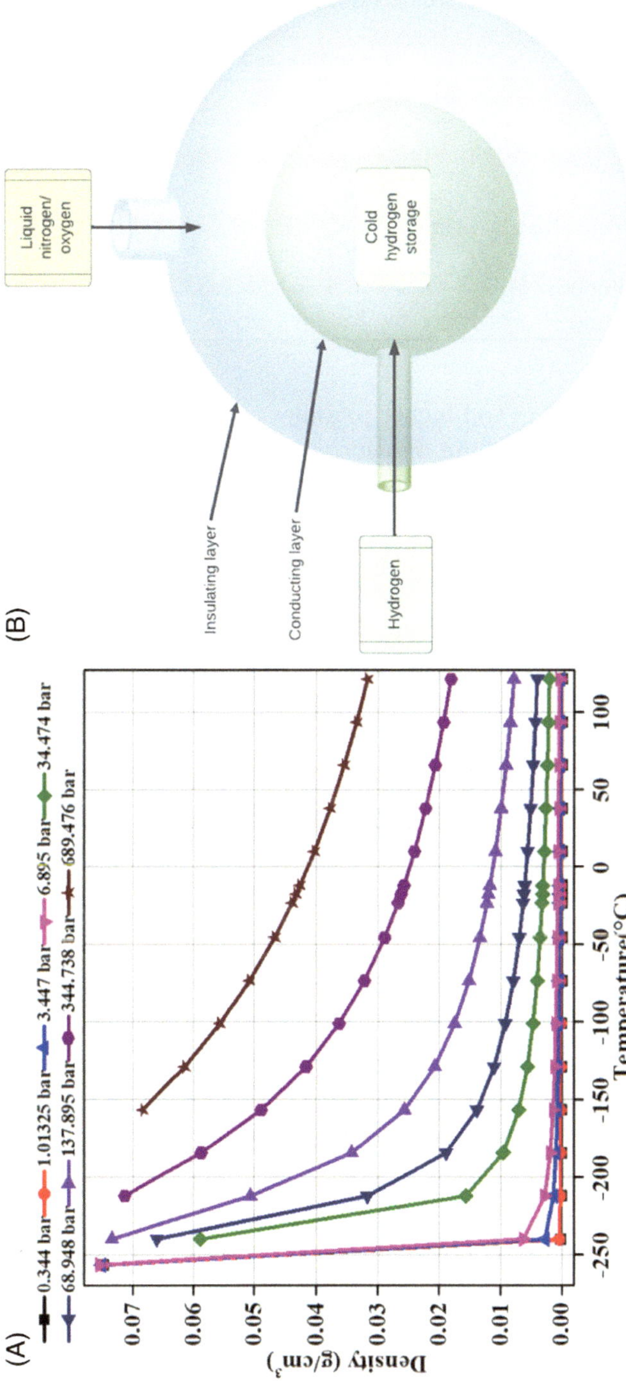

Figure 4.2.7 (A) Variation in density with the temperature at different pressures and (B) assembly for energy-efficient cold hydrogen gas storage.

Table 4.2.2 Comparison between compressed and cryo-compressed hydrogen vessels [29,59].

Storage method	Compressed H_2	Cryo-compressed H_2
Storage pressure, bar	350	350
Liner	HDPE	2 mm-SS
Storage temperature, K	288	62
Storage density, g/L	24	71
Gravimetric storage capacity, wt%	5.4	7.5
Volumetric storage density, g/L	17.7	35.5
Carbon fiber composite, kg	61.9	20
Cost, $/kWh	13	11

(-253°C) temperatures and higher pressures of up to 350 bar as depicted in Fig. 4.2.8B. Therefore, the insulation part maintains cryo temperature while high pressure is borne by the type III vessel. However, there is limited research has been conducted to study the feasibility and safety of this storage technique. Also, cylinders storing cryo-compressed hydrogen exhibit lower weight performance.

The performance of cryo-compressed hydrogen storage strongly depends on the vessel characteristics, the driving frequency, the liquid hydrogen pump performance, the heat transfer from the environment, and the conversion rate of para to ortho hydrogen.

Few technical challenges during the operation of type III cryo-compressed hydrogen vessels have been identified [25]. Significant vacuum space is required that reduces the volumetric performance of the storage system, and a wide operating range from cryogenic to room temperature creates issues with composite material performance due to thermomechanical loading conditions [26]. The insulation degrades under continuous operation and the fatigue life/aging reduces due to thermal cycling and constant hydrogen exposure [23].

4.2.2.5 Cryo liquid storage

The liquid storage tanks are made of typical steel and aluminum super-insulated cryogenic tanks. They have been discovered and utilized for ages by many commercial manufacturers present around the world as shown in Fig. 4.2.9. The gas is cooled down to -253°C by adiabatic expansion [27]. Taming liquid hydrogen was one of the significant technical achievements which led to several advances in the aeronautical industry. However, a significant amount of hydrogen is lost in the form of boil-off when stored

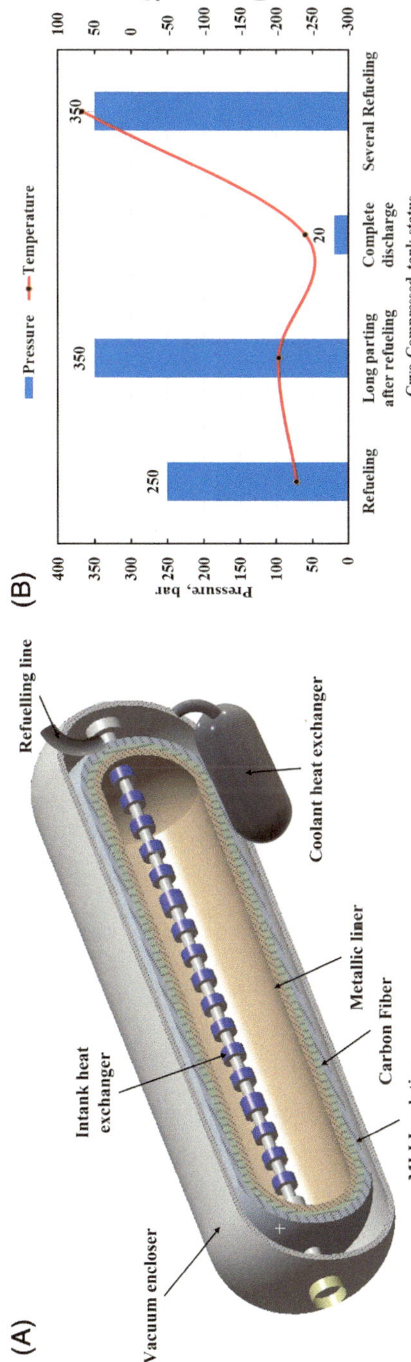

Figure 4.2.8 Cryo-compressed hydrogen (CcH$_2$) vessel (A) type III CcH$_2$ tank schematic and (B) operating range of CcH$_2$ tank [24].

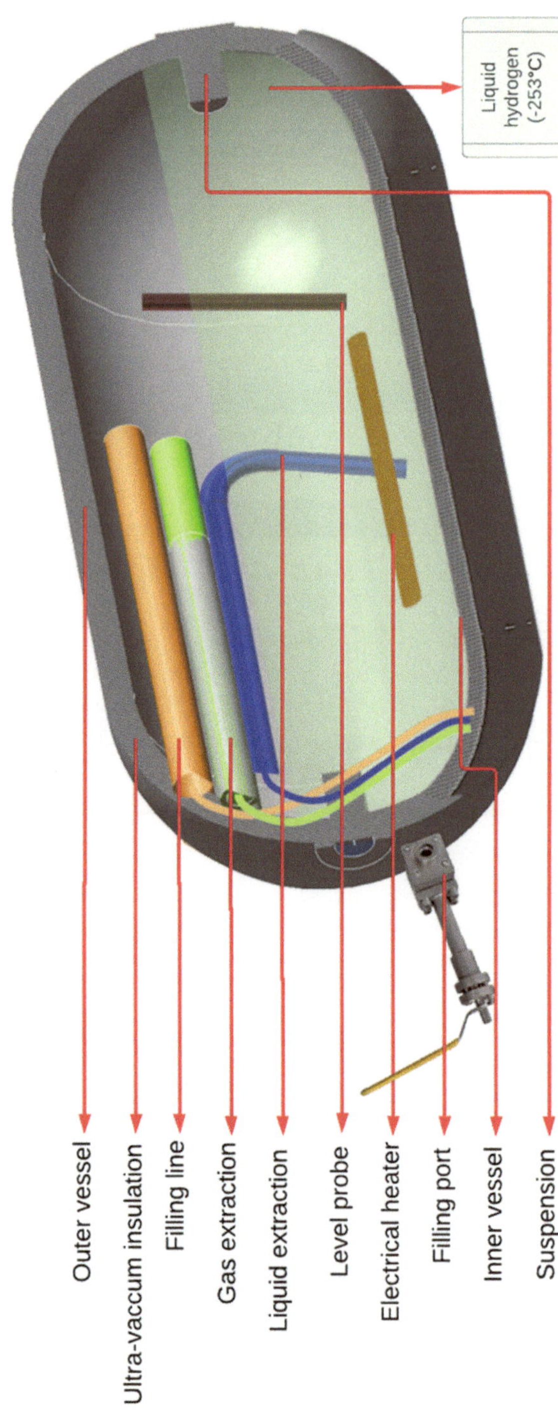

Figure 4.2.9 Schematic of a cryogenic tank for storing liquid hydrogen at $-253°C$.

in the liquid state. Therefore, commercial utilization involving the large-scale daily deployment of liquid hydrogen fuel is the only viable routine usage frequency for efficient storage. The density of hydrogen in the liquid state at ambient pressure of 1 atm is 70.9 kg/m^3, while the purity index is close to 100% [28].

4.2.2.6 Cryo-compressed liquid storage

Cryo-compressed liquid storage is the most compact form of physical hydrogen storage. The liquid hydrogen stored at cryogenic temperatures is pressurized to a higher pressure of up to 350 bar to further increase its density as seen in Fig. 4.2.8A. The density of liquid hydrogen at 1 atm is 70.9 kg/m^3 which is increased to 87 kg/m^3 upon pressurizing the system to 240 bar [28]. It was found that the cryo-compressed hydrogen tank storing liquid hydrogen at 500 bar exhibited superior performance when compared to a compressed gas tank storing hydrogen gas at 350 bar of pressure. It was installed in a fuel cell–operated bus to check the performance characteristics while on-board operation [29]. A 91% enhancement in the gravimetric volume capacity and a 21% decrease in the cryo-compressed system cost were observed as compared to compressed gas cylinders. A cryo-compressed vessel is a typical type III cylinder used for compressed gas storage enclosed in a super-insulated vacuum jacket. The dormancy losses including the liquid boil-off percentage in the case of cryo-compressed storage are also less compared to liquid storage [30].

4.2.3 Hydrogen distribution in gaseous or liquid forms

Hydrogen storage capacities of less than 50 kg are supplied in compressed gas cylinders, whereas for deliveries over 50 kg, gaseous tube trailers or liquid tankers are used [31]. Hydrogen is transported via a dedicated pipeline for continuous demands. Currently, pure hydrogen is transported to fertilizer industries and petroleum refineries via pipelines from hydrogen-producing facilities close to these large customers. For a continuous supply of hydrogen to remote locations, hydrogen can be combined with natural gas and then separated and purified according to the end user's requirements.

4.2.3.1 Distribution of compressed hydrogen gas and its blends

Pressurized hydrogen tube trailers and a large-scale pipeline distribution network can be used for gaseous hydrogen distribution to remote locations.

Tube trailers are mainly used for early markets and small stations, whereas pipeline networks can be used for mature and large markets [32]. As of now, the direct use of hydrogen for long-distance pipeline distribution has not been established; instead, hydrogen is being blended in small amounts with compressed natural gas in many countries, and efforts are being made to enhance the hydrogen blending ratio.

4.2.3.1.1 *Transportation in tube trailers*

Small to medium-sized volumes of gaseous hydrogen can be delivered by truck in compressed gas containers. However, for larger volume deliveries, long pressurized gas cylinders or tubes are assembled on CGH_2 tube trailers. The gaseous hydrogen is compressed into the gas cylinders to 180 bar or higher pressures. The substantial tubes are packed up inside a protective frame. The tubes often have a high net weight and are constructed of steel. Transportation constraints relating to bulk could result from this. The most recent pressurized storage systems use composite storage containers with 560–900 kg of hydrogen per trailer and are lighter for truck transportation [33].

Through transmission pipelines, hydrogen generated at a central plant is delivered to a loading facility for tube trailers. The gas terminal and production facility are situated together when production is semi-centralized. A compressor raises the hydrogen pressure at the gas terminal to the operating pressure for the tube trailers, and storage is included to help with demand fluctuations. The tube trailers are directly connected to the loading bays where the distribution gas terminal fills them. They are then delivered to the refueling stations and attached to the compressor. The compressor is used to transfer the hydrogen from the tube trailer to a high-pressure buffer storage system, which later distributes the hydrogen to the vehicle's fuel tanks. Low-pressure storage is used as a buffer for varying supply and demand of hydrogen but this is not required for tube trailer deliveries [31].

The pressure vessels must be mounted inside an ISO-certified container that has a length of 40 ft, a width of 8 ft, and 8 ft in height as per the U.S. Department of Transportation, Federal Highway Administration (1994) Federal Size Regulations for Commercial Motor Vehicles. Code of Federal Regulations (CFR), 23 CFR Part 658. Steel pressure vessels are constrained due to their higher density and correspondingly lower weight performance. The maximal hydrogen payload is limited to around 250 kg due to the characteristics of steel and weight restrictions [31].

4.2.3.1.2 Distribution through pipelines

For large-scale and widespread usage of hydrogen resources, a pipeline network is the best option. However, pipelines demand a significant upfront expenditure, which may be profitable but only with proportionally huge volumes of hydrogen. Local/regional networks also called micro-networks are potential sites for creating pipeline networks for the distribution of hydrogen [34]. These could then be linked to form transregional networks. Approximately 4500 km long hydrogen pipelines are already in use worldwide, most of which are run by hydrogen producers to move hydrogen between facilities in the chemical process industry. The United States' Louisiana and Texas states have the longest operating pipelines, followed by Belgium and Germany [35].

Pipeline delivery is the most cost-effective option when the hydrogen demand is so high that it can support the building of dedicated transmission and distribution pipelines. A significant obstacle to extending the hydrogen pipeline delivery infrastructure is the high initial investment expenses of new pipeline construction. However, once established, they are thought to be the least expensive solution for supplying hydrogen to major refueling stations within a city or locality. Hydrogen is received from the production plant and delivered via pipeline. When there is any outage in the hydrogen production plant, an underground storage facility acts as a backup. The hydrogen is pressurized from the plant and underground storage using a compressor and then sent to the transmission pipeline. The hydrogen is further transported to the city gate and distribution terminal through these transmission pipelines, where it is further supplied to the refueling station via a distribution pipeline network, as shown in Fig. 4.2.10. If the hydrogen production plant is near the city gate or the distribution terminal, it is called a semi-central plant, and the hydrogen produced there is directly pressurized using a compressor and then sent to the distribution pipeline network. A significant and sustained demand must recover the expense of building a pipeline [31].

Today's research focuses on addressing technical pipeline transmission issues, such as hydrogen embrittlement of the steel and welds used to construct the pipelines, the need to stop hydrogen leaks and permeation, and the demand for more affordable, dependable, and long-lasting hydrogen compression technology. Fiber-reinforced polymer (FRP) pipelines for the distribution of hydrogen are one possible remedy. Due to the availability of far longer FRP sections than steel, which reduces the need for welding, the installation costs for FRP pipelines are around 20% lower than those for steel pipelines [36].

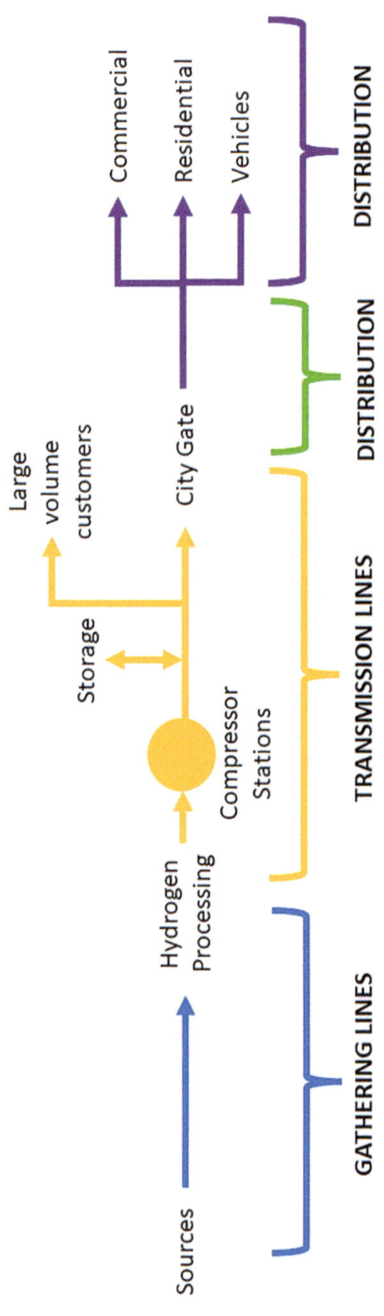

Figure 4.2.10 Hydrogen distribution pipeline network.

4.2.3.1.3 Distribution of hydrogen blends

Mixing hydrogen into natural gas pipeline networks has been suggested for supplying pure hydrogen to the markets. According to the needs of the end application, we can either directly use the blend or separate the hydrogen and purify it from the blends. Blending and utilizing the widespread natural gas pipeline infrastructure at this early stage of economic growth will help in decreasing the cost of building separate hydrogen-dedicated pipelines and delivery infrastructure. It can improve the energy sector scenario by expanding hydrogen usage and reducing environment-related problems such as global warming by decreasing CO_2 emissions. The advantages also include an increase in combustion rate, a reduction in the width of the flame, and an extension of the ignition limits. The blends can be used as a direct fuel for power plants giving high efficiency and being environmentally friendly. Instead of liquid fuels, gaseous ones result in complete combustion with fewer carbon deposits and toxic byproducts [37].

There are still limitations to this method because of the low gravimetric and volumetric energy density of the hydrogen gas. It may limit the energy delivery by the grid, and the additional pressure drops due to transporting hydrogen may not be entirely compensated by the installed compressors at the grid.

Along with blending hydrogen with natural gas, we should also focus on the end user's appliances, for which this blend could be an alternative and cheaper option. In literature, it has been discussed that blending hydrogen up to 10% has no significant effect on its utilization [38], whereas increasing the concentration to 20% needs to focus more on the compressor capabilities [39]. The grid properties must be varied if we want to improve it from 20% to 50%. In many countries, hydrogen and natural gas blends are successfully used in automobiles with internal combustion engines also, hydrogen of up to 20% is used for personal/public services, and for running domestic appliances like cooking burners [40]. Methane–hydrogen blend is used as a fuel in the automotive sector consisting of 10–30% vol. hydrogen called hythane [41]. The presence of hydrogen improves the engine's efficiency and reduces CO_2 and NOx emissions.

4.2.3.2 Liquid hydrogen transportation

Alternatively, liquid hydrogen can be carried using trucks or other vehicles. Since the density of liquid hydrogen is higher than its gaseous state, special LH_2 trailers are used for its transportation. At atmospheric pressure, hydrogen can exist in a liquid state below 20 K, hence vacuum-jacketed stainless-steel tanks are required for storage [31]. Obtaining such low temperatures is both energy-intensive and costly. The density of liquid hydrogen, which is about 800 kg/m^3, is very low compared to the available liquid fuels, so a relatively small volume of hydrogen can be carried. At a loading volume of 50 m^3, approximately 3500 kg of liquid hydrogen can be transported at a density of 70.8 kg/m^3. Hydrogen's volumetric mass and energy densities are significantly increased through liquefaction. The tanker can hold roughly 4 metric tons of liquid, which is 5–6 times more than a composite tube trailer and 15–20 times more than a steel tube trailer [31]. Since a liquid hydrogen tank can carry significantly more hydrogen compared to a pressurized gas tank, it is typically more economical to transport hydrogen over longer distances in this form.

The hydrogen is placed into insulated cryogenic tanks for liquid transportation. An LH$_2$ trailer's range is roughly 4000 km [34]. Heat loss through the tank walls causes liquid hydrogen to vaporize or boil off. Any pressure build-up in the storage tank can be prevented by recovering or releasing the boil-off into the atmosphere. A high volume-to-surface ratio can help to reduce the tank's boil-off rate [16]. A significant boil-off loss also occurs when the fuel is unloaded at the refueling station during the pumping of hydrogen to the on-site cryogenic tank from the liquid tanker. Limiting the number of deliveries and planning the delivery route appropriately will help to reduce these boil-off losses. LH$_2$ can be carried by ship or rail like truck transport, provided that the necessary waterways, railroad lines, and loading terminals are accessible. Transmission pipelines may be used to transmit hydrogen created at a central production facility to the distribution terminal, which is loaded into liquid tankers after liquefaction. Hydrogen is created and liquefied when production is semi-central at the distribution facility. A pump transfers liquid hydrogen to liquid tankers from the cryogenic storage tank. The liquid tanker is later unloaded at the refueling facility into another cryogenic storage tank. Vehicles are refueled using liquid hydrogen at the refueling storage tank [31].

4.2.4 Thermodynamics of hydrogen phase change

The accurate estimation of the thermodynamics of hydrogen storage and distribution is essential for the development of efficient hydrogen infrastructure. It mainly focuses on energy efficiency, the ratio of energy stored per consumption or loss of energy for that storage.

4.2.4.1 Pressurization energy for compressed gas storage

The storage of hydrogen in compressed gaseous form requires compression of the produced hydrogen from normal temperature and pressure into the storage system at normal temperature but high pressure. Considering the use of a 350–700 bar Type III/IV pressure vessel for hydrogen storage, the amount of work done to isothermally compress the gas from 1.01 bar to 350 bar of pressure at 288.15 K of temperature using a standard mechanical piston is calculated to be 138.216 kJ/mol using Eq. (4.2.1) [42].

$$\Delta G = RT. \ln \left(\frac{p}{p_0} \right) \tag{4.2.1}$$

where R denotes the universal gas constant with a value of 8.314 J/mol/K, p/p_0 is the end pressure and the starting pressure, respectively, and T is the absolute temperature.

4.2.4.2 Specific heat capacity for cold storage

The specific heat of hydrogen at 25°C is 14.304 J/gK and it is calculated by using Eq. (4.2.2) [43].

$$Q = m.C_p.\Delta T \tag{4.2.2}$$

where Q is the amount of energy in joule, m is the mass of stored hydrogen in g, and $.\Delta T$ is the temperature difference in K.

4.2.4.3 Latent heat of phase change for liquid storage

The minimum theoretical work required to thermodynamically express the liquefication process is expressed as a point function for commercial scale conversion and is defined using Eq. (4.2.3) [27].

$$A_i - A_f = H_i - H_f - T_0\left(S_i - S_f\right) \tag{4.2.3}$$

where H and S are the enthalpies and entropies of the hydrogen in the initial and final phases, respectively. For the liquefaction process, the liquid state is taken as the final state, and T_0 is the heat sink temperature. The electrical energy consumption for this process is calculated to be 39.0 MJ/kg [27].

During utilization, the liquified hydrogen is again converted to gas. Therefore, the latent heat of liquefication for different temperatures is calculated by Eq. (4.2.4), i.e., the Clausius–Clapeyron equation [44].

$$\ln \frac{P_2}{P_1} = \frac{H_v}{R} \left(\frac{1}{T_1} - \frac{1}{T_2} \right) \tag{4.2.4}$$

Table 4.2.3 Purity of pure and ultra-pure hydrogen gas commercially available in compressed gas cylinders.

Component	Unit	High purity hydrogen	Pure hydrogen
H_2	%	99.998	99.995
N_2	ppm	13.097	29.778
O_2	ppm	0.49	4.82
Moisture	ppm	1.7	5.43

where H_v is the latent heat of liquefication, P_1 and P_2 are pressure in the liquid and gaseous phase in Pa, T_1, and T_2 are the temperatures of the liquid and gaseous phase in K, and R is the universal gas constant. The amount of energy consumed in changing the liquid hydrogen from $-253°C$ and 1 MPa pressure to the gas phase at $25°C$ and 35 MPa per kg of hydrogen is -2810.92 kJ/kg or 0.78 kWh.

4.2.4.4 Logistics summary of storage and distribution solutions

The purity of hydrogen in commercial compressed gas cylinders plays a vital role in affecting the energy efficiency of the storage system. If the purity of hydrogen is reduced by 1%, it increases the windage loss by 12%. Therefore, a reduction in the gas purity from 99% to 92% reduces the generator output by around 2100 kW [45]. However, purifying hydrogen is also a costly process. A common consensus between the utilization site requirements and the production cost of the fuel is established to develop economically feasible physical storage solutions as shown in Table 4.2.4. The content of hydrogen gas supplied in different grades in commercially available cylinders is shown in Table 4.2.3.

Liquid hydrogen has a high purity index for which purifying technologies are quite essential for the continuous operation of the production unit. The presence of impurities may hamper the process flow leading to system shutdown. The operational cost of various storage techniques has been estimated based on the operational conditions in Table 4.2.4 [46].

4.2.5 Way forward towards completing the hydrogen energy square

The logistics summary of the commercially utilized storage solutions has been discussed in Section 4.2.4.4. The offered hydrogen purity index, the operating temperature and pressure, the density of the gas at those conditions, and the operating cost have been discussed. Fig. 4.2.11 illustrates the

Table 4.2.4 Summary of the storage solutions.

Storage technique	H_2 purity index	Operating temperature	Operating pressure	Density of gas	Operational cost/kWh	Lifetime
Underground gas storage	99%	Ambient	50–200 bar	4.5–11.4 kg/m^3	–	50 years
Compressed gas storage	99.995%	Ambient	350–700 bar	23–70 kg/m^3	Rs. 1183.17	10–15 years
Cold gas storage at low temperatures and normal pressure	99%	−50°C to −155°C	1.013–70 bar	7.09–13.85 kg/m^3	–	–
Cold-compressed gas storage	99.995%	−155°C to −253°C	20–350 bar	52–70 kg/m^3	Rs. 848.18	–
Cryo liquid storage	~100%	≤−253°C	1.013 bar	71 kg/m^3	Rs. 1311.47	3000 life cycles
Cryo-compressed liquid storage	~100%	≤−253°C	240–500 bar	87–90 kg/m^3	–	–

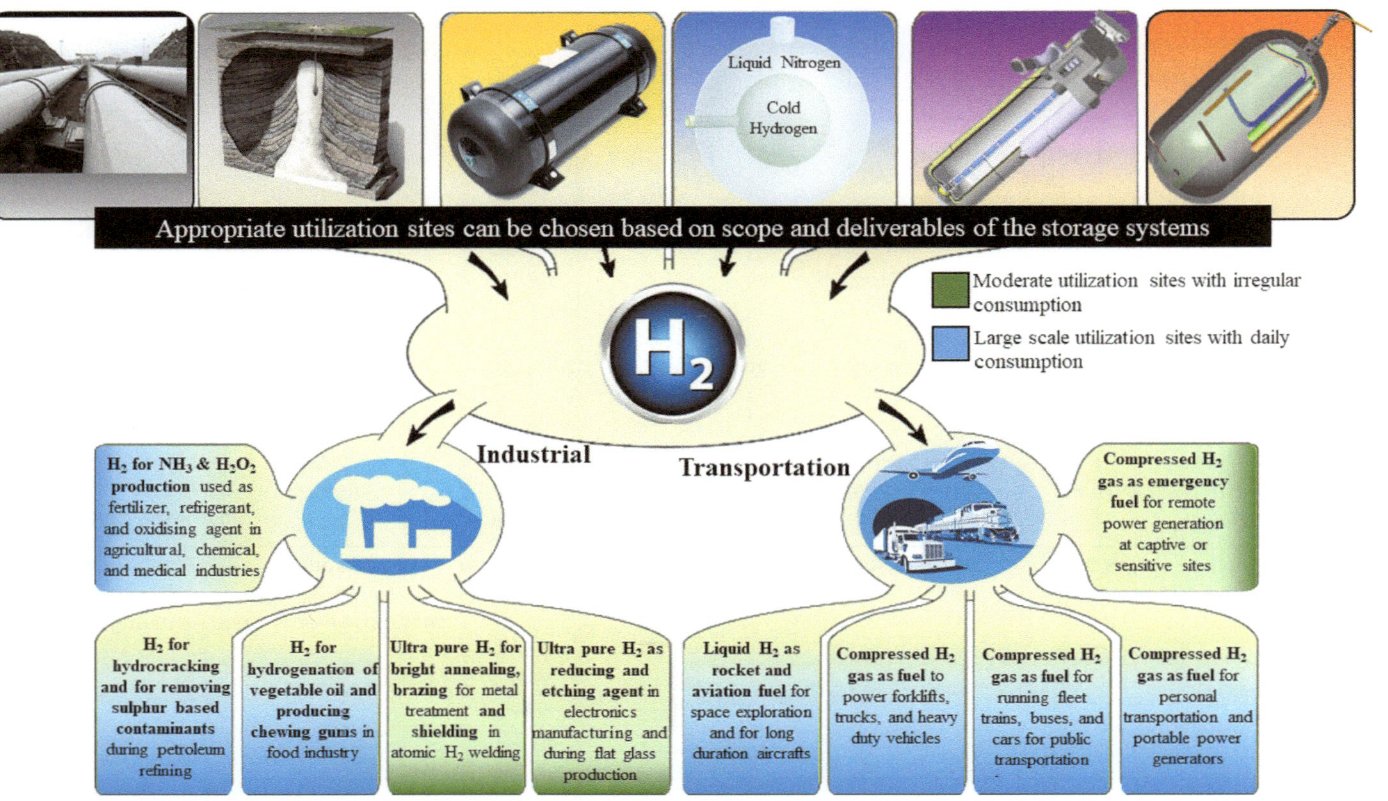

Figure 4.2.11   Available storage systems and characteristics of the utilization sites.

requirements of the utilization sites. Appropriate storage and distribution solutions can be chosen based on the consumption frequency, purity requirements, and the operating conditions of the utilization sites for maximum fuel efficiency in terms of energy output as well as economic feasibility.

The characteristic requirements of the utilization sites are analyzed, and the corresponding storage and distribution solutions are identified. The details of the utilization sites with the matched capacity of the storage/distribution units have been discussed in this section. The possible hydrogen utilization sites in the transportation sector are discussed below. Based on the state of fuel utilization, refiling frequency, the required purity index, and required performance of the storage system, the storage/distribution units are matched as follows:

1. **Public and personal transportation**: Compressed hydrogen gas storage cylinders are gaining popularity in the automotive sector for powering vehicles. Many automobile manufacturers produce mid-size fuel-cell powered cars for personal transportation and fleet buses and trains are also being adopted after the successful operation of the Hydrail project. The purity index of compressed hydrogen gas is suitable for the chosen utilization site and has already been tested in many case studies. The compressed gas storage tanks can be filled at the fuel dispensing stations. To meet the daily fuel demands of the transportation sector, hydrogen gas pipelines are laid between the production site and the city gate as illustrated in Fig. 4.2.10. Distribution service lines are used to distribute hydrogen from the city gate to the fuel dispensing stations. For dispensing stations located in remote areas with occasional consumption, buffer storage systems can be used to save pipeline installation costs.

2. **Aviation**: Large amounts of aviation fuel are consumed daily by the aviation industry to fly air vehicles. Liquid hydrogen or cryo-compressed hydrogen can be used to power the aviation industry. Liquid hydrogen dispensing stations are used for distribution. Cryotanks transfer liquid hydrogen from the production site to the dispensing station. While other maintenance operations are fulfilled using gaseous hydrogen. The boil-off hydrogen gas can be stored along with compressed hydrogen gas for producing electricity to improve the interior of aircraft.

3. **Global transportation logistics**: Compressed hydrogen gas tanks are used to run fuel-cell-powered forklift trucks. The storage and distribution system is similar to that proposed for automotive applications.

4. **Space exploration**: Liquid hydrogen has been successfully deployed for space exploration. They require high-purity fuel for which liquid or cryo-compressed hydrogen can be used. The storage/distribution system for spacecraft applications is similar to that deployed for the aviation industry.

5. **Marine**: Submarines, boats, and ships demand storing more fuel in low volumetric capacity. The case studies indicate that compressed hydrogen gas was utilized for marine applications. However, liquid/cryo-compressed hydrogen should be used to improve the sailing range.

6. **Defence**: Police fleets, autonomous air vehicles, etc. have been powered using compressed gas hydrogen. For land transportation, the gaseous form should be used while liquid/cryo-compressed powered vehicles can cover longer durations for defence applications. Also, the utilization rate is high in aircraft applications so liquid boil-off can be controlled giving better energy output.

7. **Power generation**: Compressed hydrogen gas is used for power generation in sensitive sites. An underground storage solution should be adopted to store the hydrogen as the utilization site will only be operational for emergency purposes and have no volumetric requirements for stational storage. Underground storage is a naturally formed geological location with the ability to store hydrogen gas with the least construction and maintenance cost.

8. **Ammonia production**: Gaseous hydrogen reacts with nitrogen at high pressure and temperature in the Haber–Trosch process. The daily consumption is quite high so compressed gas tanks are the emergency storage options while gas pipelines can be used for distribution from hydrogen-producing sites to the ammonia production factory.

9. **Hydrogen peroxide production**: Hydrogen gas is adsorbed over anthraquinone catalyst to react with oxygen and form hydrogen peroxide. A similar solution as proposed for ammonia production can be adopted for this industry also.

10. **Reducing agent**: Hydrogen acts as a reducing agent in a gaseous state with large daily consumption. The distribution/storage system should be the same as the ammonia production industry.

11. **Provide inert, reducing, shielding, or etching atmosphere**: A smaller amount of hydrogen is required for these cases so gas grid systems or compressed gas tanks can be deployed to meet the fuel requirements [47–57].

References

[1] M. Ball, M. Wietschel, The future of hydrogen: opportunities and challenges, Int. J. Hydrogen Energy 34 (2) (2009) 615–627. https://doi.org/10.1016/j.ijhydene.2008.11.014.

[2] A. Demirbas, Future hydrogen economy and policy, Energy Sourc. Part B: Eco. Plan. Policy 12 (2) (2017) 172–181.

[3] A. Kovač, M. Paranos, D. Marciuš, Hydrogen in energy transition: a review, Int. J. Hydrogen Energy 46 (16) (2021) 10016–10035.

[4] A. Midilli, M. Ay, I. Dincer, M.A. Rosen, On hydrogen and hydrogen energy strategies I: current status and needs, Renew. Sustain. Energy Rev. 9 (3) (2005) 255–271. https://doi.org/10.1016/j.rser.2004.05.003.

[5] F. Dawood, M. Anda, G.M. Shafiullah, Hydrogen production for energy: an overview, Int. J. Hydrogen Energy 45 (7) (2020) 3847–3869. https://doi.org/10.1016/j.ijhydene.2019.12.059.

[6] DOE. (2022a). *Hydrogen storage.* https://www.energy.gov/eere/fuelcells/hydrogen-storage.

[7] S. Sharma, S.K. Ghoshal, Hydrogen the future transportation fuel: from production to applications, Renew. Sustain. Energy Rev. 43 (2015) 1151–1158.

[8] E. Rivard, M. Trudeau, K. Zaghib, Hydrogen storage for mobility: a review, Materials 12 (12) (2019a) 1973.

[9] F. Schüth, Challenges in hydrogen storage, Europ. Phys. J.: Spec. Top. 176 (1) (2009) 155–166. https://doi.org/10.1140/epjst/e2009-01155-x.

[10] M.C. Kane (2008). Permeability, solubility, and interaction of hydrogen in polymers: an assessment of materials for hydrogen transport. SRNL, WSRC-STI-2008-00009, Rev. 0. https://doi.org/10.2172/927901, https://www.osti.gov/biblio/927901.

[11] Y.S.H. Najjar, Hydrogen safety: the road toward green technology, Int. J. Hydrogen Energy 38 (25) (2013) 10716–10728. https://doi.org/10.1016/j.ijhydene.2013.05.126.

[12] Y. Feng, H. Wang, F. Hou, H. Chen, P. Tan, Mechanism analysis of hydrogen blisters on the surface of a gas tank and research on its preventive measures, Proceed. 2015 Int. Conf. Mater. Environ. Biol. Eng. 10 (2015) 578–582. https://doi.org/10.2991/mebe-15.2015.133.

[13] S. Lynch, Hydrogen embrittlement phenomena and mechanisms, Corros. Rev. 30 (3–4) (2012) 105–123. https://doi.org/10.1515/corrrev-2012-0502.

[14] K. Mazloomi, C. Gomes, Hydrogen as an energy carrier: prospects and challenges, Renew. Sust. Energy Rev. 16 (5) (2012) 3024–3033. https://doi.org/10.1016/j.rser.2012.02.028.

[15] H. Barthelemy, M. Weber, F. Barbier, Hydrogen storage: recent improvements and industrial perspectives, Int. J. Hydrogen Energy 42 (11) (2017) 7254–7262. https://doi.org/10.1016/j.ijhydene.2016.03.178.

[16] E. Rivard, M. Trudeau, K. Zaghib, Hydrogen storage for mobility: a review, Materials 12 (12) (2019b). https://doi.org/10.3390/ma12121973.

[17] C. Sambo, A. Dudun, S.A. Samuel, P. Esenenjor, N.S. Muhammed, B. Haq, A review on worldwide underground hydrogen storage operating and potential fields, Int. J. Hydrogen Energy 47 (2022) 22840–22880.

[18] M. Panfilov, G. Gravier, S. Fillacier, Underground storage of H_2 and H_2-CO_2-CH_4 mixtures, in: ECMOR X-10th European Conference on the Mathematics of Oil Recovery, cp-23, European Association of Geoscientists & Engineers, 2006. https://doi.org/10.3997/2214-4609.201402474.

[19] S.R. Thiyagarajan, H. Emadi, A. Hussain, P. Patange, M. Watson, A comprehensive review of the mechanisms and efficiency of underground hydrogen storage, J. Energy Storage 51 (2022) 104490.

[20] P. Krawczak, Composite high pressure tanks, Engineering Technique: Plastic and Composite, Tech. Ing., Plast. Compos, 4th Edition TIP100WEB (2002) 1–11. https://doi.org/10.51257/A-V1-AM5530.

[21] Wong, J. (2009). CNG & hydrogen tank safety, R&D, and testing. Powertech, 1–33. https://www.energy.gov/eere/fuelcells/articles/cng-and-hydrogen-tank-safety-rd-and-testing.

[22] L. Zhou, Y. Zhou, Y. Sun, Enhanced storage of hydrogen at the temperature of liquid nitrogen, Int. J. Hydrogen Energy 29 (3) (2004) 319–322. https://doi.org/10.1016/S0360-3199(03)00155-1.

[23] A. Sarkar, R. Banerjee, Net energy analysis of hydrogen storage options, Int. J. Hydrogen Energy 30 (8) (2005) 867–877. https://doi.org/10.1016/j.ijhydene.2004.10.021.

[24] O. Kircher, G. Greim, J. Burtscher, T. Brunner, Validation of cryo-compressed hydrogen storage (CcH_2): a probalistic approach, 4th International Conference on Hydrogen Safety, 2011. http://conference.ing.unipi.it/ichs2011/papers/258.pdf.

[25] R.K. Ahluwalia, J.-K. Peng, T.Q. Hua, Cryo-compressed hydrogen storage, Compendium of Hydrogen Energy, Elsevier Ltd, 2016. https://doi.org/10.1016/b978-1-78242-362-1.00005-5.

[26] P. Sharma, H.H. Sardar, S. Neogi, Thermomechanical processing of type-4 composite cylinders under static load, J. Energy Stor. 55 (2022) 105465. https://doi.org/10.1016/j.est.2022.105465.

[27] C.R. Baker, R.L. Shaner, A study of the efficiency of hydrogen liquefaction, Int. J. Hydrogen Energy 3 (3) (1978) 321–334.

[28] O.Y. Cheremnykh, Development of a method and means for transporting high-purity liquid hydrogen, Chem. Pet. Eng. 54 (11) (2019) 827–830.

[29] R.K. Ahluwalia, J.K. Peng, H.S. Roh, T.Q. Hua, C. Houchins, B.D. James, Supercritical cryo-compressed hydrogen storage for fuel cell electric buses, Int. J. Hydrogen Energy 43 (22) (2018) 10215–10231.

[30] R.K. Ahluwalia, J. Peng, T.Q. Hua, *Cryo-compressed hydrogen storage: performance and cost review* (2011). https://www.energy.gov/eere/fuelcells/articles/cryo-compressed-hydrogen-storage-performance-and-cost-review.

[31] K. Reddi, M. Mintz, A. Elgowainy, E. Sutherland, Challenges and opportunities of hydrogen delivery via pipeline, tube-trailer, LIQUID tanker and methanation-natural gas grid, Hydrogen Sci. Eng.: Mater. Proc. Syst. Technol. 2 (2016) 849–874. https://doi.org/10.1002/9783527674268.CH35.

[32] G. Hu, C. Chen, H.T. Lu, Y. Wu, C. Liu, L. Tao, Y. Men, G. He, K.G. Li, A review of technical advances, barriers, and solutions in the power to hydrogen (P2H) roadmap, Engineering 6 (12) (2020) 1364–1380.

[33] DOE. (2022b). *Hydrogen tube trailers.* https://www.energy.gov/eere/fuelcells/hydrogen-tube-trailers.

[34] J. Adolf, C.H. Balzer, J. Louis, U. Schabla, M. Fischedick, K. Arnold, A. Pastowski, D. Schüwer, *Energy of the future? Sustainable mobility through fuel cells and H2; shell hydrogen study* (2017). https://epub.wupperinst.org/frontdoor/index/index/docId/6786.

[35] Hydrogen Europe. (2022). *Hydrogen Europe.* https://hydrogeneurope.eu/.

[36] IEA. (2019). The future of hydrogen: analysis: IEA. https://www.iea.org/reports/the-future-of-hydrogen.

[37] J. Wang, Z. Huang, J. Zheng, H. Miao, Effect of partially premixed and hydrogen addition on natural gas direct-injection lean combustion, Int. J. Hydrogen Energy 34 (22) (2009) 9239–9247.

[38] I.A. Gondal, Hydrogen integration in power-to-gas networks, Int. J. Hydrogen Energy 44 (3) (2019) 1803–1815.

[39] F. Bainier, R. Kurz, Impacts of H_2 blending on capacity and efficiency on a gas transport network, Proc. ASME Turbo Expo: Power Land, Sea, Air 58721 (2019) V009T27A014.

[40] M. Sun, X. Huang, Y. Hu, S. Lyu, Effects on the performance of domestic gas appliances operated on natural gas mixed with hydrogen, Energy 244 (2022) 122557.

[41] B. Mahant, P. Linga, R. Kumar, Hydrogen economy and role of hythane as a bridging solution: a perspective review, Energy Fuels 35 (19) (2021) 15424–15454.

[42] G. Sdanghi, G. Maranzana, A. Celzard, V. Fierro, Review of the current technologies and performances of hydrogen compression for stationary and automotive applications, Renew. Sustain. Energy Rev. 102 (2019) 150–170. https://doi.org/10.1016/j.rser.2018.11.028.

[43] J. Liu, L. Teng, B. Liu, P. Han, W. Li, Analysis of hydrogen gas injection at various compositions in an existing natural gas pipeline, Front. Energy Res. 9 (2021) 1–13. https://doi.org/10.3389/fenrg.2021.685079.

[44] A.V. Rusanov, V.V. Solovey, M.V. Lototskyy, Thermodynamic features of metal hydride thermal sorption compressors and perspectives of their application in hydrogen liquefaction systems, J. Phys. Energy 2 (2) (2020). https://doi.org/10.1088/2515-7655/ab7bf4.

[45] H. Gürbüz, The effect of H_2 purity on the combustion, performance, emissions and energy costs in an SI engine, Therm. Sci. 24 (2018) 315. https://doi.org/10.2298/TSCI180705315G.

[46] A. Małachowska, N. Łukasik, J. Mioduska, J. Gębicki, Hydrogen storage in geological formations: the potential of salt caverns, Energies 15 (14) (2022) 1–19. https://doi.org/10.3390/en15145038.

[47] P.H. Berry, A.G. Lipman, book & media reviews, J. Pain Palliative Care Pharmacother. 9 (3) (2001) 89–94. https://doi.org/10.1300/j088v09n03_08.

[48] H. de Vries, A.V. Mokhov, H.B. Levinsky, The impact of natural gas/hydrogen mixtures on the performance of end-use equipment: interchangeability analysis for domestic appliances, Appl. Energy 208 (2017) 1007–1019.

[49] A. Haratian, S.E. Meybodi, M. Dehvedar, A comprehensive review on the feasibility of largescale underground hydrogen storage in geological structures with concentration on salt caverns a comprehensive review on the feasibility of large-scale underground hydrogen storage in geological structures (2022). https://www.researchgate.net/publication/359191528.

[50] K. Nanthagopal, R. Subbarao, T. Elango, P. Baskar, K. Annamalai, Hydrogen enriched compressed natural gas: a futuristic fuel for internal combustion engines, Therm. Sci. 15 (4) (2011) 1145–1154. https://doi.org/10.2298/TSCI100730044N.

[51] D. Onyango, Britain's gas grid to accept 20% hydrogen blend by 2023, preparations underway, Pipeline Technol J (2022). https://www.pipeline-journal.net/news/britains-gas-grid-accept-20-hydrogen-blend-2023-preparations-underway.

[52] M.D. Paster, R.K. Ahluwalia, G. Berry, A. Elgowainy, S. Lasher, K. McKenney, M. Gardiner, Hydrogen storage technology options for fuel cell vehicles: well-to-wheel costs, energy efficiencies, and greenhouse gas emissions, Int. J. Hydrogen Energy 36 (22) (2011) 14534–14551. https://doi.org/10.1016/j.ijhydene.2011.07.056.

[53] R.R. Ratnakar, N. Gupta, K. Zhang, C. van Doorne, J. Fesmire, B. Dindoruk, V. Balakotaiah, Hydrogen supply chain and challenges in large-scale LH_2 storage and transportation, Int. J. Hydrogen Energy 46 (47) (2021) 24149–24168.

[54] M. Balat, M. Balat, Political, economic and environmental impacts of biomass-based hydrogen, Int. J. Hydrogen Energy 34 (9) (2009) 3589–3603.

[55] A.M. Swanger, K.M. Jumper, J.E. Fesmire, W.U. Notardonato, Modification of a liquid hydrogen tank for integrated refrigeration and storage, IOP Conf. Ser. Mater. Sci. Eng. 101 (1) (2015) 12080.

[56] R. Tarkowski, Underground hydrogen storage: characteristics and prospects, Renew. Sustain. Energy Rev. 105 (2019) 86–94.

[57] M.R. Withers, R. Malina, C.K. Gilmore, J.M. Gibbs, C. Trigg, P.J. Wolfe, P. Trivedi, S.R.H. Barrett, Economic and environmental assessment of liquefied natural gas as a supplemental aircraft fuel, Prog. Aerosp. Sci. 66 (2014) 17–36.

[58] UAC. (2022). CNG, biogas and hydrogen storage and transportation solutions. https://www.uac.no/download/brochure_cng-biogas-and-hydrogen-storage-and-transportation-solutions/.

[59] J. Adams, K.L. Simmons, Cold and Cryo-Compressed Hydrogen Storage R&D and Applications, Department of Energy, USA (2020). https://www.energy.gov/sites/default/files/2020/08/f77/hfto-webinar-cryogenic-h2-july2020.pdf.

Metal hydride hydrogen storage: A systems perspective

Mahvash Afzal
Department of Mechanical Engineering, Islamic University of Science and Technology, Kashmir, Jammu and Kashmir, India

4.3.1 Introduction

As the transition from fossil fuels to renewables takes place globally, hydrogen is emerging as a key energy alternative in the scenario. Hydrogen, although not a source of energy, is a promising energy carrier with its high energy density and benign environmental impact. The hydrogen energy structure is comprised of mainly three components, i.e., hydrogen production, storage and distribution, and application. The storage of hydrogen can be done either in the compressed gas state, liquid state, or as a hydride in the solid state. As hydrogen is a very light molecule with very low density, the gaseous and liquefied hydrogen storage is highly energy intensive. Hydrogen is also highly inflammable, therefore, storing hydrogen in the compressed gas form or liquid form entails multiple safety concerns as well. Hydrogen storage in the form of a hydride is relatively easier without a huge energy penalty or any safety concerns. Hydrogen at high pressure is injected into a bed of powdered metal/alloy. It reacts with the metal or alloy to form a metal hydride. This process is accompanied by the release of energy. If we require the hydrogen back in the gaseous form, heat is to be supplied to the metal hydride to obtain hydrogen. This makes the storage pathway a very safe one.

$$M + \frac{x}{2}H_2 \leftrightarrow MH_x + HEAT$$

The exothermic nature of the hydrogenation reaction implies that we need to continuously withdraw heat from the reaction for the reaction to keep progressing. The reverse reaction also requires the supply of heat to the reaction vessel. The MHHST has to be designed to ensure quick and efficient transport of heat to/from the reaction bed. Low thermal conductivity of the alloy powder (approx. 1 W/mK) aggravates the problem. From a design perspective, after taking into account the allowable stresses

Towards Hydrogen Infrastructure: Advances and Challenges in Preparing for the Hydrogen Economy.
DOI: https://doi.org/10.1016/B978-0-323-95553-9.00004-2

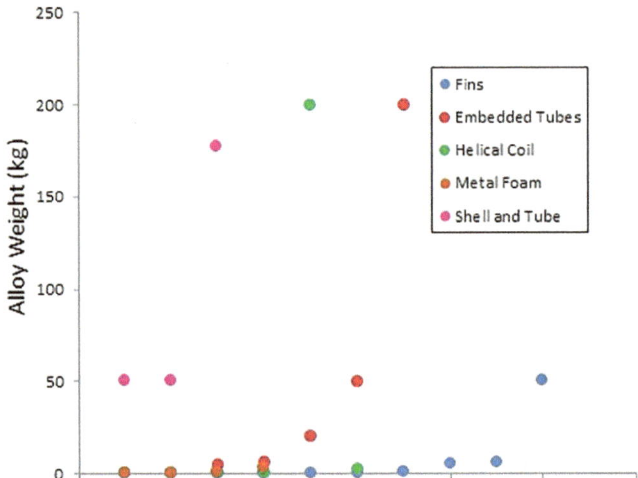

Figure 4.3.1 Graphical Representation of system scale correlated with thermal management system.

due to the design pressure and expansion/contraction of the alloy lattice, heat transfer forms the cornerstone of the MH hydrogen storage tank (MHHST) design. It is imperative that a robust thermal management system be designed for the MHHST to ensure efficient hydrogenation/ dehydrogenation.

The designs for MHHSTs have been many and varied, ranging from the most common cylindrical tanks [1–4] to rectangular or plate-type storage reactors [5–7], to multi-tubular [8], to elliptical reactors [9], etc. Each of these configurations employs different heat transfer enhancements such as fins [10–17], internal or external; cooling tubes [10,18–24] which may be straight or coiled; metal foam [11,16,25–28], expanded natural graphite (ENG) compacts [29–31], phase change materials (PCM) [32–37], etc. The choice of heat transfer enhancements depends heavily on system scale, application whether stationary or vehicular, availability of heat transfer fluid, cost of alloy, etc. Fig. 4.3.1 shows the type of thermal management system correlated with system scale. It may be observed that metal foam is restricted mostly to small-scale systems with alloy weight of the order of a few kgs while shell and tube type arrangement is mostly used for larger systems. Pertinent to mention here that data related to the quantity of alloy used is generally not available in the literature. Fig. 4.3.1, therefore, is a representative of the investigations carried out. It is also evident that the number of lab-scale studies outnumbers studies carried out on a larger scale.

4.3.2 Classification of metal hydride hydrogen storage tanks

Hydrogen storage tanks may be classified on the basis of scale, design, and end-use. Although there are no clear categories for the classification of MHHSTs in terms of scale, a broad distinction between lab-scale and large-scale systems can still be made. In terms of end-use MHHSTs can be designed for stationary or mobile applications. The end-use can strongly influence the design and will determine the type of heat exchanger/enhancement configuration employed in the tank.

4.3.2.1 Classification on the basis of system scale

On the basis of system scale, MHHSTs may be classified as lab-scale and large-scale. Systems that are designed for theoretical studies in labs for investigation of influence of operating or design parameters on reactor behavior may be classified as lab-scale. These are generally a few cm in height and diameter. The quantity of MH alloy ranges from a few grams to a few kilograms. Lab-scale systems are not designed with any specific application in mind. The majority of the studies conducted on MHHSTs comprise lab-scale systems. A comprehensive understanding of system operation and a foundation for application-oriented studies is built on the basis of these studies.

Multiple studies on assessing the importance of using three-dimensional models for numerical investigations were conducted. It was established that two-dimensional models were of sufficient accuracy while three-dimensional models could increase the computational without any significant advantage especially in the case of symmetrical cylindrical systems [38]. The optimum H/R (height to radius) ratio for MH reactors has been established such that it should be neither H/R $>>$ 2 nor H/R $<<$ 2 [39]. It is also a good practice to limit the expansion volume to 20% of the reactor volume; although this maybe lesser in larger reactors. The effectiveness of external and internal fins, concentric cooling tubes has been amply demonstrated in literature. Numerical and experimental studies on the effect of operating pressure, addition of fins, heat transfer coefficient, and cooling fluid and flow rate were conducted by Chung et al., Ye et al., Muthukumar et al., Souahlia et al., and Bao et al. [12,40–45]. These studies established that increasing supply pressure, addition of cooling tubes, and heat transfer coefficients improved the reactor performance significantly. These studies established the importance of heat transfer parameters in improving reactor performance. In addition to the importance of the aforementioned operating parameters, multiple lab-scale studies have also been conducted to determine the effect of thermophysical properties of the alloy such as thermal conductivity, activation energy, and reaction constant on the performance of hydrogen storage alloys [46–49]. These studies have established that thermal conductivity is of pivotal importance in improving the performance of metal hydride hydrogen storage reactors.

While the lab-scale studies are of colossal importance in establishing the character of reactions occurring in the absorption/desorption of hydrogen in metal hydrides, it is imperative to underline that large-scale systems have unique challenges associated with them due to their large size. While transverse fins and metal foams may be employed in small reactors without difficulty, in large-scale systems where scores of kilograms of the alloy have to be filled, homogeneous distribution of alloy within the reactor would pose a challenge if transverse fins or foam is used. Similarly, limitations in heat transfer resulting from large bed radius arise. This has resulted in the development of a multitubular shell and tube-type reactor with alloy housed in multiple tubes of small radius surrounded by a shell containing the heat transfer fluid [8,17,23,50–53]. Restricting the radius ensures that the heat transport is not affected by the size of the reactor. Alternatively, many designs have used embedded cooling tubes in a bed of metal alloy [20–22,54]. The idea is to reduce the conduction pathway between the cooling surfaces as it has been established that effective heat transfer can take place if the conduction pathway is limited to 10 mm [55].

4.3.2.2 Classification on the basis of tank configuration

The mechanical design of MHHSTs is based on the ASME (American Society of Mechanical Engineers) Pressure vessel code [56]. The code governs the relationship between the thickness of the high-pressure vessel and the maximum internal pressure that it is subjected to. In the case of welded joints or flanges, the efficiency of the joints can also be taken into account while determining the suitable thickness of the vessel on the basis of the internal pressure. As most of the MHHSTs have a cylindrical geometry, the mechanical design tends to be straightforwardly guided by the ASME pressure vessel code. The heat transfer design, on the other hand, results in a multitude of reactor configurations based on the heat transfer enhancement that is chosen.

4.3.2.2.1 Fins

Perhaps the most common heat transfer enhancement that has been employed to improve conduction heat transport in MHHSTs are fins. Extensive studies have been carried out to optimize the dimensions of fins that have been used in different reactors. Fins may be external or internal [12,57], transverse or longitudinal [58], and connected to a heat exchanger or the reactor body itself [15,59]. As mentioned earlier, the use of transverse fins could lead to the problem of homogeneous distribution of alloy especially with large quantities of alloy. To counter this problem, unique fin designs have been suggested such as fins with perforations [59], conical fins [60] that facilitate the filling of alloy in the reactor. To increase the surface area and reach of fins in reactors, bifurcating longitudinal fins were investigated and optimized [61]. The optimization of fins is critical to ensure an efficiency of at least greater than 85%. Otherwise, the parasitic weight added to the system would be a major drawback to contend with.

4.3.2.2.2 Embedded tubes, coils, and cooling jacket

As ubiquitous as the fins, embedded cooling tubes are also extensively used for augmenting heat transport to/from metal hydride beds. These maybe used in the form of a single concentric tube, multiple tubes, helical coils, or even mini-channels [7,62]. Often times, embedded tubes may be used in conjunction with fins to improve their effectiveness [58,63]. The diameter of the tubes, their number, and the pitch in the case of helical tubes can be optimized on the basis of the reactor for which it is used [3,64,65]. Some interesting designs that have led to a significant improvement in heat transport include the branched channel heat exchanger in which the main longitudinal cooling tubes branch into transverse tubes that empty into the external water jacket as shown in Fig. 4.3.2 [66]. Embossed plate heat exchangers (EPHX) with parallel and serpentine flow fields were compared with helical coils. The hydrogen absorption rate with serpentine EPHX was found to be similar to that of the helical coil [67,68]. The appropriate heat transfer mechanism may be selected taking into account factors such as end-use, availability of manufacturing facilities, heat transfer fluid etc.

While cooling tubes/coils offer an effective heat transport solution from the metal hydride bed, the fabrication is an intricate task. In case of multiple embedded tubes, welding of tubes of diameters of a few mm can be challenging. Any lapse in welding can result in leakage of the heat transfer fluid into the alloy bed. In addition, it is imperative to maintain the purity of the heat transfer fluid. Presence of impurities can lead to clogging of

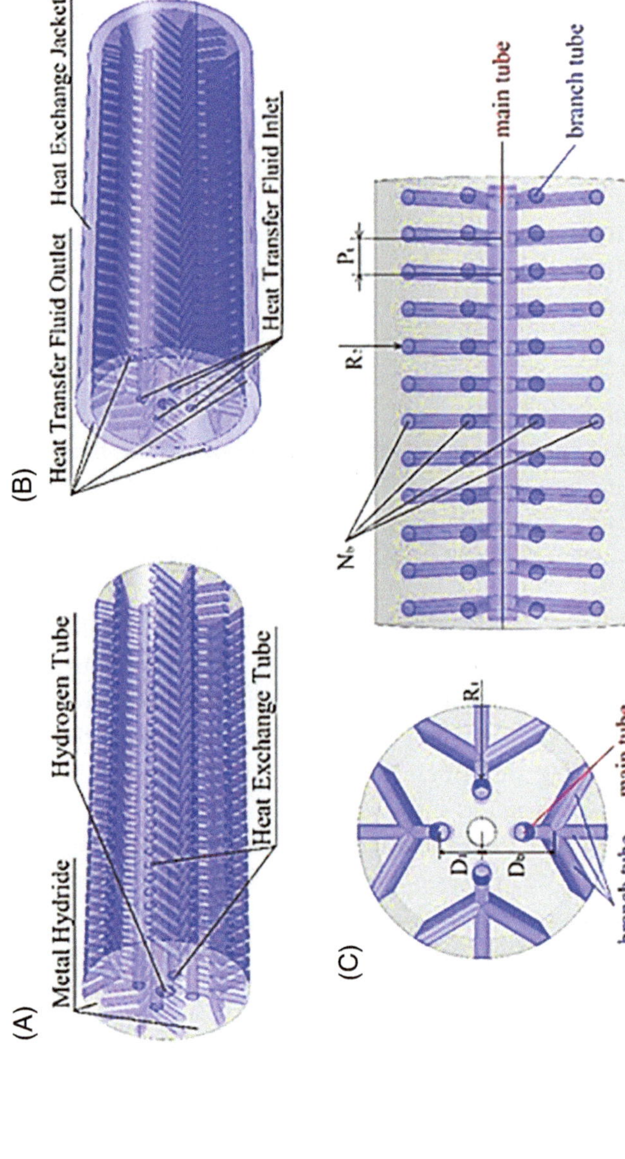

Figure 4.3.2 Graphical representation of the branched channel reactors (A) without external jacket (B) with jacket [66].

Figure 4.3.3 Magnesium compacts before and after 1000 cycles [29].

tubes especially in case of smaller tubes and coils. If water is used as a heat transfer fluid, the possibility of corrosion of tubes exists over long periods of exposure.

4.3.2.2.3 Expanded natural graphite and metal foam

Thermal conductivity of the MH bed is very low ranging from 0.1 to 1 W/mK [69]. High thermal conductivity materials such as metal shavings and expanded natural graphite (ENG) may be mixed with the alloy to improve the thermal conductivity. This can result in an increased effective thermal conductivity of over 10 W/mK. As early as 1980, a metal matrix composite was developed where in addition of copper and aluminum was done to form an alloy of $LaNi_5$ [70]. The resulting composite was pelletized by compaction. The thermal conductivity of the pellet was observed to be 32 W/mK with rapid hydrogen absorption.

Advancements in material science have led to the identification and development of newer heat transfer materials. Expanded natural graphite (ENG) is produced by the chemical treatment of natural graphite flakes. It has high thermal conductivity and stability. ENG 5–10% by volume is mixed with metal alloy and compacted under a pressure of around 75 MPa [29,44,71]. These compacts show promising hydrogenation/dehydrogenation characteristics. Over a few hundred cycles, they may begin to disintegrate resulting in a decreased thermal conductivity. Fig. 4.3.3 shows the pictures of Mg compacts before and after 1000 cycles. While the disintegration of compacts after hundreds of cycles may not be very significant it still depends upon the operating pressure and temperature, the material in question, and the compaction pressure. In order to further reduce the possibility of disintegration, compacts may be reinforced with a mesh matrix for mechanical stability [72].

Metal foam has also been used as an effective heat transfer augmentation technique [11,73,74]. The effective thermal conductivity after the addition of metal foam with porosity of around 90% increases up to 10 W/mK. The effective thermal capacity per unit volume and effective thermal conductivity are expressed as:

$$\left(\rho C_p\right)_{eff} = \rho_{foam} C_{p_{foam}} + \left(\rho_g + \rho_{MH}\right) C_{p_{MH}} + \rho_g C_{p_g} \tag{4.3.1}$$

$$k_{eff} = k_{foam} + \varepsilon_{foam} \varepsilon_{MH} k_g + \varepsilon_{foam} k_{MH} \tag{4.3.2}$$

The foam porosity also has a bearing on the absorption/desorption in the reactor. It was observed that a foam porosity of <20 ppi (pores per inch) adversely affected the rate of absorption/desorption [74]. A smaller pore (cell) size would also lead to accumulation of alloy and hinder free flow of alloy within the foam.

4.3.2.2.4 Adjunct heat transfer enhancement techniques

The use of phase change materials (PCM) in metal hydride–based hydrogen storage systems has gained traction in the last few years [33–36]. PCM such as paraffins are materials that absorb large amounts of heat when going through a change in phase and release the same amount of heat when the process is reversed. The heat released during hydrogen absorption is taken up by the PCM and is expended as latent heat of fusion of the PCM. During desorption, the PCM solidifies while supplying the latent heat of fusion to the metal hydride bed. PCM add tremendous amount of additional weight to the MHHST, therefore, they are not recommended for mobile/vehicular hydrogen storage applications. An appropriate thermal coupling between the hydride and PCM is essential for the effective functioning of the PCM, i.e., the hydriding temperature should be greater than the melting point of the PCM.

A thermal coupling between metal hydride reactors and the associated fuel cell is also an effective route of thermal management. The heat released by the fuel cell can be diverted to the metal hydride reactor during dehydrogenation. The heat may be diverted directly by diverting the hot air through a duct in case of open-cathode fuel cells or alternatively an addition of heat pipes may be undertaken [75–77]. Heat pipes are heat transfer devices that work on the principle of phase change and capillary action. Working fluid contained in the inner tube of the heat pipe takes up the heat from the fuel cell and rises as a vapor towards the metal hydride reactor where dehydrogenation is taking place. It loses the heat to the reactor and condenses to liquid phase and returns back via capillary action through the outer annular space. It was observed that of the heat generated during the operation of a 500 W fuel cell only 40% was sufficient to maintain the required supply of hydrogen from the reactor.

4.3.2.3 Classification on the basis of end-use

On the basis of end-use, the MHHSTs may be classified as stationary or vehicular. The use of metal hydride based hydrogen storage systems has been tested for vehicular applications under static and dynamic loading and the results have been found to be promising [78,79]. A metal hydride based hydrogen powered bus by the name of H2FUEL bus was operated in Augusta and was observed to meet all the performance requirements. The MH tank was equipped with a U-tube heat exchanger and aluminum foam [80]. The MH tank was stored on the bus floor. In case of MHHSTs designed for smaller vehicles, system weight has to be as low as possible which rules out the use of heat transfer enhancement techniques like PCM or the use of shell and tube type reactors where heat transfer fluid (HTF) weight contributes significantly to the system. This leads to air-cooled MHHSTs as a favorable option for vehicular applications [81]. A fuel cell powered scooter was observed to operate successfully with a mileage three times that of the existing figures for gasoline powered scooters [82].

In terms of stationary applications, system weight is not a major inhibiting factor in the selection of heat exchanger configurations. A totalized hydrogen energy utilization system (THEUS) was developed as a collaborative effort of the United States and Japan [83,84]. It was equipped with a solar powered electrolyzer for production of hydrogen, a metal hydride-based storage unit and a fuel cell. A resistive load of 0.5 kW was operated for over 5 hours. In addition, the fuel cell was also used to operate the auxiliary systems. The metal hydride tank which was equipped with double helical coil heat exchanger indicated good thermal performance. Using a total of 99 embedded cooling tubes, La-based metal hydride tank was designed and tested for industrial applications. It was capable to storing over 500 gm of hydrogen at a supply pressure of a moderate 20 bar and HTF flow rate of up to 30 Lpm [20]. A 50 kg La-based reactor fitted with hexagonal shaped heat transfer enhancements was designed to provide back-up of 10 kWh [85]. It was equipped with a water jacket where the flow rate of water as heat transfer fluid could be regulated.

There is a great diversity in the types of metal hydride hydrogen storage tanks based on the materials, applications, and scale. The challenge lies in the low system gravimetric capacity of the metal hydride. Most metal hydrides have a gravimetric capacity below 2 wt% except for Magnesium which can go up to 4 wt% [86]. With the addition of system weight, the gravimetric capacity drops further to less than 1 wt%. Maintenance of appropriate temperature and pressure is also vital for the efficient storage and withdrawal of hydrogen. Despite these constraints, MH storage of hydrogen can be explored as an alternative to compressed gas storage for economical as well as safety reasons.

4.3.3 Safety and reliability

Hydrogen has a flammability range from 4% to 74% which makes it potentially dangerous for public use especially at higher pressures. It becomes critical to evaluate the safety and reliability of hydrogen infrastructure before it is opened for operations. While there have been multiple investigations to study the risks associated with compressed gas hydrogen storage and liquified hydrogen storage, there is a severe dearth of data on the safety of hydrogen storage in metal hydrides. This could be due to the fact that metal hydride based hydrogen storage is inherently a safe method of hydrogen storage. It offers higher safety in comparison to gaseous or liquid hydrogen storage due low storage pressures and nonspontaneous release of hydrogen.

However, some hydrogen storage materials such AlH_3 and Ti-based alloys may become pyrophoric after activation and react violently with air or water [87–89]. Activation is the cyclic absorption and desorption of hydrogen in metal alloy powder till the maximum capacity of the alloy is reached. Utmost care has to be taken in these cases to prevent exposure to air or water and using corrosion resistant tank materials. Water based cooling/heating solutions which could lead to corrosive deterioration of joints, stress spots should be avoided in these cases.

4.3.4 Present status

Hydrogen powered fuel cell vehicles currently in operation use compressed gas hydrogen housed at 700 bar pressure in carbon fiber reinforced tanks. These tanks, known as type-IV tanks, are cost intensive both in terms of structure and operation since the pressure is high. Storing hydrogen in metal hydrides for light vehicular purposes is challenging because of system weight and the response of the MHHST to the variation in load. The issue of developing hydrogen refilling infrastructure where hydrogen can be filled on the lines of diesel or gasoline also exists. However, Toyota has come up with a unique plug and play hydrogen cartridge which can be fitted into the vehicle fully charged and replaced after discharging [90]. This resolves the issues associated with longer fill times and simplifies the refueling infrastructure. Many metal hydride based hydrogen storage solutions have already been commercialized catering mostly to stationary applications. Several industries such as MAHYTEC and Pragma industries in France, GKN in Europe and the United States, Hystorsys in Norway, and Hbank in Taiwan are some of the leading suppliers of metal hydride-based hydrogen storage solutions. While the complete commercialization of hydrogen as an energy carrier is yet to be realized, at the same time that market penetration of hydrogen based energy solutions has increased.

References

[1] P. Muthukumar, M.P. Maiya, S.S. Murthy, Experiments on a metal hydride based hydrogen compressor, Int. J. Hydrogen Energy 30 (2005) 879–892. https://doi.org/10.1016/j.ijhydene.2004.09.003.

[2] S. Mellouli, F. Askri, H. Dhaou, A. Jemni, S. Ben Nasrallah, Numerical study of heat exchanger effects on charge/discharge times of metal-hydrogen storage vessel, Int. J. Hydrogen Energy 34 (2009) 3005–3017. https://doi.org/10.1016/j.ijhydene.2008.12.099.

[3] Z. Bao, F. Yang, Z. Wu, S. Nyallang Nyamsi, Z. Zhang, Optimal design of metal hydride reactors based on CFD-Taguchi combined method, Energy Convers. Manag. 65 (2013) 322–330. https://doi.org/10.1016/j.enconman.2012.07.027.

[4] S. Anbarasu, P. Muthukumar, S.C. Mishra, Tests on LmNi$_{4.91}$Sn$_{0.15}$ based solid state hydrogen storage device with embedded cooling tubes – part A: absorption process, Int. J. Hydrogen Energy 39 (2014) 3342–3351. https://doi.org/10.1016/j.ijhydene.2013.12.090.

[5] T. Oi, K. Maki, Y. Sakaki, Heat transfer characteristics of the metal hydride vessel based on the plate-fin type heat exchanger, J. Power Sources 125 (2004) 52–61. https://doi.org/10.1016/S0378-7753(03)00822-X.

[6] M. Botzung, S. Chaudourne, O. Gillia, C. Perret, M. Latroche, A. Percheron-Guegan, et al., Simulation and experimental validation of a hydrogen storage tank with metal hydrides, Int. J. Hydrogen Energy 33 (2008) 98–104. https://doi.org/10.1016/j.ijhydene.2007.08.030.

[7] X. Meng, Z. Wu, Z. Bao, F. Yang, Z. Zhang, Performance simulation and experimental confirmation of a mini-channel metal hydrides reactor, Int. J. Hydrogen Energy 38 (2013) 15242–15253. https://doi.org/10.1016/j.ijhydene.2013.09.056.

[8] M. Raju, S. Kumar, Optimization of heat exchanger designs in metal hydride based hydrogen storage systems, Int. J. Hydrogen Energy 37 (2012) 2767–2778. https://doi.org/10.1016/j.ijhydene.2011.06.120.

[9] C. Veerraju, M.R. Gopal, Heat and mass transfer studies on elliptical metal hydride tubes and tube banks, Int. J. Hydrogen Energy 34 (2009) 4340–4350. https://doi.org/10.1016/j.ijhydene.2009.03.022.

[10] H. Dhaou, A. Souahlia, S. Mellouli, F. Askri, A. Jemni, S. Ben Nasrallah, Experimental study of a metal hydride vessel based on a finned spiral heat exchanger, Int. J. Hydrogen Energy 35 (2010) 1674–1680. https://doi.org/10.1016/j.ijhydene.2009.11.094.

[11] S. Ferekh, G. Gwak, S. Kyoung, H.G. Kang, M.H. Chang, S.H. Yun, et al., Numerical comparison of heat-fin-and metal-foam-based hydrogen storage beds during hydrogen charging process, Int. J. Hydrogen Energy 40 (2015) 14540–14550. https://doi.org/10.1016/j.ijhydene.2015.07.149.

[12] B.D. MacDonald, A.M. Rowe, Impacts of external heat transfer enhancements on metal hydride storage tanks, Int. J. Hydrogen Energy 31 (2006) 1721–1731. https://doi.org/10.1016/j.ijhydene.2006.01.007.

[13] S. Garrier, A. Chaise, P. De Rango, P. Marty, B. Delhomme, D. Fruchart, et al., MgH2 intermediate scale tank tests under various experimental conditions, Int. J. Hydrogen Energy 36 (2011) 9719–9726. https://doi.org/10.1016/j.ijhydene.2011.05.017.

[14] A. Singh, M.P. Maiya, S.S. Murthy, Effects of heat exchanger design on the performance of a solid state hydrogen storage device, Int. J. Hydrogen Energy 40 (2015) 9733–9746. https://doi.org/10.1016/j.ijhydene.2015.06.015.

[15] V.K. Kukkapalli, S.W. Kim, Metal hydride reactor design optimization for hydrogen energy storage, Key Eng. Mater. 708 (2016) 85–93. https://doi.org/10.4028/www.scientific.net/KEM.708.85.

[16] X.S. Bai, W.W. Yang, X.Y. Tang, F.S. Yang, Y.H. Jiao, Y. Yang, Hydrogen absorption performance investigation of a cylindrical MH reactor with rectangle heat exchange channels, Energy 232 (2021) 121101. https://doi.org/10.1016/J.ENERGY.2021.121101.

[17] M. Afzal, N. Sharma, N. Gupta, P. Sharma, Transient simulation studies on a metal hydride based hydrogen storage reactor with longitudinal fins, J. Energy Storage 51 (2022) 104426. https://doi.org/10.1016/j.est.2022.104426.

[18] M. Visaria, I. Mudawar, Coiled-tube heat exchanger for High-pressure metal hydride hydrogen storage systems: part 1. experimental study, Int. J. Heat Mass Transf. 55 (2012) 1782–1795. https://doi.org/10.1016/j.ijheatmasstransfer.2011.11.035.

[19] S. Mellouli, F. Askri, H. Dhaou, A. Jemni, S. Ben Nasrallah, Numerical simulation of heat and mass transfer in metal hydride hydrogen storage tanks for fuel cell vehicles, Int. J. Hydrogen Energy 35 (2010) 1693–1705. https://doi.org/10.1016/j.ijhydene.2009.12.052.

[20] A. Kumar, N.N. Raju, P. Muthukumar, P.V. Selvan, Experimental studies on industrial scale metal hydride based hydrogen storage system with embedded cooling tubes, Int. J. Hydrogen Energy 44 (2019) 13549–13560. https://doi.org/10.1016/j.ijhydene.2019.03.180.

[21] P. Muthukumar, M.S. Patil, N.N. Raju, M. Imran, Parametric investigations on compressor-driven metal hydride based cooling system, Appl. Therm. Eng. 97 (2016) 87–99. https://doi.org/10.1016/j.applthermaleng.2015.10.155.

[22] N.N. Raju, P. Muthukumar, P.V. Selvan, K. Malleswararao, Design methodology and thermal modelling of industrial scale reactor for solid state hydrogen storage, Int. J. Hydrogen Energy 44 (2019) 20278–20292. https://doi.org/10.1016/j.ijhydene.2019.05.193.

[23] C. Na Ranong, M. Höhne, J. Franzen, J. Hapke, G. Fieg, M. Dornheim, et al., Concept, design and manufacture of a prototype hydrogen storage tank based on sodium alanate, Chem. Eng. Technol. 32 (2009) 1154–1163. https://doi.org/10.1002/ceat.200900095.

[24] S. Tiwari, P. Sharma, Simulations of hydrogen-storage system integrated with sensible storage system, Nanomater. Energy 8 (2019) 33–41. https://doi.org/10.1680/jnaen.18.00016.

[25] H. Wang, A.K. Prasad, S.G. Advani, Hydrogen storage systems based on hydride materials with enhanced thermal conductivity, Int. J. Hydrogen Energy 37 (2012) 290–298. https://doi.org/10.1016/j.ijhydene.2011.04.096.

[26] S. Levesque, M. Ciureanu, R. Roberge, T. Motyka, Hydrogen storage for fuel cell systems with stationary applications—I. transient measurement technique for packed bed evaluation, Int. J. Hydrogen Energy 25 (2000) 1095–1105. https://doi.org/10.1016/S0360-3199(00)00023-9.

[27] Y. Zhuo, S. Jung, Y. Shen, Numerical study of hydrogen desorption in an innovative metal hydride hydrogen storage tank, Energy Fuels 35 (2021) 10908–10917. https://doi.org/10.1021/acs.energyfuels.1c00666.

[28] Y. Chen, C.A.C. Sequeira, C. Chen, X. Wang, Q. Wang, Metal hydride beds and hydrogen supply tanks as minitype PEMFC hydrogen sources, Int. J. Hydrogen Energy 28 (2003) 329–333. https://doi.org/10.1016/S0360-3199(02)00064-2.

[29] M. Dieterich, C. Pohlmann, I. Bürger, M. Linder, L. Röntzsch, Long-term cycle stability of metal hydride-graphite composites, Int. J. Hydrogen Energy 40 (2015) 16375–16382. https://doi.org/10.1016/j.ijhydene.2015.09.013.

[30] K. Herbrig, L. Röntzsch, C. Pohlmann, T. Weißgärber, B. Kieback, Hydrogen storage systems based on hydride-graphite composites: computer simulation and experimental validation, Int. J. Hydrogen Energy 38 (2013) 7026–7036. https://doi.org/10.1016/j.ijhydene.2013.03.104.

[31] Y. Madaria, E.A. Kumar, P. Maiya, S.S. Murthy, Simulation of effective thermal conductivity of metal hydride packed beds, Heat Transf. Eng. 37 (2016) 616–624. https://doi.org/10.1080/01457632.2015.1066653.

[32] S. Garrier, B. Delhomme, P. De Rango, P. Marty, D. Fruchart, S. Miraglia, A new MgH$_2$ tank concept using a phase-change material to store the heat of reaction, Int. J. Hydrogen Energy 38 (2013) 9766–9771. https://doi.org/10.1016/j.ijhydene.2013.05.026.

[33] H. Ben Mâad, F. Askri, S. Ben Nasrallah, Heat and mass transfer in a metal hydrogen reactor equipped with a phase-change heat-exchanger, Int. J. Therm. Sci. 99 (2016) 271–278. https://doi.org/10.1016/j.ijthermalsci.2015.09.003.

[34] A.A.R. Darzi, H.H. Afrouzi, A. Moshfegh, M. Farhadi, Absorption and desorption of hydrogen in long metal hydride tank equipped with phase change material jacket, Int. J. Hydrogen Energy 41 (2016) 9595–9610. https://doi.org/10.1016/j.ijhydene.2016.04.051.

[35] Y. Ye, J. Lu, J. Ding, W. Wang, J. Yan, Numerical simulation on the storage performance of a phase change materials based metal hydride hydrogen storage tank, Appl. Energy 278 (2020) 115682. https://doi.org/10.1016/j.apenergy.2020.115682.

[36] S. Mellouli, F. Askri, E. Abhilash, S. Ben Nasrallah, Impact of using a heat transfer fluid pipe in a metal hydride-phase change material tank, Appl. Therm. Eng. 113 (2017) 554–565. https://doi.org/10.1016/j.applthermaleng.2016.11.065.

[37] Y. Ye, J. Lu, J. Ding, W. Wang, J. Yan, Numerical simulation on the storage performance of a phase change materials based metal hydride hydrogen storage tank, Appl. Energy 278 (2020) 115682. https://doi.org/10.1016/J.APENERGY.2020.115682.

[38] K. Aldas, M.D. Mat, Y. Kaplan, A three-dimensional mathematical model for absorption in a metal hydride bed, Int. J. Hydrogen Energy 27 (2002) 1049–1056. https://doi.org/10.1016/S0360-3199(02)00010-1.

[39] F. Askri, Prediction of transient heat and mass transfer in a closed metal–hydrogen reactor, Int. J. Hydrogen Energy 29 (2004) 195–208. https://doi.org/10.1016/S0360-3199(03)00089-2.

[40] C.A. Chung, C.J. Ho, Thermal-fluid behavior of the hydriding and dehydriding processes in a metal hydride hydrogen storage canister, Int. J. Hydrogen Energy 34 (2009) 4351–4464. https://doi.org/10.1016/j.ijhydene.2009.03.028.

[41] J. Ye, L. Jiang, Z. Li, X. Liu, S. Wang, X. Li, Numerical analysis of heat and mass transfer during absorption of hydrogen in metal hydride based hydrogen storage tanks, Int. J. Hydrogen Energy 35 (2010) 8216–8224. https://doi.org/10.1016/j.ijhydene.2009.12.086.

[42] P. Muthukumar, M.P. Maiya, S.S. Murthy, Experiments on a metal hydride-based hydrogen storage device, Int. J. Hydrogen Energy 30 (2005) 1569–1581. https://doi.org/10.1016/j.ijhydene.2004.12.007.

[43] A. Souahlia, H. Dhaou, F. Askri, S. Mellouli, A. Jemni, S. Ben Nasrallah, Experimental study and characterization of metal hydride containers, Int. J. Hydrogen Energy 36 (2011) 4952–4957. https://doi.org/10.1016/j.ijhydene.2011.01.074.

[44] Z. Bao, F. Yang, Z. Wu, X. Cao, Z. Zhang, Simulation studies on heat and mass transfer in high-temperature magnesium hydride reactors, Appl. Energy 112 (2013) 1181–1189. https://doi.org/10.1016/j.apenergy.2013.04.053.

[45] C.A. Chung, C.S. Lin, Prediction of hydrogen desorption performance of Mg_2Ni hydride reactors, Int. J. Hydrogen Energy 34 (2009) 9409–9423. https://doi.org/10.1016/j.ijhydene.2009.09.061.

[46] Y. Madaria, E. Anil Kumar, S. Srinivasa Murthy, Effective thermal conductivity of reactive packed beds of hydriding materials, Appl. Therm. Eng. 98 (2016) 976–990. https://doi.org/10.1016/j.applthermaleng.2016.01.013.

[47] V.K. Sharma, E.A. Kumar, Measurement and analysis of reaction kinetics of La-based hydride pairs suitable for metal hydride e based cooling systems, Int. J. Hydrogen Energy 39 (2014) 19156–19168. https://doi.org/10.1016/j.ijhydene.2014.09.083.

[48] M. Valizadeh, M.A. Delavar, M. Farhadi, Numerical simulation of heat and mass transfer during hydrogen desorption in metal hydride storage tank by Lattice Boltzmann method, Int. J. Hydrogen Energy 41 (2016) 413–424. https://doi.org/10.1016/j.ijhydene.2015.11.075.

[49] A. Chibani, S. Merouani, C. Bougriou, L. Hamadi, Heat and mass transfer during the storage of hydrogen in $LaNi_5$-based metal hydride: 2D simulation results for a large scale, multi-pipes fixed- bed reactor, Int. J. Heat Mass Transf. 147 (2019) 118939. https://doi.org/10.1016/j.ijheatmasstransfer.2019.118939.

[50] M. Raju, J.P. Ortmann, S. Kumar, System simulation model for high-pressure metal hydride hydrogen storage systems, Int. J. Hydrogen Energy 35 (2010) 8742–8754. https://doi.org/10.1016/j.ijhydene.2010.05.024.

[51] M. Afzal, P. Sharma, Design and computational analysis of a metal hydride hydrogen storage system with hexagonal honeycomb based heat transfer enhancements-part A, Int. J. Hydrogen Energy 46 (2021) 13116–13130. https://doi.org/10.1016/j.ijhydene.2021.01.135.

[52] M.V. Lototskyy, I. Tolj, M.W. Davids, Y.V. Klochko, A. Parsons, D. Swanepoel, et al., Metal hydride hydrogen storage and supply systems for electric forklift with low-temperature proton exchange membrane fuel cell power module, Int. J. Hydrogen Energy 41 (2016) 13831–13842. https://doi.org/10.1016/j.ijhydene.2016.01.148.

[53] M. Lototskyy, I. Tolj, Y. Klochko, M. Wafeeq, D. Swanepoel, V. Linkov, Metal hydride hydrogen storage tank for fuel cell utility vehicles, Int. J. Hydrogen Energy 45 (2019) 7958–7967. https://doi.org/10.1016/j.ijhydene.2019.04.124.

[54] P. Muthukumar, K. Singh Patel, P. Sachan, N. Singhal, Computational study on metal hydride based three-stage hydrogen compressor, Int. J. Hydrogen Energy 37 (2012) 3797–3806. https://doi.org/10.1016/j.ijhydene.2011.05.104.

[55] M. Visaria, I. Mudawar, T. Pourpoint, Enhanced heat exchanger design for hydrogen storage using high-pressure metal hydride: part 2. Experimental results, Int. J. Heat Mass Transf. 54 (2011) 424–432. https://doi.org/10.1016/j.ijheatmasstransfer.2010.09.028.

[56] B.S. Thakkar, S.A. Thakkar, Design of pressure vessel using ASME code, section VIII, division 1, Int. J. Adv. Eng. Res. Stud. 1 (2012) 228–234.

[57] F. Askri, M. Ben Salah, a. Jemni, S. Ben Nasrallah, Optimization of hydrogen storage in metal-hydride tanks, Int. J. Hydrogen Energy 34 (2009) 897–905. https://doi.org/10.1016/j.ijhydene.2008.11.021.

[58] S.L. Garrison, B.J. Hardy, M.B. Gorbounov, D.A. Tamburello, C. Corgnale, B.A. Vanhassel, et al., Optimization of internal heat exchangers for hydrogen storage tanks utilizing metal hydrides, Int. J. Hydrogen Energy 37 (2012) 2850–2861. https://doi.org/10.1016/j.ijhydene.2011.07.044.

[59] M. Melnichuk, N. Silin, H.A. Peretti, Optimized heat transfer fin design for a metal-hydride hydrogen storage container, Int. J. Hydrogen Energy 34 (2009) 3417–3424. https://doi.org/10.1016/j.ijhydene.2009.02.040.

[60] S. Chandra, P. Sharma, P. Muthukumar, S. Sarma, V. Tatiparti, Modeling and numerical simulation of a 5 kg LaNi$_5$-based hydrogen storage reactor with internal conical fins, Int. J. Hydrogen Energy 45 (2020) 8794–8809. https://doi.org/10.1016/j.ijhydene.2020.01.115.

[61] A. Mallik, P. Sharma, Modeling and numerical simulation of an industrial scale metal hydride reactor based on CFD-Taguchi combined method, Energy Storage 3 (2021) e227. https://doi.org/10.1002/est2.227.

[62] B. Satya Sekhar, M. Lototskyy, A. Kolesnikov, M.L. Moropeng, B.P. Tarasov, B.G. Pollet, Performance analysis of cylindrical metal hydride beds with various heat exchange options, J. Alloys Compd. 645 (2015) S89–S95. https://doi.org/10.1016/j.jallcom.2014.12.272.

[63] A. Souahlia, H. Dhaou, F. Askri, M. Sofiene, A. Jemni, S. Ben Nasrallah, Experimental and comparative study of metal hydride hydrogen tanks, Int. J. Hydrogen Energy 36 (2011) 12918–12922. https://doi.org/10.1016/j.ijhydene.2011.07.022.

[64] S. Tiwari, P. Sharma, Design of metal hydride reactor with embedded helical coil using optimization methods, Eng. Res. Express 3 (2021) 025022. https://doi.org/10.1088/2631-8695/abf9e5.

[65] Z. Wu, F. Yang, Z. Zhang, Z. Bao, Magnesium based metal hydride reactor incorporating helical coil heat exchanger: Simulation study and optimal design, Appl. Energy 130 (2014) 712–722. https://doi.org/10.1016/j.apenergy.2013.12.071.

[66] D. Wang, Y. Wang, F. Wang, S. Zheng, S. Guan, L. Zheng, et al., Hydrogen storage in branch mini-channel metal hydride reactor: Optimization design, sensitivity analysis and quadratic regression, Int. J. Hydrogen Energy 46 (2021) 25189–25207. https://doi.org/10.1016/j.ijhydene.2021.05.051.

[67] P. Chippar, S.D. Lewis, S. Rai, A. Sircar, Numerical investigation of hydrogen absorption in a stackable metal hydride reactor utilizing compartmentalization, Int. J. Hydrogen Energy 43 (2018) 8007–8017. https://doi.org/10.1016/j.ijhydene.2018.03.017.

[68] S.D. Lewis, P. Chippar, Numerical investigation of hydrogen absorption in a metal hydride reactor with embedded embossed plate heat exchanger, Energy 194 (2020) 116942. https://doi.org/10.1016/j.energy.2020.116942.

[69] V.I. Borzenko, D.O. Dunikov, S.P. Malyshenko, Crisis phenomena in metal hydride hydrogen storage facilities, High Temp. 49 (2011) 249–256. https://doi.org/10.1134/S0018151X11010019.

[70] M. Ron, D. Gruen, M. Mendelsohn, I. Sheft, Preparation and properties of porous metal hydride compacts, J. Less Common Met. 74 (1980) 445–448. https://doi.org/10.1016/0022-5088(80)90183-6.

[71] A.R. Sánchez, H.-P. Klein, M. Groll, Expanded graphite as heat transfer matrix in metal hydride beds, Int. J. Hydrogen Energy 28 (2003) 515–527. https://doi.org/10.1016/S0360-3199(02)00057-5.

[72] Y. Madaria, E. Anil Kumar, Effect of heat transfer enhancement on the performance of metal hydride based hydrogen compressor, Int. J. Hydrogen Energy 41 (2016) 3961–3973. https://doi.org/10.1016/j.ijhydene.2016.01.011.

[73] M. Afzal, P. Sharma, Design of a large-scale metal hydride based hydrogen storage reactor: Simulation and heat transfer optimization, Int. J. Hydrogen Energy 43 (2018) 13356–13372. https://doi.org/10.1016/j.ijhydene.2018.05.084.

[74] F. Laurencelle, J. Goyette, Simulation of heat transfer in a metal hydride reactor with aluminium foam, Int. J. Hydrogen Energy 32 (2007) 2957–2964. https://doi.org/10.1016/j.ijhydene.2006.12.007.

[75] R. Omrani, H.Q. Nguyen, B. Shabani, Thermal coupling of an open-cathode proton exchange membrane fuel cell with metal hydride canisters: an experimental study, Int. J. Hydrogen Energy 45 (2020) 28940–28950. https://doi.org/10.1016/j.ijhydene.2020.07.122.

[76] A.P. Tetuko, B. Shabani, J. Andrews, Thermal coupling of PEM fuel cell and metal hydride hydrogen storage using heat pipes, Int. J. Hydrogen Energy 41 (2016) 4264–4277. https://doi.org/10.1016/j.ijhydene.2015.12.194.

[77] C.A. Chung, S.W. Yang, C.Y. Yang, C.W. Hsu, P.Y. Chiu, Experimental study on the hydrogen charge and discharge rates of metal hydride tanks using heat pipes to enhance heat transfer, Appl. Energy 103 (2013) 581–587. https://doi.org/10.1016/j.apenergy.2012.10.024.

[78] G. Capurso, B. Schiavo, J. Jepsen, G.A. Lozano, O. Metz, T. Klassen, et al. Metal hydride-based hydrogen storage tank coupled with an urban concept fuel cell vehicle: off board tests. Adv. Sustain. Systems 2 (2018) 1800004. https://doi.org/10.1002/adsu.201800004.

[79] G. Capurso, B. Schiavo, J. Jepsen, G. Lozano, T. Klassen, M. Dornheim, Development of a modular room-temperature hydride storage system for vehicular applications, Appl. Phys. A 236 (2016) 1–11. https://doi.org/10.1007/s00339-016-9771-x.

[80] L.K. Heung, On-board hydrogen storage system using metal hydride, in: Hypothesis II Hydrogen Power, Theoretical, and Engineering Solutions, International Symposium at Grimstad, Norway, 1997.

[81] M.W. Davids, M. Lototskyy, M. Malinowski, D. Van Schalkwyk, Metal hydride hydrogen storage tank for light fuel cell vehicle, Int. J. Hydrogen Energy 44 (2019) 29263–29272. https://doi.org/10.1016/j.ijhydene.2019.01.227.

[82] B. Lin, Conceptual design and modeling of a fuel cell scooter for urban Asia, J. Power Sources 86 (2000) 202–213.

[83] A. Nakano, T. Maeda, H. Ito, M. Masuda, Y. Kawakami, M. Tange, et al., Study on absorption/desorption characteristics of a metal hydride tank for boil-off gas from liquid hydrogen, Int. J. Hydrogen Energy 37 (2012) 5056–5062. https://doi.org/10.1016/j.ijhydene.2011.12.021.

[84] A. Nakano, T. Maeda, H. Ito, T. Motyka, J.M. Perez-Berrios, S. Greenway, Experimental study on a metal hydride tank for the totalized hydrogen energy utilization system. Energy Procedia 29 (2012) 463–468. https://doi.org/10.1016/j.egypro.2012.09.054.

[85] M. Afzal, N. Gupta, A. Mallik, K.S. Vishnulal, P. Sharma, Experimental analysis of a metal hydride hydrogen storage system with hexagonal honeycomb-based heat transfer enhancements-part B, Int. J. Hydrogen Energy 46 (2021) 13131–13141. https://doi.org/10.1016/j.ijhydene.2020.11.275.

[86] J.C. Crivello, R.V. Denys, M. Dornheim, M. Felderhoff, D.M. Grant, J. Huot, et al., Mg-based compounds for hydrogen and energy storage, Appl. Phys. A Mater. Sci. Process 122 (2016) 1–17. https://doi.org/10.1007/s00339-016-9601-1.

[87] S. Flueckiger, T. Voskuilen, T. Pourpoint, T.S. Fisher, Y. Zheng, In situ characterization of metal hydride thermal transport properties, Int. J. Hydrogen Energy 35 (2010) 614–621. https://doi.org/10.1016/j.ijhydene.2009.11.001.

[88] M. Visaria, I. Mudawar, T. Pourpoint, Enhanced heat exchanger design for hydrogen storage using high-pressure metal hydride: part 1. Design methodology and computational results, Int. J. Heat Mass Transf. 54 (2011) 413–423. https://doi.org/10.1016/j.ijheatmasstransfer.2010.09.029.

[89] J. Huot, Metal hydrides, in: M. Hirscher (Ed.), Handbook of Hydrogen Storage: New Materials for Future Energy Storage, Wiley-VCH, Weinheim, 2009, pp. 81–116. https://doi.org/10.1002/9783527629800.

[90] Green Car Congress. Toyota and Woven Planet develop portable hydrogen cartridge prototype 2022. https://www.greencarcongress.com/2022/06/20220602-hydrogencartridge.html. (Accessed 26 September 2022).

Hydrogen sensors for safety applications

Orhan Sisman[a], Mustafa Erkovan[b,c] and Necmettin Kilinc[d]

[a]Center for Functional and Surface Functionalized Glass, Alexander Dubcek University of Trencin, Trencin, Slovakia
[b]Instituto de Engenharia de Sistemas E Computadores—Microsistemas e Nanotecnologias (INESC MN), Lisbon, Portugal
[c]Instituto Superior Tecnico (IST), Universidade de Lisboa, Lisbon, Portugal
[d]Department of Physics, Faculty of Science & Arts, Inonu University, Malatya, Türkiye

5.1.1 Introduction

Hydrogen is a green energy source and the most abundant element in the universe, approximately 75%. Hydrogen is a clean, efficient, sustainable energy source, and is considered an ideal energy source in the future [1]. The use of and the demand for hydrogen are increasing day by day because hydrogen is used not only as an energy source but also has found many uses in industry. Hydrogen plays a significant role in the modern world due to its wide range of application fields, such as the chemical industry for the production of ammonia, the refining industry for the creation of petroleum products and methanol, the food industry for hydrogenation of fats and oils, the semiconductor industry for using as etching, processing, annealing and reducing gas in the production of electronic devices, the medical industry for fabrication of hydrogen peroxide, the metal industry for annealing, welding and heat treating, the space industry for rocket engines, the energy industry for clean transportation, etc. [2–4].

Hydrogen is the lightest element with one proton and one electron and exists as a diatomic molecule in the universe generally. The density of hydrogen gas is very low as 0.0899 kg/m^3 under standard conditions, the melting and boiling points of hydrogen are 14.1 K and 20.93 K, respectively [5]. Hydrogen has a high diffusion coefficient (0.61 cm^2/s in air), an ignition temperature of 560°C, a high burning and propagation velocity, high heat of combustion (142 kJ/g), a low minimum ignition energy (0.017 MJ), and a high permeability through many materials [6]. Hydrogen is highly flammable in a wide range from 4% (i.e., lower explosive limit) up to 75% (i.e., upper explosive limit), but hydrogen is not toxic. In addition, hydrogen

Towards Hydrogen Infrastructure: Advances and Challenges in Preparing for the Hydrogen Economy.
DOI: https://doi.org/10.1016/B978-0-323-95553-9.00061-3

275

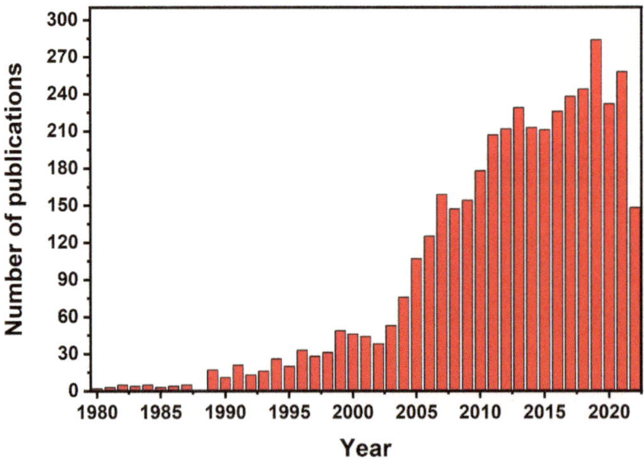

Figure 5.1.1 The number of scientific publications since 1980 according to an inquiry "hydrogen sensor, detection and sensing" keywords in all fields in September 2022 (Web of Science, Clarivate Analytics).

also acts as a strong reducing agent for many elements and has a high permeability through many materials, which demands special precautions in certain applications.

During the usage of hydrogen, there are several risks and hazards such as physiological (asphyxiation, hypothermia, overpressure injury, …), physical (component failures and embrittlement), and chemical (flammability, ignition, explosion, …). The detection of hydrogen is vital because human senses cannot detect hydrogen and hydrogen is the lightest element, nontoxic, odorless, colorless, flammable, tasteless, and explosive above 4% concentrations in any ambient. Hydrogen sensors are used for real-time quantitative analysis, safety issues, leak detection, and recently the other important usage of hydrogen sensors is the disease diagnosis in the human digestive system [7–9].

There are various types of hydrogen sensors commercially available or in development. Fig. 5.1.1 gives the enhanced publication number about hydrogen sensors from the Web of Science in September 2022. The rising numbers reflect the growing demand for hydrogen sensors in widespread applications. The common trend is to develop reliable, portable, selective, reversible, fast, energy-friendly, user-friendly, and cheap hydrogen sensors. Until now, various types of hydrogen sensors have been extensively studied and hydrogen sensors are commonly assorted into nine categories according to the evident physico-chemical principles of the detection mechanism:

(1) resistor-based (semiconducting metal–oxide and metallic resistor), (2) work function based (metal–semiconductor, metal–insulator–semiconductor), (3) catalytic (pellistor and thermoelectric), (4) electrochemical (amperometric and potentiometric), (5) optical (interferometric-based, surface plasmon resonance, intensity-based, fiber grating-based), (6) thermal conductivity (calorimetric), (7) mechanical, (8) acoustic (surface acoustic wave, quartz crystal microbalance), and (9) magnetic [10–17].

5.1.2 Risks and hazards of hydrogen usage

During the usage of hydrogen in industries, there are several risks and hazards such as physiological (asphyxiation, hypothermia, overpressure injury, …), physical (component failures and embrittlement), and chemical (flammability, ignition, explosion, …) [5,18]. Up to now, there have been quite a few severe accidents in the industrial and transport sectors of hydrogen usage. The first known accident of hydrogen usage was the Hindenburg disaster which was the biggest airship Zeppelin in 1937, in Manchester Township, New Jersey, United States. The origin of this accident was the fire of the ignition of hydrogen and as a result, 36 people died [5]. Another accident occurred in 1989 when a hydrogen gas cloud exploded at a polyethylene factory in Pasadena, Texas, resulting in 23 deaths and 314 injuries. The main causes of accidents are over-pressurization, increased embrittlement of storage tanks, material or mechanical failure, expanding vapor explosion during transitioning from liquid to gas phase, corrosion, human error, and others.

Hydrogen is not a toxic gas and the inhaled hydrogen gas causes a high-pitched voice and sleepiness. On the other hand, if the hydrogen concentration increases in the environment and the oxygen in the environment falls below 19% by volume, it causes asphyxiation which is the one of physiological hazards. Hydrogen fire, like other hydrocarbon fires, has risks and hazards. If a person is exposed to hydrogen fire, burns of varying degrees can occur. Burns from hydrogen fire depend on many factors such as the size of the burning surface, burning rate, exposure time, heat of combustion, and atmospheric conditions. There are also risks and hazards during the usage of liquid hydrogen. If a person contacts with ultra-cold liquid and vessel, it causes cryogenic burns or frostbite. The damage of frostbite is freezing the tissues because of the formation of ice crystals inside cells. In addition, the spills of high liquid hydrogen exposure can cause hypothermia. During the explosion of hydrogen, blast waves affect humans directly by increasing pressure and this causes damage to the ears and lungs that organs are sensitive to high pressure. The blast waves also affect the humans indirectly through objects or parts such as fragments, shrapnel, and debris resulting from hydrogen explosions.

The lightest and smallest molecule, hydrogen can diffuse in many metals and this causes component failures and embrittlement during hydrogen storage and transportation. Hydrogen embrittlement is defined as brittle and small cracks in the metal due to exposure to hydrogen. This occurs over a long period of time and can lead to liquid hydrogen spillage or hydrogen gas leaks. Hydrogen embrittlement depends on many parameters such as hydrogen pressure, purity, concentration, exposure time, and temperature of the environment, and the stress state, physical and mechanical properties, microstructure, surface conditions, etc.

5.1.3 Hydrogen sensors

A gas sensor is a device that measures or detects the amount and the presence of an indicated gas in an area. Mainly a gas sensor consists of three parts: analyte, sensing material, and transducer as seen in Fig. 5.1.2A. In general, a gas sensor is called with the type of gas such as hydrogen sensor, carbon monoxide sensor, oxygen sensor, volatile organic compound sensor, etc. Also, a gas sensor can be classified by sensing material type, transducer type, physico-chemical detection mechanism, signal type, etc. This chapter provides hydrogen sensors that are classified with physico-chemical detection mechanisms.

Hydrogen sensors will be classified into nine categories depending on the evident physico-chemical principles of the detection mechanism as given in Fig. 5.1.2B. These categories are catalytic, electrochemical, resistor-based, work function-based, mechanical, optical, acoustic, thermal conductivity, and magnetic. The details about these categories will be presented in subsections. In order to evaluate the performance of hydrogen sensors, it is useful to introduce some basic sensor parameters. These are sensitivity, selectivity, lifetime, response time, recovery time, power consumption, sensor size, stability, cost, measurement range, etc.

5.1.3.1 Resistive hydrogen sensors

A resistive sensor detects the changes in the resistance or the conductivity of the sensing layer due to the ad/absorption of hydrogen gas. Generally, an interdigital electrodes (IDE) with a back-side heater are used for measuring the

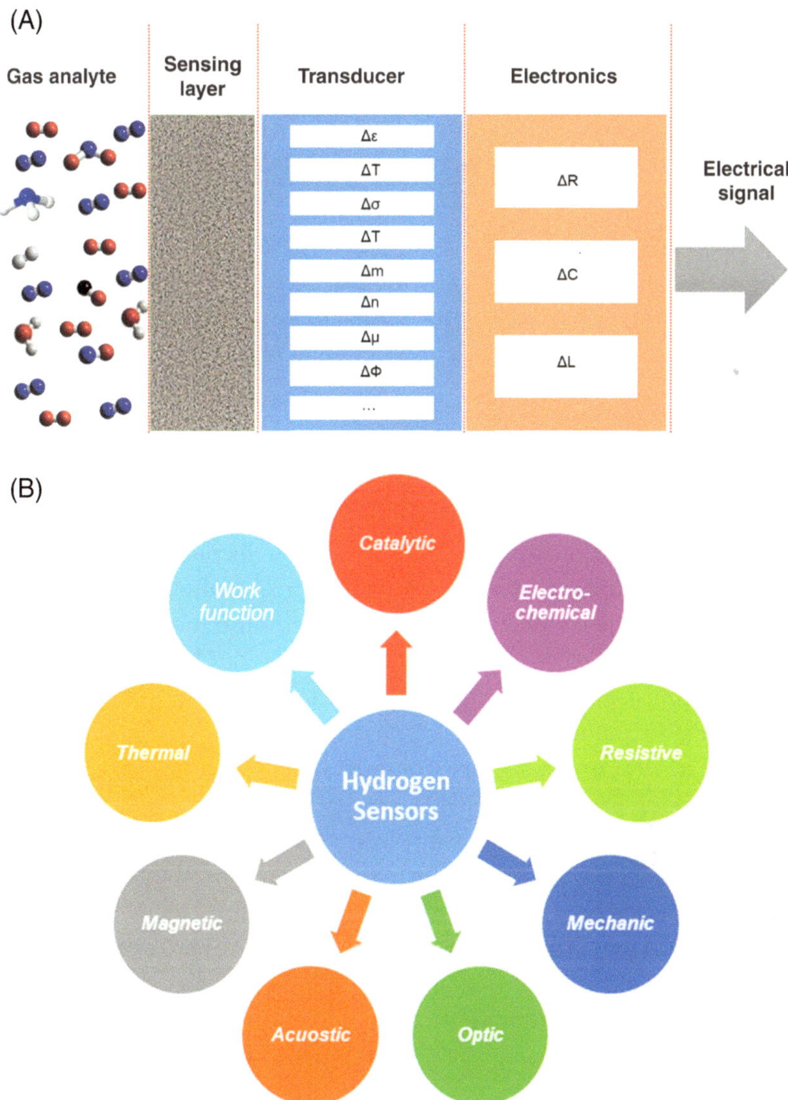

Figure 5.1.2 A schematic illustration for a gas sensor (A) and a smart art illustration of hydrogen sensor categories (B).

resistance or the conductivity of the sensing layer that is coated on the IDE as seen in Fig. 5.1.3. Gold (Au) and platinum (Pt) are mostly used for fabricating the IDE and wire heater with the photolithography method respectively. The resistive hydrogen sensors could be classified into two parts depending on sensitive material type such as semiconductor and metallic. The semi-conducting and metallic resistive hydrogen sensors are examined in detail.

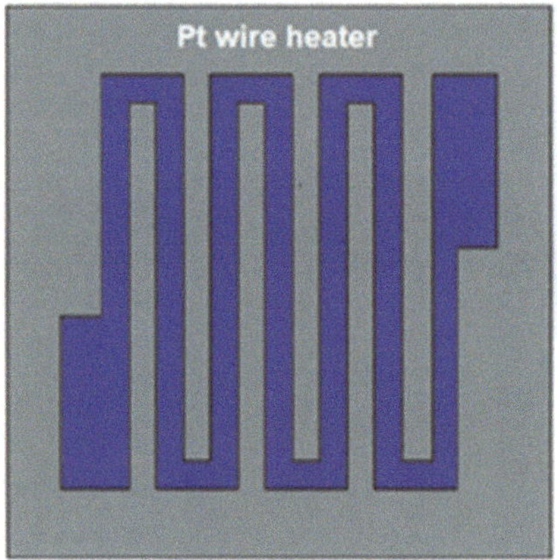

Figure 5.1.3 A schematic view of interdigital electrodes with a backside heater.

5.1.3.1.1 *Semiconducting resistive hydrogen sensors*

Semiconducting metal oxides (SMOX) have been the pioneer semiconducting materials for resistive gas sensors with their special electron configurations partially occupied d shells since the first demonstration of the influence of hydrogen on the electrical conductivity of the ZnO surface by Heiland in 1957 [19]. Just after the first reports by Heiland [19] and Seiyama et al. [20], Taguchi introduced the first commercial MOX-based electrochemical gas sensors in 1971 under the first commercial company "Figaro" [21].

The working principle of SMOX sensors starts with the oxygen adsorption and desorption processes with increased temperature [22]. In the range of 100–200°C, the oxygen molecules in the air were adsorbed on the SMOX surface by capturing electrons from the conduction band of SMOX as shown in the Eq. (5.1.1). At higher temperatures (>200°C), the oxygen molecules dissociate into oxygen ions (singly or doubly negative electric charges) by trapping more electrons from the conduction band as shown in the Eqs. (5.1.2) and (5.1.3).

$$O_{2(gas)} + e^- \leftrightarrow O^-_{2(adsorbed)} \tag{5.1.1}$$

$$\frac{1}{2}O_2 + e^- \; k_{Oxy} \Leftrightarrow \; O^-_{ads} \tag{5.1.2}$$

$$\frac{1}{2}O_2 + 2e^- \; k_{Oxy} \Leftrightarrow \; O^{2-}_{ads} \tag{5.1.3}$$

where the k_{gas} refers to the reaction rate constant [23].

Until reaching thermal equilibrium, a charge trade occurs between the bulk and the surface due to the acceptor and donor states in semiconductors

[24]. In the equilibrium, a space charge region emerges as a consequence of charge trades. The characteristics of space charge regions change according to the physical properties of the SMOX surface. For instance, oxygen ionosorptions establish an electron depletion region for the n-type and a hole accumulation region for the p-type SMOX surface. The electrons held by ionosorbed oxygen return to conduction when the SMOX surface interacts with the hydrogen. Hydrogen is classified as a "reducing gas" for n-type SMOXs since the released electrons reduce the electron depletion region and resistivity. On the contrary, the released electrons reduce the hall accumulation region which increases the resistivity of p-type SMOXs.

The net charge carrier concentration (n) can be extracted by considering the sum of charge exchanges during oxygen ion adsorption and desorption in Eq. (5.1.4) [25]. The sensitivity, ratio of resistance change under air and hydrogen atmosphere, is defined by extending the resistance definition with surface charge carrier concentration changes as given in Eq. (5.1.5),

$$n = \Gamma_t k_{gas} \left[O_{ads}^{ion} \right]^b [H]^b + n_0 \tag{5.1.4}$$

$$S_g = \frac{R_a}{R_g} = \frac{\Gamma_t k_{gas} \left[O_{ads}^{ion} \right]^b [H]^b}{n_0} + 1 \tag{5.1.5}$$

where Γ_t constant (time-dependent), $\left[O_{ads}^{ion} \right]$ adsorbed oxygen ion concentration, b is the valance value, $[H]$ hydrogen ion concentration.

At the beginning of the 90s, the developments in nanotechnology boosted the gas sensor studies. Yamazoe reported that the smaller crystallite sizes led to remarkable performances for SMOX gas sensors [26]. Nanostructures promoted gas sensing by higher aspect ratio, lower detection limit and short response time for SMOX-based gas sensors [24,27–29]. The increased surface area over volume ratio in nanomaterials triggers more electronic interactions between the SMOX and O_2 molecules. This enhances the sensitivity by increasing number of the oxygen ions $\left[O_{ads}^{ion} \right]$ on the surface with the dimension reduction.

Semiconductors have characteristic space charge region relying on their donor concentrations, which is called L_d (Debye length) and defined in Eq. (5.1.6) [26]. It depends only on the carrier concentration at a constant temperature.

$$L_d = \left(\frac{\varepsilon k_B T}{q^2 n} \right)^{1/2} \tag{5.1.6}$$

where ε is a static dielectric constant, q is an electrical charge of a carrier, and n is a carrier concentration. For instance, the L_d value of SnO_2 can change around 10–130 nm depending on the temperature changes from 400 K to 700 K [30]. In the case of one-dimensional nanowires, the relation between the sensitivity, diameter of nanowires (D), and Debye length (L_d) can be derived as given in Eq. (5.1.7) [23]. The sensitivity can be changed according to the relation between D, L_d, and w (depletion depth) [31,32].

$$S_{L_d} = \left(\frac{\Gamma_t k_{gas}[O]^b H^b}{n_0} \right) \frac{D^2}{(D - 2L_d)^2} + 1 \qquad (5.1.7)$$

Mathematical interpretation of the sensitivity helps to comprehend the sensing mechanism and the relevant parameters for hydrogen sensing.

The favorable hydrogen sensing SMOXs are the TiO_2, ZnO, and SnO_2 due to their superior surface properties, wide–band gaps, and higher charge carrier mobilities. However, they suffer from high operation temperatures, lack of selectivity, interference with humidity or high resistance values which creates needs for advanced electronic units, and set-ups for sensors. The best solution to overcome these handicaps is metal doping or loading on the surface of SMOXs. There are two different scenarios for doped and loaded surfaces in terms of gas sensing kinetics, electronic sensitization, and chemical sensitization.

In the case of electronic sensitization, the additive dopants are taking place in SMOX's lattice by increasing the charge carrier concentrations, shifting the band–gap energy values, and decreasing the adsorption energy of hydrogen. Doping has an impact on both surface and bulk properties [33]. The effect of Cr^{2+}, Cu^{2+}, and Pd^{2+} on hydrogen sensing properties of SnO_2 nanocomposites was compared both theoretically and experimentally [34]. The Pd^{2+} doped SnO_2 surface had the lowest adsorption energy among the reported ions ($E_{ads} = -0.28$ eV) [34]. In another study, it was demonstrated that the adsorption of hydrogen on the Ni-doped TiO_2 nanotubes surfaces caused a significant change in the band structure since the generated impurity level could give a migration pathway for the hydrogen to overcome the energy barrier consequently led to a higher sensitivity [35]. Besides the amount of dopants that can invert the conduction type from n-type to p-type or p-type to n-type, the excessive amount of dopants may lead to deactivation of the surface due to high concentration of surface states [36].

In the case of chemical sensitization, the loaded metals take a role mainly in the surface reactions. The noble metals (Ag, Au, Pd, and Pt) are preferred as loading materials for hydrogen detection due to their catalytic effect on H_2 dissociations reactions. The size and coverage of the loaded metal are important for the sensitization effect. Therefore, the small size of additives is better for more surface reactivity.

5.1.3.1.2 *Metallic resistive hydrogen sensors*

In general, platinum (Pt), palladium (Pd), and their alloy are used as a sensitive material for metallic resistive hydrogen sensors. The sensing mechanism of Pd and Pt resistive hydrogen sensors differs from each other. The adsorption and diffusion of hydrogen in Pd could be explained as follows: adsorption of molecular hydrogen on the Pd, dissociation of the molecule leading to the formation of H atoms, migration of H atoms into the Pd, and reaction between H and Pd atoms leading to the formation of a hydride (PdH_x) [37–39].

Palladium is a unique transition metal for hydrogen absorption since it can absorb up to $\sim 10^3$ volumes of H_2 corresponding to atomic ratio $PdH_{0.7}$ to 8320 cm^3 of H_2 per 100 gm of Pd under the normal pressure of H_2 at room temperature. This brings an excellent H_2 storage capacity of Pd and causes a huge volume expansion in the Pd layer.

Sievert's law quantifies the solubility of gases in metals by proposing a mathematical relation between partial pressure and adsorbed H atoms on the Pd surface as given in Eq. (5.1.8).

$$\left(\frac{H}{Pd}\right)_{at} = K_s\sqrt{P_{H_2}} \tag{5.1.8}$$

where $(H/Pd)_{at}$ is the atomic ratio of H and Pd, K_s the Sievert's constant, and P_{H_2} the H_2 partial pressure (*Pa*).

During PdHx formation, the volume of the structure is expanded and different phases are occurred depending on hydrogen concentration. At low hydrogen concentration (x = H/Pd < 0.015) and at high concentration (x > 0.6), PdHx is in α phase and β phase respectively. While the hydrogen concentration is between these values (0.015 < x < 0.6) the α to β phase transition occurs and at this concentration range α and β phases coexist. The lattice parameters of pure Pd, α phase PdHx (x = 0.015 that contains maximum hydrogen in this phase), and β phase PdHx (x = 0.6 that contains minimum hydrogen in this phase) are $a_0 = 3.890$ Å, $a_{\alpha max} = 3.895$ Å, and $a_{\beta min} = 4.025$ Å, respectively. Although the volume expansion rate of Pd is a maximum of 0.38% in the α phase, this rate is around 10.8% at the minimum in the β phase [37–39]. Therefore, during the PdHx phase transition from the α phase to the β phase, the volume of the structure expands seriously.

Depending on the continuity of Pd nanostructure, the sensing mechanism of Pd based resistive hydrogen sensor could be explained in two categories: continuous Pd based resistor and nanogap based Pd resistor [15,16,40]. For continuous Pd based resistive hydrogen sensors, the resistance of the sensor increases during exposure to hydrogen gas. In general, upon absorption of hydrogen, the PdHx phase is formed and the resistance of Pd increases due to the increase in the number of scatterings of charge carriers. Besides, for nanogap based Pd resistive hydrogen sensor, the sensitive layer Pd nanostructure have cracks or nanogaps. The resistance of Pd structure with nanogaps decreases during the absorption of hydrogen because the swelling of the Pd nanostructure causes the closing of nanogaps. The nanogap based resistive hydrogen sensors are also well known as on–off type or switch-type hydrogen sensors. The nanogap based resistive hydrogen sensors have hydrogen concentration limitation because the huge volume expansion of PdHx ($0.015 < x < 0.6$) occurs during the phase transition from the α phase to the β phase. Fast response, high sensitivity, and selectivity are the main advantages of nanogap based resistive hydrogen sensors. Fig. 5.1.4 shows a schematic illustration of continuous and discontinuous Pd nanoparticles covered on two electrodes and resistance change under dry air and hydrogen ambient conditions.

On the other hand, there are fewer studies on Pt-based resistive hydrogen sensors compared to Pd based resistive hydrogen sensors. The increase and the decrease in the resistance of Pt nanostructures while exposed to hydrogen are explained by different detection mechanisms. The decrease in the resistance of the Pt based sensor during exposure to hydrogen is explained by the scattering of electrons from the Pt surface as seen in a schematic diagram of Fig. 5.1.5B [41–44]. Fig. 5.1.5A shows the hydrogen sensor response of Pt nanowire array fabricated with PDMS molds by Yoo et al. and is observed that the resistance of Pt nanowire array decreases meanwhile hydrogen exposure [42]. Fig. 5.1.5B gives a schematic description for the hydrogen sensing mechanism of Pt nanostructures with the surface scattering phenomenon [42]. On the contrary, the increase in the resistance of the Pt nanostructure in the hydrogen environment is explained in three ways. The increase in the resistance of Pt nanostructure upon exposure to hydrogen is explained by the formation of PtHx hydride [45], by the scattering of electrons from defects in Pt [46], and from the grain boundary of the Pt nanostructure [47].

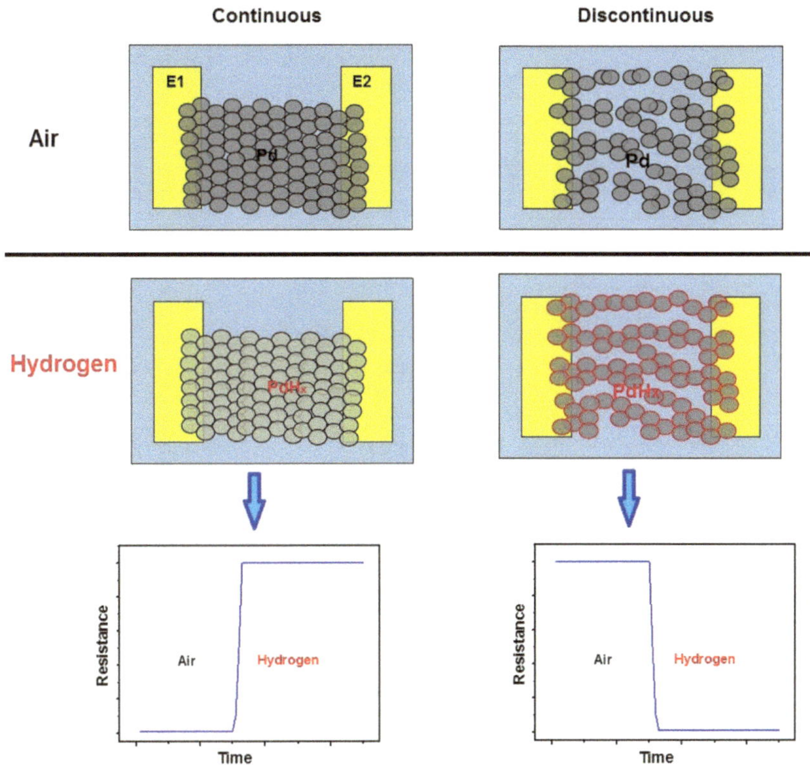

Figure 5.1.4 A schematic illustration of continuous and discontinuous Pd nanoparticles covered on two electrodes and resistance change under dry air and hydrogen ambient conditions.

5.1.3.2 Work function based hydrogen sensors

The work function is defined as the minimum energy required to remove an electron from a solid material to the vacuum level at that level the electron is free. The sensing mechanism of the work function-based sensors is related to the variation of the energy band diagram at the interface of the junction between metal, isolator, and semiconductor [10,48]. There are three main structures for work function based hydrogen sensors as given in Fig. 5.1.6: Schottky diode, capacitor, and transistor [10,48,49]. The Schottky diode structure could be metal-semiconductor (MS) or metal–insulator–semiconductor (MIS) or metal–insulator–metal (MIM) in which the isolator is generally an oxide material. The capacitor and transistor structures are generally in the form of metal-oxide-semiconductor (MOS) capacitors and metal-oxide-semiconductor field effect transistors (MOSFET) respectively.

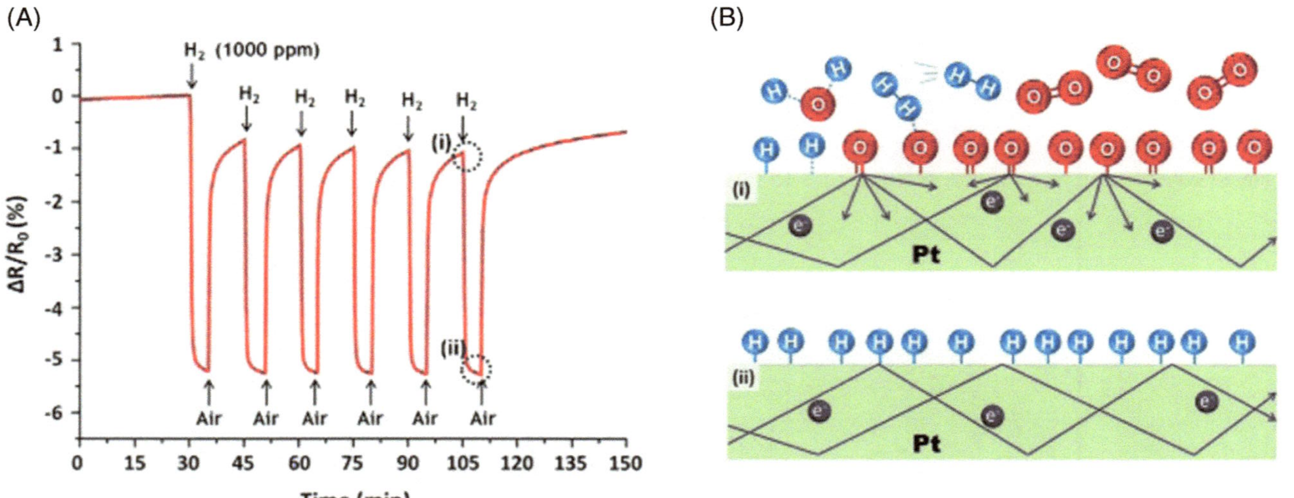

Figure 5.1.5 The sensor response of Pt nanowire array for 1000 ppm hydrogen in a cyclic test under dry air conditions (A). A schematic description for hydrogen sensing mechanism of Pt nanostructures with the surface scattering of electrons (B). A Pt—H surface (ii) greatly reduces surface scattering as compared to a Pt—O surface (i), resulting in a decreased resistance to hydrogen ad/absorption. *(Reprinted with permission from reference [42]. From: American Chemical Society, 2015).*

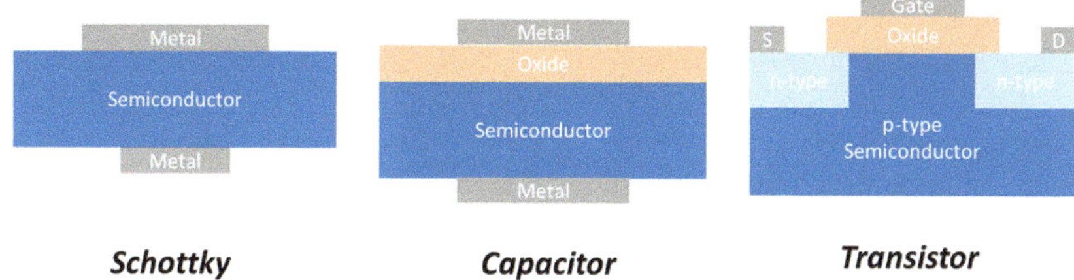

Figure 5.1.6 Cross-sectional schematic illustrations for Schottky diode, capacitor, and transistor-based hydrogen sensor that modified from reference [49]. *(Reprinted from T. Hubert, L. Boon-Brett, V. Palmisano, M.A. Bader, Developments in gas sensor technology for hydrogen safety, Int. J. Hydrogen Energy, 39, 20474–20483. From: Elsevier, 2014).*

Generally, TiO_2, La_2O_3, SiO_2, ZnO, HfO_2, GaO_x, WO_3, etc. materials are used as an oxide or an isolator in work function-based hydrogen sensors and Si, SiC, GaN, InP, GaAs, etc. materials are utilized as a semiconductor [48].

The top metal is crucial for hydrogen sensors and is usually chosen from catalytic metals such as Pd, Pt, Au, Ni, their alloy, etc. [48]. During the ab/adsorption of hydrogen at the top metal surface, hydrogen molecules dissociates into hydrogen atoms and then the atoms diffuse through the metal and adsorb at the interface between the metal and isolator or semiconductor or oxide layers. The hydrogen atoms at the interface causes a depletion layer due to the polarization, and this changes the work function of the metal and also the energy band diagram of the diode or the capacitor of the transistor structure. So, the ab/adsorption of hydrogen on work function based sensor changes the voltage or the current or the capacitance of the sensor. Schottky diode, capacitor, and transistor-based hydrogen sensors are separately examined in detail as given below.

5.1.3.2.1 *Metal–semiconductor (Schottky) diode*

The electrical properties of Schottky diode are accurately described with the thermionic-emission model and the relationship between current and voltage is given in the following Eq. (5.1.9) [50]:

$$I = I_0 \left[\exp\left(\frac{qV}{nkT} - 1 \right) \right] \tag{5.1.9}$$

where, k is the Boltzmann constant, q is the electron charge, T is the temperature in Kelvin, n is the ideality factor, and I_0 is the saturation current defined as:

$$I_0 = AA^* T^2 \exp\left(\frac{q\Phi_b}{kT} - 1\right) \qquad (5.1.10)$$

where A^* is the Richardson constant, A is the device junction area, and Φ_b is the effective Schottky barrier height (SBH). According to the thermionic-emission model, SBH is examined as following Eq. (5.1.11) by rearranging the above Eq. (5.1.10):

$$\Phi_b = \frac{kT}{q} ln\left(\frac{AA^* T^2}{I_0}\right) \qquad (5.1.11)$$

When a contact or a junction is formed between the metal and the semiconductor, the Fermi level of the metal and the semiconductor is aligned with dominating the fermi level of the metal and the energy band diagram of the MS structure bends. The SBH is the difference between the work functions of metal and semiconductor or isolator. When a Schottky diode is exposed to hydrogen, hydrogen molecules dissociated at the surface of the catalytic metal electrode and the hydrogen atoms diffused in the metal, and this causes the chance in the work function of the metal and as a result, SBH is differed. Therefore, this variation could be measured with current–voltage characteristic.

5.1.3.2.2 *Capacitor based hydrogen sensor*

A capacitor-based hydrogen sensor is in the structure form of metal–oxide–semiconductor (MOS) so similar to the structure of the previously described Schottky diode device. But the thickness of the oxide layer in the MOS capacitor is higher to prevent current conduction and also to build charges easily on both sides. The total capacitance of MOS capacitor could be explained with the following Eq. (5.1.12) [51].

$$C_T = \left(\frac{t_{ox}}{\varepsilon_{ox}\varepsilon_0 A} + \frac{X_d}{\varepsilon_s \varepsilon_0 A}\right)^{-1} \qquad (5.1.12)$$

where $\varepsilon_0, \varepsilon_{ox}$, and ε_s are the relative permittivity of free space, oxide film, and the semiconductor substrate, respectively. A is the area of the metal electrode, X_d is the width of the depletion layer, and t_{ox} is the thickness of the oxide film. The capacitance voltage $(C–V)$ characteristic of the capacitor based hydrogen sensor is varied during exposure to the hydrogen. Because the diffused hydrogen atoms at the interface of the metal and the oxide causes dipole layer formation. This layer results in a flat band voltage shift (ΔV_{FB})

and ΔV_{FB} could be represented in the equation below [48,51]:

$$\Delta V_{FB} = \frac{W_{MS}}{q} - \frac{(Q_{it} + Q_f + Q_m + Q_{ot})t_{ox}}{\varepsilon_{ox}\varepsilon_0 A} \tag{5.1.13}$$

where Q_m, Q_{it}, Q_{ot}, and Q_f are mobile ionic, interface trapped, oxide trapped, and fixed oxide charges, respectively. q is the charge of the electron and W_{MS} is the work function difference between the metal and silicon.

5.1.3.2.3 Transistor based hydrogen sensor

The basic schematic structure of a MOSFET in the npn semiconductor is given in the Fig. 5.1.6 and the MOSFET structure could be pnp structure. Two n-type parts are generally fabricated on the p-type semiconductor substrate with the ion implantation method. An isolating oxide material is coated on a substrate between two n-type parts of the semiconductor. Metal electrodes are coated to obtain source (S), drain (D), and gate (G) electrodes. For an ideal MOSFET, when the gate voltage is zero or increases to a certain value that called as threshold voltage in one direction, there is not any current between D and S by applying a voltage to these electrodes. But, when the gate voltage is greater or equal to the threshold voltage, a significant current flows from S to D. The drain current of a MOSFET could be explained with the following equation in the linear region of the transistor [48,50]:

$$I_D = \frac{\mu C}{L}\left[(V_G - V_T)V_D - \frac{V_D^2}{2}\right] \tag{5.1.14}$$

where C is the capacitance of the oxide layer, μ is the mobility of the charge carriers within the channel, L is the length of the channel, V_G is the gate voltage, V_D is the drain voltage, and V_T is the threshold voltage. Catalytic metals are used as a gate electrode material for transistor based hydrogen sensor. Lundström et al. first reported MOSFET based hydrogen sensors by using Pd as a gate electrode [52]. During ab/adsorption of hydrogen on the catalytic gate metal, the dissociated hydrogen atoms creates a dipole layer, this causes the change in the threshold voltage, and as a result, the I–V characteristic of the MOSFET is varied.

5.1.3.3 Catalytic hydrogen sensors

Catalytic hydrogen sensors measure the evolved energy during the combustion of hydrogen on the catalyst surface. The standard heat value from hydrogen combustion is 141.9 kJ/g [6]. There are two different set-ups, pellistor and thermoelectric-based catalytic sensors.

5.1.3.3.1 Pellistor

In pellistor design, two platinum coils are integrated into a pellet by mounting in a Wheatstone bridge circuit for easy benchmarking of resistance values. One of the coils is used as a reference resistance thermometer in its bead [53]. The other bead is activated with a suitable hydrogen catalyst (Pd, Pt, etc.). The coils are heated above 300°C to trigger chemisorption in the ambient atmosphere. With the injection of hydrogen, adsorbed oxygen on the catalyst combust the hydrogen. The raising temperature in the bead cause to variation in the resistance of the active coil.

The pellistor type catalytic hydrogen sensors can operate in a wide range of atmospheric conditions, at temperatures from 20°C to 70°C, under pressures from 70 kPa to 130 kPa, in relative humidity 5–95%. The disadvantages are the poor selectivity towards other type of combustion gases like, hydrocarbons or carbon monoxide [10].

5.1.3.3.2 Thermoelectric

Similar to pellistor type catalytic hydrogen sensors, thermoelectric catalytic hydrogen sensors convert the catalytic hydrogen combustion to the electrical signal. In thermoelectric sensors, the electrical signal is derived by the Seebeck effect, a voltage difference occurs due to the temperature gap between two points of a conductor or semiconductor. Basically, hydrogen combustion creates a voltage difference between the reference and active sensor. The amount of oxidized hydrogen can be quantified by the following equation [10]:

$$U = \alpha \cdot \Delta T \approx \alpha \cdot \left\{ const. \cdot exp\left(\frac{-E_a}{RT}\right) \right\} \cdot \Delta H \qquad (5.1.15)$$

where α is the Seebeck coefficient, ΔT is the temperature difference active, and cold sensors. $const. \cdot exp\left(\frac{-E_a}{RT}\right)$ is the reaction rate of the combustion and ΔH is the heat of the hydrogen combustion.

The main advantages of the thermoelectric type are lower working temperature, low cross–sensitivity to other combustion gases compared to pellistor type sensors.

5.1.3.4 Electrochemical hydrogen sensors

Electrochemical hydrogen sensors are devices for signal transduction of electrochemical interaction between a protective interface and hydrogen gas. Electrochemistry is dealing with ion diffusion, ion transport, and electron transfer reactions between an electrode and solution interfaces. Usually a three-electrode measuring system is used, consisting of the sensing or working electrode, the counter electrode and the reference electrode. Laconti and Maget proposed the first electrochemical based hydrogen sensor [54]. A few reviews related to electrochemical hydrogen sensor have been published in recent years [14,55,56]. There are two main measurement techniques for electrochemical hydrogen sensors: potentiometric and amperometric.

5.1.3.4.1 Amperometric based hydrogen sensor

Generally, a constant external voltage applies to an amperometric based hydrogen sensor, and the current between electrodes measures continuously. The magnitude of this current is proportional to the concentration of hydrogen gas present in the environment, and can be measured using a multimeter. By measuring this current, the electrochemical hydrogen sensor is able to detect the presence of hydrogen gas in the surrounding environment. The current of the sensor is proportional to equation below [14],

$$i = n \cdot F \cdot Q \cdot C \tag{5.1.16}$$

here n is the number of electrons per molecule in the reaction, F is the faraday constant (9.648×10^4 C/mol), Q is the gas consumption rate in the unit of m^3/s, and C is hydrogen concentration in the unit of mol/m^3.

5.1.3.4.2 Potentiometric based hydrogen sensor

In a potentiometric based hydrogen sensor, no current flows and the potential between the electrodes is measured. The voltage is related to the concentration of hydrogen gas in the air, so by measuring the voltage, the sensor can determine the concentration of hydrogen gas. The potential can be expressed by the following equation [14],

$$E = E^0 + \left[\frac{RT}{nF}\right]\ln(a) \tag{5.1.17}$$

where E is the potential of the electrode, E^0 is standard electrode potential, R is the universal gas constant, T is the absolute temperature, a is the chemical activity of hydrogen, n is the number of electrons per molecule in the reaction, and F is the faraday constant.

5.1.3.5 Optical hydrogen sensors

The optical hydrogen sensors track the changes in the optical properties of the materials during interaction with light under hydrogen purged atmospheres. After the primary demonstrations of the optical hydrogen sensors [57,58], there have been proposed various optical hydrogen sensors working in different set-ups mainly with fibers. Despite the variety of the experimental designs like reflectivity on micromirrors, interferometric measurements, surface plasmon resonance, evanescent field interaction, fiber Bragg gratings, and others, the sensing materials are Pd, Pd based bimetallic coatings, chemochromic materials (WO_3, YH_2, etc.) and, their noble metal loaded derivates.

5.1.3.5.1 *Reflectivity measurements on micromirrors*

A thin layer of sensing material coated on the micromirror is placed at the end of an optical fiber. The hydrogen exposure modulates the intensity of the reflected light which is affected by the sensing material on the micromirror. The advantages of the micromirrors optical fiber hydrogen sensors are their low cost and practical use. On the other side, the dependency on the light paths reduces their stability. Their multiplexing capability is limited due to compatibility for point measurements [59].

5.1.3.5.2 *Interferometric measurements*

The interferometric hydrogen sensors read out the phase difference in the light due to refractive index change in the sensing layers. The amount of phase differences are correlated with the amount of hydrogen. The interferometry design can be Mach-Zehnder, Fabry–Perot, Michelson, and other types for hydrogen detection. In Mach–Zehnder interferometers, the interference pattern of two light beams propagating through the core and cladding parts of the fiber was investigated towards hydrogen injections. Despite the same lengths of the core and cladding, the optical paths are different because the refractive index of the cladding is smaller which also leads to the excitation of multi modes [60]. The wavelength domain converts to the spatial frequency domain by Fourier transformation to find the effective modes for interference pattern formation. The interference of coupling waves can be formulated as follows [61],

$$I_{total} = I_{core} + I_{clad} + 2\sqrt{I_{core}I_{clad}}\ cos\phi \qquad (5.1.18)$$

where the I_{core} and I_{clad} are the intensities of core and cladding modes. Φ is the phase angle that occurred traveling of modes through a length L. It can be extracted by using the formula given below.

$$\phi = \frac{2\pi}{\lambda}\Delta n_{eff}L \qquad (5.1.19)$$

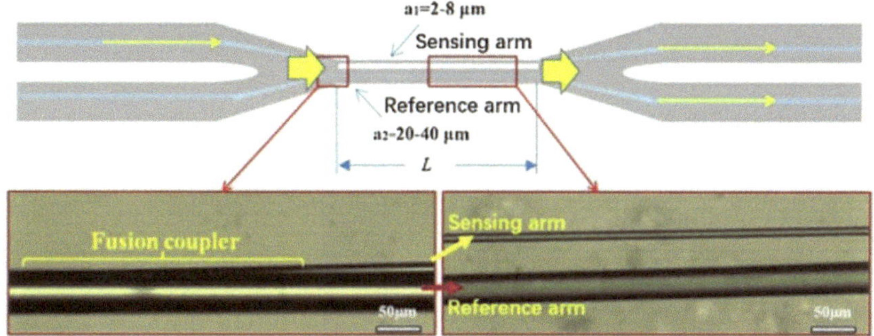

Figure 5.1.7 A schematic figure of Mach–Zehnder set-up and the microscope photographs of different regions [63]. *(Reprinted from M.S. Chen, H.T. Dang, J. Zhang, Y. Wang, J. Li, Mach-Zehnder interferometer refractive index sensing probe based on dual microfiber coupler, Optik, 228, 166181. From: Elsevier, 2021).*

where Δn_{eff} is the effective index difference between the core and the cladding mode. Here, regarding the modal dispersions, differential modal group index (Δm_{eff}) can be introduced for phase angle by redefining λ as its shift $\Delta\lambda$ from the center by using Taylor expansion with the first order at the λ_0 [62].

$$\phi = \frac{2\pi\,\Delta\lambda}{\lambda_0^2}\,\Delta m_{eff}L \tag{5.1.20}$$

with

$$\Delta m_{eff} = \Delta n_{eff} - \lambda_0\frac{\partial}{\partial\lambda}\Delta n_{eff} \tag{5.1.21}$$

The Fourier-transformed spatial frequency spectrum can show the peaks at phase angles which can be presented via the following mathematical relation.

$$\phi = 2\pi\xi\Delta\lambda \tag{5.1.22}$$

By using the Eq. (5.1.20) and spatial frequency peaks, one can find Δm_{eff} in terms of L, empirical ξ, and constant λ_0.

A hydrogen measurement set-up with a Mach–Zehnder interferometer was given in Fig. 5.1.7. The other interferometer gas sensors work with similar principles tracking phase angle differences that occurred due to the hydrogen interactions. Despite their high sensitivity, high accuracy, and reliability, interferometric optical hydrogen sensors are highly affected by temperature fluctuations. They need stable environmental conditions.

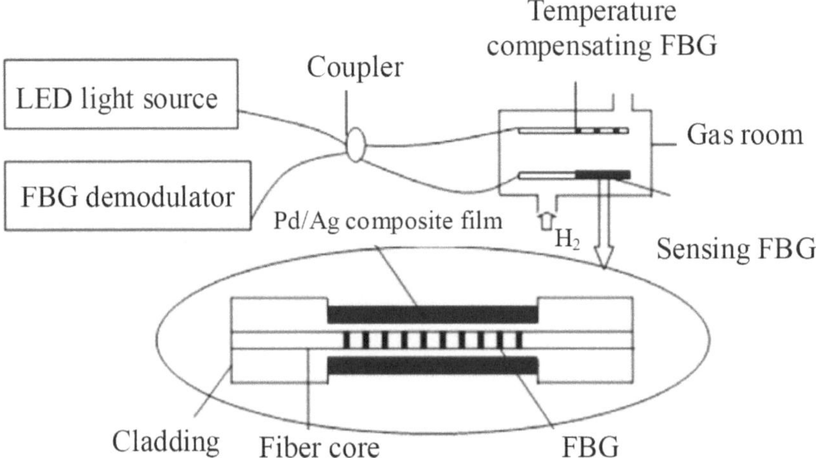

Figure 5.1.8 A schematic configuration for a Pd/Ag sensing layer deposited FBG hydrogen sensor [65]. *(Reprinted from M.H. Yang, J.X. Dai, Fiber optic hydrogen sensors: A review, Photonic Sens., 4, 300–324. From: Springer Nature, 2014).*

5.1.3.5.3 *Fiber Bragg gratings*

One of the favorite designs is the fiber Bragg gratings-based sensors that detect the hydrogen through the shifts in the Bragg wavelength of the refracted lights from the gratings in the core. The multiple gratings improve the sensor response towards hydrogen while avoiding thermal effects with the operation of a reference grating. In FBG sensors, the effective refractive index (n_{eff}) and grating pitch (Λ) are decisive in reflection wavelength (λ_B) as described below [64]:

$$\lambda_B = 2\, n_{eff}\, \Lambda \tag{5.1.23}$$

In the case of Pd or Pd-based bimetallic coatings, the sensing layers can thicken with the formation of PdHx. The FBG wavelength alters due to the grating pitch variations by the thickening in the sensing layer. FBG sensors work by correlating the hydrogen amount with the magnitude of wavelength shifts. A schematic configuration for an FBG sensor is given below in Fig. 5.1.8.

5.1.3.5.4 *Measurement of evanescent field interaction*

Developing material process technologies like ion etching and lithography techniques simplified the fabrication of optical fiber evanescent field and SPR hydrogen sensors. The evanescent field is the electromagnetic field formed at the boundary of the core of an optical fiber exponentially correlated with the distance from the core [10]. Therefore, the cladding is removed and the sensing layer is deposited directly on the core to detect the hydrogen by attenuation change in the evanescent field due to a change in the refractive index. Tailoring the shape of the fiber like D-shape designs and decorating the boundaries with nanostructured sensing materials enhanced the sensitivity, sensor response times, and reversibility of these sensors effectively [59].

5.1.3.5.5 Surface plasmon resonance

The interaction of light with metal/dielectric surfaces creates electromagnetic waves that propagate parallel to the surface (surface plasmons). To excite surface plasmons is possible at a certain angle of incidence for noble metals like Au, Ag, and Pd. The special interaction of the Pd with hydrogen allows for designing SPR hydrogen sensors with different techniques.

5.1.3.6 Thermal conductivity hydrogen sensors

The heat conduction capabilities of elements and their physical forms (solid, liquid, gas, and plasma) are different. As the smallest and lightest element, hydrogen has higher thermal conductivity (TC) approximately seven times than the other atmospheric gases (O_2, N_2) at room temperature [66]. To measure TC value gases, a temperature difference is created between a hot and a cold element that leads to a thermal conduction through the investigated gas [67]. The hot element, a resistor is heated up until its resistance deviates from the limit of Ohm's low [68]. The resistance change in the heated filament is an indicator of the thermal conductivity of the surrounding atmosphere. TC sensors are quite favorable due to wide working ranges (1–100%), low power consumption, low dimensionality, no requirement for carrier or background gas, and less contamination risk. Therefore, they are highly commercial and in use compared to other types of hydrogen sensors.

There are several models to formulate the TC values with hydrogen concentrations. In the case of two parallel (hot–cold) plates heat can be dissipated by the radiation, heat flow over contact points, and heat flow over gases. The radiation and contact point losses could be compensated with micromachining technologies. Therefore, the heat flow rate (Q) correlated to the temperature gradient and thermal conductivity (λ) of the gas between the plates as follows:

$$\dot{Q} = -\lambda \cdot grad\ (T) \tag{5.1.24}$$

The temperature dependence of thermal conductivity of a gas is expressed with a fourth-order approximation below.

$$\frac{\lambda}{W\,m^{-1}K^{-1}} = A + B\frac{T}{K} + C\left(\frac{T}{K}\right)^2 + D\left(\frac{T}{K}\right)^3 + E\left(\frac{T}{K}\right)^4 \qquad (5.1.25)$$

The coefficients are substance-dependent and defined in VDI heat atlas [69]. For hydrogen, $A = 0.651 \times 10^3$, $B = 0.767 \times 10^3$, $C = -0.687050 \times 10^6$, $D = 0.506510 \times 10^9$, $E = -0.138540 \times 10^{12}$ respectively.

In the jump model, the thermal conductivity is correlated with the pressure-dependent material property as given below,

$$\lambda = \frac{\varepsilon \cdot d \cdot p}{1 + \gamma \cdot d \cdot p} \qquad (5.1.26)$$

where d is the distance between the filament and the heat sink, p is the pressure, ε and γ are constants in the kinetic gas theory. $\varepsilon \propto \sqrt{m}$ and $\gamma \propto \Lambda^{-1}$ where m is the particle mass and Λ is the mean free path of gas [66]. These sensors are operated in a periodic pulsed mode which causes two different evaluation options, transition state and steady state. In the transition state, the time constant τ of the pulse is inversely correlated with the thermal conductivity of the surrounding gas as given below in Eq. (5.1.27).

$$\frac{1}{\tau} = \frac{\lambda\frac{A}{L} - \alpha I^2 R_0}{\rho V C_p} \qquad (5.1.27)$$

where α is the temperature coefficient resistance (TCR). $I^2 R_0$ is the initial input power and $\rho V C_p$ is the total thermal capacitance of the system [66].

In the steady state, considering the total temperature shift at the end of the pulse. Similar to frequency, the temperature shift is inversely correlated with the thermal conductivity Eq. (5.1.28).

$$\frac{1}{T - T_0} = \frac{\lambda}{P_0}\left(\frac{A}{L}\right) - \alpha \qquad (5.1.28)$$

For the gas mixtures, the thermal conductivity of the gases can be calculated by the mixing rule regarding individual gas components Eq. (5.1.29) [70].

$$\lambda_{mix} = \sum_{i=1}^{n} \frac{\lambda_i}{\frac{1}{x_i} \cdot \sum_{j=1}^{n} A_{ij} x_j} \qquad (5.1.29)$$

The A_{ij} parameter is related to molar masses and dynamic viscosities.

5.1.3.7 Mechanical hydrogen sensors

Generally, cantilevers are operated in two modes as dynamic and static mode. An external influence is required to actuate the cantilever for the dynamic mode. The actuation types can be magnetic, electrical, piezoelectric, thermal noise, etc. The resonance frequencies of cantilevers are measured by different readout methods such as optics, piezoelectric, impedance, etc. The resonance frequency of a cantilever operated in dynamic mode is expressed as:

$$f = \frac{1}{2\pi} \sqrt{\frac{k}{m^*}} \tag{5.1.30}$$

where k is the spring constant of the cantilever and m^* is the effective mass of the cantilever. The resonance frequency of the cantilever in a vacuum environment can be expressed as [71],

$$f_{vac} = \frac{t}{4\pi} \frac{3.52}{L^2} \sqrt{\frac{E}{3\rho}} \tag{5.1.31}$$

where, E, t, L, and ρ represent the elastic constant, thickness, length, and density of the cantilever, respectively. As can be understood from this expression, the material of the cantilever and the dimensions of the cantilever affect the resonance frequency. On the other hand, sensor parameters of the cantilever depend on many conditions such as the temperature of the environment, the sensing material coated on the cantilever, the interference analytes in the environment, light, pressure, and quality factor.

When any biological or chemical analyte is ab/adsorbed onto the cantilever, the overall mass can be expressed as:

$$m_t = m^* + m_a \tag{5.1.32}$$

where m_a is the total mass of the ad/absorbed analytes. Thus, the resonance frequency of the cantilever under any analyte environment can be written as by taking into account the spring constant of the cantilever [72].

$$f_a = \frac{1}{2\pi} \sqrt{\frac{Ewt^3}{4L^3(m_a + 0.24\rho wtL)}} \tag{5.1.33}$$

where, w represents the width of the cantilever.

Otherwise, the amount of bending in the cantilever is measured in static mode. The bimetallic cantilevers are bent with temperature change due to the difference the thermal expansion coefficients of these metals. Bimetallic cantilevers are generally used as temperature sensor. In addition, the cantilever is bent with the stress created by the analytes ab/adsorbed on the cantilever or sensitive layer coated cantilever. In general, the optical, piezoelectric, and capacitive methods are used to measure the deflection of the cantilever. The Stoney's equation is the most commonly used formula to relationship between cantilever deflections and surface stresses [73,74].

$$\Delta\sigma = \frac{Et^2}{3L^2\,(1-\nu)}\Delta z \qquad (5.1.34)$$

where $\Delta\sigma$, ν, and Δz are the variation in the stress, Poisson ratio, and the amount of the deflection respectively. This equation is for assumption of a circular shape of the cantilever bend with small deflection. Up to now cantilever has been coated with Pd thin films and hydrogen detection of Pd coated cantilever investigated [75,76].

The hydrogen–Pd interaction is given in details in the metallic resistive hydrogen sensor part. While hydrogen ad/absorbed by Pd, physical properties of Pd changes as hydrogen dissociates, occupies interstitial sites in Pd lattice causing lattice expansion. Generally this lattice expansion principle explains the Pd coated cantilever based hydrogen sensors. During the ad/absorption of hydrogen with Pd film, the expansion of the Pd film induces stresses on cantilever that Pd coated and this causes curvature or bending of the cantilever. There are fewer works about mechanical hydrogen sensor compared to other hydrogen sensor types.

Volume expansion during the absorption of hydrogen creates deformations in Pd layers due to internal stress. The emerged strain δ in the Pd film can be calculated depending on the $(H/Pd)_{at}$ by using the relation as given below,

$$\delta = \frac{\Delta a}{a_0} = C_{Pd} * (H/Pd)_{at} \qquad (5.1.35)$$

where a_0 is the lattice constant and C_{Pd} is the linear expansion coefficient of Pd with H. The working principles of the mechanical hydrogen sensor rely on measuring the deformations (bending, stretching, etc.) in the structure correlating to the amount of adsorbed hydrogen. Mechanical hydrogen sensors can be useful for a fuse design for safety applications.

5.1.3.8 Acoustic hydrogen sensors

5.1.3.8.1 QCM based hydrogen sensors

Quartz crystal microbalance (QCM) is a sensor device quantify the amount of mass change by shifts in the bulk crystal resonance frequency. The real-time monitoring of the mass changes on the surface QCM electrodes make it a strong tool for gas detection. The Saurbery introduced the mathematical relation between the mass change and the resonance frequency as seen in Eq. (5.1.36) [77]

$$\Delta f = -\frac{2 f_0^2}{A \sqrt{\mu_Q \rho_Q}} \Delta m \qquad (5.1.36)$$

where ρ_Q *and* μ_Q are the density and shear modulus of quartz, f_0 is the base frequency.

QCM systems work at low temperatures to avoid the temperature effect on the oscillations. Therefore, the sensing mechanism is mainly based on the adsorption and desorption processes of the analyte gas on sensing layer. Besides some noble metal doped SMOX nanostructures [78], the use of organic sensing materials and blends are more favorable for QCM sensors.

Due to the conductive nature of electrodes, electrochemical deposition, and nanomaterial synthesis techniques are more compatible with QCM. For instance, TiO_2 nanotubes were electrochemically grown on electrodes for H_2 sensing [79]. The QCM sensors can be combined with optical setups to monitor adsorption and desorption surface kinetics of the hydrogen [80].

5.1.3.8.2 *Surface acoustic waves (SAW) based hydrogen sensors*

The surface acoustic waves based sensors track the changes on the modulated waves after their propagation on the sensitive layer. Surface waves transform the acoustic information due to physical and chemical interactions between analyte gas and sensing layer to electrical signal. Compared to QCM, there are more hydrogen sensor applications with SAW sensors. In addition to noble metal films, or noble metal loaded SMOX nanostructures, graphene loaded, and organic sensing materials have been studied extensively [81–89].

SAW gas sensors work at stable conditions constant temperature and pressure. Due to the piezoelectricity of the substrates, finite element (FE) simulations can be used for modelling hydrogen sensing [90]. The perturbation theory defines the mass loading effect, elastic loading effect and acoustoelectric loading effect as major effects cause the negative frequency shift of SAW gas sensors [91]. The relationship between the center frequency changes (Δf) of SAW sensor and these perturbation parameters can be defined as:

$$\frac{\Delta v}{V_0} = -C_m f_0 \Delta(\rho_s) + C_e f_0 h_\Delta \left(\left(\frac{4\mu}{V_0^2}\right) \times \left(\frac{\mu + \lambda}{\mu + 2\lambda}\right) \right)$$

$$- \frac{K^2}{2} \times \Delta \left(\frac{1}{1 + (V_0 C_s / \sigma_s)} \right) \qquad (5.1.37)$$

where f_0 is the initial center frequency, Δf is the frequency shift, k is the ratio of the length of the sensitive film to the center to-center distance between IDTs, ΔV and V_0 are the velocity shift and base velocity of SAW, ρ_s is the mass surface density (mass per unit area) of the sensitive film, C_m is the coefficient of mass sensitivity, C_s is the capacitance per unit length of the SAW substrate material ($C_s = \varepsilon_s + \varepsilon_0$, where ε_s and ε_0 are the permittivities of the substrate and free pace, respectively), C_e is the coefficient of elasticity sensitivity, μ and λ are the shear and bulk modulus of elasticity, h is the thickness of the sensitive films, σ_s is the sheet conductivity of the film ($\sigma_s = \sigma_h$, where σ is the film bulk conductivity), and K^2 is the electromechanical coupling coefficient.

5.1.3.9 Magnetic hydrogen sensors

A magnetic hydrogen gas sensor is a relatively new concept and is receiving more attention in the scientific world. Over the past few years, a number of researchers have been studying the magnetic detection of gases since they discovered that some gases, like hydrogen, being able to alter the magnetic properties of certain materials [92]. It has been shown that hydrogen interacts with ferromagnetic structures containing Pd in a way that affects their structural, electronic, optical, and magnetic properties [92]. In terms of magnetic properties, notable changes were observed in susceptibility, magnetization, magnetic anisotropy, and ferromagnetic resonance in Co/Pd multilayers [93,94], Pd/Co/Pd tri-layers [95,96], Pd/Fe, Pd/Co and Pd/Ni bilayers [97,98], and in Pd-rich CoPd alloy films [93,99]. There are also several other materials whose magnetic properties are observed to change systematically in their magnetic properties when hydrogen is present in the system, such as Fe/Nb [100] and Fe/V superlattices [101,102], and $SnFeO_2$ ferrites [103]. All of the effects mentioned above were detected using laboratory magnetometry techniques including superconducting quantum interference devices (SQUID), vibrating magnetometers, optical Kerr effect setups, ferromagnetic resonance, and also some synchrotron-based techniques like polarized neutron reflectivity, X-ray resonant magnetic scattering (XRMS), and X-ray magnetic circular dichroism (XMCD). A formidable challenge awaits these techniques when they are adapted to field conditions. There has been some work done on using magnetic materials to sense gas, but it hasn't been implemented in practical devices as of yet.

The first study to deal with hydrogen-related magnetic properties manipulation was published by D. Sander in 2004 [104]. In their study, 8 mL Ni epitaxial films were grown on a Cu (0 0 1) single crystal substrate, and the magneto optical Kerr effect was studied to determine magnetic properties. The innovation part of their study was the way in which they exposed hydrogen gas to manipulate the magnetic properties of Ni films. It has been observed that the magnetic properties of Ni films have changed from in plane to out of plane with hydrogen exposure. After hydrogen exposure stopped, the magnetic properties of Ni films were returned to the initial state. This effect is totally reversible. Several different types of magnetic hydrogen gas sensors have been studied after this study. Magnetization based hydrogen gas sensors what the using magnetic properties changing to sense hydrogen gas are of four types: (1) Hall effect based, (2) Kerr effect based, (3) Ferromagnetic Resonance Based, and (4) Magnetostatic surface spin-wave (MSSW) oscillator based magnetic hydrogen gas sensors. As an alternative to these types of magnetic based gas sensors, there may also be the possibility of using various types of magnetic based sensors in the near future. In terms of hydrogen sensors, the magnetic based ones described above are still considered fundamental research projects.

A basic understanding of magnetic based hydrogen sensors and their recent advancements will be presented in this section.

5.1.3.9.1 Hall effect based magnetic hydrogen sensors

Hall effect is the occurrence of a voltage difference across an electrical conductor generated by an electron current flowing through the conductor and a perpendicular magnetic field applied. It has been found that there are two types of Hall effects based on where they originate in the magnetic material as the ordinary Hall effect and the extraordinary Hall effect. Both types of effects have advantageous properties, which make them suitable for the fabrication of magnetic gas sensors. The Hall voltage changing can be measured with and without hydrogen gas.

It has been reported that Gerber and colleagues [105] have conducted an investigation of the magnetic gas detection system using extraordinary Hall effects (EHE). Their approach allowed them to simultaneously assess two independent variables at the same time, either resistivity or magnetization which were both manipulated by the target gas and with a method that was compatible with existing conductometric gas detection techniques. As a way to demonstrate the feasibility of the approach, thin CoPd films were used in the sensor as a means of detecting hydrogen at low concentrations. A hydrogen/nitrogen atmosphere containing hydrogen concentrations below 0.5% has a Hall effect sensitivity of over 240% per 104 ppm. This is compared to conductance detection, the accuracy of this method is at least two orders of magnitude more accurate.

5.1.3.9.2 *Kerr effect based magnetic hydrogen sensors*

A magneto-optic Kerr effect, also known as MOKE, measures the change in polarization and intensity of light that is reflected from a magnetized surface. It can be employed in the detection of gas sensing applications. With its high surface-sensitive properties, MOKE is a very efficient method of generating signals. In other words, diamagnetic and paramagnetic additives from the substrate do not affect the signal. It is therefore a very efficient tool for analyzing the magnetic properties of magnetic surfaces with very small variations in their properties. This approach can thus be used to monitor changes in the reflected light beam as well as the magnetic properties of materials by monitoring these changes. As a result of rotating a polarized light beam within a gas atmosphere, we are able to measure the Ms, Hc, and squareness (Mr/Ms) of a hysteresis loop as well as its perpendicular and horizontal directions (in-plane) [92].

5.1.3.9.3 *FMR based magnetic hydrogen sensors*

Measurement of magnetic properties via ferromagnetic resonance (FMR) involves measuring the processional motion of magnetization in ferromagnetic samples. Electron paramagnetic resonance is therefore related to this technique. A macroscopic perspective is that in the presence of the static magnetic field H_0, the total magnetic moment processes around the direction of the local field H_{eff}, before relaxation processes dampen this precession and cause it to align with the local field. By irradiating the sample with a transverse radio frequency field (typically microwaves ranging from 1 GHz to 35 GHz) and if the RF frequency coincides with the processional frequency, resonance occurs, and microwave energy is absorbed.

A coplanar waveguide (CPW) sits over the sample to investigate the hydrogen effects on magnetic samples with FMR. It is possible to control hydrogen or nitrogen gas through the cell. With this type of gas sensor device, the change in perpendicular magnetic anisotropy (PMA) is measured using the FMR method. Using this approach, the frequency value of the input signal controls the FMR constant, and when gas is present, the FMR peak position of the magnetic layer moves to the lower applied magnetic field H, which is a result of the PMA decreasing when the analyte gas molecule is present. A sensor's sensitivity is determined by the magnitude of the FMR peak shift [97].

5.1.3.9.4 Magnetostatic surface spin-wave (MSSW) oscillator based magnetic hydrogen sensors

Magnetostatic or exchange interactions cause spin waves which are eigen disturbances in magnetic moments to propagate within magnetic materials, including ferrimagnets, ferromagnets, and antiferromagnets. It is well known that magnetostatic spin waves propagate in three distinctly different directions, depending on their relative orientation with respect to static magnetization (M). It can be categorized as (1) magnetostatic surface waves (MSSW), (2) backward volume mode (BVM), and (3) forward volume mode (FVM) [106–110]. In MSSW and BVM, both k and $M\rho$ are in the film plane. According to MSSW, k is perpendicular to M, but according to BVM, k is parallel with M. In FVM, k lies in the plane of the magnetic film, while M lies out of the plane. MSSW oscillators have been demonstrated to be useful in a variety of novel devices and applications [111–114]. These considerations suggest that a new generation of chemical sensors can be designed by combining an MSSW oscillator as a magnetic field detector with a layer of magnetic nanoparticles of $CuFe_2O_4$ as a gas sensor. Using a MSSW oscillator, one measures the change in magnetic properties to create a chemical sensor, which it can call a "magnonic gas sensor." A sensitive magnetic material has been used in the form of nanoparticles in order to create a very large surface-to-volume ratio that can significantly increase sensitivity.

5.1.4 Hydrogen sensor market

A significant increase has been observed in the worldwide market for hydrogen gas sensors in recent years. There have already been a number of commercial markets that have adopted hydrogen as a fuel, including stationary power systems, public transportation (train, ferry, bus, etc.), and industrial trucks [115,116]. Increasing demand for hydrogen in these markets is driving the development of hydrogen infrastructure, including transportation and production capabilities. There are three main segments of the hydrogen gas sensor market: technology, end-user industry, and geographical region. In terms of technology with their percentages, the market can be segmented into electrochemical (76.2%), catalytic (12.5%), thermal conductivity (8.9%), and metal oxide semiconductor (2.3%), or other types (2.5%). As well as this, the market is separated by end-user industries, such as automotive, oil and gas, healthcare, aerospace and defense, and others. Depending on the region, the hydrogen sensor market research can be segmented into North America,

Latin America, Europe, East Asia, South Asia, Middle East, and other regions [117].

The demand for hydrogen sensors increased by 1.5% CAGR between 2017 and 2021. Currently, the global hydrogen market is valued at USD 314.4 million as of 2022, and it is expected to increase at a compound annual growth rate (CAGR) of 6.7% to reach USD 599 million by the year 2032. As marketing increases in phases over the next 10 years, there will be three distinct phases: a short-term marketing phase (2022–2025), a medium-term marketing phase (2025–2028), and a long-term marketing phase (2028–2032) [115]. A growing power generation industry as well as the automotive industry are likely to impact the market in the short term for hydrogen sensors. The market for hydrogen sensors is expected to expand significantly in the medium term due to the increasing awareness of green hydrogen as a cleaner, newer source of energy. A long-term will be activated with IoT-enabled sensors supplying accurate real-time information regarding the presence of hydrogen, the prevention of equipment malfunctions, and the avoidance of preventable accidents [117].

A comparison of hydrogen sensor performances and as well as their working conditions of various hydrogen gas sensors that are currently being used on the market can be found in Table 5.1.1 [10]. Electrochemical hydrogen gas sensors have several benefits, including their ability to detect very low concentrations of hydrogen gas, quick response to changes in gas concentration, and ease of use. However, these sensors may not always provide highly accurate measurements, may have a shorter lifespan compared to other types of sensors, and may be affected by the presence of other gases.

Catalytic hydrogen gas sensors are a popular choice for detecting and measuring hydrogen gas due to their high sensitivity, fast response time, and ease of use. However, these sensors also have some limitations. For example, they may have a shorter lifespan compared to other types of sensors, may not always provide highly accurate measurements, and may only be able to operate within a limited temperature range. Additionally, they may be affected by the presence of other gases, which can impact their accuracy and reliability.

Thermal conductivity hydrogen gas sensors have several advantages, including their high sensitivity, fast response time, and simple design. However, they also have some disadvantages, including a potentially shorter lifespan compared to other types of hydrogen gas sensors, sensitivity to other gases, and a limited temperature range within which they can operate.

Table 5.1.1 An overview of the commercial hydrogen sensors with their performances [10].

Sensor type	Principle/device	Performance					
		Measuring range (vol%)	Accuracy (% of indication)	Response time (t_{90} − s)	Power consumption (mW)	Gas environment	Lifetime (years)
Catalytic	Pellistor	Up to 4	<±5	<30	1000	−20–70°C 5–95% RH 70–130 kPa	5
Thermal conductivity	Calorimetric	1–100	±0.2	<10	<500	0–50°C 0–95% RH 80–120 kPa	5
Electrochemical	Amperometric	Up to 4	≤±4	<90	2–700	−20–55°C 5–95% RH 80–110 kPa	2
Resistance based	Semiconducting metal–oxide	Up to 2	±10–30	<20	<800	−20–70°C 10–95% RH 80–120 kPa	>2
	Metallic resistor	0.1–100	≤±5	<15	>25	0–45°C 0–95% RH Up to 700 kPa	<10
Work function based	Capacitor	Up to 5	<±7	<60	4000	−20–40°C 0–95% RH 80–120 kPa	10
	MOS field effect transistor	Up to 4.4	<±7	<2	700	−40–110°C 5–95% RH 70–130 kPa	10
Optical	Optrode	0.1–100	±0.1	<60	1000	−15–50°C 0–95% RH 75–175 kPa	>2

(Reprinted from: T. Hubert, L. Boon-Brett, G. Black, U. Banach, Hydrogen sensors: a review, Sens. Actuators B: Chem. 157 (2011) 329–352 with permission from Elsevier.)

Resistive hydrogen gas sensors have the ability to detect hydrogen gas with high sensitivity, are easy to use due to their simple design, and have a fast response time. However, it is important to consider their potential limitations as well, such as potential inaccuracies in measurement, a limited temperature range in which they can be used, a shorter lifespan compared to other types of sensors, and the potential for interference from other gases.

Work function-based hydrogen gas sensors are able to detect hydrogen gas with a high level of sensitivity and have a fast response time and a simple design. However, there are also some drawbacks to consider. These sensors may not always provide highly accurate measurements, have a shorter lifespan compared to other types of sensors, be affected by the presence of other gases, and have a limited temperature range within which they can operate effectively.

There are a variety of prices for commercial hydrogen sensors, ranging from \$10 to \$2700. According to reports, the cheapest detectors are work function-based detectors that are small and provide fast and accurate results between 100 ppm and 1000 ppm when used in certain restricted conditions. A resistance-based detector has a high hydrogen sensitivity between 10 ppm and 1000 ppm and a response time of 15 seconds and is slightly cheaper (around \$30). While they are highly sensitive and have an acceptable lifetime, they exhibit poor selectivity. There are also catalytic and electrochemical gas sensors in the mid-price range, and some models are not specifically constructed to detect hydrogen gas, but they have a long lifetime and can detect multiple gases. The thermal conductivity type of sensor is quite expensive but provides a wide detection range that works at slightly higher temperatures and pressures, but their sensing abilities diminish as temperatures rise [117].

5.1.5 Future trend in hydrogen sensors

As a clean energy source compared to fossil fuels, hydrogen will take more place in industry, transport, and residential heating. Besides the available technologies, a full transformation to hydrogen will take time to adapt the production, storage, distribution, transfer, and consumption stages [118]. Regarding all these stages, the process control and safety concerns, there is an emerging demand for advanced hydrogen sensors work in various applications and under different conditions. Some of the desired specifications for hydrogen sensors are high sensitivity in the operational concentration ranges (0.01–10% for safety and 1–100% full cell) [119], stability, selectivity, reliability, robustness, rapid sensor feedbacks, energy, and cost efficiency.

New sensing materials, their combinations and nanomaterial designs with novel heterostructured MOXs [120], 2D materials [121,122], and metal–organic frameworks (MOFs) seem attractive research fields for future advanced H_2 sensors targeting to reach the listed requirements.

Heterostructures MOXs nanomaterials always keep the interest on electrochemical hydrogen sensing due to adjustable sensing properties, low cost, low dimensional designs, and easy integration to electrical devices. The core-shell MOXs nanostructures can present higher sensing properties by modulating heterojunction layer, surface depletion regions. The controlling the thickness and depletion layers by atomic layer deposition systems (ALD) provide high control on sensing mechanisms. In addition, surface loading, and ternary MOX structures can present exceptional sensing mechanisms due to synergetic effects.

The 2D materials have been favorable with the widespread ALD and CVD systems. The layer by layer material designs allow to discover new electrical interactions and surfaces by graphene, metal dichalcogenides, etc.

The most of the hydrogen production (>40%) is based on natural gas reforming technology [123]. In these steam reforming plants with carbon capture, utilization, and storage (CCUS) are operational. Besides the sensor technology, the membrane and filter technologies are emerging for gas capturing, storage, transport, and separation processes. The discovery of metal-organic-frameworks (MOFs) boosted these technologies by their unique crystalline structures with metal ions bounded organic linkers [124]. Their tunable pore sizes gave them huge advantage for molecular sieving while proposing a high surface area/volume ratio for gas storage applications, etc. [125]. Today there are more than 70,000 structural combinations with the diversity of metal ions and organic linkers in available MOFs [126]. With its lowest molecular size and weight, hydrogen can be an essential material for MOF electroactive membrane applications for sensing [127,128].

References

[1] S.A. Sherif, D.Y. Goswami, E.K. Stefanakos, A. Steinfeld, Handbook of Hydrogen Energy, CRC Press, 2014.

[2] J.O. Abe, A.P.I. Popoola, E. Ajenifuja, O.M. Popoola, Hydrogen energy, economy and storage: review and recommendation, Int. J. Hydrogen Energy 44 (2019) 15072–15086.

[3] M. Momirlan, T.N. Veziroglu, Current status of hydrogen energy, Renew. Sustain. Energy Rev. 6 (2002) 141–179.

[4] N. Sazali, Emerging technologies by hydrogen: a review, Int. J. Hydrogen Energy 45 (2020) 18753–18771.

[5] F. Rigas, P. Amyotte, Hydrogen Safety, CRC Press, 2018.

[6] W.M. Haynes, CRC Handbook of Chemistry and Physics, CRC Press, 2014.

[7] T.S. King, M. Elia, J.O. Hunter, Abnormal colonic fermentation in irritable bowel syndrome, Lancet North Am. Ed. 352 (1998) 1187–1189.

[8] A. Rezaie, M. Buresi, A. Lembo, H. Lin, R. McCallum, S. Rao, M. Schmulson, M. Valdovinos, S. Zakko, M. Pimentel, Hydrogen and methane-based breath testing in gastrointestinal disorders: the North American consensus, Am. J. Gastroenterol. 112 (2017) 775.

[9] W. Shin, Medical applications of breath hydrogen measurements, Anal. Bioanal. Chem. 406 (2014) 3931–3939.

[10] T. Hubert, L. Boon-Brett, G. Black, U. Banach, Hydrogen sensors: a review, Sens. Actuators, B 157 (2011) 329–352. https://doi.org/10.1016/j.snb.2011.04.070.

[11] S.K. Arya, S. Krishnan, H. Silva, S. Jean, S. Bhansali, Advances in materials for room temperature hydrogen sensors, Analyst 137 (2012) 2743–2756. https://doi.org/10.1039/c2an16029c.

[12] H. Hashtroudi, P. Atkin, I.D.R. Mackinnon, M. Shafiei, Low-operating temperature resistive nanostructured hydrogen sensors, Int. J. Hydrogen Energy 44 (2019) 26646–26664. https://doi.org/10.1016/j.ijhydene.2019.08.128.

[13] J.X. Dai, L. Zhu, G.P. Wang, F. Xiang, Y.H. Qin, M. Wang, M.H. Yang, Optical fiber grating hydrogen sensors: a review, Sensors-Basel 17 (2017). https://doi.org/10.3390/s17030577.

[14] G. Korotcenkov, S. Do Han, J.R. Stetter, Review of electrochemical hydrogen sensors, Chem. Rev. 109 (2009) 1402–1433. https://doi.org/10.1021/cr800339k.

[15] N. Kilinc, Resistive hydrogen sensors based on nanostructured metals and metal alloys, Nanosci. Nanotechnol. Lett. 5 (2013) 825–841. https://doi.org/10.1166/nnl.2013.1653.

[16] R.M. Penner, A nose for hydrogen gas: fast, sensitive H_2 sensors using electrodeposited nanomaterials, Acc. Chem. Res. 50 (2017) 1902–1910. https://doi.org/10.1021/acs.accounts.7b00163.

[17] J.S. Noh, J.M. Lee, W. Lee, Low-dimensional palladium nanostructures for fast and reliable hydrogen gas detection, Sensors-Basel 11 (2011) 825–851. https://doi.org/10.3390/s110100825.

[18] A. Kotchourko, T. Jordan, Hydrogen safety for energy applications: engineering design, Risk Assessment, and Codes and Standards, Elsevier Science, 2022.

[19] G. Heiland, Zum Einfluß von adsorbiertem Sauerstoff auf die elektrische Leitfähigkeit von Zinkoxydkristallen, Z. Med. Phys. 138 (1954) 459–464.

[20] T. Seiyama, A. Kato, K. Fujiishi, M. Nagatani, A new detector for gaseous components using semiconductive thin films, Anal. Chem. 34 (1962) 1502–1503.

[21] Taguchi, N. Gas detection semiconductor element, Japanese Patent Application, S45-38200, 1962.

[22] O. Sisman, Strategies to Enhance the Performances of Metal Oxide Gas Sensors, University of Brescia, 2020.

[23] S. Choopun, N. Hongsith, E. Wongrat, Metal-oxide nanowires for gas sensors, in: X. Peng (Ed.), Nanowires, IntechOpen, Rijeka, 2012, Ch. 1.

[24] N. Barsan, U. Weimar, Conduction model of metal oxide gas sensors, J. Electroceram. 7 (2001) 143–167.

[25] N. Hongsith, E. Wongrat, T. Kerdcharoen, S. Choopun, Sensor response formula for sensor based on ZnO nanostructures, Sens. Actuators B 144 (2010) 67–72.

[26] N. Yamazoe, New approaches for improving semiconductor gas sensors, Sens. Actuators B: Chem. 5 (1991) 7–19.

[27] N. Barsan, D. Koziej, U. Weimar, Metal oxide-based gas sensor research: How to? Sens. Actuators B 121 (2007) 18–35.

[28] E. Comini, Metal oxide nano-crystals for gas sensing, Anal. Chim. Acta 568 (2006) 28–40.

[29] E. Comini, C. Baratto, G. Faglia, M. Ferroni, A. Vomiero, G. Sberveglieri, Quasi-one dimensional metal oxide semiconductors: preparation, characterization and application as chemical sensors, Prog. Mater Sci. 54 (2009) 1–67.

[30] E. Comini, G. Faglia, G. Sberveglieri, Solid State Gas Sensing, Springer US, 2008.

[31] E. Wongrat, N. Hongsith, D. Wongratanaphisan, A. Gardchareon, S. Choopun, Control of depletion layer width via amount of AuNPs for sensor response enhancement in ZnO nanostructure sensor, Sens. Actuators B 171 (2012) 230–237.

[32] N. Yamazoe, K. Shimanoe, Theory of power laws for semiconductor gas sensors, Sens. Actuators B 128 (2008) 566–573.

[33] M. Hübner, N. Bârsan, U. Weimar, Influences of Al, Pd and Pt additives on the conduction mechanism as well as the surface and bulk properties of SnO_2 based polycrystalline thick film gas sensors, Sens. Actuators B 171 (2012) 172–180.

[34] W. Zeng, T. Liu, D. Liu, E. Han, Hydrogen sensing and mechanism of M-doped SnO_2 (M = Cr^{3+}, Cu^{2+} and Pd^{2+}) nanocomposite, Sens. Actuators B 160 (2011) 455–462.

[35] Z. Li, D. Ding, Q. Liu, C. Ning, X. Wang, Ni-doped TiO_2 nanotubes for wide-range hydrogen sensing, Nanoscale Res. Lett. 9 (2014) 1–9.

[36] Z. Zhang, J.T. Yates Jr, Band bending in semiconductors: chemical and physical consequences at surfaces and interfaces, Chem. Rev. 112 (2012) 5520–5551.

[37] Y. Fukai, The Metal-Hydrogen System: Basic Bulk Properties, Springer, Berlin Heidelberg, 2010.

[38] F.A. Lewis, A. Aladjem, Hydrogen Metal Systems I, Trans Tech Publications Limited, 1996.

[39] F.A. Lewis, A. Aladjem, Hydrogen in Metal Systems II, Trans Tech Publications Limited, 2000.

[40] H.S. Lee, J. Kim, H. Moon, W. Lee, Hydrogen gas sensors using palladium nanogaps on an elastomeric substrate, Adv. Mater. 33 (2021) 2005929.

[41] F. Yang, K.C. Donavan, S.-C. Kung, R.M. Penner, The surface scattering-based detection of hydrogen in air using a platinum nanowire, Nano Lett. 12 (2012) 2924–2930.

[42] H.-W. Yoo, S.-Y. Cho, H.-J. Jeon, H.-T. Jung, Well-defined and high resolution Pt nanowire arrays for a high performance hydrogen sensor by a surface scattering phenomenon, Anal. Chem. 87 (2015) 1480–1484.

[43] N. Kilinc, Palladium and platinum thin films for low-concentration resistive hydrogen sensor: a comparative study, J. Mater. Sci.-Mater. Electron. 32 (2021) 5567–5578. https://doi.org/10.1007/s10854-021-05279-w.

[44] E. Sennik, S. Urdem, M. Erkovan, N. Kilinc, Sputtered platinum thin films for resistive hydrogen sensor application, Mater. Lett. 177 (2016) 104–107. https://doi.org/10.1016/j.matlet.2016.04.134.

[45] A. Abburi, N. Abrams, W.J. Yeh, Synthesis of nanoporous platinum thin films and application as hydrogen sensor, J. Porous Mater. 19 (2012) 543–549.

[46] F. Cao, P. Zhao, Z. Wang, X. Zhang, H. Zheng, J. Wang, D. Zhou, Y. Hu, H. Gu, An ultrasensitive and ultraselective hydrogen sensor based on defect-dominated electron scattering in pt nanowire arrays, Adv. Mater. Interfaces 6 (2019) 1801304.

[47] K. Rajouâ, L. Baklouti, F. Favier, Platinum for hydrogen sensing: surface and grain boundary scattering antagonistic effects in Pt@ Au core–shell nanoparticle assemblies prepared using a Langmuir–Blodgett method, Phys. Chem. Chem. Phys. 20 (2018) 383–394.

[48] T. Sahoo, P. Kale, Work function-based metal-oxide-semiconductor hydrogen sensor and its functionality: a review, Adv. Mater. Interfaces 8 (2021) 2100649. https://doi.org/10.1002/admi.202100649.

[49] T. Hubert, L. Boon-Brett, V. Palmisano, M.A. Bader, Developments in gas sensor technology for hydrogen safety, Int. J. Hydrogen Energy 39 (2014) 20474–20483.

[50] S.M. Sze, Y. Li, K.K. Ng, Physics of Semiconductor Devices, Wiley, 2021.

[51] E.H. Nicollian, J.R. Brews, MOS (Metal Oxide Semiconductor) Physics and Technology, Wiley, 2002.

[52] I. Lundström, S. Shivaraman, C. Svensson, L. Lundkvist, A hydrogen sensitive MOS field effect transistor, Appl. Phys. Lett. 26 (1975) 55–57.

[53] P.S. Chauhan, S. Bhattacharya, Hydrogen gas sensing methods, materials, and approach to achieve parts per billion level detection: a review, Int. J. Hydrogen Energy 44 (2019) 26076–26099. https://doi.org/10.1016/j.ijhydene.2019.08.052.

[54] A.B. LaConti, H.J.R. Maget, Electrochemical detection of H_2, CO, and hydrocarbons in inert or oxygen atmospheres, J. Electrochem. Soc. 118 (1971) 506.

[55] M.K. Hossain, R. Chanda, A. El-Denglawey, T. Emrose, M.T. Rahman, M.C. Biswas, K. Hashizume, Recent progress in barium zirconate proton conductors for electrochemical hydrogen device applications: a review, Ceram. Int. 47 (2021) 23725–23748. https://doi.org/10.1016/j.ceramint.2021.05.167.

[56] E.G. Lee, S.-W. Jung, Y.E. Jo, H.R. Yoon, B.K. Yoo, S.H. Choi, J.W. Choi, J.S. Jang, S.-Y. Lee, Electrochemical hydrogen sensor assembly for monitoring high-concentration hydrogen, Phys. Status Solidi (A) 219 (2022) 2100782.

[57] M.A. Butler, Optical fiber hydrogen sensor, Appl. Phys. Lett. 45 (1984) 1007–1009.

[58] K. Ito, T. Kubo, Gas detection by hydrochromism, Proceedings of the 4th Sensor Symposium, Tokyo, Japan (1984) 153–156.

[59] G.P. Wang, J.X. Dai, M.H. Yang, Fiber-optic hydrogen sensors: a review, IEEE Sensors J. 21 (2021) 12706–12718. https://doi.org/10.1109/JSEN.2020.3029519.

[60] Y.H. Kim, M.J. Kim, B.S. Rho, K.H. Kwack, B.H. Lee, Mach-Zehnder interferometric hydrogen sensor based on a single mode fiber having core structure modification at two sections, in: IEEE SENSORS 2010 Conference, 2010, pp. 1483–1486. https://doi.org/10.1109/Icsens.2010.5690361.

[61] H.Y. Choi, M.J. Kim, B.H. Lee, All-fiber Mach-Zehnder type interferometers formed in photonic crystal fiber, Opt. Express 15 (2007) 5711–5720. https://doi.org/10.1364/Oe.15.005711.

[62] B.H. Lee, J. Nishii, Dependence of fringe spacing on the grating separation in a long-period fiber grating pair, Appl. Optics 38 (1999) 3450–3459. https://doi.org/10.1364/Ao.38.003450.

[63] M.S. Chen, H.T. Dang, J. Zhang, Y. Wang, J. Li, Mach-Zehnder interferometer refractive index sensing probe based on dual microfiber coupler, Optik (2021) 228. https://doi.org/10.1016/j.ijleo.2020.166181.

[64] A.D. Kersey, M.A. Davis, H.J. Patrick, M. LeBlanc, K.P. Koo, C.G. Askins, M.A. Putnam, E.J. Friebele, Fiber grating sensors, J. Lightwave Technol. 15 (1997) 1442–1463. https://doi.org/10.1109/50.618377.

[65] Yang, M.H., Dai, J.X. Fiber optic hydrogen sensors: a review. Photonic Sens. 2014, 4, 300–324, https://doi.org/10.1007/s13320-014-0215-y.

[66] D. Berndt, J. Muggli, F. Wittwer, C. Langer, S. Heinrich, T. Knittel, R. Schreiner, MEMS-based thermal conductivity sensor for hydrogen gas detection in automotive applications, Sens. Actuators A 305 (2020) 111670.

[67] I. Simon, M. Arndt, Thermal and gas-sensing properties of a micromachined thermal conductivity sensor for the detection of hydrogen in automotive applications, Sens. Actuators A 97 (2002) 104–108.

[68] S. Foorginezhad, M. Mohseni-Dargah, Z. Falahati, R. Abbassi, A. Razmjou, M. Asadnia, Sensing advancement towards safety assessment of hydrogen fuel cell vehicles, J. Power Sources 489 (2021) 229450.

[69] Gesellschaft, V.D.I. *VDI Heat Atlas*, Springer Berlin Heidelberg: 2010.

[70] A.L. Lindsay, L.A. Bromley, Thermal conductivity of gas mixtures, Ind. Eng. Chem. 42 (1950) 1508–1511.

[71] B.N. Johnson, R. Mutharasan, Biosensing using dynamic-mode cantilever sensors: A review, Biosens. Bioelectron. 32 (2012) 1–18.

[72] T. Thundat, P.I. Oden, R.J. Warmack, Microcantilever sensors, Microscale Thermophys. Eng. 1 (1997) 185–199.

[73] A. Loui, S. Elhadj, D.J. Sirbuly, S.K. McCall, B.R. Hart, T.V. Ratto, An analytic model of thermal drift in piezoresistive microcantilever sensors, J. Appl. Phys. 107 (2010) 054508.

[74] G.G. Stoney, The tension of metallic films deposited by electrolysis, in: Proceedings of the Royal Society of London. Series A, Containing Papers of a Mathematical and Physical Character, 82, 1909, pp. 172–175.

[75] H. Li, Y. Li, K. Wang, L. Lai, X. Xu, B. Sun, Z. Yang, G. Ding, Ultra-high sensitive micro-chemo-mechanical hydrogen sensor integrated by palladium-based driver and high-performance piezoresistor, Int. J. Hydrogen Energy 46 (2021) 1434–1445.

[76] J.T. Gurusamy, G. Putrino, R.D. Jeffery, K.D. Silva, M. Martyniuk, A. Keating, L. Faraone, MEMS based hydrogen sensing with parts-per-billion resolution, Sens. Actuators B 281 (2019) 335–342.

[77] G. Sauerbrey, Verwendung von Schwingquarzen zur Wägung dünner Schichten und zur Mikrowägung, Z. Med. Phys. 155 (1959) 206–222.

[78] S. Ozturk, A. Kosemen, Z.A. Kosemen, N. Kilinc, Z.Z. Ozturk, M. Penza, Electro-chemically growth of Pd doped ZnO nanorods on QCM for room temperature VOC sensors, Sens. Actuators, B 222 (2016) 280–289. https://doi.org/10.1016/j.snb.2015.08.083.

[79] G. Neri, A. Bonavita, G. Micali, G. Centi, S. Perathoner, R. Passalacqua, M.-G. Willinger, N. Pinna, Synthesis, characterization and sensing applications of nanotubular TiO_2-based materials, Sensors and Microsystems, Springer, 2011, pp. 151–154.

[80] K.J. Palm, J.B. Murray, J.P. McClure, M.S. Leite, J.N. Munday, In situ optical and stress characterization of alloyed Pd_xAu_{1-x} hydrides, ACS Appl. Mater. Interfaces 11 (2019) 45057–45067.

[81] A. D'Amico, A. Palma, E. Verona, Surface acoustic wave hydrogen sensor, Sens. Actuators 3 (1982) 31–39.

[82] Y.-S. Huang, Y.-Y. Chen, T.-T. Wu, A passive wireless hydrogen surface acoustic wave sensor based on Pt-coated ZnO nanorods, Nanotechnology 21 (2010) 095503.

[83] D. Sil, J. Hines, U. Udeoyo, E. Borguet, Palladium nanoparticle-based surface acoustic wave hydrogen sensor, ACS Appl. Mater. Interfaces 7 (2015) 5709–5714.

[84] R. Arsat, M. Breedon, M. Shafiei, P.G. Spizziri, S. Gilje, R.B. Kaner, K. Kalantar-zadeh, W. Wlodarski, Graphene-like nano-sheets for surface acoustic wave gas sensor applications, Chem. Phys. Lett. 467 (2009) 344–347. https://doi.org/10.1016/j.cplett.2008.11.039.

[85] L. Yang, C. Yin, Z. Zhang, J. Zhou, H. Xu, The investigation of hydrogen gas sensing properties of SAW gas sensor based on palladium surface modified SnO_2 thin film, Mater. Sci. Semicond. Process. 60 (2017) 16–28. https://doi.org/10.1016/j.mssp.2016.11.042.

[86] S.J. Ippolito, S. Kandasamy, K. Kalantar-Zadeh, W. Wlodarski, Layered SAW hydrogen sensor with modified tungsten trioxide selective layer, Sens. Actuators B 108 (2005) 553–557. https://doi.org/10.1016/j.snb.2004.11.048.

[87] L. Al-Mashat, H.D. Tran, W. Wlodarski, R.B. Kaner, K. Kalantar-zadeh, Polypyrrole nanofiber surface acoustic wave gas sensors, Sens. Actuators B 134 (2008) 826–831. https://doi.org/10.1016/j.snb.2008.06.030.

[88] R. Arsat, X.F. Yu, Y.X. Li, W. Wlodarski, K. Kalantar-zadeh, Hydrogen gas sensor based on highly ordered polyaniline nanofibers, Sens. Actuators B 137 (2009) 529–532. https://doi.org/10.1016/j.snb.2009.01.028.

[89] D.-T. Phan, G.-S. Chung, Surface acoustic wave hydrogen sensors based on ZnO nanoparticles incorporated with a Pt catalyst, Sens. Actuators B 161 (2012) 341–348. https://doi.org/10.1016/j.snb.2011.10.042.

[90] M.Z. Atashbar, B.J. Bazuin, M. Simpeh, S. Krishnamurthy, 3D FE simulation of H_2 SAW gas sensor, Sens. Actuators B 111-112 (2005) 213–218. https://doi.org/10.1016/j.snb.2005.06.054.

[91] V.B. Raj, H. Singh, A.T. Nimal, M.U. Sharma, V. Gupta, Oxide thin films (ZnO, TeO_2, SnO_2, and TiO_2) based surface acoustic wave (SAW) E-nose for the detection of chemical warfare agents, Sens. Actuators B 178 (2013) 636–647. https://doi.org/10.1016/j.snb.2012.12.074.

[92] K. Munbodh, F.A. Perez, D. Lederman, Changes in magnetic properties of Co/Pd multilayers induced by hydrogen absorption, J. Appl. Phys. 111 (2012) 123919.

[93] W.-C. Lin, C.-J. Tsai, H.-Y. Huang, B.-Y. Wang, V.R. Mudinepalli, H.-C. Chiu, Hydrogen-mediated long-range magnetic ordering in Pd-rich alloy film, Appl. Phys. Lett. 106 (2015) 012404.

[94] K. Munbodh, F.A. Perez, C. Keenan, D. Lederman, M. Zhernenkov, M.R. Fitzsimmons, Effects of hydrogen/deuterium absorption on the magnetic properties of Co/Pd multilayers, Phys. Rev. B 83 (2011) 094432.

[95] W.-C. Lin, C.-J. Tsai, B.-Y. Wang, C.-H. Kao, W.-F. Pong, Hydrogenation induced reversible modulation of perpendicular magnetic coercivity in Pd/Co/Pd films, Appl. Phys. Lett. 102 (2013) 252404.

[96] S. Okamoto, O. Kitakami, Y. Shimada, Enhancement of magnetic anisotropy of hydrogenated Pd/Co/Pd trilayers, J. Magn. Magn. Mater. 239 (2002) 313–315.

[97] C.S. Chang, M. Kostylev, E. Ivanov, Metallic spintronic thin film as a hydrogen sensor, Appl. Phys. Lett. 102 (2013) 142405.

[98] W.-C. Lin, C.-S. Chi, T.-Y. Ho, C.-J. Tsai, Hydrogen absorption induced reversible effect on magneto-optical property of Pd/Fe, Pd/Co and Pd/Ni bilayers, Thin Solid Films 531 (2013) 487–490. https://doi.org/10.1016/j.tsf.2013.01.022.

[99] S. Akamaru, T. Matsumoto, M. Murai, K. Nishimura, M. Hara, M. Matsuyama, Sensing hydrogen in the gas phase using ferromagnetic Pd–Co films, J. Alloys Compd. 645 (2015) S213–S216. https://doi.org/10.1016/j.jallcom.2015.01.055.

[100] F. Klose, C. Rehm, D. Nagengast, H. Maletta, A. Weidinger, Continuous and reversible change of the magnetic coupling in an Fe/Nb multilayer induced by hydrogen charging, Phys. Rev. Lett. 78 (1997) 1150.

[101] A. Remhof, G. Nowak, H. Zabel, M. Björck, M. Pärnaste, B. Hjörvarsson, V. Uzdin, Remote control of the Fe magnetic moment in magnetic heterostructures, Europhys. Lett. 79 (2007) 37003.

[102] D. Labergerie, C. Sutter, H. Zabel, B. Hjörvarsson, Hydrogen induced changes of the interlayer coupling in $Fe_{(3)}/V_{(x)}$ superlattices (x = 11–16), J. Magn. Magn. Mater. 192 (1999) 238–246. https://doi.org/10.1016/S0304-8853(98)00546-0.

[103] A. Punnoose, K.M. Reddy, J. Hays, A. Thurber, M.H. Engelhard, Magnetic gas sensing using a dilute magnetic semiconductor, Appl. Phys. Lett. 89 (2006) 112509.

[104] D. Sander, W. Pan, S. Ouazi, J. Kirschner, W. Meyer, M. Krause, S. Müller, L. Hammer, K. Heinz, Reversible H-induced switching of the magnetic easy axis in Ni/Cu(001) thin films, Phys. Rev. Lett. 93 (2004) 247203.

[105] A. Gerber, G. Kopnov, M. Karpovski, Hall effect spintronics for gas detection, Appl. Phys. Lett. 111 (2017) 143505.

[106] R.W. Damon, J.R. Eshbach, Magnetostatic modes of a ferromagnetic slab, J. Appl. Phys. 31 (1960) S104–S105.

[107] A.A. Serga, A.V. Chumak, B. Hillebrands, YIG magnonics, J. Phys. D Appl. Phys. 43 (2010). https://doi.org/10.1088/0022-3727/43/26/264002.

[108] S.O. Demokritov, B. Hillebrands, A.N. Slavin, Brillouin light scattering studies of confined spin waves: linear and nonlinear confinement, Phys. Rep. 348 (2001) 441–489. https://doi.org/10.1016/S0370-1573(00)00116-2.

[109] D.D. Stancil, Theory of Magnetostatic Waves, Springer, New York, 2012.

[110] G. Srinivasan, A.N. Slavin, High frequency processes in magnetic materials, World Scientific (1995).

[111] R. Marcelli, E. Andreta, G. Bartolucci, M. Cicolani, A. Frattini, A magnetostatic wave oscillator for data relay satellite, IEEE T. Magn. 36 (2000) 3488–3490. https://doi.org/10.1109/20.908869.

[112] N. Qureshi, O.V. Kolokoltsev, C.L. Ordonez-Romero, G. Lopez-Maldonado, An active resonator based on magnetic films for near field microwave microscopy, J. Appl. Phys. 111 (2012). https://doi.org/10.1063/1.3672081.

[113] D. Matatagui, O.V. Kolokoltsev, N. Qureshi, E.V. Mejia-Uriarte, J.M. Saniger, A novel ultra-high frequency humidity sensor based on a magnetostatic spin wave oscillator, Sens. Actuators B: Chem. 210 (2015) 297–301. https://doi.org/10.1016/j.snb.2014.12.118.

[114] D. Matatagui, O.V. Kolokoltsev, N. Qureshi, E.V. Mejia-Uriarte, J.M. Saniger, A magnonic gas sensor based on magnetic nanoparticles, Nanoscale 7 (2015) 9607–9613. https://doi.org/10.1039/c5nr01499a.

[115] U. Bossel, Hydrogen economy: what future? in: A. Iulianelli, A. Basile (Eds.), Current Trends and Future Developments on (Bio-) Membranes, Elsevier, 2020, pp. 421–451.

[116] A. Demirbas, Future hydrogen economy and policy, Energy Sourc. Part B: Econ. Plan. Pol. 12 (2017) 172–181.

[117] https://www.factmr.com/report/3401/hydrogen-sensor-market.

[118] I.E. Agency, Global Hydrogen Review 2022. https://www.iea.org/reports/global-hydrogen-review-2022.

[119] Hübert, T., Boon-Brett, L., Palmisano, V., Frigo, G., Hellstrand, Å., Kiesewetter, O., May, M. Trends in gas sensor development for hydrogen safety. International Conference on Hydrogen Safety, Brussels, Belgium, 2013.

[120] D. Zappa, V. Galstyan, N. Kaur, H.M.M. Munasinghe Arachchige, O. Sisman, E. Comini, "Metal oxide-based heterostructures for gas sensors": a review, Anal. Chim. Acta 1039 (2018) 1–23. https://doi.org/10.1016/j.aca.2018.09.020.

[121] M. Donarelli, L. Ottaviano, 2D materials for gas sensing applications: a review on graphene oxide, MoS_2, WS_2 and phosphorene, Sensors-Basel 18 (2018) 3638.

[122] S. Yang, C. Jiang, S.-h. Wei, Gas sensing in 2D materials, Appl. Phys. Rev. 4 (2017) 021304.

[123] M.N. Uddin, V.V. Nageshkar, R. Asmatulu, Improving water-splitting efficiency of water electrolysis process via highly conductive nanomaterials at lower voltages, Energy Ecol. Environ. 5 (2020) 108–117.

[124] H. Furukawa, K.E. Cordova, M. O'Keeffe, O.M. Yaghi, The chemistry and applications of metal-organic frameworks, Science 341 (2013) 1230444.

[125] A. Knebel, J. Caro, Metal–organic frameworks and covalent organic frameworks as disruptive membrane materials for energy-efficient gas separation, Nat. Nanotechnol. 17 (2022) 911–923.

[126] P.Z. Moghadam, A. Li, S.B. Wiggin, A. Tao, A.G.P. Maloney, P.A. Wood, S.C. Ward, D. Fairen-Jimenez, Development of a Cambridge Structural Database subset: a collection of metal–organic frameworks for past, present, and future, Chem. Mater. 29 (2017) 2618–2625.

[127] Y. Li, A.-S. Xiao, B. Zou, H.-X. Zhang, K.-L. Yan, Y. Lin, Advances of metal–organic frameworks for gas sensing, Polyhedron 154 (2018) 83–97.

[128] D.-K. Nguyen, J.-H. Lee, T.L.-H. Doan, T.-B. Nguyen, S. Park, S.S. Kim, B.T. Phan, H_2 gas sensing of Co-incorporated metal-organic frameworks, Appl. Surf. Sci. 523 (2020) 146487.

Hydrogen safety/standards (national and international document standards on hydrogen energy and fuel cell)

Sanjay Kumar[a], Anilia Nanan-Surujbally[b], Davinder Pal Sharma[b] and Dinesh Pathak[b]
[a]Department of Physics, Faculty of Science, University of South Bohemia, České Budějovice, Czech Republic
[b]Department of Physics, The University of the West Indies, St. Augustine, Trinidad & Tobago

5.2.1 Introduction

5.2.1.1 Hydrogen energy

Energy is one of the most needed entities of this era. The demand for it is very high due to growing economies and populations. This need has caused the energy sectors of the world to lean on multiple sources that are not friendly to nature, such as fossil fuels. Hence, the world is facing the consequences in the form of global warming and climate change. Today we are looking for alternative ways to reduce the number of greenhouse gases in the atmosphere and hydrogen shows a promising future. Hydrogen can be produced using multiple processes and sources, depending on the process and source it is denoted by color. Grey hydrogen is produced using fossil fuels like coal and natural gas, the excess carbon dioxide is released into the atmosphere making it less environmentally friendly, and turquoise hydrogen is produced through the pyrolysis of methane the by-products are carbon in the solid phase and hydrogen. In cases where renewables are used to power the reactor, the process is carbon neutral. Additionally, green hydrogen uses renewable energy sources, and the process of electrolysis is applied. Electrolysis is the procedure by which water molecules are separated into hydrogen and oxygen only. Green hydrogen can also be produced by steam reforming of biomethane and pyrolysis of biogenic feedstock. This process has zero emissions and is considered the most environmentally safe way to produce hydrogen. Moreover, blue

Towards Hydrogen Infrastructure: Advances and Challenges in Preparing for the Hydrogen Economy.
DOI: https://doi.org/10.1016/B978-0-323-95553-9.00011-X

hydrogen is formed using steam reforming. Natural gas is split into hydrogen and carbon dioxide and the carbon dioxide is captured and stored. The carbon capture technology mitigates the environmental impact. Furthermore, yellow hydrogen is produced in a similar way to green hydrogen but the only power source used is solar power and as well as pink hydrogen is produced using electrolysis and the power source is nuclear energy [1,2].

5.2.1.1.1 *The physical properties of hydrogen and basic safety issues*

Hydrogen exists in the liquid and gaseous phases. The element has a very low boiling and melting point and hence is considered a cryogenic liquid or gas. The boiling and melting points of the substance are very crucial for transporting and storing the element. This factor plays a crucial role in the future of hydrogen. Hydrogen in its purest form is tasteless, odorless, and colorless. If a hydrogen leak were to occur in broad daylight its presence is almost invisible. Moreover, when hydrogen gas is mixed with sufficient air the oxygen molecules can be displaced making it very toxic. In comparison to other gases, hydrogen has a very low vapor and liquid density. Furthermore, the volume proportion of liquid hydrogen to gaseous hydrogen is 1:848, and hydrogen is very flammable when exposed to a certain amount of air. Its flammability range is 4–75% which is greater than many other fuels at atmospheric pressure [3,4]. Table 5.2.1 summarizes some of the physical properties of hydrogen [3,4].

Hydrogen can be stored in both gaseous and liquid phases. Hydrogen can be stored for longer periods in the gas phase. Existing natural gas storage systems can be adopted. Hydrogen can be stored in underground caverns with built-in salt domes with depts of 1000 meters near electrolytes. Metal hydrides can be used as a storage possibility since the hydrogen molecules can bond with metal compounds at atmospheric pressure in low-pressure systems and remain in a stable state [5,6]. Hydrogen storage is important since it holds the highest energy per mass of any fuel. It can be stored in liquid and gas phases, but specific equipment and conditions are necessary due to its physical and chemical properties. High-pressure tanks are needed with psi ranging from 5000 to 10,000, and cryogenic temperatures are needed since at one atmospheric pressure it boils at $-252.8°C$. Hydrogen is distributed using three methods.

Table 5.2.1 Physical properties of hydrogen.

S. no.	Properties	U.S. units	SI units
1	Autoignition temperature	752°F	400°C
2	Boiling point at 14.69 psia (101.283 kPa)	-423.0°F	-252.8°C
3	Chemical formula	H_2	H_2
4	Critical temperature	-399.8°F	-239.9°C
5	Critical pressure	188 psia	1296.212 kPa, abs
6	Critical density	1.88 lb/ft^3	30.12 kg/m^3
7	Cp	3.425 Btu/(lb)(°F)	14.34 kJ/(kg)(°C)
8	Cv	2.418 Btu/(lb)(°F)	10.12 kJ/(kg)(°C)
10	The density of the gas at boiling point and 1 atm	0.083 lb/ft^3	1.331 kg/m^3
11	The density of the liquid at boiling point and 1 atm	4.23 lb/ft^3	67.76 kg/m^3
12	Flammable limits in air	4–75%	
13	Latent heat of fusion at the triple point	24.97 Btu/lb	58.09 kJ/kg
14	Latent heat of vaporization at boiling point	191.7 Btu/lb	446.0 kJ/kg
15	Molecular weight	2.016	2.016
16	The melting point at 14.69 psia (101.283 kPa)	-434.5°F	-259.2°C
17	The ratio of specific heats	1.42	1.42
18	Solubility in water vol/vol at 60°F (15.6°C)	0.019	0.019
19	The specific gravity of the liquid at boiling point and 1 atm	0.0710	0.0710
20	The specific gravity of the gas at 32°F and 1 atm (air $=1$)	0.0696	0.0696
21	The specific volume of the gas at 70°F (21.1°C) and 1 atm	192.0 ft^3/lb	11.99 m^3/kg
23	Vapor pressure at -423°F (-252.8°C)	14.69 psia	101.283 kPa

1. Pipelines are a safe cost friendly medium to distribute but it requires more infrastructure since places like the United States only have 2575 km of pipelines the volume transferred is limited. In the Gulf Coast and California, the pipelines are situated near petroleum refineries and chemical plants.
2. High-pressure tube trailers are used to transport hydrogen within a 321-km radius. The tube trailers are transported by trucks and ships. This medium is very costly with limitations.
3. Liquefied hydrogen tanks are considered the most expensive of all three since the hydrogen is cooled to a liquid state before it is transported. The phase conversion is costly and time-consuming. However, this method is most efficient since the hydrogen can be transported over long distances without any hindrance. It is possible during transportation the hydrogen can boil off or evaporate so the consumption rates must be corresponding [7].

5.2.1.1.2 *Fuel cell*

Fuel is an electrochemical apparatus that converts chemical energy to electrical energy without combustion and the fuel or oxidant is supplied outward. This technology was developed in 1839 by Sir William Grove, later in the 20th century, the technology was adopted in the Gemini and Apollo space mission. In comparison to internal combustion engines and batteries, fuel cells are considered environmentally friendly when it runs purely on hydrogen, its only by-products are water and heat. When it runs on a hydrogen-heavy reformate gas mix, some greenhouse gasses are released into the atmosphere, but the quantity of emissions is smaller in comparison to combustion engines using fossil fuels. Fuel cells have higher power-generating efficiency in comparison to gasoline-electric, steam and gas turbines, and diesel–electric. Since it operated at low temperatures it is ideal for automotive applications. Fuel cell systems do not require turning and recharging. However, since no major infrastructure for this technology is not put in place the drawback is storage and how expensive it is to produce. Fuel cell applications are abundant for example they can be used in stationary power plants, submarines, buses, cars, and portable power systems [8]. The five different types of fuel cells are polymer electrolyte membrane, alkaline, phosphoric acid, molten carbonate, and solid oxide. Table 5.2.2 shows the comparison of different fuel cell technologies [9].

Table 5.2.2 Comparison of different fuel cell technologies.

S. no.	Fuel cell type	Common electrolyte	Operating temperature	Typical stack size	Electrical efficiency
1	Alkaline (AFC)	Aqueous potassium hydroxide soaked in a porous matrix, or alkaline polymer membrane	<100°C	10–100 kW	60%
2	Molten carbonate (MCFC)	Molten lithium, sodium, and/or potassium carbonates, soaked in a porous matrix	600–700°C	300 kW–3 MW, 300 kW module	50%
3	Phosphoric acid (PAFC)	Phosphoric acid soaked in a porous matrix or imbibed in a polymer membrane	150–200°C	5–400 kW, 100 kW module	40%
4	Polymer electrolyte membrane (PEM)	Perfluorosulfonic acid	<120°C	<1 kW–100 kW	60% transportation, 35% stationary
5	Solid oxide (SOFC)	Yttria stabilized zirconia	500–1000°C	1 kW–2 MW	60%

5.2.1.1.3 Hydrogen safety codes, standards, and regulations

Hydrogen safety codes and standards are a vital project since it ensures that hydrogen is handled and used in a very safe and cautious manner. The National Renewable Energy Laboratory (NREL) is currently testing the safety sensor's efficiency in their safety Sensor Testing Laboratory. The engineers and scientists at this organization are also collaborating with partners in other industries and organizations to further develop and test hydrogen sensor technology. NREL has validated a memorandum of agreement (MOA) with the European Commission's Joint Research Centre (JRC) Institute of energy and transport in The Netherlands. In recent times codes and standards are being established for hydrogen systems for transportation, commercial and residential applications. Hydrogen is becoming a more significant energy carrier and fuel with the advancement of new codes and standards. Standards development organizations (SDOs) are responsible for formulating codes and standards that would assist with the commercial applications of fuel vehicle technology [10–12]. Table 5.2.3 shows codes and standards development organizations and their roles.

5.2.2 Hazards, incidents, and preventions

5.2.2.1 Hydrogen leaks and fires

Hydrogen leaks are the foundation of all gaseous hydrogen hazards. Generally, hydrogen has a low molecular weight and scientifically it is the smallest molecule. The element has high buoyancy since the density is low. The element is tasteless, colorless, and odorless, when the concentration is high it acts as an asphyxiant. Leaks are difficult to avoid since the element molecule is very small. Using hydrogen as a fuel has a lot of precautions attached to it, beginning with the design and the type of material that would be used when manufacturing pipes and storage cylinders since hydrogen is known for embrittling certain metals. Some of the precautions taken by the transport manufacturing industry were to run fuel lines outside the passenger section of the vehicle to avoid leaks inside the enclosed vehicle. Hydrogen is stored in the hoods of transit buses in high-pressured tanks. This placement allows the gas to dissipate rapidly and vertically in the air with minimum impedance in case of a leak. Hydrogen is stored in the fuel cell at low pressure, in small amounts when the engine is operating and when the engine is shut off no hydrogen is present in the fuel cell. This

Table 5.2.3 Codes and standards development organizations and their roles.

S. no.	Code	Organization	Responsibility
1	AGA	American Gas Association	Material analysis standard
2	API	American Petroleum Institute	Apparatus standards for petroleum production, storage, and handling
3	ASHRAE	American Society of Heating, Refrigerating, and Air-Conditioning Engineers	Advocating sustainability through research, standards writing, publishing, and education
4	ASTM	American Society for Testing and Materials	Practical test standard for materials, products, and systems
5	ASME	American Society of Mechanical Engineers	Automated and multidisciplinary engineering, design codes, and standards
6	CGA	Compressed Gas Association	Equipment design and performance standards for compressed gas systems and components
7	CSA	CSA Standards	U.S. and Canadian equipment standards
8	GTI	Gas Technology Institute	Supplies technical support and training for the energy industry
9	ICAO	International Civil Aviation Organization	Promotes understanding and security through cooperative aviation regulation
10	ICC	International Code Council	Family of model building codes, including the international fire code
11	IEA	International Energy Agency	Energy policy advisor to 28 member countries ensuring clean energy for their citizens
12	ISO	International Organization for Standardization	Supports U.S. concerns at key international codes and standards
13	OIML	International Organization of Legal Metrology	Intergovernmental treaty organization

(continued on next page)

Table 5.2.3 Codes and standards development organizations and their roles—cont'd

S. no.	Code	Organization	Responsibility
14	IPHE	International Partnership for Hydrogen and Fuel Cells in the Economy	Facilitating commercial utilization activities related to hydrogen and fuel cell technologies
15	IEEE	Institute of Electrical and Electronics Engineers	Electrical standards
16	NFPA	National Fire Protection Association	Model codes and standards, including the national electric code
17	NIST	National Institute of Standards and Technology	Measurement standards
18	SAE	SAE International	Vehicle standards
19	UL	Underwriters Laboratory	Equipment and performance testing standards
20	UN	United Nations	Developing friendly relations among nations to promote social progress and human rights

initiative is to reduce exposure and mishaps. When hydrogen is used in a fuel cell the air and the hydrogen do not mix directly, this is achieved by using seals. It is understood that hydrogen seep is not a hazard by itself, but it can become a risk if it is mixed with highly concentrated air. If the seep causes the hydrogen to dissipate from the air molecules it can become an asphyxiation issue [13,14].

Hydrogen can be detected using thermal conductivity (TC) and gas chromatography (GC). The response time using these methods is rapid, but its major disadvantage is being able to detect other gases like nitrogen and carbon dioxide that are in the presence of hydrogen. Mass spectrometry (MS) is another method that can be used to detect hydrogen gas. Both methods of detection require maintenance, skilled operations personnel, and an abundance of gases for calibration purposes, which can be very costly and composite. Laser gas analyzers are very efficient and simple to use, the apparatus uses light frequencies to detect concentrations of hydrogen, but the apparatus is very expensive, costing roughly

US$100,000 [15]. Recently, many different types of hydrogen sensors have surfaced, most using the alloy of palladium. One differs from the other based on the characteristics such as high sensitivities, wide range concentrations, stability standing, response and recovery time, and low energy consumption.

1. Electrochemical sensors (EC) detect the presence of hydrogen gas. The hydrogen gas oxidizes on the platinum-coated electrode, signaling a current or voltage depending on the levels of concentration. These devices are available for commercial uses but only give warranties for up to two years. The advantages are it is sensitive hence the amount of power needed for operations is very small making it a good product for the transport industry. However, it is dependent on environmental pressures and the temperature range is narrow since the liquid electrolyte is used.

2. Metal–oxide sensors (MOX), are manufactured using a wide band gap semiconductor material. MOX is not affected by environmental temperature fluctuations. The limitation includes long response and recovery time since the sensor is not linear. In addition, the presence of silicone compounds can permanently damage the sensors. This sensor requires the presence of oxygen to give stable responses. The sensor has a metal-oxide-semiconductor, with a metal layer, an insulating layer, and a semiconductor layer. These layers enable the sensor to work in the absence of oxygen.

3. Catalytic gas sensors (CGS), are comprised of double platinum wires that sit on a ceramic block, connected in a Wheatstone bridge. The ceramic block is heated at 773–823 K, causing oxidization on the block's surface to occur which leads to a rise in the resistance of the platinum filament. Hydrogen concentration and the Wheatstone bridge are linearly related. These sensors are used to test hydrogen concentration in the air from 0% to 10%. Since the sensor requires high temperatures to operate a lot of electrical energy is required, and this limitation is weighted heavily in terms of energy conservation and efficiency.

4. Thermal conductivity sensors (TC) deal with the change in temperature in an electrical heating element when susceptible to gases. The sensor is not raised to a temperature causing combustion but instead is a result of deviations in Ohm's Law. TCS is very reliable and stable and has a lower chance of being contaminated. The sensor gives rapid results and is time efficient. The amount of energy required to run this sensor

is smaller in comparison to the others. However, the limitation of this sensor includes environmental factors such as humidity, pressure, and temperature.

5. Optical sensors (OS) are not very common since hydrogen does not absorb ultraviolet and infrared. Some devices react to the properties of the palladium films whilst other reacts to chemical changes such as colors when in contact with hydrogen gas. These sensors are very sensitive and simple to use, providing long-term stability [15].

5.2.2.1.1 *Leaks repairs*

Hydrogen leaks can be repaired and the principle it stands by is sealing the leak or replacing the damaged component or fittings. Precautions must be taken when attempting to fix a leak. These precautions begin with limiting the number of persons in the contaminated area, ensuring the surrounding is free of sources that can start a fire, and prohibiting smoking. For example, if a leak exists in a vehicle, the location of the leak must be first identified, and leak detection apparatus and sensors can facilitate this task. Leaks take place when the circuit component is pressurized, and the circuit remains this way even if the ignition is turned off. Hence, it is important to take the battery off to ensure the vehicle is electrically dead before attempting to fix the leaks. Once all the precautions are taken, the fittings can be tightened but if the leak continues the fixtures must be replaced. Tightening and replacing fittings can only take place when the ambient pressure is low else physical injury can occur. It is advised that if a leak occurs when the pressure is high proper ventilation methods should be administered before attempting to fix it [13,15].

5.2.2.1.2 *Fires*

Hydrogen generates electrostatic charges that can potentially cause sparks, it can burn without any smoke or flame and the flammability range is very wide. Hydrogen fires potentially start once released hydrogen is mixed with air and becomes ignited. The flame speed is 230 cm/s and hence the fire spreads rapidly. The fire can occur in broad daylight and be unnoticeable because it is invisible. Leaks in an enclosed space are hazardous since explosions can occur. In open spaces, the risk factor is smaller since the gas can dissipate and diffuse quickly. Fires can be avoided if precautionary techniques are applied from the early stages of manufacture and design. For example, static charge accumulation can be avoided if the materials are fire-resistant and well-set. Storage tanks can be built with pressure-release valves and pipes to reduce high-pressure build-ups that can lead to explosions. Persons responsible for dealing with hydrogen must be properly educated and skilled in handling the gas and its technology, putting safety and precision first. The danger involved in a hydrogen fire is related to the amount of pressure that was built up. Noticing a hydrogen fire is very difficult, but in most cases, you can identify it by smelling smoke, heat waves, or being alerted by a fire suppression system. The fire inhibition system is a set of instruments (sensors) that are in connection with the power system in a vehicle, when the heat level and temperature rise, an alert is set off. The driver will be informed, and the vehicle would then automatically power off and retardants would be expelled.

The method of extinguishing a hydrogen fire is like that of any other fire. Persons can either locate the source and put a stop or let the fire run its course while ensuring the safety of persons and paraphernalia. The size of the fire determines the extinguishing techniques that would be considered. If the fire is small, it can be extinguished with halon, carbon dioxide, or retardant but for a larger fire, the fuel supply must be turned off. If the supply of hydrogen cannot be powered off then the fire must run its course while wetting and cooling the area to reduce damage can remove flammable objects to reduce mishaps and explosions. People must keep a 450 m distance away from the fire and ensure not to inhale the smoke since it can be toxic [13–15].

5.2.2.1.3 Temperature-related hazards

Temperature and heat are not the same, temperature is the degree of hotness of a substance while heat is the energy released from the motion of molecules. Heat is energy and temperature is a unit of measure. The apparatus used to measure temperature are thermometers, thermocouples, and temperature transducers. Furthermore, hydrogen is usually used in gaseous form however to transport it from one place to another it is done in liquid form. When converting gaseous hydrogen to liquid hydrogen ($H_{2(l)}$), 33% is lost, and the process is not very efficient. The characteristic of liquid hydrogen is the same as gaseous hydrogen since it is odorless, colorless, and transparent. Hydrogen has a boiling point of minus 252.87°C at 1 atmospheric pressure, hence at low temperatures when in a liquid state the hazards involved are secondary fires and frostbites. Frostbites can occur to persons or any surface it encounters. This occurs from the cold boil-off gas that is produced from the liquid [16–18].

Rapid phase transitions (RPT) can cause explosions, this is a reaction when hot $H_{2(l)}$ encounters cold liquids at a temperature greater than the boiling point of the cold liquids. This reaction is physical and not chemical, it is very famous with liquid nitrogen gas (LNG). No reports have surfaced when come to $H_{2(l)}$, but the possibility still exists. The requirement for storing $H_{2(l)}$ is it must be in perfectly isolated tanks such as cryogenic hydrogen containers. This technology is reserved for very particular cases like space travel, the tanks in the Ariane launcher hold $H_{2(l)}$ that will fuel the engine, and the casing is only 1.3 mm thick. A cryogenic tank doesn't function well at high-pressure levels, when the pressure is high a liquid vapor is formed which can expand at a rate of 845 times. Therefore, when manufacturing the tanks precaution like applying release vents are placed. At low temperatures, most gases condense and solidify. If the hydrogen solidifies in the seals and gaskets this can cause serious plugging which is a great hazard. This causes a reaction known as a cryopump, the plugging decreases the volume of gas that can flow through the pipes and the pipes start to operate like a vacuum sucking air in. If this reaction continues for long periods, the liquefied gas (LG) will be displaced. If the oxygen-to-hydrogen ratio is one to one the mixture can become flammable at the pressure levels. The issue can be resolved by purging the LG tanks with helium since at low temperatures helium does not solidify or condense. In addition, if $H_{2(l)}$ is spilled into the air and purging is not sufficient, pools of hydrogen are left in the atmosphere, and it is shock-sensitive and can behave as an explosive. It is advised by professional safety personnel to never open or release any attachment of components under high pressure since it can be shot with great force and never try to close or tighten an attachment under high pressure since it can shatter and cause bodily harm. Due to the dangers involved in the pressure and temperature of hydrogen vessels needs to have certification to operate and inspections are a must. Hence manufacturing precautions must be considered and codes, standards, and regulations must be set, this section will be further discussed in the next section [16–18].

5.2.2.1.4 Chemical hazards

The chemical hazard involved in fuel cells are not very serious, but precautions must be taken. People maintaining the fuel cell filters must wear protective goggles to keep their eyes safe. When de-ironizing resin to remove cations and anions from the coolant, vapor rises that can be hazardous to eye health causing irritation and discomfort. Coolant circuits contain ethylene glycol which is a very toxic chemical that can cause major health issues if ingested. These health issues include heart and kidney failure, central nervous system failure, and even death. Symptoms to be aware of are eye discomfort, contortion, hypertension, rapid heart palpitations, and great respiration. If the vapor of ethylene glycol is inhaled this cause less serve health issues like headache and cough. If persons are exposed to these vapors for long periods, the severity increase and can lead to deep unconsciousness. Additionally, ethylene glycol is amicable with other chemicals, this characteristic allows it to ignite at 100°C. It is double the weight of water hence it should not be exposed to water or chemicals with higher humidity levels. If it encounters other metals like lead, the metals should be disposed of since it is considered contaminated. In the event of ethylene glycol flames, it should be smothered with carbon dioxide, dry chemicals, or water or alcohol sprays. People putting the fire out should be dressed in full protective wear [19,20]. Moreover, purple K Dry Chemical (PKDC) is a retardant that is used to extinguish chemical fires. The PKDC is too dangerous in low concentrations but in high concentrations, it operates asphyxiation. In the event of a set off the chemical is emitted in the vehicle as white power, during the release a loud noise is heard. The chemical properties avoid it from absorbing into human skin so slight surface irritation is observed. If the chemical remains on the skin's surface for a long time, the skin can become very dry, and cracking can occur. A person should not inhale or ingest PKDC since it can pose health problems like vomiting and upset stomach. If the chemical enters your nose, eyes, or mouth persons should rise thoroughly for roughly 20 min and do not eat, drink, or smoke when in contact. This chemical is very toxic to the environment and aids in thinning of the ozone layer hence it should not be burnt. The local authorities should be informed when disposing of so proper protocol would be observed.

5.2.2.1.5 Physical hazards

The fuel cell comprises belts and fans that are always rotating. These components are covered during normal conditions but should in case one is left open the engine can still operate and here is where the risk lies. Persons should be extremely careful and avoid wearing roomy clothes that can get trapped in the spinning components. Should tragedy occur it can result in the loss of limbs or even loss of life. A fuel cell-powered bus weighs 1000 kg this is much more than a compressed natural gas (CNG) bus weighs. So, if difficulty arises and this equipment must be moved by a crane. Heavy equipment being moved by a crane sometimes slip; hence proper precautionary measures must take place to prevent these mishaps. A vehicle of this weight slipping off can lead to loss of possessions and life. Tables 5.2.4 and 5.2.5 show hydrogen-related incidents and prevention respectively [21–25].

5.2.3 Regulations, codes, and standards

The versatility of hydrogen and fuel cell applications in our world today calls for advancements in minimizing and mitigating dangers related to its use as a fuel. To understand hydrogen as a fuel, awareness of its properties, safety measures, engineering controls, rules, standards, and regulations are key. This understanding is needed as the hydrogen and fuel cell industry heightens and since fuels have levels of danger associated with them. A standard is a file that has requirements and prerequisites for a specific item, component, service, or procedure. A code is a file that describes the results of what a product should stand for rather than exist as. Codes and standards are optional but regulations are compulsory. Table 5.2.6 shows the different characteristics of standards and the legal requirements.

Codes and standards are necessary to ensure that each hydrogen and fuel cell system and facility is manufactured, maintained, and handled safely. Codes and standards will also ensure that safety measures both locally and internationally are equal and uniform. Building codes and equipment standards allow persons to accurately measure the amount of risk involved in a project and it also gives insurance to the client, public, and fire safety officials. Codes are established by a jurisdiction, whilst standards are an agreement where safety and compatibility are a priority. Despite, hydrogen being known for its chemical component and application it is mostly used as a chemical and not a fuel. This is the result of low confidence in the fuel and panic over its safety. With proper implementation of hydrogen and fuel cell codes and standards, the uses can become commercialized. The Fuel Cell and Hydrogen Energy Association (FCHEA) is a state-run association comprising over seventy organizations and companies. FCHEA deals

Table 5.2.4 Hydrogen-related incidents.

S. no.	Years	Location	Damage
1	1937	Naval Air Station Lakehurst	A fire, triggered a hydrogen cell, causing it to rapture nearby cells and fall to the ground. The fire traveled to the stern and the remainder of the cells was ignited.
2	1986	Kennedy Space Centre	Seven astronauts were killed aboard the Space Shuttle Challenger when a liquid hydrogen tank ruptured.
3	1999	Hanau Germany	Hydrogen storage and the chemical tank exploded.
4	2007	Muskingum River Coal Plant	One person die when the explosion of compressed hydrogen occurred.
5	2011	Japan	A hydrogen explosion caused damage to a few reactor buildings.
6	2015	Taiwan (Formosa Plastics Group)	Chemical plant eruption.
7	2018	Los Angeles (Diamond Bar)	A hydrogen-carrying truck caught fire while transporting 24 compressed hydrogen tanks. Evacuation for a one–mile radius was initiated.
8	2018	Veridian El Cajon, CA	A hydrogen-carrying motor vehicle caught fire at the Veridian Manufacturing Plant.
9	2019	Waukegan Illinois	Four workers dies and one was badly injured in an explosion.
10	2019	Santa Clara CA (Ail Products and Chemicals)	The tanker truck exploded.
11	2019	Norway	Uno-X fuelling station had an explosion.
12	2019	Waukesha, Wisconsin	One person was injured, and two hydrogen storage tanks were affected after the explosion.
13	2020	North Carolina (OneH$_2$)	And explosion affected nearby homes and building for miles around the explosion.
14	2020	Texas (Praxair)	The hydrogen production tank has an explosion.
15	2020	Changhua City, Taiwan	The driver of a hydrogen tanker was killed in a crash.
16	2021	South Africa (Medupi Power Station)	A plant explosion in one unit.
17	2022	Detroit MI	Hydrogen yank exploded in a pickup truck.

Table 5.2.5 Prevention methods that can per applied.

S. no.	Prevention mechanism	Application
1	Inserting and purging	A standard safety procedure to apply when transfer and hydrogen are too inert or purge the pipelines. The physical and chemical properties of hydrogen require these safety measures.
2	Ignition source management	The amount of energy required for hydrogen to be ignited is the lowest amount of other elements. Hence ignition can occur rapidly. The best prevention would be to use a relief valve and pressure relief systems instead of rupture discs.
3	Mechanical integrity and reactive chemistry	The four main chemical properties of hydrogen when in contact with other materials must be noted: (1) the interaction of hydrogen is unique from other chemicals and when reacted together with different elements the byproducts are different; (2) the compatibility of hydrogen with steel (building material); (3) when exposed to high temperatures the outcomes are different; and (4) hydrogen diffuses at different rates in comparison to other gases.
4	Leak and flame detection system	Hydrogen pipes should be placed above with proper labels to avoid leaks. This position is necessary since the gases are very light and buoyant. Hydrogen sensors can also be implemented to detect leaks, should in case it occurs. Flame detectors should be implemented since hydrogen fires are invisible to the naked eye.
5	Ventilation and flaring	Two concerns with hydrogen are its flammability and its ability to act as an asphyxiant gas. Proper ventilation is necessary to deal with both issues. Ventilation areas should be on the ceilings or facilities since hydrogen rises to the ceiling and the peaks of the structure. In instances where the hydrogen may flare it is necessary to note that this can cause damage to the vehicles.
6	Inventory management and facility spacing	In an ideal case fires and explosions do not occur but buildings must be designed so should in case damages are minimal. Hydrogen units should have minimal separation distances and proper venting should in case an explosion occurs.
7	Cryogenics	$H_{2(l)}$ has different chemistry to cryogenic chemicals, when it mixes with air it behaves like trinitrotoluene which is an explosive. Hence hydrogen requires proper storage technology. Thermally insulated containers are necessary when dealing with cryogenic bodies.
8	Human factors	Safety training and project safety procedures must be implemented. The testing of high points should be employed with a guide for persons working in the field of hydrogen.

Table 5.2.6 The different characteristics of standards and the legal requirements.

S. no.	Characteristics	Standards	Legal requirements
1	Purpose	Assist with the open exchange of goods and services	Keeps the public, environment, employees, and valuable materials safe from danger
2	Source	Opens agreements by persons interested	Legal teams, governments political teams, and technical experts
3	Legal character	A principle, not a mandate, that is considered applicable	Law and ordinance

with the advancement of clean and reliable energy sectors like hydrogen and fuel cell. They represent the comprising companies and organizations giving them a voice to policymakers and regulators. Its mission is to increase marketing for fuel cells and hydrogen energy and commercialize it [26–31].

5.2.3.1 FY 2021 merit review and peer evaluation report

The Safety, Codes, and Standards (SCS) is a portion of the Technology Acceleration portfolio. Their function is to Research, Development, and Demonstration (RD&D), the science behind the safety and technicality of codes and standards. The advancement of hydrogen and fuel cell technology gives rise to the development of codes and standards. Many codes and standards are already in existence at different stages, but revision and improvement are needed. Safety-related barriers exist for both the short term and long term and this can be a hindrance in executing proper regulations, codes, and standards [32].

5.2.3.1.1 Short-term barriers

1. Shortage of component failure information.
2. Irregularity with fuel quality assurance.
3. The development of safety sensors is inadequate.
4. The restrictive separation distance for $H_{2(l)}$ storage.

5.2.3.1.2 Long-term barriers

1. Unknown regulatory processes for emerging applications, such as those for bulk transport of hydrogen as cargo.
2. The insufficient number of standards for high throughput fuelling for lusty applications like marine and rail, etc.

3. Partial, codes, and standards for bulk storage of hydrogen.

In the fiscal year 2020–2021, many codes and standards were revised and published. Some updates were done on the hydrogen incident, hydrogen safety plan, and streamlined safety planning for low-volume hydrogen and fuel cell projects. A total of $10 million was budgeted and shared between codes and standards, component RD&D, and the behavior of hydrogen and risk [32].

5.2.3.2 Project #SCS-005: research and development for safety, codes, and standards: material and component compatibility

The main objective of this project is to facilitate technology implementation by executing and applying introductory research toward the growth and expansion of scientific codes and standards that enable the deployment of hydrogen technologies [32].

The technical barriers at the beginning of the project in 2003:

1. Limited access and availability to safety information.
2. A lack of technical information to revise and revisit current standards.
3. A shortage of compatible regulations, codes, and standards to authorize international and national markets.

The following points show how the procedure can be standardized for selecting materials for high-pressure vehicle fuel systems and stationary pressure vessels.

5.2.3.2.1 *The performance-based procedure for high-pressure vehicle fuel system*

1. Set up materials performance metrics.
2. Contemplate the technicality of the service condition.
3. Investigate different relevant systems and determine presiding conditions.

5.2.3.2.2 *Design-based procedure for stationary pressure vessel*

1. Consider and quantify reliable design data.
2. Determine bounding behavior for environment and mechanics; balance between evaluating the efficiency and valid data.
3. Evaluate data in clusters to establish global behavior.

During this approach, all barriers were tackled and are valid but the limitations attached were the complexity of the materials such as

Table 5.2.7 Strengths and weaknesses of the project.

S. no.	Project strengths	Project weaknesses
1	The project was very organized, comprehensive, and well-presented.	More research and communication need to be established concerning material capability.
2	It provided industries and SDOs with information, codes, and standards.	The project lacks a test matrix to show what is being tested and determined in comparison to what is being generated.
3	A balanced approach was given with progress, accomplishments, and partnerships.	Participation from the United States is insufficient.

the metals and nonmetals. This was with regard to the material definition, tensile characteristics, and fatigue behavior. In the sectors of transportation, heating, power, and industrial uses many accomplishments were made. Performance requirements were established for SAE 72579. Safety, Codes, and Standards 2021, and ASME code Case 2938 were approved for stationary pressure vessels, leaving a few unanswered questions.

1. **SAE 72579**: Standard for fuel systems in fuel cells and other hydrogen vehicles. The objective of this file is to portray, model, construct, operate, and maintain the requirements for hydrogen fuel storage and handling systems in vehicles. It was established in 2008 and last revised in 2018.

2. **ASME code Case 2938**: Fatigue crack growth rate design curves were developed at Sandia National Laboratories and accepted into ASME Section VIII Division 3 code Case 2938 in 2018. The curves are for steels used in high-pressure hydrogen storing vessels generally mounted at hydrogen refueling stations (HRS) [32]. Table 5.2.7 summarizes the strengths and weaknesses of the project.

5.2.3.3 Project #SCS-007: fuel quality assurance research and development and impurity testing in support of codes and standards

This project aims to concentrate on polymer electrolyte membrane fuel cell examination and work with the American Society for Testing and Materials (ASTM) to build standards and create an electrochemical analyzer to measure contaminations in the fuel stream. This project was initiated in

Table 5.2.8 Shows the strengths and weaknesses of the project.

S. no.	Project strengths	Projects weaknesses
1	Many international collaborations and very current.	This project needs more publications on the status and development of the work.
2	A well-planned project with common goals to that of the Hydrogen and Fuel Cell Technologies Office.	Inline detection of contaminants needs to be better introduced and developed.
3	Good teamwork in improving the quality and efficiency of the detector system.	The testing methods of the fuel need to be clearer.

2006 and is intended to be completed by September 2022. A budget of 6.3 million dollars is to be used. The main collaborator for the project is H2Frontier, SKYRE, NREL, and ONEH2 [32]. The barriers existing are:

1. A lack of mechanical information to modify standards.
2. Insufficient consistent codification plan and process for synchronization of research and development (R&D) and code development.

The project was outlined into two categories of hydrogen development and deployment (HCD) which are the offline and inline HDC deployment and development status. From these approaches, good progress was made, and a cost-friendly method of detecting fuel contaminants was developed. The offline approach was more successful than the inline approach. Two standards were developed [32].

1. **SAE J2719:** This standard deals with related expertise in hydrogen fuel quality standards for commercial proton exchange membrane (PEM) fuel cell vehicles. The standard was initiated in 2005 and the latest revision took place in 2015.
2. **ISO 14687:** This standard deals with the minimum value attributes of hydrogen fuel as transported for implementation in vehicular and fixed services. This standard was published in 2019. Table 5.2.8 shows the strengths and weaknesses of the project.

5.2.3.4 Project #SCS-010: research and development for safety, codes, and standards: hydrogen behavior

This mission aims to perform research and development scientific and engineering foundation for the release detonation and combustion behavior of hydrogen across its range and use. In addition to creating models and apparatus to measure levels of safety and risk of hydrogen systems. This project started in 2003 and ended in 2020. The main partners and collaborators are H_2 scale CRADA, LLNL, NREL, CGA 5.5 testing task force, Fuel Cells and Hydrogen Joint Undertaking (EU), and NFPA 2 code committee. The barriers present are a lack of availability and access to safety data and inadequate technical information to revise standards [32].

During this project, many accomplishments and progress were made including calculations related to leaking flow and separation distance. Scientific tables for $H_{2(l)}$ exposure were created with NFPA 2 (which will be discussed in the future). Hazard distances were reduced, and the work was properly presented. The codes and standards process can benefit from this with the production and authentication of the models. However, the codes and standards divisions will be able to function better if more planning and presenting are done soon [32].

5.2.3.5 Project #SCS-011: hydrogen quantitative risk assessment

This project aims to build a thorough scientific and manufacturing foundation for evaluating the safety risk of hydrogen structures and facilitate the use of that data for revising regulations, codes, and standards for promising hydrogen technologies. This project commenced in 2003 and was completed in 2018, using a budget of $325K. The collaborators and partners were Linde, First Element, PNNL, NREL, Air Liquide, Quong & Associates, and over 40 organizations using HyRAM, NFPA 2, H2USA, CaFCP, and FPRF [32]. The following points are showing the Barriers from 2015 in this project.

1. Insufficient availability and access to safety data and information.
2. Consistent regulations, codes, and standards are needed to enable national and international markets.
3. Inconsistencies in mapwork for codes and process for synchronization of research and development and code formulation.
4. Restrictions to use and access tunnels and parking structures, etc.

The project was a success for $H_{2(l)}$ and it is in agreement with the proposed deadlines and milestones for 2021. Progress was made with regard to the scientific gaseous separation distance and the NFPA 55/2 was updated in 2020. This code enables safety from physiological, over-pressurization, unstable, and flammability hazards associated with compression gases and cryogenic fluids. The revision of this code made it more acceptable and applicable to hydrogen infrastructure. During this project, barriers were decreased [32].

5.2.3.6 Project #SCS-019: hydrogen safety panel, safety knowledge tools, and first responder training resources

The aim is to allow the safe and well-timed switch to hydrogen and fuel cell technologies due to distinctive and impactful safety means. This project started in 2003 and ended in 2019 with multiple partners such as the Panel member organization, California Fuel Cell Partnership (CaFCP), NREL, NFPA, and California Energy Commission (CEC) [32]. The barriers to this project were:

1. Restricted access to and accessibility of safety data and information.
2. Security is not always regarded as a continuous process.
3. Shortage of hydrogen expertise by authorities.
4. Inadequate technical data to revise standards.

The project was approached from three positions hydrogen safety panels, hydrogen tools, and emergency response training resources. The accomplishments at the end were very outstanding and partnering with AIChE was a good decision since it opened door for new exposure. The training was very beneficial, and it can go a long way since it can be reapplied to other projects.

5.2.3.7 Project #SCS-021: hydrogen sensor testing laboratory

Radars are a vital hydrogen safety component and will enable the safe execution of the hydrogen structure. The NREL Sensor Testing Laboratory tests and authenticates sensor performance for companies, developers, and end users. This project aids in the development of rules for the utilization of hydrogen safety sensors under an array of conditions and uses [32]. The barriers to this project were:

1. Inadequate technical data to revise standards.
2. Consistent regulations, codes, and standards are needed to enable national and international markets.
3. Safety is not always handled as a continuous process.

NREL R&D achievements have encouraged developers, industry, and SDOs by providing individual third base party assessments of execution. The Characterization of On-Board Vehicular Hydrogen Sensors J3089_201810 was issued in 2018 because of the project. the report on this standard shows methods for evaluating and observing hydrogen sensors for hydrogen and fuel cell vehicles. The test accommodated multiple environmental conditions and possible scenarios within a vehicle. It is not a standard to use onboard hydrogen sensors, but it is advised and it relevant for the future. The positive outcomes of this project are numerous and relevant since it is observed that leaks are a major cause of injury and detecting them can prevent causalities and injury.

5.2.3.8 Project #SCS-022: fuel cell and hydrogen energy association codes and standards support

This project aims to support and build and create essential codes and standards that will assist in the deployment of market entry of hydrogen and fuel cell technologies. FCHEA in collaboration with Oak Ridge National Laboratory, are members of regulation, code, and standards technical committees to achieve the aim. The barriers addressed in this project were:

1. Consistent regulations, codes, and standards are needed to enable national and international markets.
2. Harmonization of national codes and standards is lacking.
3. Limited participation in the business code development process.

The project was considered well accomplished since it was in line with the Department of Energy (DOE) goals. FCHEA Has been very effective in supporting the improvement of codes and standards. The following codes and standards were addressed since it is crucial for commercialization CSA, SAE, ISO/TC 197, IEC/TC 105, ICC, and NFPA [32]. These codes and standards are discussed further in the next subheading.

5.2.3.9 Project #SCS-029: point-of-use hydrogen purification and impurity reporting systems that utilize metal-organic frameworks

This project, sponsored through the Small Business Innovation Research program, the goal is to build filtration methods that can remove pollution from hydrogen fuel streams at the point of use. The existing hydrogen supply chain has the capacity for the introduction of impurities into the hydrogen stream at multiple points, making point-of-production purification ineffective at eliminating impurities. By facilitating hydrogen purification at the point of use, project outcomes will reduce the risk of damage to fuel cells due to contaminations. The hydrogen safety panel is a resource of this project. This project is very applicable to guaranteeing the advancement of the hydrogen economy and applications of fuel cells [32].

5.2.4 Global technical regulations—2022

5.2.4.1 Proposal for amendments 1 to global technical regulation, no. 13 (gtr13), phase 2 of hydrogen and fuel cell vehicles (HFCV)

Phase one of this project was established in 2010, the Global Technical Regulations (GTRs) specified that each hydrogen-fuelled vehicle will undergo the existing national crash test of contracting parties, with a quantity of hydrogen filled at the maximum allowable volume. Phase two of the project is currently under development with December 2022 being the extended deadline. During this phase, new technologies will be added and harmonized by looking at the component, subsystem, and vehicle-level requirements [32]. IWG will be looking at the following:

1. Keeping the original items constant.
2. Applying the scope to different classes of vehicles.
3. The capability of materials with regard to hydrogen embrittlement.
4. Requirements for the fuelling receptacle.
5. Editorials for the various test procedures requirements.
6. Review the conformable tanks and other new types of containers.
7. Decrease the nominal working pressure (NWP) to 200%.

5.2.4.2 Organizational updates on codes and standards—2022

5.2.4.2.1 Institute of Electrical and Electronics Engineers (IEEE)

1. **IEEE P1547.2**: This is in support of the IEEE Std 1547–2018, by characterizing several types of distributed energy resource skills and their shortcomings. It offers tips and hacks to assist with the understanding of IEEE Std 1547–2018 related to Distributed Energy Resources (DERs) project implementation [32].
2. **IEEE P1547.3**: Guidelines relevant to cybersecurity and DER interconnection with electric power systems (EPS). Its inactive reserve standard is the IEEE guide for monitoring, information exchange, and control of distributed resources interconnected with EPS [32].
3. **IEEE P1547.9**: Approved draft guide to using IEEE Standard 1547 for interconnection of energy storage distributed energy resources with electric power systems. It gives light to sample interconnections and guidance on the technical approaches, the guide considers additional

energy storage factors that are not yet fully addressed in IEEE Std 1547–2018. This standard was approved by the Board on June 6, 2022 [32].

4. **IEEE P2800:** Standard for interconnection and interoperability of inverter-based resources interconnecting with associated transmission EPS. The standard pays particular attention to the performance requirements when the inverter-based resources are connected to the bulk power system. It provides a technical minimum requirement that is constant for interconnection capability and performance when integration occurs. this standard was published on May 22, 2022 [32].

5.2.4.2.2 *International Electrotechnical Commission (IEC)*

IEC TC 105: The scope is to develop international standards for fuel cell technologies proposed by Korea. The standard covers all the various applications. Applications regarding the field of road vehicles work hand in hand with ISO TC 22. The TC 105 work program has over 18 projects and publications with the most recent being in June 2022; PNW 105-901 ED1—fuel cell technologies, performance test of fuel cell-based tri-generation system for combined cell cooling, heat, and power generation. The next stage of this project is said to commence in July 2022 (IEC, https://www.iec.ch/homepage). This item is applies to the hydrogen-fuelled fuel cell, $H_{2(l)}$-fuelled fuel cell systems that have a heat input based on a lower heating value of less than or equal to 70 kW. The test procedure considered is electric power output, the heat and electrical efficiency of heating and cooling, and environmental performance like noise and water quality [32].

5.2.4.2.3 *International Standards Organization (ISO)*

ISO/TC 197: the aim is to provide standardization in the field systems and apparatus for the production, distribution, storage, measurement, and application of hydrogen [32]. Table 5.2.9 shows published and developing standards under ISO/TC 197.

5.2.4.3.3 *National Fire Protection Association (NFPA)*

NFPA 2: The purpose is to provide fundamental precautions for the generation, installation, storage, piping, use, and handling of hydrogen cryogenic liquid form. In this form, hydrogen is kept in its liquid state at low temperatures [32]. Table 5.2.10 shows recently published projects.

Table 5.2.9 Published and developing standards under ISO/TC 197.

S. no.	Published standards	Developing standards
1	ISO 13984:1999 liquid hydrogen for in land fuelling systems.	ISO/AWI 14687 hydrogen fuel quality.
2	ISO 13985:2006 liquid hydrogen for inland fuel tanks.	ISO/AWI TR 15916 safety of hydrogen systems.
3	ISO 14687: 2019 hydrogen fuel quality specification of product.	ISO/AWI 17268 land refueling connection device.
4	ISO/TR 15916:2015 safety consideration for hydrogen systems.	ISO/AWI 198805-5 gaseous hydrogen—fuelling stations—P 5.
5	ISO 16110-1:2007 hydrogen generators using fuel processing technologies—P 1.	ISO/CD 19880-6 gaseous hydrogen—fuelling stations—Part 6.
6	ISO 16110-2:2010 hydrogen generators using fuel processing technologies—P 2.	ISO/AWI 19880-7 gaseous hydrogen—fuelling stations—P 7.
7	ISO 16111:2018 transportable gas storage devices. Hydrogen is absorbed in reversible metal hydride.	ISO/AWI-8 gaseous hydrogen—fuelling stations—P 8.
8	ISO 17268:2020 gaseous hydrogen for land vehicle refueling connection devices.	ISO/AWI 19880-9 gaseous hydrogen—fuelling stations—P 9.
9	ISO 19880-1:2020 Gaseous hydrogen for fuelling stations—P 1.	ISO/AWI 19881 land vehicle fuel containers for gaseous hydrogen.
10	Iso 19880-3:2018 gaseous hydrogen for fuelling stations—P 3.	ISO/AWI 19882 gaseous hydrogen for thermal pressure relief devices.

(continued on next page)

11	ISO 19880-8:2019 gaseous hydrogen for fuelling stations—P 8.	ISO/WD 19884 gaseous hydrogen—cylinders and tubes for stationary storage.
12	ISO 19880-8: 2019/AMD1:2021 gaseous hydrogen, fuelling stations—P 8: fuel quality control, Amendment 1: alignment with Grade D of ISO 14687.	ISO/AWI 19885-1 gaseous hydrogen—fuelling protocols for hydrogen-fuelled vehicles—P 1: design and development process for fuelling protocols.
13	ISO 19882:2018 gaseous hydrogen—thermally activated pressure relief devices for compressed hydrogen vehicle fuel containers.	ISO/AWI 19885-3 gaseous hydrogen—fuelling protocols for hydrogen-fuelled vehicles—Part 3: high flow hydrogen fuelling protocols for heavy-duty road vehicles.
14	ISO/TS 19883:2017 safety of pressure swing adsorption system for purification and separation of hydrogen.	ISO/AWI 22734-1 hydrogen generators using water electrolysis—industrial, commercial, and residential applications—Part 1: general requirements, test protocols, and safety requirements.
15	ISO 22734:2019 hydrogen generators using electrolysis for industrial, residential, and commercial applications.	ISO/AWI TR 22734-2 Hydrogen generators using water electrolysis—Part 2: testing guidance for performing electricity grid service.
16	ISO 26142:2010 hydrogen detection sensors for stationary applications.	

Table 5.2.10 Showing recently published projects.

S. no.	TSC	Title	Scope/status
1	HGV 4.3	Test methods for hydrogen fuelling parameter evaluation	This project is a revision of an existing standard and will include content related to the MC formula.
2	HGV 4.2	Hoses for dispensing compressed gaseous hydrogen	This project is a revision of an existing standard and will update to align with current hose technology and remove requirements for on-board vehicle hoses (content will be transferred to HGV 3.1.
3	HGV 5	Compact hydrogen fuelling systems	This project is to develop a NEW standard for compact hydrogen fueling systems (HGV 5.2). The TSC completed content development.
4	HGC 3	Onboard vehicle components for hydrogen gas vehicles	A revision of the current standard.
5	HGV 2	Compressed hydrogen gas vehicle fuel containers.	A revision of the current standard.
6	HGV 4.1	Priority and sequencing equipment for hydrogen vehicle fuelling.	The aim is to reinstate and update the priority and sequencing standard.
7	C22.2 NO. 22734	Hydrogen generators using water electrolysis	The CSA technical subcommittee continues to work on a binational adoption of ISO 22734.

5.2.5 Conclusion

Energy is one of the most needed entities of this era. The demand for it is very high due to growing economies and populations. Hydrogen can be produced using multiple processes and sources, depending on the process and source it is denoted by color. Due to the physical and chemical properties of hydrogen, the process by which it is produced, distributed, and stored is crucial. Fuel is an electrochemical apparatus, which converts chemical energy to electrical energy without combustion and the fuel or oxidant is supplied outward. Hydrogen safety codes and standards are a very important project since it ensures that hydrogen is handled and used in a very safe and cautious manner.

Hazards due to hydrogen and fuel cell technology occur due to leaks and fires, temperature-related factors, electrical shock, chemicals, and physical mishaps. The first accounted hazard was in 1937 at the Naval Air Station in Lakehurst and the last account was in February of 2022 in Detroit when a hydrogen tank exploded in a pickup truck. Some precautions that can be implemented are safety training, proper storage standards, well-situated ventilation mechanisms, leak, and flame detection systems, and regular inerting and purging of lines.

To understand hydrogen as a fuel, awareness of its properties, safety measures, engineering controls, rules, standards, and regulations are key. This understanding is needed as the hydrogen and fuel cell industry heightens and since fuels have levels of danger associated with them. In the fiscal year 2020–2021, many projects were conducted with hydrogen and fuel cells, from these projects awareness of the current barriers in the industry was brought to light. Each project has been accomplished where the nature of the barriers was resolved or even lessen. The project groups worked with national and international partners and collaborators introducing new codes and standards and revising old ones to better suit the new technologies.

In 2022, proposal for Amendments 1 to GTR13, Phase 2 of HFCV. The focus will be on keeping the original item constant, whilst applying the scope to different classes of vehicles. The issue of hydrogen embrittlement was analyzed, etc. The different organizations updated codes and standards such as the IEEE, IEC, ISO, NFPA, CDS, and CGA. Shortly, ASTMA is discussing specifications for propulsion technology and ASME is conducting meetings to discuss facilitating deployment, regulatory matrix hydrogen fuelling stations in California, and material compatibility of hydrogen with regards to metals and polymers.

References

[1] P.P. Edwards, V.L. Kuznetsov, W.I.F. David, Hydrogen Energy, Philos. Trans. A Math. Phys. Eng. Sci. 365 (2007) 1043–1056.
[2] M. Momirlan, T.N. Veziroglu, Current status of hydrogen energy, Renew. Sustain. Energy Rev. 6 (1-2) (2002) 141–179.
[3] W. Jolly Lee, Hydrogen, Encyclopedia Britannica, 2022.

[4] M. Momirlan, T.N. Veziroglu, The properties of hydrogen as fuel tomorrow in sustainable energy system for a cleaner planet, Int. J. Hydrogen Energy 30 (7) (2005) 795–802.

[5] E.L.V. Eriksson, Gray E. MacA, Optimization and integration of hybrid renewable energy hydrogen fuel cell energy systems: a critical review, Appl. Energy 202 (2017) 348–364.

[6] S. Amrouche Ould, D. Rekioua, T. Rekioua, S. Bacha, Overview of energy storage in renewable energy systems, Int. J. Hydrogen Energy 41 (45) (2016) 20914–20927.

[7] M. Yue, H. Lambert, E. Pahon, R. Roche, S. Jemei, D. Hissel, Hydrogen energy systems: a critical review of technologies, applications, trends and challenges, Renew. Sustain. Energy Rev. 146 (2021) 111180.

[8] O.Z. Sharaf, M.F. Orhan, An overview of fuel cell technology: fundamentals and applications, Renew. Sustain. Energy Rev. 32 (2014) 810–853.

[9] S. Mekhilef, R. Saidur, A. Safari, Comparative study of different fuel cell technologies, Renew. Sustain. Energy Rev. 16 (1) (2012) 981–989.

[10] W. Vielstich, A. Lamm, H.A. Gasteiger, Handbook of Fuel Cells: Fundamentals, Technology, Applications, 4 Volume Set, Wiley (2003) ISBN: 978-0-471-49926-8.

[11] J. Hord, Hydrogen safety: an annotated bibliography of regulations, standards and guidelines, Int. J. Hydrogen Energy 5 (6) (1980) 579–584.

[12] C.S. Marchi, E.S. Hecht, I.W. Ekoto, K.M. Groth, C. LaFleur, B.P. Somerday, R. Mukundan, T. Rockward, J. Keller, C.W. James, Overview of the DOE hydrogen safety, codes and standards program, part 3: advances in research and development to enhance the scientific basis for hydrogen regulations, codes and standards, Int. J. Hydrogen Energy 42 (11) (2017) 7263–7274.

[13] V.H. Dayan, R.L. Proffit, B. Rosen, Hydrogen Leak and Fire Detection: A Survey, NASA, Environmental Science (1970).

[14] R. Wei, J. Lan, L. Lian, S. Huang, C. Zhao, Z. Dong, J. Weng, A bibliometric study on research trends in hydrogen safety, Process Saf. Environ. Prot. 159 (2022) 1064–1081.

[15] S.H.N. Yousef, S. Mashareh, Hyddrogen leakages sensing and control, Biomed. J. Sci. Tech. Res. 21 (5) (2019) 16228–16240.

[16] E. Derempouka, T. Skjold, O. Njå, H. Haarstad, The role of safety in the framing of the hydrogen economy by selected groups of stakeholders, Chem. Eng. Trans. 90 (2022) 757–762.

[17] S.Y. Jeong, D. Jang, M.C. Lee, Property-based quantitative risk assessment of hydrogen, ammonia, methane, and propane considering explosion, combustion, toxicity, and environmental impacts, J. Energy Storage 54 (2022) 105344.

[18] K. Verfondern, Hydrogen fundamentals, Hydrogen Safety for Energy Applications, Butterworth-Heinemann, 2022, pp. 1–23.

[19] M.L. Blaylock, L.E. Klebanoff, Hydrogen gas dispersion studies for hydrogen fuel cell vessels I: vent Mast releases, Int. J. Hydrogen Energy 47 (50) (2022) 21506–21516.

[20] Y. Manoharan, S.E. Hosseini, B. Butler, H. Alzhahrani, B.T. Senior, T. Ashuri, J. Krohn, Hydrogen fuel cell vehicles; current status and future prospect, Appl. Sci. 9 (11) (2019) 2296.

[21] F. Rigas, P. Amyotte, Hydrogen Safety, CRC Press, 2012.

[22] C. Kirchsteiger, A.V. Arellano, E. Funnemark, Towards establishing an international hydrogen incidents and accidents database (HIAD), J. Loss Prev. Process Ind. 20 (1) (2007) 98–107.

[23] N.R. Mirza, S. Degenkolbe, W. Witt, Analysis of hydrogen incidents to support risk assessment, Int. J. Hydrogen Energy 36 (18) (2011) 12068–12077.

[24] D. Melideo, E.W. Ronnefeld, F. Dolci, P. Morett, HIAD – Hydrogen Incident and Accident Database, In: 53rd ESReDA Seminar, 14–15 November 2017, Ispra, 53rd ESReDA Seminar, 2017, p. 326–336, JRC108666. (Publisher: esreda).

[25] M. Fischer, Safety aspects of hydrogen combustion in hydrogen energy systems, Int. J. Hydrogen Energy 11 (9) (1986) 593–601.

[26] C.H. Rivkin, R.M. Burgess, W.J. Buttner, Regulations, codes, and standards (RCS) for large scale hydrogen systems 2017

[27] C. San Marchi, E.S. Hecht, I.W. Ekoto, K.M. Groth, C. LaFleur, B.P. Somerday, R. Mukundan, T. Rockward, J. Keller, C.W. James, Overview of the DOE hydrogen safety, codes and standards program, part 3: advances in research and development to enhance the scientific basis for hydrogen regulations, codes and standards, Int. J. Hydrogen Energy 42 (11) (2017) 7263–7274.

[28] S. Foorginezhad, M. Mohseni-Dargah, Z. Falahati, R. Abbassi, A. Razmjou, M. Asadnia, Sensing advancement towards safety assessment of hydrogen fuel cell vehicles, J. Power Sourc. 489 (2021) 229450.

[29] J.M. Ohi, C. Moen, J. Keller, R. Cox, Risk assessment for hydrogen codes and standards, H_2 Hydrogen tools (2005).

[30] C. Rivkin, C. Blake, R. Burgess, W.J. Buttner, M.B. Post, A national set of hydrogen codes and standards for the United States, Int. J. Hydrogen Energy 36 (3) (2011) 2736–2741.

[31] I. MacIntyre, A.V. Tchouvelev, D.R. Hay, J. Wong, J. Grant, P. Bénard, Canadian hydrogen safety program, Int. J. Hydrogen Energy 32 (13) (2007) 2134–2143.

[32] S. Satyapal, US Department of Energy Hydrogen Program: 2021 Annual Merit Review and Peer Evaluation Report; June 7–11, 2021. National Renewable Energy Lab (NREL), Golden, CO; 2022.

Power to gas 'pathway'

Potential of hydrogen in powering mobility and grid sectors

Biswajit Ghosh
The Neotia University, 24 Paragons (S), West Bengal, India

6.1.1 Introduction

According to National Aeronautics Space Administration (NASA), the average surrounding temperature of the earth's atmosphere is about 15°C [1] where human beings are living. It is further reported that there is increasing in temperature of the earth's atmosphere resulting in climate change. The possible cause for the earth's climate change is due to the increase in so-called greenhouse gases (GHGs) that are acting as the blanket to the earth's surroundings not allowing the infrared originating from man-made activities in the sectors like, industry, transportation, and agriculture through the burning of fossil fuels. It was reported that the highest temperature 70.7°C was observed in the Lut Desert of Iran and the coldest temperature was about −89.2°C at Vostok, Antarctica [1]. The solar radiation hit the equator directly at the right angle and hit the polar region at a shallower angle. The geometrical exposition of solar radiation is causing the creation of the hottest region as well as the coldest region of the earth's surface. The researchers reported that since the 1880s, the average surface temperature of the earth has risen by 0.8°C with the highest increase occurring in the past few decades which may cause in melting of polar ice lands and consequently rising of the sea levels and many areas of the earth's surface will go under the sea level. This calls for reducing the emission of GHGs and in the last 27th Conference of Parties (COP 27) of the United Nations Framework Convention on Climate Change (UNFCCC) it was warned that the average temperature of the earth should not exceed more than 1.5°C from its present value in maintaining sustainability. This has now worldwide sustainable development (SD) is a burning issue and is addressed in the last COP 27.

Sustainable development (SD) addresses the concerns about the relationships between the surroundings of human beings along with their

Towards Hydrogen Infrastructure: Advances and Challenges in Preparing for the Hydrogen Economy.
DOI: https://doi.org/10.1016/B978-0-323-95553-9.00063-7

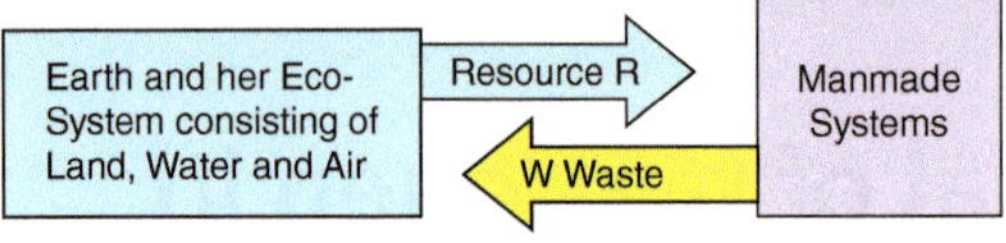

Figure 6.1.1 Application of the Carnot cycle in the eco-systems, Carnot efficiency $\eta_{eco} = [1 - W/R] \times 100\%$, and $[1 - W/R]$ represents the Carnot factor (CF).

interactions. This can be correlated with the thermodynamic principle of mechanical systems. The earth's ecological system and its surroundings can be represented in terms of the source as well as the sink as presented in Fig. 6.1.1.

In Fig. 6.1.1 it was perceived that human activities are represented in the box of manmade systems that are operating with the input resources "R" from the earth's ecosystems. In its operation, the manmade systems rejected the waste part "W" again into the earth's ecosystems. If the total system is superposed in the Carnot principle, then the ecological efficiency can be represented as $\eta_{eco} = [1 - W/R] \times 100\%$, with $[1 - W/R]$ as the Carnot factor (CF). The value of CF is confined between 0 and 1, i.e., $0 < CF < 1$, and the value of "W" is a perturbing element in maintaining sustainability, and with time "W" is loading the earth's ecosystem.

In the case of power production using fossil fuel the waste "W" consists of the GHGs containing oxides of carbon as well as black carbon/shoot and it is very difficult for the earth's ecosystem to digest/neutralize them. The wastes are triggering back the ecological systems and with time they are exceeding the earth's carrying capacity and destabilizing the ecosystems. In the use of H_2 in power production the waste "W" is the H_2O vapor which has no adverse effect on earth's eco-system. Thus, the use of H_2 in power production is one of the possible sustainable solutions.

It is well known that electricity is not only the secondary form of energy, but also it is the critical input for productivity and it is one of the central parameters of global economic development. Access to electricity has increased worldwide, particularly in developing nations. It was further reported that by 2030 power consumption in the agricultural sectors has enhanced by more than 18–20%. Despite this growth, about 700 million people about 90% of them from the sub-Saharan African region, will still lack access to electricity by 2030 [2]. This is indicating that the demand for electricity will continue to increase day by day and to match with sustainable solutions H_2 will take a possible crucial role.

At present, the major part of electricity is coming from the conversion of fossil and nuclear fuels. Fossil fuels are the major threat to sustainable development as they are a source for emitting GHGs and nuclear power remains unpopular due to public safety issues. Electricity from solar photovoltaic (PV) power plant is another option for obtaining C-free energy, but PV power will not be the substitute for fossil fuel and nuclear power plants with several limitations. In this context, for distributed power generation utilizing C-neutral energy resources H_2 is proposed to be a possible solution. Nevertheless, the H_2 can make a meaningful contribution to the power budget if its (1) availability, (2) accessibility, (3) affordability, and (4) adoptability fulfills the requirements for power production.

6.1.1.1 Hydrogen

According to the periodic table H_2 is regarded as a rogue element and its position in the periodic table till today is not satisfactory. Although it has been placed on the top of alkali metals on the basis of electronic configuration, it is a nonmetal and also resembles the halogens of a group with many characteristics. It is unique as its nucleus does not have any neutrons and is the only element to lack neutrons. As it has only one electron orbiting around its nucleus, but it frequently donates this and becomes a single, positively charged proton. Thus, sometimes it exists as an element and sometimes it is the proton.

It is the lightest gaseous element in the earth's atmosphere and at standard conditions, it is available in the molecular form. It is colorless, odorless, tasteless, nontoxic, and highly combustible gas as well as the most abundant element in the universe, constituting roughly 70–75% of all normal matters. It is acting as the fuel in the stars like sun to energize them through their fusion with the formation of helium ($_2He^4$) as its ashes. Most of the H_2 on earth's surface exists in molecular form, those are embedded in water and many other organic compounds. Combining with C it forms food, fuel, and fiber and their calorific value depends on number of H_2 within the compound which represent as C_mH_n. In case of wood the value of m is comparatively greater than n and in case of coal value of m further reduces and that of n increases and in case of oil value of m further reduces and n is further increased and gas like methane (CH_4) have the higher calorific value as m is reduced to 1 and n is increased to 4 and at zero "m" the calorific value enhanced further. This indicated the influences of H_2 in enhancing the calorific value. There is no specific scientific fact about how H_2 originated in the universe. One of the plausible logic behind its origin is from the Big Bang theory, which indicated about its nuclei occurred from the plasma cooling effect with the infusion of electrons with the bounding with a proton. The emergence of neutral H_2 atoms throughout the universe occurred about

370,000 years later during the recombination epoch when the plasma had cooled enough for electrons to remain bound to protons [3].

Except under very high pressures, it is nonmetallic in nature and readily forms a single covalent bond with the majority of nonmetallic elements to create molecules like water and almost all organic compounds. Because acid–base interactions frequently entail the exchange of protons between soluble molecules, H_2 plays a crucial role in these reactions. Species denoted by the symbol H^+ represent the proton in aqueous solutions and in ionic compounds involve screening of its electric charge by nearby polar molecules or anions, where it is known as a hydride. In ionic compounds, it can take the form of a positively charged cation or a negatively charged anion. The study of hydrogen's energetics and chemical bonding has been crucial in the development of quantum mechanics as it is the only neutral atom for which the Schrödinger equation can be analytically solved.

6.1.1.2 Hydrogen production

In the 16th century Sir Henry Cavendish first produced H_2 artificially from the reaction of acids on metals and when it is burned it turned into water. As a result, in Greek, hydrogen means the "water-former." For large-scale industrial applications, it is produced mainly from the gasification of coal as well as steam reforming of hydro-carbonaceous items, oil, and natural gas. Its extraction from the electrolysis of water is an energy-intensive process. At industrial production sites, particularly in ammonia (NH_3) production, the source for H_2 is the cracking of fossil fuel. In storage and transportation one of its major problems is the embrittlement of many metals and that welcomes the complications in designing the pipelines as well as the storage tanks. Various methods for H_2 production along with various aspects in presented below.

6.1.1.3 Steam reformation

Steam reformation is a process for producing syngas, a mixture of carbon monoxide (CO), and H_2, with the reaction of hydrocarbon (C_mH_n) with water vapor. The natural gas contains methane (CH_4) that can be used to produce H_2 using a steam-methane reforming process. It is a high-temperature process in which the gaseous mixture of hydrocarbon and steam is processed at a high temperature in the range of 700–1000°C and high pressure 3–12 bar in the presence of nickel (Ni) catalyst convert hydrocarbon to H_2, CO, and a relatively small amount of carbon dioxide (CO_2). As the reaction is an endothermic one, a high temperature is required to maintain the reactions. The process further initiates the water gas shift reaction to enrich H_2. It is basically a two steps reaction process, e.g., $C_nH_m + nH_2O \rightarrow nCO + mH_2$ and $CO + H_2O \rightarrow CO_2 + H_2$. The advantage of the process is that the reaction produces the highest yield of H_2 and the disadvantage is the increase in heat loading capacity resulting from the large endothermic reaction and the continuous supply of heat to keep the reaction on.

6.1.1.4 Gasification

Gasification is a technological process that can convert any carbon-based raw material such as coal, oilcake, biomass, etc. into fuel gas, also known as synthesis gas or syngas. Gasification occurs in a gasifier in the presence of little amount of air/O_2 and steam is directly contacted with the feed like coal, oilcake, or biomass and initiates a series of chemical reactions to convert the feed masses into syngas and ash or charcoal. The syngas can be further converted to H_2 and CO_2 by reacting it with steam over a catalytic surface in a water gas shift reactor. The advantage of the process is a large-scale industrial supply source and the disadvantage is that it emitted some GHGs into the atmosphere.

6.1.1.5 Electrolysis

It is the electrochemical process for splitting of water into H_2 and O_2 using electricity. It is a promising method for H_2 production using carbon-free resource when the electricity is generating either from renewables or nuclear resources. The electrochemical reaction takes place in the unit called electrolyzer. The electrolyzers consist of an anode and a cathode separated by an electrolyte and their function depends upon the type of electrolyte material involved and the ionic species it conducts. Normally there are three types of electrolyzers, e.g., polymer electrolyte membrane (PEM), alkaline, and solid state. Both PEM and alkaline electrolyzers operate in between 70°C and 90°C on the other hand for operation of solid state electrolyzers at high temperatures between 500°C and 600°C. The advantage of this process is that the raw material is water and is available easily. The disadvantage is that the production cost of H_2 is a function of the cost of electricity used for electrolysis.

6.1.1.6 Solar method

Via the splitting of water, solar radiation can also serve as an agent in the synthesis of hydrogen. Hydrogen is extracted from split water via photobiological, photoelectrochemical, and thermochemical processes that are powered by the sun. The production of hydrogen by bacteria and green algae during their normal photosynthetic processes is known as photobiological reactions. Semiconductors are used in the photo-electrochemical processes to separate H_2O into H_2 and O_2. Concentrated solar energy is used in the solar thermochemical H_2 generation process to fuel H_2O splitting processes, which are occasionally mediated by metallic oxides. High-temperature heat (500–2000°C) is used in thermochemical water-splitting systems to power a series of chemical reactions that yield hydrogen. Each cycle's chemical input is recycled, resulting in a closed loop that uses only water and generates hydrogen and oxygen. In photolytic biological systems, sunlight is absorbed by tiny organisms like cyanobacteria or green microalgae, which then divide water into oxygen and hydrogen ions. Direct or indirect combinations of the hydrogen ions might result in the release of hydrogen gas. Low hydrogen production rates and the fact that splitting water also produces oxygen, which quickly stymies the hydrogen production reaction and may be dangerous when coupled with hydrogen in certain amounts, are challenges for this method.

6.1.2 Hydrogen as fuel

In comparison to other fuels, H_2 exhibits extremely high flame propagation rates within the engine cylinder over a wide range of temperatures and pressure. Even for extremely lean combinations that are far beyond the domain of stoichiometric mixtures, these rates continue to be adequate. As a result of the rapid energy release, combustion times are frequently brief, which results in high power output efficiency and rapid pressure rise after spark ignition. When hydrogen is used as fuel in a spark ignition engine, the lean operational limit mixture is substantially lower than when other common fuels are used. In H_2-fueled engines, this enables stable lean mixture operation and management. High output efficiency values are produced by operating on lean mixtures in conjunction with the fast combustion energy release rates at a top dead center that is associated with the extremely rapid burning of H_2-air mixtures.

Being connected with fewer unfavorable exhaust emissions than operation on other fuels is one of the most significant characteristics of H_2 ICE operation. Regarding the H_2 fuel's contribution to emissions, there aren't any unburned hydrocarbons, CO, CO_2 or sulfur dioxide, oxides, smoke, or particles. In well-maintained engines, the lubricating oil usually makes up a very small portion of these emissions. The only combustion byproducts that are released are oxides of N_2 and water vapor. Moreover, NOx levels during lean operation are frequently much lower than during operations utilizing other fuels. H_2's fast-burning properties make high-speed engine operating much more enjoyable. As a result, there would be less of a penalty for operating in a lean mixture, increasing power output [4].

It is further observed that specific physical characteristics of H_2 fuel are quite different from those common fuels. Some of those properties make H_2 potentially less hazardous, while other characteristics of H_2 could theoretically make it more dangerous in certain situations even when using in IC engine they introduce knocking impulse in the engine which degrade the engine efficiency. However, in case of using syngas with enriched with H_2 reduction in knocking phenomenon observed. Thus, use of H_2-enriched syngas is the possible fuel for particularly for operation of IC engine as observed in field experiments.

6.1.3 Power conversion methodologies

Reduction of GHGs in power conversion is one and recommended items in sustainable development goal (SDG). Thus, it is recommended that there needs a fundamental paradigm shift in the power generation industry from conventional fossil fuel to syngas and H_2 rich syngas fuel. This is the direct result of shifting an acceleration in the installed capacity of renewable power sources, including solar and wind. Nevertheless, the power generation using H_2 and its blending with the power from other renewable sources may fulfil the substitution of power generation using fossil fuel. In addition to this reduction in decentralized consumption of fossil fuel may also be substituted by H_2. These efforts may have a positive impact in addressing SD goals. Several commercial power plants have extensive experience employing a wide range of H_2 concentrations to replace the fossil fuel used in gas turbine power generation. Gas turbines running on H_2 might therefore support the grid firming while also emitting a lot less greenhouse gases. Every time an energy carrier is created, transformed, or consumed, efficiency losses occur, including with fossil fuels. These losses in the case of H_2 might add up across the value chain at several points. Using an electric fuel cell to extract H_2 from H_2O, store it, transport it, and then turn it back into electricity, the supplied energy may be less than 30% of the initial electric input. Because of this, producing H_2 is more "expensive" than using electricity or natural gas to do it. It also offers a case for reducing the number of energy carrier transformations along any value chain. Having said that, as long as the intensity of CO_2 emissions is assessed and there are no restrictions on the supply of energy. Efficiency can essentially be viewed as a function, to be taken into account at the level of the entire value chain, while taking economic involvement into account. This is crucial because H_2 has the potential to be produced with no GHG emissions and can be used in some applications with considerably higher efficiency. An H_2 fuel cell is about 58–60% efficient when used in a car, compared to a gasoline IC engine's 18–20% efficiency, and a modern coal-fired power plant's 43–45% efficiency, plus another 10–15% or more due to electric power line losses. Using internal combustion engines (ICE), gas turbines, and fuel cells to produce electricity from hydrogen with virtually no negative emissions and possibly high efficiency at the point of use is interesting due to the range of ways in which it can be produced. According to a recent study, the output of final fuel per hectare of land for various fuels obtained from biomass and of the H_2 from photovoltaics or wind power [5] is superior to that of biofuels. The findings demonstrate that when land is used to capture wind or solar energy, its energy yield is significantly higher. In terms of volumetric and gravimetric energy storage density, employing H_2 as an energy carrier is superior to using electricity. When using H_2 as an energy carrier, there are significant obstacles to overcome. Even when compressed to 700 bar or liquefied, both of which result in significant energy losses, it has a very low density compared to the fuels now in use, despite being better at storing than batteries. As a result, storage on board vehicles, bulk storage, and distribution are severely affected. Also, it is important to watch out for the well-to-wheel greenhouse gas emission reduction of hydrogen-fueled automobiles compared to hydrocarbon fuel [6]. Nonetheless, the benefits that hydrogen offers are substantial enough to justify investigating its potential. However, several questions arise on the use of H_2 for energy conversion through the ICEs. Many research reports discussed about the "best" fuel as the energy carrier in powering mobility sectors by optimizing hydrocarbon fueled engines, biofuels, electricity, H_2, etc. It is difficult to predict who will perform best because there are always a wide range of factors to be considered, including well-to-wheel or cradle-to-grave primary energy use, GHG emissions, tailpipe emissions related to local pollution, cost, practicality, customer acceptance, etc. These factors are obviously not all easily scored and ranked. Furthermore, it appears that there is no magic solution, and a wide range of options justifies a thorough investigation into the benefits and limitations. Researchers [6] have compared gasoline and electric vehicles to hydrogen fuel cell (H2FC) and hydrogen internal combustion engine (H2ICE) vehicles based on well-to-wheel CO_2 emissions and primary energy use. The researchers came to the further conclusion that, when compared to vehicles powered by gasoline and natural gas, H2FC vehicles could reduce primary energy use and GHG emissions, but H2ICE vehicles could increase both. Moreover, ICEs can be powered by a combination of H_2 and gasoline, supporting increased fuel station density and autonomy requirements. This might speed up the development of H_2 infrastructure, which would allow fuel cell-powered vehicles to immediately benefit from the knowledge gained in transportation, fueling, and storage. These investigations also came to the following conclusions: (1) H_2 produced using alternative, cleaner production methods and from a more varied range of energy sources can be used in existing applications. (2) Hydrogen can replace present fuels and inputs in a variety of new uses, or it can work in tandem with increased electricity use in these applications. H_2 can be utilized in its pure form in these applications for transportation, heating, making steel, and electricity, or it can be transformed into hydrogen-based fuels such as synthetic methane (CH_4), synthetic liquid fuels, ammonia (NH_3), and methanol (CH_3OH). In order to increase energy supply resilience and reduce end users' reliance on specific energy sources, hydrogen is installed alongside electrical infrastructure. Electricity can be converted to hydrogen and back again, as well as to other fuels. H_2 has the capacity to strengthen and link various components of the energy system in both ways. Renewable electricity can be used in situations where chemical fuels work better by generating H_2. Low-carbon energy can be transported over very long distances, and electricity can be stored to balance supply and demand on a weekly or monthly basis. Furthermore, it can be argued that in a highly electrified, low-carbon world, H_2 could offer an additional method of storing and emerging as a strategic energy reserve. In addition to lowering regional and local air pollution, using H_2 in place of carbon-containing fuels could also improve environmental and health consequences. Also, if H_2 can be utilized to cut industrial emissions cheaply without causing any manufacturing to move, that will aid in keeping local jobs. Similar to this, some fossil fuel resources may still be utilized if carbon capture, utilization, and storage (CCUS) was used to lower the CO_2 intensity of producing hydrogen from fossil fuels. Using the infrastructure, resources, and talents already in place can make transition pathways easier to follow and less expensive [7]. Improvements in the integrated electrolyzer, hydrogen storage, and fuel cell designs have created opportunities for off-grid H_2 generating and storage systems. Containerized systems, which can be used in conjunction with off-grid energy sources, are being developed to offer backup power for crucial buildings like hospitals and electricity storage for longer periods of time than battery-based systems or the natural gas used to generate electricity. It also offers a case for reducing the number of energy carrier transformations along any value chain. Having said that, as long as the intensity of CO_2 emissions is assessed and there are no restrictions on the supply of energy. Efficiency can essentially be viewed as a function, to be taken into account at the level of the entire value chain, while taking economic involvement into account. This is crucial because H_2 has the potential to be produced with no GHG emissions and can be used in some applications with considerably higher efficiency. An H_2 fuel cell is about 58–60% efficient when used in a car, compared to a gasoline IC engine's 18–20% efficiency and a modern coal-fired power plant's 43–45% efficiency, plus another 10–15% or more due to electric power line losses. Using internal combustion engines (ICE), gas turbines, and fuel cells to produce electricity from hydrogen with virtually no negative emissions and possibly high efficiency at the point of use is interesting due to the range of ways in which it can be produced. According to a recent study, the output of final fuel per hectare of land for various fuels obtained from biomass and of the H_2 from photovoltaics or wind power [5] is superior to that of biofuels. The findings demonstrate that when land is used to capture wind or solar energy, its energy yield is significantly higher. In terms of volumetric and gravimetric energy storage density, employing H_2 as an energy carrier is superior to using electricity. When using H_2 as an energy carrier, there are significant obstacles to overcome. Even when compressed to 700 bar or liquefied, both of which result in significant energy losses, it has a very low density compared to the fuels now in use, despite being better at storing than batteries. As a result, storage on board vehicles, bulk storage, and distribution are severely affected. Also, it is important to watch out for the well-to-wheel greenhouse gas emission reduction of hydrogen-fueled automobiles compared to hydrocarbon fuel [6]. Nonetheless, the benefits that hydrogen offers are substantial enough to justify investigating its potential.

6.1.4 Safety issues

Safety is the biggest aspect while using any fuel or other energy resources used for energizing the machines. When H_2 is used as the energy carrier it poses risks if not properly used with safety devices/systems. Therefore, one must consider relative risks to the common fuels such as gasoline, propane or natural gas, and with H_2. Since H_2 has the smallest molecule, it has a greater tendency to escape through small bores/hole/openings than other liquid or gaseous fuels. Based on properties of H_2 such as density, viscosity, and diffusion coefficient in air, the propensity of H_2 to leak through holes or joints of low-pressure fuel lines may be only 1.26 in case of laminar flow and 2.8 for turbulent flow. It is many times faster than a natural gas leak through the same hole and not 3.8 times faster as frequently assumed based solely on diffusion coefficients. Since natural gas has over three times the energy density per unit volume the natural gas leak would result in more energy release than a H_2 leak. For very large leaks from high pressure storage tanks, the leak rate is limited by sonic velocity. Due to higher sonic velocity to the order of 1308 m/s, H_2 would initially escape much faster than natural gas as the sonic velocity of natural gas is about 440–449 m/s. Again, since natural gas has more than three times the energy density than hydrogen, a natural gas leak will always contain more energy.

If any leak originated for whatever reason H_2 will disperse much faster than any other fuel, as a result it reduced the hazard levels. In fact, H_2 is both more buoyant and more diffusive than either gasoline, propane, or natural gas. The H_2 and the air mixture ratio can burn in relatively wide volume ratios between 4% and 75% of H_2 in air. Other fuels have much lower flammability ranges, natural gas 5.3–15%, propane 2.1–10%, and gasoline 1.2–6%. However, the range has little practical value. In many normal leak situations, the key parameter that determines if a leak would ignite is the lower flammability limit, and hydrogen's lower flammability limit is four times higher than that of gasoline, 1.9 times higher than that of propane, and slightly lower than that of natural gas [8].

Hydrogen has a very low ignition energy about 0.02 mJ, about one order of magnitude lower than other fuels. The ignition energy is a function of air–fuel ratio, and for H_2 it reaches to a minimum of about 25–30% H_2 content with air. In normal conditions, the H_2 has a flame velocity six times faster than that of natural gas or gasoline. Thus, the H_2 flame would therefore be more likely to progress to a detonation than other fuels. However, the likelihood of a detonation depends in a complex manner on the exact fuel/air ratio, the temperature, and particularly the geometry of the confined space [9].

The H_2 detonation in the open atmosphere is an unlike phenomenon. The lower deniability of air–fuel ratio for H_2 is about 13–18%, which is two times higher than that of natural gas and 12 times higher than that of gasoline [10]. Since the lower flammability limit is 4% an explosion is possible only under the most unusual scenarios, which is, H_2 would first have to accumulate and reach 13% concentration in a closed space without ignition, and only then an ignition source would have to be triggered. At that condition if an explosion occurs, and has the lowest explosive energy per unit stored energy in the fuel, a given volume of H_2 would have 22 times less explosive energy than the same volume filled with gasoline vapor.

The flame of H_2 is nearly invisible, which may be dangerous, because people in the vicinity of a H_2 flame may not even know there is any fire phenomenon. This may be overcome by adding some chemical additives that will provide the necessary luminosity. The low emissivity of H_2 flames means that nearby materials and people will be much less likely to ignite and/or hurt by radiant heat transfer. The fumes and soot from a gasoline flame pose a risk to anyone inhaling the smoke, while H_2 fires produce only water vapor unless secondary or additive materials begin to burn in conjunction with H_2. The liquid H_2 has another set of safety issues, such as the risk of cold burns, and the increased duration of leaked cryogenic fuel. A large spill of liquid H_2 has some characteristics like a gasoline spill, however it will dissipate at a much faster rate. Some potential danger may occur due to the violent explosion of a boiling liquid expanding vapor due to the failure of a pressure relief valve [11].

In onboard conditions a H_2 vehicle may pose a safety hazard. The hazards should be considered in situations when the vehicle is in immobile condition when the vehicle is in normal operational mode, and in collisions with other objects. The possible potential hazards are due to fire, and explosion of toxicity. The latter can be ignored since neither the H_2 nor its fumes in case of fire are toxic. The possible source of fire or explosion due to H_2 use may come from the fuel storage, or from the fuel supply lines. The largest amount of H_2 at any given time is present in the tank may develop some casualties. It was reported that several tank failure modes may be considered in both normal operation and collision, such as catastrophic rupture, due to a manufacturing defect in the storage tank, a defect caused by abusive handling of the tank or stress fracture, puncture by a sharp object, and external fire combined with a failure of pressure relief device to open. The massive leak, due to the above causes or any other incidence may induce openings in the fuel line connection [12].

The possible failure modes as discussed earlier may be either be avoided or their occurrence and consequences minimized by leak prevention through a proper system design, selection of adequate materials for equipment, allowing for tolerance of shocks and vibrations, locating a pressure relief device vent, protecting the high-pressure lines, installing a normally closed solenoid valve on each tank feed line, leak detection by either a leak detector, or by adding an odorant to the H_2 fuel. A possible problem for fuel cell's ignition prevention, through automatically disconnecting battery bank, thus eliminating source of electrical sparks which are the cause of 80–85% gasoline fires after a collision and by designing the fuel supply lines so that they are physically separated from all electrical devices, batteries, motors and wires to the maximum extent is possible. By designing the system for both active and passive ventilation such as an opening to allow the H_2 to escape upward directions.

6.1.5 Engine characteristics for hydrogen fuel

Engine is the heart of the mobility systems and it operates with the burning of fuels. Burning of fuel is the function of chemical components and its quality. When H_2 is used as the fuel in operating engine its burning process will differ from the conventional hydrocarbon fuels. Absence of C may welcome some other undesirable phenomenon those require to address before use it as fuel for engine. Fuel composition in cycle variations is one of the issues while using H_2 than with other fuels, even for very low mixture of other component while is in operation. The low C concentration may lead to a reduction in GHGs emissions, enhance efficiency, reduce knocking effect, and perform smoother engine operation. The high effective octane number of H_2 initiates high burning rates and maintain slow preignition reactivity. The excellent additive quality of H_2 even at low concentrations may form other fuel components like CH_4 and that may lead to an efficient burning process. Another advantage of H_2 is that even at a gaseous state in low temperatures it can generate flame until it reaches a temperature around 20 K. Due to this property and fast burning rate, the H_2 used as fuel is more appropriate for high-speed engine operation. This condition also helps in having better efficiency and improved power output at less sparking conditions, reduces, less heat loss, and reduces emission rates in comparison to the other conventional fuels used in the IC engines. Even at a lean mixture ratio with air, the engine performs its better output with higher efficiency at high compression ratio operation.

While H_2 is used as the fuel for the engine there are possibilities for the formation of H_2O vapor which consumes some of its internal energy and at the same time, it introduces some additional thermal load on the engine itself. This phenomenon reduces engine efficiency and at the same time, the H_2O vapor introduces a corrosion effect inside the engine components. To overcome this phenomenon some preventive measures like engine and piston require coating with an anticorrosive protective layer in overcoming this issue.

It is mentioned in the above section that during H_2 burning in the IC engine there is the formation of H_2O vapor that enhance the heat load of the system. This property is very much useful in cogeneration applications. The generated heat may be extracted from the engine body surface and may be integrated with some other applications also. Apart from all other conventional hydrocarbon fuels H_2 is a pure fuel having well-known characteristics as a result performance optimization of the engine is possible. The pure H_2 may pass through the catalytic agents to have the optimum performance. This optimization process will help in improving the combustion process and develop the process for treating the waste emission from the engine. Normally H_2O and C are the waste emission from the engine and treating this $C + H_2O$ composite may develop another fuel like kerosene [13].

The thermodynamic characteristics of H_2 indicate its heat transfer methodologies. The characteristics indicated that the burning of H_2 has the tendency to have a high compression temperature. This high compression temperature contributes to improving engine efficiency even at a low and meager ratio of air–fuel mixture. The high burning rates with appropriate air–fuel mixture indicated the engine performance is independent of the shape of the engine and its related components. The independent shape and size have no influence on the turbulence level as well as the intake charge swirling effect.

Normally burning of H_2 takes place at the ICEs with wider ranges of air–fuel mixtures than the other hydrocarbon-based fuel like gasoline. The wider flammability range of H_2 limits the higher flame speed makes and makes it more efficient in starting and stopping and it can tolerate better even in the presence of fuel diluents. As a result, the engine can run even at higher air–fuel ratio or even high diluted fuel mixing state. On the other hand, it can operate even with O_2 rich air–fuel ratio. The O_2-enriched fuel gas is highly diffusive and buoyant that makes quick dispersal of mixed fuel gas even if any leaks generate. This aspect reduces the explosion hazards associated with the operation of ICEs with H_2 as the primary fuel. However, for efficient operation and maintaining all the safety aspects as well as hazards there needs modification of standard ICEs for H_2 fuels.

The standard ICEs are designed for burning of the gaseous fuels like propane (C_3H_8), CH_4, and H_2, etc. to convert mechanical power. For efficient operation and maintaining safety at a large air–fuel ratio the carburetors of standard ICEs need modifications. The knocking and backfire-safety issues at wider ranges of air fuel mixture are the main aspects for engine structure modifications. The storage of the fuel at non operation condition is another important issue while operating the engine with H_2. Another aspect for engine modification along with cylinder specification need to addressed while storing liquid gasoline (C_8H_{18}) as its low energy per unit volume produces less energy in the cylinder. The standard ICEs fueled with H_2 produces less power than that of C_8H_{18} and supercharging may help in overcoming this issue by compressing the incoming air fuel mixture before it enters into the cylinder. Additional weight and complexity are added to the modified ICEs engine for their successful efficient operation. However, the effective power gain and backfire resisting property can be controlled by cooling the cylinder with more air to compensates the mentioned limitations.

The spray nozzles for water spraying over the engine surface in developing ambience is essential in providing backfire free operation. The spray nozzle design for supplying of appropriate water volume for cooling purpose is the function of mechanical load, engine speed, and heat formation during the ignition process. There are the recommendation for use of liquid H_2 at low temperature for reducing the thermal effect and backfiring effect. However, the material selection for the liquid fuel injectors, fuel supply connecting line, storage tank, and metering devices must be made according to the efficient ignition process and delivering mechanical power to the external load. Space research systems have introduced a number of modifications in their ICEs that are using liquid H_2 as fuel for space vehicles. Nevertheless, space systems may not fulfill all the requirements in the case of small vehicles operated using liquid H_2 fuel [14].

Sometimes there is the origination of a hotspot in the piston during the need for quick burning at low ignition temperature. The origin of the hotspot initiates backfire, preignition, as well as a knocking phenomenon. It was observed that these phenomena occurred particularly at high air–fuel mixtures. There is the possibility of power loss due to uncontrolled preignition resists that initiate upward compression stroke of the piston. Proper closing of the intake valves delayed the injection that sometimes initiates fuel detonation. This is helpful in reducing backfiring as well as the knocking phenomenon. To overcome the backfiring and knocking phenomenon there needs timely controlled manifold injection system that can overcome the problems of backfiring in ICEs that use H_2 as the fuel [15] for engine operation. The knocking phenomenon sometimes occurs in engine systems while injecting H_2 in the ignition systems and the piston failed to return back to its exact position. Researchers also reported that 70% conversion of syngas into H_2 reduces the knocking phenomenon possibly due to the presence of CO_2 and H_2O and these are acting as the additive shock absorber and initiate delayed ignition. However, this requires detailed studies for efficient fuel utilization.

The fast-burning properties of H_2 can also be controlled by keeping the air and fuel in separate containers and mixing can be done during the ignition time. Sometimes the low flammability limits the requirements for ignition of H_2 and that causes the preignition as well as backfiring effect. The ignition occurs when air–fuel mixture ignites in the combustion chamber before the intake valve remains in closed conditions. Sometimes the preignition can also cause the backfiring phenomenon when ignited fuel-air mixture explodes back into the intake systems before entering into the ignition chamber. The preignition of H_2 is not only a necessary precursor to the backfiring and knocking in the ICEs but probably not occurs under normal circumstances at moderate compression with an appropriate equivalence ratio. Low volumetric energy content, higher compression ratios, or high fuel delivery pressures are needed to avoid reduced power being delivered to the mechanical systems. During the supercharging spark ignition process the engine compresses the air–fuel mixture before being injected into the ignition cylinder. Direct fuel injection involves mixing the fuel with air inside the combustion chamber before sparking is introduced. External mixing of air–fuel sometimes helps the carburetor for efficient injection in the ICEs systems as was observed in the mixing of syngas with diesel fuel.

Sometimes the liquid/cryogenic conditions of the fuel mixture have some beneficial effect on combustion process as well as appropriate mixture formation. For an efficient burning process H_2 can be mixed with 30% of air and these are controlled from the air–fuel injecting systems. It is different from the burning of other hydrocarbon-based conventional fuels where mixing can be done up to 50% for complete burning as well as to avoid the formation of CO, black C, and other oxide ingredients. The volumetric heat value of the air H_2 mixture results in a corresponding power loss at the engine compared to the conventional fuel. The wide flammability range of air H_2 mixtures enables very lean operation with substantially reduced NOx emissions much more easily than that of hydrocarbon conventional fuels. Even, H_2 offers a considerable reduction of air throttle and cylinder charge intake flow losses. This is very different from other gaseous fuels such as C_3H_8 or C_8H_{18} normally used in running ICEs. The NOx emissions from the burning process can also be controlled by allowing the waste gas through a catalytic bed or even passing through H_2O spray chamber. For efficient engine operation, several additional measures are necessary to prevent uncontrolled preignition, knocking, and backfiring effects into the intake manifold. Supercharging is an additional measure to be taken in compensating for the loss in power output and that is related to the lean air–fuel mixture concepts. Due to the low energy content in the exhaust gas, there is larger partial load efficiency and lower volumetric heat value, exhaust gas turbo-charging is less suitable with H_2 operation than with conventional fuel despite un-throttled air supply. Due to the emission of lower exhaust gas volume along with lower temperature the H_2 operation under partial load meager energy density is available from the exhaust gas for charging up and improving torque resulting with the formation of turbocharger hole. The turbocharger hole can be effectively bridged with an additional centrifugal compressor driven directly by the engine via a high-speed transmission gear. Although the recognized turbocharger deficiency can be diminished through the reduction in flow through the orifice of the ignition chamber of ICEs, this results in increased choking of the exhaust gas and welcome additional problems with uncontrolled preignition process. The emission of hot residual gas could basically be reduced through an increase of the compression ratio in the range of either 10:1 or 11:1. Although this measure is contradictory to supercharging, the ICEs are sometimes then coupled with turbine and alternator in generating electrical power. The power generated using this method is decentralized one and can use in meeting the local demand as well as can be blended with the grid power like blending of power from other renewable sources as biomass, solar PV, or wind. For blending purpose additional measures require to be taken for matching of voltage, phase, and frequency through smart controlling systems.

For small system operation with electric power, the H_2 is coupled with the fuel cell systems for producing electricity. With proper conversion and controlling systems the power can also be deliver to the load. This is going to be the upcoming perception for electrical vehicles considering the fuel cells systems could have a lower material and C footprint than that of lithium–ion batteries (LiB). The captive vehicle fleets can help in overcoming the challenges of low utilization of refueling stations. For long-distance and heavy-duty vehicles like marine vessels or high-speed bullet trains H_2 is an attractive option. This will put a positive impact on the utilization of refueling stations in terms of ecology, economy, and empowerment. However, on the other hand, an increase in refilling stations needed appropriate storage facilities for H_2 and that may have an impact on the price of the fuel value also. In order to make flexible storage there is the option for converting H_2 into appropriate chemicals for its applications. The increase in stages from energy carriers to chemicals and its reconverting again for powering the wheel reduce the overall efficiencies. One of such option is to convert H_2 into Ammonia (NH_3) and which is used as fuel [16]. Nevertheless, the NH_3 has additional advantages in terms of the burning process, storage, and safety issues.

In powering mobility sectors using NH_3 as the fuel have several limitations also. The alkaline nature of NH_3 welcomes caustic hazards particularly to the operators as well as in the engine systems also. Although some experimentation on the use of NH_3 was conducted in heavy vehicles like marine vessels, trucks, and buses but no conclusive decision was indicated on the longevity as well engine performance with time. The NH_3 is also used in some marine vessels as demonstration projects with small ships and onboard power supply in larger vessels. The GHGs emission reduction strategy over the sea surface welcomes the use of H_2-based fuels as well as NH_3. It is projected that marine freight activities will increase by 45–50% more than that of 2030 as a result H_2 and NH_3 are the potential candidates for both national action on domestic shipping decarbonization. According to the International Maritime Organization (IMO) the GHGs reduction strategy gives limitations on the use of other C-free fuels and the use of H_2 is given priority [17] to reduce the photochemical reactions between the marine air and continental air in smog forming over the sea surface. Comparative studies on the properties of hydrogen with most used hydrocarbon fuels is presented in Table 6.1.1 and that of engine characteristics used in marine vessels is presented in Table 6.1.2.

Further studies indicated that the storage cost of H_2 is comparatively higher than other hydrocarbon fuels as a result the cargo volume reduces due to the introduction of storage facilities for H_2. Some demonstration projects were also carried out on the use of H_2 in rail transportation. Preliminary results indicated that the trains run by H_2 can be most competitive in rail freight. Rail in most of the countries are operating in electrified transport mode and electric lines are energized from burning of fossil fuels.

Table 6.1.1 Comparison of properties of hydrogen with most used hydrocarbon fuels [32].

Properties	Syngas	Hydrogen	Methane	Gasoline
Density (kg/m^3) at NTP	1.04	0.082	0.717	5.11
Vol% Stoichiometric compression in air	2.07	29.53	9.48	1.65
Combustion of mol nos.	1.66	0.85	1.00	1.058
Lower heating value (MJ/kg)	7.47	119.7	46.72	44.79
Stochiometric combustion energy/kg (MJ)	2.02	3.37	2.56	2.79
Limit to flammability (%)	1.0–3.5	4–75	5.3–15.0	1.2–6.0
Ignition energy (mJ)	0.18	0.02	0.28	0.25
Flame speed at NTP (m/s)	0.28–0.35	1.90	0.38	0.37–0.43
Temperature for autoignition (K)	898	858	813	500–750

Table 6.1.2 Engine characteristics for using syngas, hydrogen engine, and diesel fuel.

Engine characteristics	Syngas	Hydrogen	Diesel
Speed (RPM)	625	775	600
Ratio of fuel pressure	15	9	13
Fuel energy (MJ/kg)	40	130	43
Bore (cm)	34	32	34
Stroke (cm)	42	42	42
Power (kW)	2700	2700	2700
Thermal efficiency (%)	45	30	47
Fuel consumption (g/kWh)	180	93.5	179
Cylinder nos.	6	6	6
Effective pressure (bar)	27	10	26
Compression pressure (bar)	85	76	84
Combustion pressure (bar)	125	132	124

The decentralized H_2 and battery-operated electric trains with partial line electrification are both options to be replace nonelectrified operations, which are substantial in many regions in the developing countries. As a result, there is higher possibility for emerging H_2 as the fuel for surface, water, and air transportation.

The theoretical potential for future use of H_2 in road transport is very large. Any road transport mode can technically be powered using H_2, either directly using fuel cells or via H_2-based fuels in ICEs.

Large vehicles utilized for long distances typically have the highest energy usage per kilometers in terms of total cost of ownership. This means that for heavier vehicles and those with high utilization rates, such as trucks, intercity buses, and fleets of commercial vehicles, fuel expenditures typically represent a larger portion of total costs. It will be critical to reduce the cost of fuel cell systems and H_2 storage tanks to attain cost competitiveness with other options, as the capital cost of a car ranges from 70% to 95% of the total cost of ownership, depending on the vehicle. For trucks, the situation is somewhat different because capital expenses account for 40–70% of overall ownership costs, therefore cost reductions for delivered hydrogen are equally crucial.

Sulfur (S) and greenhouse gas emissions for maritime transportation have been reduced by methods put in place by the IMO. Scrubber installation, fuel switching to LNG, and the use of very low sulfur fuel oil (VLSFO) are potential solutions to the problem of lowering S emissions, although these solutions will only partially contribute to the 50% greenhouse gas reduction target by 2050 compared to 1990. Limits on S emissions are anticipated to increase refinery demand for H_2 rather than shipping fuel demand. Advanced biofuels, H_2, and NH_3, as well as synthetic liquid fuels based on H_2, are all options for meeting the GHG emissions target for reaching the SDG. The decision to switch fuels depends on infrastructure deployment that is out of the ship owners' direct control. The development of bunkering facilities would be necessary for LNG, H_2, and NH_3, although only LNG and NH_3 could rely on the current distribution system. The intense rivalry from other sectors for a limited supply of sustainable biomasses, which are the sources for biofuels, has led to a new concern about the availability and pricing of advanced biofuels. Some nations are aiming for low GHG emissions and low-carbon options for domestic marine transportation. A decarbonization strategy for the water transportation sectors has also been developed by industry leaders, and it comprises demonstration projects, technological adoption, openness, and knowledge sharing on the subject of low GHG emissions.

It is clear from the discussion above that H_2 and its derivatives will play a significant role in the decarbonization of sectors like heavy industry, shipping, aviation, and heavy-duty transportation where emissions are difficult to reduce and alternative solutions are either unavailable or difficult to implement. These projected advancements, however, fall short of what is required to meet the "Net Zero Emissions by 2050 Scenario" [18]. There is a need for swifter action to increase demand for low-emission fuels like H_2 hydrogen and to release investments that can hasten production with scale-up and infrastructure deployment as needed.

By using low-emission H_2 and H_2-based fuels, the "Net Zero Scenario" aims to achieve modest reductions in GHG emissions overall and C-based components in particular by 2030. H_2 technologies do not contribute nearly as much as other important mitigation strategies including the use of renewable energy, direct electrification, and altering human behaviors in the global environment. Yet, H_2 and H_2-based fuels can be crucial in industries where emissions are difficult to reduce and where other mitigation measures would not be accessible or would be challenging to put into place. Due to the fact that H_2-based technologies need to mature over time, their overall contribution to the power sector is also greater in the long run.

6.1.6 Hydrogen-blended syngas

Synthetic gas or syngas (SnG) produced from gasification of biomass offers another interesting as well as possible prospect for distributed power generation with the capacity to the ranges of megawatt (MW) ranges. The distributed power generation from syngas showed its potential applications in industrial sectors particularly in process industries. Major advantages of the distributed generation include flexible location, reduced costs, reliability improvement, grid congestion reduction, transmission loss reduction, water and land conservation, and reactive power improvement [19]. However, the above drawbacks are still remain including relatively higher capital costs and primary fuel prices, as compared to central generation systems [20]. During the beginning of 2000 in India a good effort has been given in using syngas both for thermal as well as electric power production purposes. In power production the syngas produced from biomass gasification integrated with conventional diesel generators operating in dual fuel mode. Integration of syngas with diesel generator can save 70–80% diesel. Enrichment of syngas with H_2 can further reduce the diesel fuel consumption while operating in dual fuel mode. The policies require be given in this direction in converting "Waste to Wealth" [20] in upcoming scenario of SDGs.

The syngas from biomass gasification containing a mixture of CO (18–22%), H_2 (8–2%), CH_4 (3–4%), CO_2, H_2O, hydrocarbons, H_2S, phenolic acid, or tar. The syngas compositions are typically consumed based on the gasifier's operational parameters, including the properties of the feedstock, the gasification medium (such as steam, air, oxygen, and CO_2), temperature ranges, pressure, and catalyst types. Syngas is currently transformed into important chemical products like H_2, NH_3, and CH_3OH depending on the input materials or feedstock.

Normally the biomass gasification occurred in four stages, e.g., drying, pyrolysis, oxidation, and reduction [8]. Drying or dehydrating of feedstock typically happens when a temperature of 120°C is reached at the input biomass store. Pyrolysis takes place in between 120°C and 700°C, and start releasing of volatile species below 500°C. Charcoal, bio-oil, and syngas are the byproducts of the pyrolysis process. Biomass produces more syngas than coal because its volatile components are substantially higher—between 70% and 86% in coal than roughly 30% in biomass. The primary components of the final solid products are char and ash. After gasification, syngas is released at temperatures of up to 1000°C under normal air conditions and 1600°C under pressure. An efficiency of 80–95% is achieved in the conversion of the C part of biomass into syngas, as a result of gasification processes is much more efficient than that of burning in local oven. In the presence of suitable catalysts, the syngas can process through water gas shift (WGS) reactions in further enriching the syngas to H_2. However, there are certain difficulties in producing syngas and enriching it to H_2, and this is the subject of current study and research.

The gasification process is significantly influenced by the characteristics of biomass. The alkalis in biomass, as opposed to the alkalis in coals, might interact with other minerals to cause fouling and slagging in the gasifier. Based on the idea that slagging develops because ash does not melt at high temperatures, slagging gasifiers are primarily made for coal gasification and can work at temperatures between 1400°C and 1600°C. Minerals or organic substances in the carbon feedstock melt at this high temperature. The three alkalis that contribute most to the creation of ash are potassium (K), calcium (Ca), and silicon (Si). When the temperature rises, the alkali species are evaporated and form eutectic mixes with silicate (SiO_2), which results in slag formation. As a result, the process temperature has a significant impact on the slag formation and ash deposition process. Ash deposition increases significantly over 600°C, and sticking effectiveness rises to 1170°C from 970°C at that temperature. Sometimes the Cl and S gas levels may accelerate the slagging process in the presence of H_2O vapor. In addition to these particular chemical components, particle viscosity is another aspect that may contribute to the slagging and fouling phenomena. For instance, small alumina-silicate particles with low kinetic energy tend to cling together, but large particles with low viscosity, like compounds based on iron, tend to slag easily.

The syngas conditioning is frequently categorized as hot, cold, or warm depending on the temperature of the syngas exiting the cleanup equipment. It is a crucial step in the gasification process, particularly when the gasifier is unable to reduce pollutants to safe levels. As previously mentioned, the syngas conditioning can also be employed to enhance particular gas components like H_2 and CO. Syngas conditioning can account for up to 30% of the overall capital cost of gasification plants at widespread scales. The degree of syngas conditioning for the abatement of various impurities, including tars and particles, is now driven by the emission standards and anticorrosion specifications of the power units. Advanced water gas shift reactors with improved designs, chemical looping, catalytic gasification, candle-based filtering, H_2/CO_2 membranes, syngas coolers, and integrated CO_2 removal are some other syngas conditioning methods.

Because electricity can be used for so many different things, prime mover technologies are continually improving. The average electrical efficiency of prime movers ranges from 10% to more than 60%. In comparison to conventional systems, these have lower capital expenditure (CAPEX) associated with grid expansion and operational expenditure (OPEX) associated with line losses when employed as a modular electric power source close to an end-user. When local generation units are connected to the grid, the two-way transactions between them and the grid increase grid capacity, lower the chances of supply interruption, and enable better energy price due to flexible use/purchase/sell choices [21]. Some new opportunities have emerged in this direction in sullying power in decentralized mode.

The H_2-enriched syngas (H_2SnG) has several advantages over the normal syngas. Researchers compared studies on the burning of SnG with H_2SnG in ICEs and other engines also. The studies indicated that H_2SnG is much more efficient in power production that that of normal SnG. In addition to this use of H_2SnG reduced knocking phenomenon as well as back firing effect in engine. Apart from burning of H_2SnG in standard ICEs the H_2SnG can be used as the input fuel for the fuel cells also for producing electrical power [22].

6.1.7 Internal combustion engines

The spark ignition engine (SIE) and compression ignition engine (CIE) are two types of internal combustion engines. A spark plug in a SIE initiates the ignition process, which ignites compressed gasoline and air in the combustion chamber. In the past few decades, ICE technology has evolved significantly. One such advancement is pressurized gain combustion, in which compressed air and syngas are burned in a pressured combustion chamber. At standard condition the electrical efficiency for ICEs are varying from 20% to 35%. The most recent ICE systems claim to have a power generation capacity of up to 6.5 MW with high economic returns and maintenance intervals of roughly 6000 hours, roughly every 1–2 years. Low noise levels of 44 dB at 3 ft. and improved emission performances also contribute to the acceptability of these systems. The ICEs are preferred for distributed power generation under 5 MW due to the positive aspects like; (1) easy to install and modify, (2) have a track record of performance, (3) can start up and shut down quickly, and (4) have less complex control strategies focused primarily on regulating the air–fuel ratio. These have high generation efficiency to the ranges of 33–41% on a low heating value (LHV) basis, and have a low capital cost to the order of \$700–1000/kW when compared to alternative prime movers using syngas. Moreover, the ICEs are more resistant to syngas impurities such as tar, particle, and alkali metals.

The thermal efficiency, specific fuel consumption (SFC), and power derating are the three main operational factors that affect an ICEs powered by syngas. Engine thermal efficiency—often referred to as "brake thermal efficiency" which is the ratio of the brake shaft power output—that is, the power available in the crankshaft rather than the power produced in cylinders—and the rate at which thermal energy from the fuel is used up. Depending on the fuel type and equivalency ratio, the maximum engine thermal efficiency varies greatly. For instance, utilizing fuels rich in CH_4 generally results in higher efficiency than using fuels rich in H_2, while high H_2 expands at the syngas' flammability range and that stabilizes its combustion conditions. When ICEs use pure CH_4 and its blend with H_2 up to 60% delivers optimal performances, engine thermal efficiencies are at their highest at equivalency ratios of approximately 0.9 [23].

The SIE system can also run on ethanol and other renewable fuels like syngas in addition to the usual fuels of gasoline and natural gas. Due to the SIEs' straightforward modification requirements, particularly in the air/fuel intake system, they are frequently employed for syngas applications.

6.1.8 Power production and power blending

The electrical power production using the syngas from small-scale biomass gasification systems is almost totally done using ICEs and implemented in several process and product industries [24]. In addition to these power from the syngas had also blended in the grid in meeting the peak load demand. In meeting the peak load demand the syngas from biomass gasifier was blended with diesel run the ICEs through appropriate controlling and blending systems. The power from this method showed big impact in rice mills and rice processing industries [25] as rice mills are captive sources for rice husk. The method for power production using syngas can also be enriched further by using it in microgas turbines (μGT) and also with fuel cells (FC) and even combination of the both to have hybrid μGT/FC power plants but these are still are in the developmental stages. Some theoretical data are available and putting these in appropriate software and with the use of artificial intelligence (AI) and machine learning (ML) successful design of μGT/FC can be made in terms of gasifier typology, gasifying agent, clean up and cooling systems, appropriate biomass, electrical and cogeneration efficiencies and of capital cost.

The figure of merit for such gasification system is presented below:

The cold gas efficiency (η_{Chem}) is the efficiency of the gasification systems. The overall electrical efficiency is defined as:

$$\eta_{\text{Elec}} = \frac{Power\ out - Power\ aux}{Syngas\ power\ from\ biomass} = \frac{Net\ available\ power}{Syngas\ power\ from\ biomass}$$
$$= \eta_{\text{Chem}} \cdot \eta_{\text{Mech}} \cdot \eta_{\text{Gen}}$$

where η_{Mech} and $\eta_{\text{Gen.}}$ are the mechanical and generator efficiency of the various components of the mechanical system and power generating system respectively.

In addition to the above the analysis provides other important information such as work temperature, fuel utilization factor and current density for the solid oxide fuel cell (SOFC) and pressure ratio as well as turbine inlet temperature for μGT.

Another important parameter is the combined heat and power (CHP) efficiency η_{CHP}, defined as:

$$\eta_{\text{CHP}} = \frac{P_{\text{out}} - P_{\text{Aux}} + Q_{\text{useful}}}{Input\ biomass}$$

with Q_{useful} means the net quantity of heat that can be used for cogeneration. Because of their well-known technology and dependability, ICE constitute one of the first initiatives to manufacture electricity from gasification producing gas. When compared to natural gas and gasoline, producer gas has a significantly lower calorific value. Thus, in order for engines to be able to run on producer gas, specific design modifications must be made [26]. The most common ones are spark ignition and diesel engines (ultimately with a small percentage of diesel in a dual-fuel mode operation [27]). Researchers also stated that the permitted particle and tar content in production gas must be less than 50–100 mg/Nm^3 for ICE functioning to be successful. The type of engine utilized in tests and its design elements should be taken into consideration when interpreting the gas quality standards as given in the literature [28]. In contrast to simulated ones, the majority of examples for ICE deal with actual plants that are used in experimental and commercial settings. Particularly, there has been considerable experience with such systems from 10 kWe to 500 kWe around the world [29].

In the rice mills of West Bengal, India several studies were conducted since 2001 on the performance of both downdrafts as well updraft biomass gasifiers range from 100 kWe to 250 kWe for operating diesel generators using dual fuel mode [30]. Some field studies were conducted by the author in 2001 in 30 rice mills. The result showed that at 80% load of a 250 kVA diesel generator consume about 52 liter of diesel per hour. Coupling the diesel generator with rice husk gasifier at an optimum condition diesel consumption reduces to 12 liter per hour. Treatment of the outcome syngas with water vapor activated with Fe_2O_3/Ni diesel consumption further reduces to 9–10 liters per hour without any knocking effect or backfiring. It was presumed that while syngas and water vapor passed through Fe_2O_3/Ni bed it enriched H_2 content resulting in enhancement of calorific value of the modified syngas gas and that has reflected in reduction of diesel consumption. Further studies require to optimize the catalytic bed and optimization of syngas mixture to utilize the maximum outcome along with minimum knocking effect and backfiring effect [31].

6.1.9 Conclusion

The H_2-enriched syngas is a possible solution in generating electricity to be used either in decentralized as well as centralized mode. The syngas can be generated using gasification technologies. The generated syngas can be enriched with H_2 by processing the syngas through water gas shift (WGS) reaction in presences of appropriate catalysts. The H_2-enriched gas can be converted into electrical power either by ICE as well as combination of ICE and FC technologies. The syngas from biomass gasification showed successful applications in process industries like rice mills as well as food processing industries. To increase the yield and quality of syngas, proper gasification with steam, indirect heat supply, and primary conditioning with catalysts may offer a potential solution. A 40% boost in electricity efficiency is possible, but doing so would increase capital costs globally. However, as technologies like fluidized beds, catalysts, and GT-fuel cells continue to advance, they will be used extensively in the transportation and mobility sectors. Yet more study is required, particularly in the development of novel catalysts and supporting materials to improve the selectivity, activity, productivity, and economy of a catalytic process for syngas cleaning and downstream applications, as well as exploration in highly energy-efficient heat and power generators like gas turbines or fuel cells. The H_2-enhanced syngas is thought to be the fuel of the future, with potential uses in the mobility industry in general and maritime transportation systems in particular.

Acknowledgment

The present study is the part of research work of the Mission Innovation Program of Department of Science and & Technology (DST), Government of India.

References

[1] T. Sharp, What is the temperature on earth? 2015. http://www.space.com/17816-earth-temperature.html.

[2] Global Electrification Database of World Bank, SDG track 7. https://data.worldbank.org/indicator/EG.ELC.ACCS.ZS?locations=ZG.

[3] Origin of the elements. https://www2.lbl.gov/abc/wallchart/chapters/10/0.html.

[4] G. Sayannal, C.M. Bagade, Hydrogen internal combustion engine, Int. J. Tech. Res. Appl. 6 (2018) 01–11.

[5] Royal Society of Chemistry Report, New approaches to biofuels, 2010. http://www.rsc.org/ScienceAndTechnology/Policy/Documents/newapproachbiofuels.asp.

[6] M. Ruth, M. Laffen, T.A. Timbario, Technical report of national renewable energy laboratory, NREL/TP-6A1-46612, 2009. https://www.nrel.gov/docs/fy10osti/46612.pdf.

[7] IEA Report, 2021. https://www.iea.org/reports/about-ccus.

[8] L.C. Cadwallader, J.S. Herring, Safety issues with hydrogen as a vehicle fuel, 1999. https://inldigitallibrary.inl.gov/sites/sti/sti/3318091.pdf.

[9] B.M. Besancon, V. Hasanov, R.I. Imbault-Lastapis, M. Benesch, M. Barrio, M.J. Mølnvik, Hydrogen quality from decarbonized fossil fuels to fuel cells, Int. J. Hydrogen Energy 34 (2009) 2350–2360.

[10] M.W. Melaina, O. Antonia, M. Penev, 'Blending hydrogen into natural gas pipeline networks: a review of key issues' Technical Report, NREL/TP-5600-51995 2013. https://www.nrel.gov/docs/fy13osti/51995.pdf.

[11] L. Dong, Q. Qi Zhang, M. Qiuju, S. Shilei Shen, Comparison of explosion characteristics between hydrogen/air and methane/air at the stoichiometric concentrations, Int. J. Hydrogen Energy 40 (28) (2015). https://doi.org/10.1016/j.ijhydene.2015.05.038.

[12] F. Yang, W. Tianze, X. Deng, J. Dangn, Review on hydrogen safety issues: incident statistics, hydrogen diffusion, and detonation process, Int. J. Hydrogen Energy 43 (61), 31467–31488. https://doi.org/10.1016/j.ijhydene.2021.07.005.

[13] V.N. Nguyen, R. Deja, R. Peters, L. Blum, D. Stolten, Study of the catalytic combustion of lean hydrogen-air mixtures in a monolith reactor, Int. J. Hydrogen Energy 43 (36) (2018) 17520–17530.

[14] S.K. Mital, J.Z. Gyekenyesi, S.M. Arnold, R.M. Sullivan, J.M. Manderscheid, P.L.N. Murthy, Review of current state of the art and key design issues with potential solutions for liquid hydrogen cryogenic storage tank structures for aircraft applications, NASA Report (2006). https://ntrs.nasa.gov/api/citations/20060056194/downloads/20060056194.pdf.

[15] P.V. Blarigan, Advanced internal combustion engine research. Proceedings of the 2000 DOE hydrogen program review, NREL/CP-570-28890. https://www1.eere.energy.gov/hydrogenandfuelcells/pdfs/28890yy.pdf.

[16] The Royal Society Report, The role of hydrogen and ammonia in meeting the net zero challenge. https://royalsociety.org/-/media/policy/projects/climate-change-science-solutions/climate-science-solutions-hydrogen-ammonia.pdf.

[17] IMO, 2020, Sulphur limit implementation – carriage ban enters into force. https://www.imo.org/en/MediaCentre/PressBriefings/Pages/03-1-March-carriage-ban-.aspx.

[18] IEA, Net zero by 2050: a roadmap for the global energy sector. https://www.iea.org/reports/net-zero-by-2050.

[19] A. Ugwu, A. Zaabout, F. Donat, G.V. Diest, K. Albertsen, C. Müller, S. Amini, Combined syngas and hydrogen production using gas switching technology, Ind. Eng. Chem. Res. 60 (9) (2021) 3516–3531. https://doi.org/10.1021/acs.iecr.0c04335.

[20] K. Gumte, P. Devi, P. Srinivas, S. Miriyala, K. Mitra, Achieving wealth from bio-waste in a nationwide supply chain setup under uncertain environment through data driven robust optimization approach, J. Cleaner Prod. 291 (2021). https://doi.org/10.1016/j.jclepro.2020.125702.

[21] Biomass gasification based power production in India, Report of Energy Alternatives India (2011). https://www.eai.in/ref/reports/biomass_gasification_based_power_production.pdf.

[22] L. Fana, Z. Tua, S.H. Chan, Recent development of hydrogen and fuel cell technologies: a review, Energy Rep. 7 (2021) 8421–8446. https://doi.org/10.1016/j.egyr.2021.08.003.

[23] L. Wang, C. Hong, X. Li, Z. Yang, S. Guo, Q. Li, Review on blended hydrogen-fuel internal combustion engines: a case study for China, Energy Rep. 8 (2022) 6480–6498. https://doi.org/10.1016/j.egyr.2022.04.079.

[24] Z. Fu, Y. Li, H. Chen, J. Du, Y. Li, W. Gao, Effect of hydrogen blending on the combustion performance of a gasoline direct injection engine, ACS Omega 7 (15) (2022) 13022–13030. https://doi.org/10.1021/acsomega.2c00343.

[25] R. Varshney, J.L. Bhagoria, C.R. Mehta, Recent research and developments in biomass gasification in India, Int. J. Adv. Sci. Technol. 3 (2) (2011) 86–116.

[26] R. Bates, K. Doelle, Syngas use in internal combustion engines: a review, Adv. Res. 10 (1) (2017) 1–8. https://doi.org/10.9734/AIR/2017/32896.

[27] E. Bocci, M. Sisinni, M. Moneti, L. Vecchione, A. Di Carlo, M. Villarini, State of art of small-scale biomass gasification power systems: a review of the different typologies, Energy Procedia 45 (2014) 247–256.

[28] J. Stewart, A. Clarke, R. Chen, An experimental study of the dual-fuel performance of a small compression ignition diesel engine operating with three gaseous fuels, Proc. Inst. Mech. Eng. Part D J. Aut. Eng. 221 (D8) (2007). https://doi.org/10.1243/09544070JAUTO458.

[29] N.N. Mustafi, R. Raine, A study of the emissions of a dual fuel engine operating with alternative gaseous fuels, SAE Technical Papers (2008). https://doi.org/10.4271/2008-01-1394.

[30] B. Ghosh, Modernize Bio-Energy, School of Energy Studies, Jadavpur University, Kolkata, 2002.

[31] B. Sudarmanta, Dual fuel engine performance using biodiesel and syn-gas from rice husk downdraft gasification for power generation, Project Report on Dual Fuel System for Diesel Engine (2015). https://doi.org/10.13140/RG.2.1.2503.4968.

[32] S. Bari, S.N. Hossain, Performance of a diesel engine run on diesel and natural gas in dual-fuel mode of operation, Energy Procedia 160 (2019), 215–222. http://doi.org/10.1016/j.egypro.2019.02.139.

Total cost of ownership analysis of fuel cell electric vehicles in India

Ujwal R. Sontakke[a], Santosh Jaju[b] and Dhiraj K. Mahajan[a]
[a]Ropar Mechanics of Materials Laboratory, Department of Mechanical Engineering, Indian Institute of Technology Ropar, Rupnagar, Punjab, India
[b]Department of Mechanical Engineering, G. H. Raisoni College of Engineering, Nagpur, Maharashtra, India

7.1.1 Introduction

For the past few centuries, the increasing energy demand in the world has been constantly met by fossil fuels. Combustion of these fuels, however, has become a major source of greenhouse gas emissions causing an increase in the global average temperature [1]. According to the objectives of the Paris Agreement on climate change, global CO_2 emissions need to be net-zero by the year 2050, which can help to limit global warming to 1.5°C [2]. To achieve the same, rapid decarbonization in all sectors is a need of the hour.

India is one of the largest importers of fossil fuels and at the same time one of the largest emitters of greenhouse gasses by burning these imported fossil fuels, as shown in Figs. 7.1.1 and 7.1.2. Therefore, self-reliance in the energy sector through clean energy is an immediate requirement. In India, both the power sector and the transportation sector are the major consumers of fossil fuels, consuming around 92% of the total requirement [3].

The transportation sector is one of the fastest-growing sources of carbon emissions in India [3]. In the fiscal year 2021, the total sale of internal combustion engine vehicles (ICEV) includes 27.11 lakh units of passenger vehicles, 151.19 lakh units of two-wheelers, 5.69 lakh units of commercial vehicles, and 2.16 lakh units of three-wheelers [4]. Globally, the automobile sector alone is responsible for 24% of CO_2 emissions while in India it contributes to 13.5% of total CO_2 emissions [5,6]. To control these emissions, several initiatives are being taken by the Government of India (GoI) among which electrification of the transport sector is the major one [7]. Electrification of the transportation sector towards net zero-emission is also an important consideration for moving the country toward a green economy.

Towards Hydrogen Infrastructure: Advances and Challenges in Preparing for the Hydrogen Economy.
DOI: https://doi.org/10.1016/B978-0-323-95553-9.00005-4

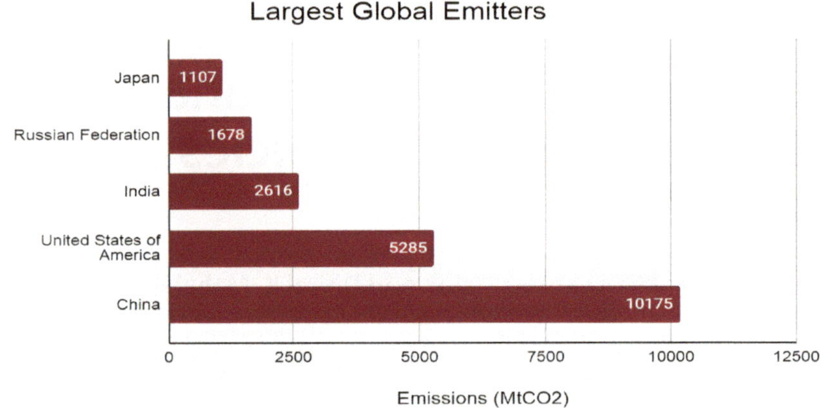

Figure 7.1.1 Top five largest global emitters of CO_2 [1].

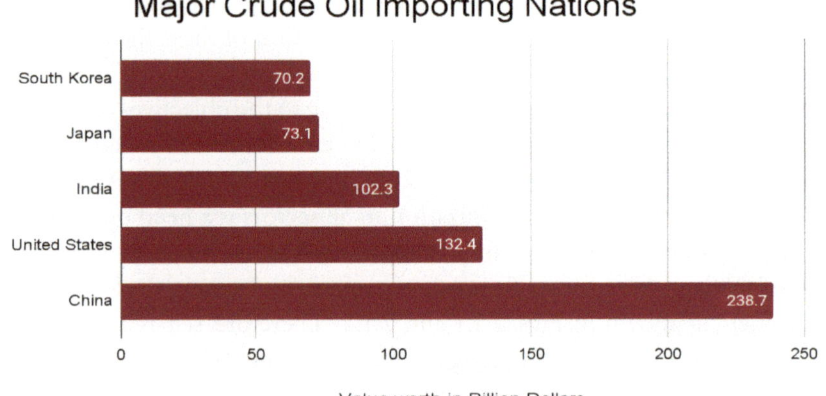

Figure 7.1.2 Top five major crude oil-importing nations in 2019 [1].

According to the United Nations Environment Program, the "green economy" is defined as a low-carbon, resource-efficient, and socially inclusive economy [8]. The green economy focuses on sustainable development and a flexible economy while providing a quality life to people by using natural and renewable energy resources [9].

India has a huge potential for renewable energy and progressively making significant development in this area. Currently, around 25% of India's total electric power share comes from the renewable sector as shown in Fig. 7.1.3 and the country aims to increase it up to 40% in upcoming years [10]. The Indian geographical scenario provides a better opportunity for the renewable energy sector to grow with its 300 clear sunny days,

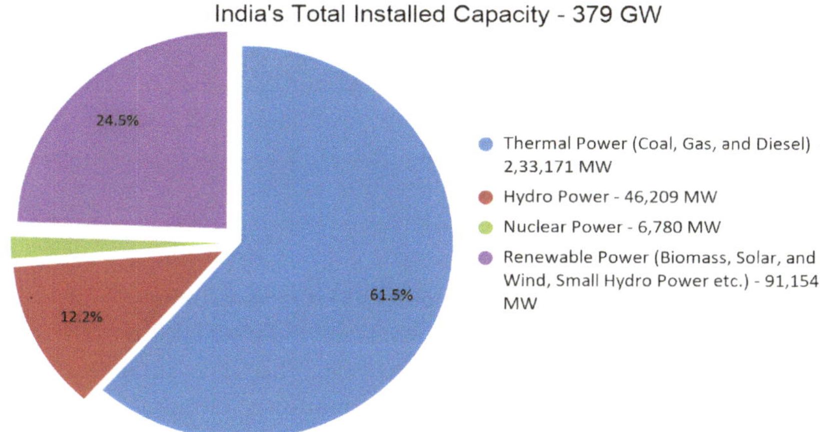

Figure 7.1.3 India's total installed power capacity (renewable and nonrenewable sources) [10].

huge landmass, long coastline, etc. [11,12]. Therefore, the government is now coming up with many initiatives and has set the target to increase its renewable energy capacity up to 227 GW in upcoming years, which includes solar-, wind-, bio-, hydro-power, etc. [11]. This increased renewable power generation capacity will provide a major advantage to the electrification of India's transport sector.

In India, as of April 2020, around 58.49% of conventional passenger rail and freight–transported rail infrastructure is already electrified [13]. However, the electrification of road transport is just at the nascent stage. In India, to decarbonize the transport sector many bold steps are required. The GoI has already announced a goal for 2030, which is to electrify all public vehicles. To achieve these targets, both battery electric vehicles (BEVs) and fuel cell electric vehicles (FCEVs) will play a crucial role in the future [6,14].

Nowadays many Indian cities are witnessing strong growth of BEVs primarily for the short and medium range of travel [15]. This growth is due to the strong support provided by GoI through the Department of Heavy Industries had launched a scheme for faster adoption and manufacturing of electric vehicles in India, that is, Faster Adoption and Manufacturing of Electric Vehicles (FAME) India which was initially launched in the year 2015. FAME India phase II, which was launched in the year 2019 is the next phase of the scheme, which is based on demand incentives, building a network of charging infrastructure, and promotional activities for publicity, information, communication, and education. Under the scheme following categories are eligible for subsidy amounting up to 40% of initial vehicle cost:

e-rikshaw, e-cart, two-wheelers, three-wheelers, and passenger four-wheeler vehicles and buses.

Insite of the immense popularity gained by BEVs, they do not provide a true zero-emission solution due to the fossil fuel-dominated energy mix of India, as shown in Fig. 7.1.3. In addition, BEVs due to their long battery charging time and limited range of travel do not meet the generic requirement of all kinds of vehicles such as heavy-duty applications and intercity buses. Therefore, BEVs do not ensure a proper solution for the transport sector required for the green economy [16]. On the contrary, FCEVs are better suited for the green economy. Well-developed and demonstrated fuel cell technologies are now successful to offer FCEVs as an alternative to BEVs. Hence, they are now grabbing the attention of the automobile sector in the international markets [16]. In addition, hydrogen can also be used with a modified internal combustion engine which can be cheaper but less efficient than FCEVs and only emits NOx [14]. To achieve a target of zero-emission, a green hydrogen-based economy can make a great impact where hydrogen is produced from clean renewable energy sources (also referred to as "green hydrogen"). In this scenario, the transport sector will be the major consumer of hydrogen, because of its features like short refueling time, long travel range, and zero carbon emissions, etc. Hence, hydrogen is being considered very crucial in the near future for heavy-duty and long-range transportation. Apart from road transport, many companies are also investing in developing hydrogen-powered boats, airplanes, marine vehicles, etc. Despite having many advantages, the hydrogen-based green economy still faces many challenges including commercialized acceptance in the Indian market, standards, regulating policies, safety, and cost [17]. With respect to fuel cell technology for vehicular applications, the main barrier is the lack of hydrogen infrastructure (e.g., hydrogen filling stations), the high initial cost of FCEVs, and low-carbon hydrogen production with cost-competency [14].

Looking at the long-term advantages offered by FCEVs, it becomes important to know the operational benefits of FCEVs in the Indian scenario. In this context, a total cost of ownership (TCO) analysis of FCEVs that considers not only their initial cost but also their operational cost over the total service life of the vehicle is required in the Indian context. To provide such insights, in this work TCO analysis of FCEVs is performed by considering the total life of the vehicle to be twelve years in comparison to BEVs and ICEVs. A realistic approach is adopted by considering the average annual distance traveled, the fuel cost, and average electrical energy consumption. The role of subsidy, as provided to BEVs under the FAME-II scheme, is also

Table 7.1.1 Cost components considered for total cost of ownership analysis of different vehicles.

S. no.	Petrol	Diesel	BEVs	FCEVs
1	Capital cost	Capital cost	Capital cost	Capital cost
2	Interest paid	Interest paid	Interest paid	Interest paid
3	Insurance price	Insurance price	Insurance price	Insurance price
4	Maintenance and repair	Maintenance and repair	Maintenance and repair	Maintenance and repair
5	Fuel cost	Fuel cost	Energy cost	Fuel cost
6	–	–	Battery replacement cost	–
7	–	–	Downtime cost	–

considered for evaluating the cost competitiveness of FCEVs in comparison to BEVs. The presented TCO analysis thus examines the best-suited vehicles for an Indian scenario considering their travel range and type. Further, the analysis is important to compare the operational cost of FCEVs with conventional vehicles to know their significance and limitations.

7.1.2 Methodology

7.1.2.1 Scope of total cost of ownership analysis

The TCO estimates the cost to own a vehicle for a particular period. In this work, TCO analysis is performed for different fuel vehicles in comparison to FCEVs, while considering their annual travel distance of 1 lakh km and service life of 12 years. The TCO for various vehicles is a function of the purchase price, interest paid on the loan taken for the purchase of a vehicle, insurance cost, battery replacement cost for EVs, maintenance and repair cost, and fueling cost over the service period, as shown in Table 7.1.1. In addition, certain assumptions are made for the TCO analysis based on information collected from relevant articles and various sources such as reports by research organizations, data from government organizations, and journals which are properly referred to in the text.

7.1.2.2 Formulation of TCO model

To get realistic estimates of the cost of ownership it is important to consider a wide range of parameters considering the capital and running cost of the vehicles. The variables considered for the TCO model along with their values are given in Table 7.1.2.

Table 7.1.2 Considered variables of the total cost of ownership calculations.

Description	Symbol	Formula	Units
Purchase price of vehicle	C_{PV}	[–]	₹
Vehicle life expectancy	L_V	[–]	years
Annual distance traveled	T_D	[–]	km
Purchase price per km	C_{PVKM}	$C_{PV}/(L_V \times T_D)$	₹/km
Fuel cell stack cost per kw	C_{FC}	[–]	₹/kW
Balance of plant cost for fuel cell system per kW	C_{BOP}	[–]	₹/kW
Battery pack cost per kWh	C_B	[–]	₹/kWh
Cost of hydrogen tank per kg	C_{HT}	[–]	₹/kg
Insurance costing	I	[–]	%
Vehicle insurance cost per km	C_I	$(I \times C_{PV})/T_D$	₹/year
Maintenance and repair cost	C_M	[–]	₹/year
Maintenance and repair cost per km	C_{MKM}	C_M/T_D	₹/km
Average electrical energy consumption per km	E_{CV}	[–]	kWh/km
Energy content of fuel per unit	E_{CF}	[–]	kWh/lit or kWh/kg
Electric powertrain/motor efficiency	ϵ_M	[–]	%
Battery efficiency	ϵ_B	[–]	%
Fuel cell efficiency	ϵ_{FC}	[–]	%
Tank to wheel efficiency of BEV	ϵ_{BEV}	ϵ_M	%
Tank to wheel efficiency of FCEV	ϵ_{FCEV}	$\epsilon_M \times \epsilon_{FC}$	%
Tank to wheel efficiency of petrol vehicles	ϵ_{PV}	[–]	%
Tank to wheel efficiency of diesel vehicles	ϵ_{DV}	[–]	%

(continued on next page)

Parameter	Symbol	Formula	Unit
Fuel cost per unit	C_F	[–]	₹/unit
Actual energy transmitted to wheel	E_{TW}	$E_{CF} \times (\in_{BEV}$ or $\in_{FCEV}$ or $\in_{PV}$ or $\in_{DV})$	kWh/lit or kWh/kg
Fuel requirement per km	F_R	E_{CV}/E_{TW}	lit/km or kg/km
Cost of fuel per km	F_C	$F_R \times C_F$	₹/km
Battery charger rating per kW	B_{CR}	[–]	kW
Battery capacity	B_C	[–]	kWh
Charging time	T_C	B_C/B_{CR}	hours
Battery price deduction per year	B_{PD}	[–]	%/year
Total cost of vehicle battery pack	C_{TBP}	$C_B \times B_C$	₹
Battery replacement cost after "N" years	C_{BR}	$C_{TBP} \times \left(1 - \frac{B_{PD}}{100}\right)^N$	₹
Total battery replacement cost per km	C_{TBR}	$C_{BR}/(L_V \times T_D)$	₹/year
Battery warranty	B_W	[–]	km
Loan interest	I	[–]	%
Loan tenure	N	[–]	years
Monthly EMI	C_{EMI}	$C_{PV} \times i \times \left(\frac{(1+i)^N}{(1+i)^N-1}\right)$	₹
The total amount paid with interest over the service period	C_{TI}	$12 \times Lv \times C_{PV} \times i \times \left(\frac{(1+i)^N}{(1+i)^N-1}\right)$	₹
Interest paid per km	C_{TIKM}	$(C_{TI} - C_{PVKM})/(L_V \times T_D)$	₹/km
FAME II subsidy for BEV	S_{BEV}	[–]	₹/kWh
Proposed subsidy for FCEV	S_{FCEV}	[–]	₹/kWh, ₹/kW
TCO of FCEV	TCO_{FCEV}	$C_{PVKM} + C_{TIKM} + C_I + C_{MKM} + F_r$	₹
TCO of BEV	TCO_{BEV}	$C_{PVKM} + C_{TIKM} + C_I + C_{MKM} + F_r + C_{TBR}$	₹
TCO of petrol vehicle	TCO_{PV}	$C_{PVKM} + C_{TIKM} + C_I + C_{MKM} + F_r$	₹
TCO of diesel vehicle	TCO_{DV}	$C_{PVKM} + C_{TIKM} + C_I + C_{MKM} + F_r$	₹

Table 7.1.3 Tank-to-wheel efficiency of vehicles.

	Efficiency (ϵ_V)	Units	References
$\epsilon_{FCEV} = \epsilon_{FC} \times \epsilon_M$	$0.36 = 0.45 \times 0.80$	[–]	[21]
$\epsilon_{BEV} = \epsilon_B \times \epsilon_M$	$0.76 = 0.95 \times 0.80$	[–]	[21]
ϵ_{PV}	0.18	[–]	[21]
ϵ_{DV}	0.202	[–]	[21]

The TCO analysis is based on all the major costs associated with owning a vehicle. The capital investment on the purchase of FCEVs is estimated based on specifications of the fuel cell system, battery pack, and hydrogen storage capacity required in equivalence to BEVs available in the Indian market of similar power rating.

$$TCO = C_{PVKM} + C_{TIKM} + C_I + C_{MKM} + F_C + C_{TBR} \tag{7.1.1}$$

$$C_{EMI} = C_{PV} \times i \times \left(\frac{(1+i)^N}{(1+i)^N - 1} \right) \tag{7.1.2}$$

$$C_{TI} = 12 \times Lv \times C_{PV} \times i \times \left(\frac{(1+i)^N}{(1+i)^N - 1} \right) \tag{7.1.3}$$

$$C_{TIKM} = (C_{TI} - C_{PVKM})/(L_V \times T_D) \tag{7.1.4}$$

$$C_I = (I \times C_{PV})/T_D \tag{7.1.5}$$

The cost of batteries is continuously declining year by year. From the year 1991 to 2021 in three decades, the price of batteries has significantly reduced by about 97% [18]. Hence, 10% price deduction per year is considered to get the battery replacement cost and calculated as,

$$C_{BR} = C_{TBP} \times \left(1 - \frac{B_{PD}}{100} \right)^N \tag{7.1.6}$$

$$C_{TBR} = C_{BR}/(L_v \times T_D) \tag{7.1.7}$$

The presented TCO model calculates the total operational cost of vehicles based on three sets of important parameters for each type of vehicle considered in this study: tank to wheel efficiency of drivetrain, as shown in Table 7.1.3; energy content of fuel per kg or liter, as shown in Table 7.1.4, and average electrical energy consumption by vehicles per km of distance traveled, as shown in Table 7.1.5. Not all the energy in the fuel is transmitted to the wheel because of drivetrain efficiencies. Hydrogen is a high energy content fuel but with a fuel cell drivetrain, it can transmit only 36% of energy

Table 7.1.4 Energy content for different fuels.

Fuel type	Energy content (E_{Cf})	Cost (C_f)	References
Hydrogen	33.6 kWh/kg	300 ₹/kg	[22]
Electricity	[–]	9 ₹/kWh	[–]
Petrol	9.3 kWh/L	105 ₹/L	[22]
Diesel	10 kWh/L	90 ₹/L	[22]

Table 7.1.5 Types of vehicles considered for the analysis and their respective average electrical consumptions.

Vehicle type	Average electrical consumption (E_{cv})	Units	References
Two-wheelers/scooters	0.033	kWh/km	[20]
Three-wheelers	0.061	kWh/km	[20]
Four-wheelers	0.106	kWh/km	[20]
Buses	1.00	kWh/km	[19]

to the wheel. However, it is still much higher than the efficiencies of ICEVs. The actual energy which is transmitted to the wheels is calculated as,

$$E_{TW} = E_{CF} \times \epsilon_V \tag{7.1.8}$$

The fuel consumption of the vehicle is calculated based on average electrical consumption in kWh/km. Average electrical consumption varies based on vehicle type, size, and capacity. For an Indian scenario, average electrical energy consumption is considered based on the values given in the literature [19,20].

$$F_R = E_{CV}/E_{TW} \tag{7.1.9}$$

$$F_C = F_R \times C_F \tag{7.1.10}$$

For effective comparison of TCO of FCEVs with vehicles already available in the Indian market, FCEVs have been compared to the same power rating BEVs with approximate hybridization of FC and battery-based power. General assumptions and major input parameters considered for TCO analysis are listed in Tables 7.1.6 and 7.1.7. The presented analysis estimates the TCO of vehicles for 1 lakh km annual average travel distance.

The plots obtained from the TCO analysis provide critical information about the cost competitiveness of FCEVs in comparison to other types of vehicles considered in the analysis.

Table 7.1.6 Assumptions made for total cost of ownership calculation.

Type of vehicles	Specifications	Two-wheeler	Three-wheeler	Four-wheeler	Buses
Fuel cell electric vehicles	Fuel cell stack (kW)	2	5	20	80
	Hydrogen storage capacity (kg)	0.5	2	3	33.6
	Battery pack (kWh)	0.15	0.3	10	50
	Fuel requirement (kg/km)	0.0027	0.0050	0.0070	0.083
Battery electric vehicles	Battery pack (kWh)	2.8	7.7	30.2	186
	Charger rating (kW)	3	5	15	50
	Energy requirement (kWh/km)	0.043	0.080	0.139	1.32
Petrol-powered vehicles	Fuel requirement (L/km)	0.020	0.036	0.063	[–]
Diesel-powered vehicles	Fuel requirement (L/km)	[–]	0.030	0.050	0.50

Table 7.1.7 General inputs of the total cost of ownership model.

Variables	Values	Units	References
L_v	12	Years	
C_B	12,000.00	₹/kWh	
C_{FC}	59,200.00	₹/kW	[23]
C_{BOP} (2 kW FC system)	308,950.00	₹/kW	[23]
C_{BOP} (5 kW FC system)	138,676.00	₹/kW	[23]
C_{BOP} (20 kW FC system)	53,539.00	₹/kW	[23]
C_{BOP} (80 kW FC system)	36,896.40	₹/kW	[23]
C_{HT}	103,600.00	₹/kg	
C_I	0.9%	of purchase price	[24]
B_{pd}	10%	% per year	
i	8%	% per year	
N	7	Years	
S_{BEV}	15,000 for light-duty vehicles 20,000 for heavy-duty vehicles	₹/kWh battery pack FAME-II subsidy	[25]
S_{FCEV}	40% of market price (considered)	₹/kW fuel cell stack ₹/kWh battery pack	

7.1.3 Results and discussion

In the adoption of electric vehicles, the subsidies given by the GoI are playing an important role. To date, more than 160,000 electric vehicles are already sold under the FAME India scheme phase II all over India. Many state governments have also announced subsidies for electric vehicles. In combination, these subsidies can go up to 60% of the purchase price of the vehicle. If a similar kind of subsidy is given to the FCEVs, it can bring the benefit of long travel range and shorter refueling time to the electric vehicles.

7.1.3.1 Two-wheelers/scooters

India is a huge market for two-wheelers. There are around 253 million motorcycles on Indian roads. From 1951 to 2019, the number of two-wheelers in India has constantly increased. In 2021, India sold around 151.19 lakhs two-wheelers. India has some of the largest two-wheeler manufacturers like Hero MotoCorp, Bajaj Auto, TVS Motor Company, Honda Motorcycle, Royal Enfield, Yamaha Motor, Mahindra Two Wheeler, etc. These manufacturers were only focusing on manufacturing ICEVs, but recently they have also come with electric two-wheelers for Indian roads. New companies like Ather Energy Pvt. Ltd., Ola Electric, and Revolt Motors have also come up with affordable and efficient electric two-wheelers for the Indian market.

Recent developments and subsidies given by the GoI on electric mobility have been partially successful to gain the attention of customers, especially of two-wheeled electric scooters. However, still, there is no development in the case of FC-based E2W vehicles. The major reason is the cost associated with FC systems, BOPs, etc., and hence FC-based E2W scooters are not an affordable option. FC-based E2W with 2 kW of FC stack and 0.15 kWh of the battery pack can cost up to 8–10 lakhs Indian rupees. Whereas conventional two-wheelers with the same rating are easily available for 80,000 Indian rupees.

Fig. 7.1.4 shows the TCO comparison of two-wheelers based on the fuel they are using. From the graph, battery-based E2W has the lowest TCO in the Indian context even when compared with petrol alternatives. The operation cost of battery-powered electric vehicles is very less. Also with incentives, E2W becomes the most affordable option. Under the FAME II scheme, purchase incentives for E2W are increased by 50%, that is, from 10,000 ₹/kWh to 15,000 ₹/kWh which contributes to up to 40% of ex-showroom price. As shown, FC-based E2W will be very costly and thus will not be comparable with electric or petrol-based two-wheelers even for long-range travel. In the current scenario, there is no scheme announced for FC-based E2W. If similar kinds of incentives are provided to the FC-based E2W, the purchase price of these vehicles will significantly reduce but still will not be comparable with battery-based E2W. However, FC-based E2W can provide a far longer range of travel with absolute zero greenhouse gas emissions.

7.1.3.2 Three-wheelers auto

India has a large fleet of three-wheelers, which are commonly passenger carriers or load carriers. In 2021, three-wheeler manufacturers sold around 2.16 lakhs vehicles in the Indian market. Bajaj Auto Limited, Mahindra & Mahindra Ltd., Atul Auto Ltd., TVS Motor Company Ltd. are some of the major players in three-wheeler manufacturing in India. Most of these vehicles, however, are based on IC engines. Along with these companies, Hero Electric Industries, Lohia Auto Industries, Saera Electric Auto Pvt. Ltd., Terra Motors, Kinetic Green Energy & Power Solutions Ltd., etc. now have their electric three-wheelers in the Indian market.

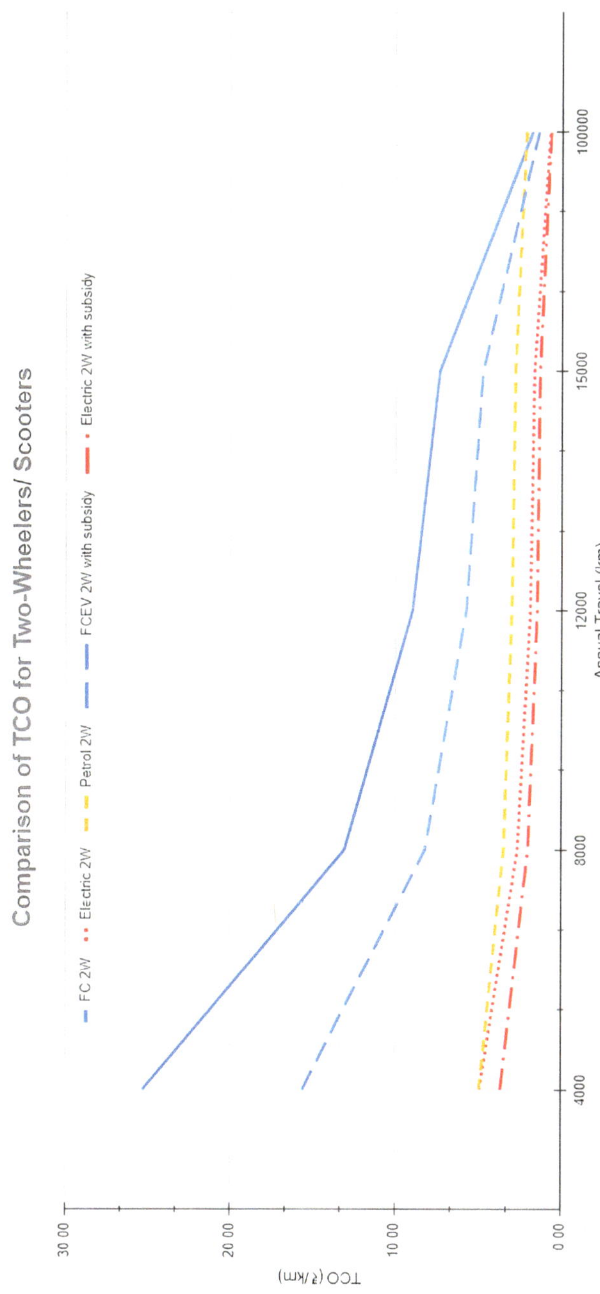

Figure 7.1.4 Total cost of ownership for two-wheelers/scooters.

Some Indian automobile manufacturers have started to work on the development of hydrogen-based three-wheelers. In the year 2012, Mahindra & Mahindra showcased its first hydrogen-based three-wheeler "HyAlfa" that was based on a hydrogen internal combustion engine and was developed under the project "DelHy 3W" [26]. Mahindra HyAlfa though was based on an IC engine can help to increase the public acceptance of hydrogen and thus can give a push to the development of fuel cell-based three-wheelers for the Indian market.

Fig. 7.1.5 shows the TCO comparison of three-wheelers. Electric and diesel-based three-wheelers come out as more affordable options as compared to petrol three-wheelers. Fuel cell three-wheelers for short travel ranges are not even comparable with battery-powered electric three-wheelers due to their high initial cost. In this scenario, the FAME subsidy by the Indian Government makes electric vehicles even more beneficial for passenger and load-carrying applications. If a similar kind of subsidy is announced for FC e3W, they can become comparable with their petrol alternatives for long-range of travel. But to attract the attention of the drivers, the cost of the FC system and BOP should be reduced drastically.

7.1.3.3 Four-wheeler passenger vehicles

As a part of climate change commitment, India wants to increase the sales of electric vehicles in the domestic market. The major automobile brands Tata Motors, Mahindra & Mahindra, Hyundai, Honda, and MG motor have launched their electric four-wheeler models in recent times. The major issue with the adoption of electric vehicles in India is its charging infrastructure. Tata Nexon eV is the most successful eV model in India. Tata has recorded the sales of 604 electric cars in July 2021 and these numbers are gradually increasing month by month. However, again these electric four-wheelers have a certain limitation of large time consumption in battery recharging. They have a limited range of travel and hence are not always suitable for taxis or intercity travel. On the contrary, FC-based electric vehicles can be the better alternative to battery-powered vehicles with their short refueling time and long range of travel. The intercity taxis can take advantage of hydrogen by opting for FC vehicles. For example, the distance between two Indian cities Delhi and Jaipur is around 280 km. By opting for FC electric vehicles, it becomes easy to travel to another city and come back on a single charge with a travel range of up to 600 km. This seems merely impossible in the current scenario with the battery-electric vehicle. Mirai is a model by Toyota

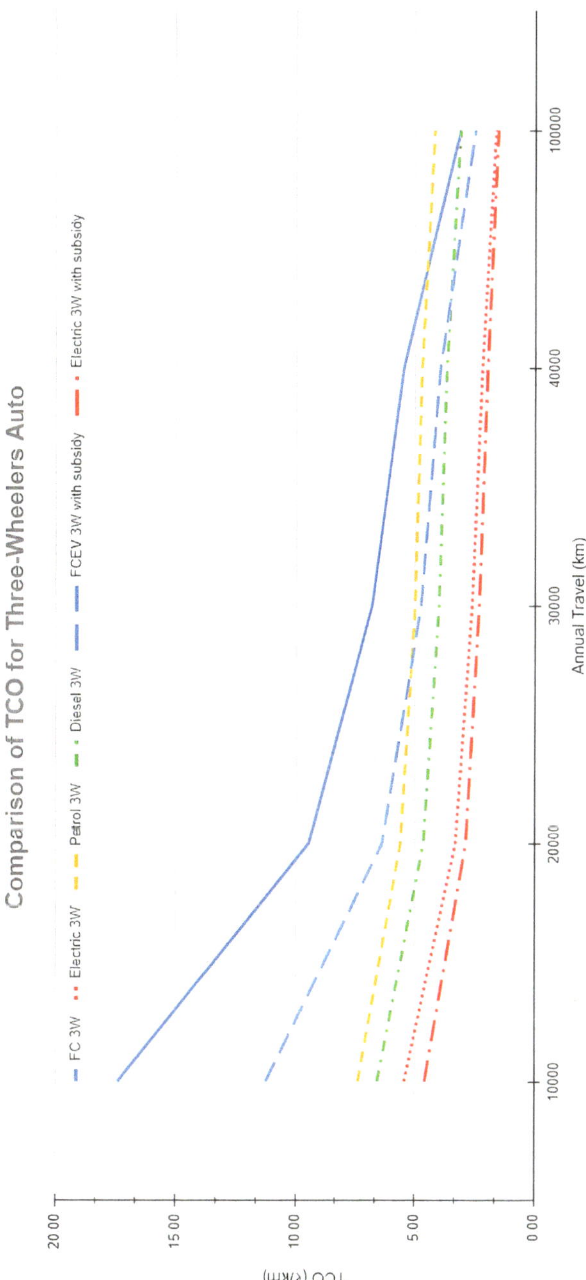

Figure 7.1.5 Total cost of ownership for three-wheelers.

Motor Company that is based on hydrogen fuel cells, and has a long range of travel up to 600 km which can be refueled within 5 min.

Fig. 7.1.6 shows that even after a higher purchase price than petrol and diesel vehicles, electric vehicles with FAME subsidies are the most affordable option. In the present scenario, the cost of the fuel cell stack and balance of the plant is very high, and hence, the purchase price of FCEVs is on the higher side as compared to conventional vehicles. However, if purchase incentives under the FAME subsidy are given, their TCO can be compatible with petrol and diesel vehicles in the longer run. With more than an annual travel distance of 50,000 km, FC-based electric vehicles become compatible with petrol alternatives. With research and developments, if the costs of the FC system and BOP drop significantly in the future, the purchase price of FCEVs can reduce significantly. This may lead to the mass acceptance of FC-based electric four-wheelers and will lead to low TCO even for a shorter range of travel.

7.1.3.4 Buses

Commercial medium and heavy-duty IC-engine-based vehicles heavily impact the environment. In India, diesel buses generate high-level of GHG emissions. Also, the increasing price of petroleum imposes a higher operational cost on diesel-based IC-engine vehicles. Buses are a very important means of both intracity and intercity public transport in India. These buses should impact as low as possible to the environment by eliminating the exhaust gasses. Battery electric buses are the best solution to the issue with low operating costs. But the issue with these buses is the long battery charging time, battery replacement period, and short range of travel. These buses can be used for intracity travel but require overnight charging. Hence, requires a larger fleet of battery-electric buses for intracity public transport.

Globally, the electrification of the bus fleet has already started. In 2017, Shenzhen city of China became the first city to electrify its bus fleet and then its taxis by 2019. The major reason behind this success was the subsidies given by both central and local governments that contribute around 60% of the total purchase cost and charging infrastructure development. When compared to diesel buses, with consideration of subsidies and warranty, the electric buses procured by Shenzhen Bus Group have 35% lower TCO than the diesel fleets [27]. Following a similar path, some Indian cities have started the electrification of bus fleets. For example, the state of Maharashtra

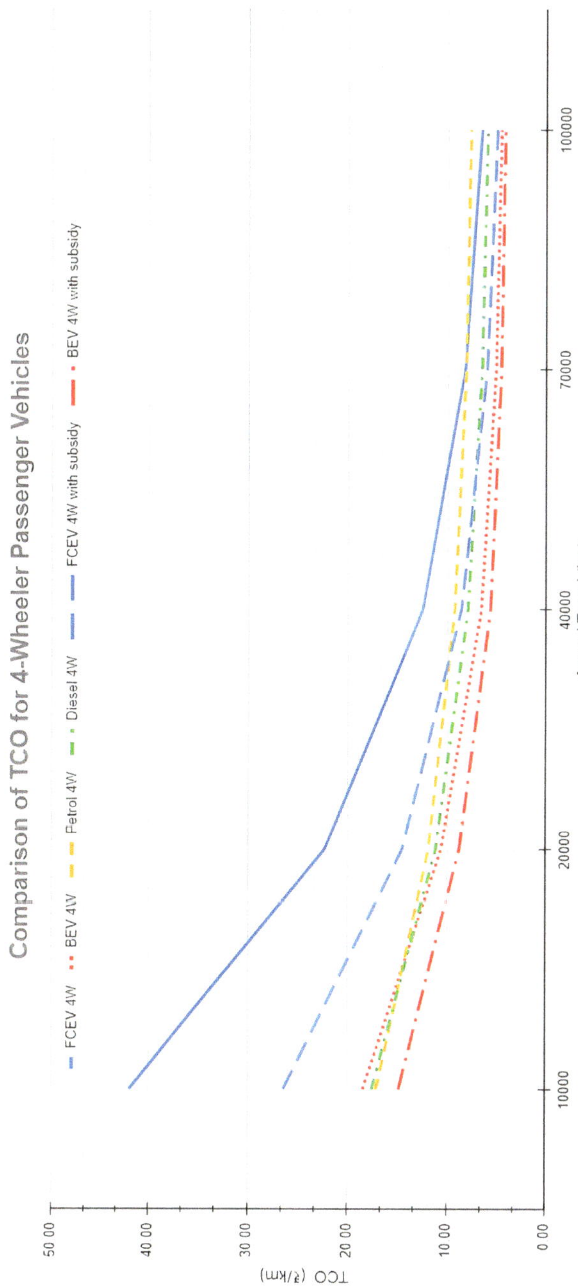

Figure 7.1.6 Total cost of ownership for four-wheeler passenger vehicles.

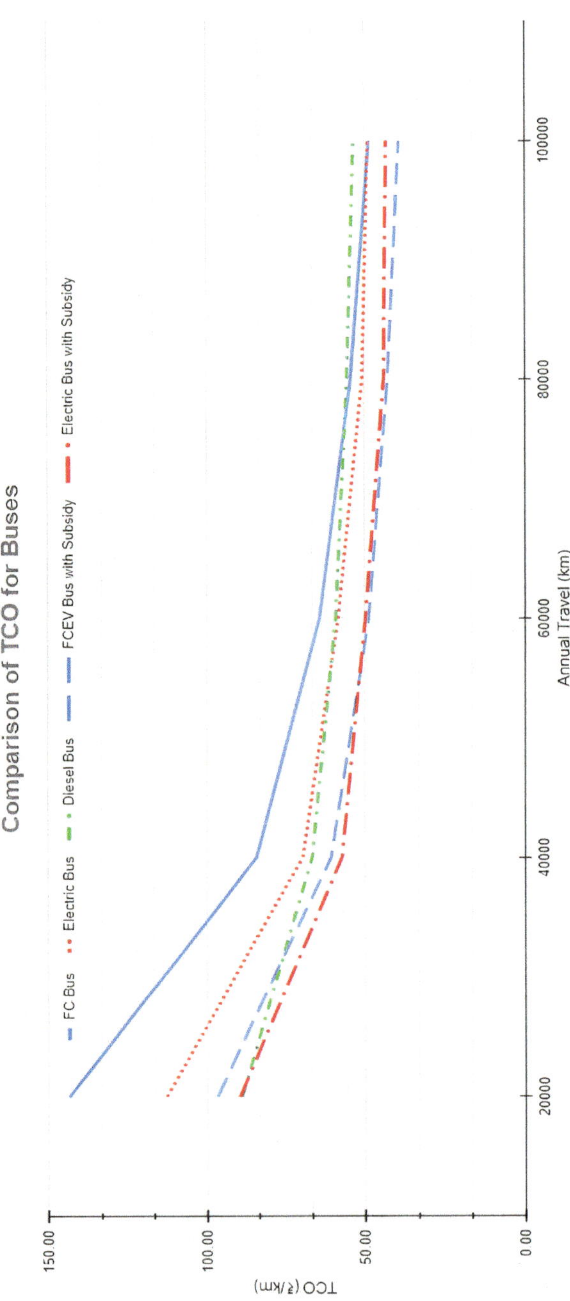

Figure 7.1.7 Total cost of ownership for buses.

has set the target to electrify 15% of public transportation by 2025. The Brihanmumbai Electric Supply & Transport Undertaking (BEST) has issued a tender for the procurement of 1900 electric buses, most of them for the city of Mumbai. Many other Indian cities including Bangalore, Kolkata, Hyderabad, Lucknow, Indore, Guwahati, etc. have floated such kinds of tenders for the procurement of electric buses. For the buses, subsidies up to 60% of the purchase price are available [28].

FC buses have lower operating costs than all other types of vehicles. However, their initial purchase cost is very high and hence suitable for medium and heavy-duty operations as well as for long-range travel. In addition, hydrogen-based FC electric vehicles have a very short refueling time and hence for both intracity and intercity travel, these hydrogen-powered buses can be the best alternative. The presented TCO analysis in Fig. 7.1.7 states that battery-powered electric vehicles are better when considering light-duty applications and short-range of travel. If the FAME subsidy is given to the fuel cell buses, the buses can be bought at significantly lower prices. These buses can give low ownership costs over diesel buses even at 40,000 annual distance travel. In the longer run, subsidized fuel cell buses will be having a lower cost of ownership than battery-powered electric buses. Fuel cell electric buses with a long range of travel of more than 500 km, short refueling time, and zero greenhouse gas emissions become an affordable option over all other types of buses. As BEVs take more time to charge the batteries and battery replacement cost is higher and hence are unaffordable in the long run. Whereas FC buses have a long range of travel and short refueling time, hence becoming a better option for heavy-duty applications. In the current scenario, the cost of fuel cell stack and balance of plants is much higher, if their cost reduces significantly FC buses will be better alternatives over other types of buses.

7.1.4 Conclusion

India has a fast-growing population and economy but has limited availability of fossil fuels to fulfill its energy demands. The consumption of fossil fuels is contributing to the heavy emission of greenhouse gasses. A large number of vehicles that are based on petroleum are the major reason for increased petroleum imports in India. To reduce the environmental pollution and petroleum imports in India, there is a need to look for an alternate fuel for the transport sector.

For the last decade, India is constantly focusing on growing its renewable energy capacity by taking advantage of its geography. But the flexibility of fossil fuels and high energy density is not available with renewables. However, hydrogen production integrated with renewables can bring this flexibility. In addition, by taking the advantage of its renewable energy scenario, India can produce and export green hydrogen to other nations. Green hydrogen when used with fuel cells can help India to significantly reduce its petroleum imports and environmental pollution. Various governmental and public agencies in India are investing in the research and development of hydrogen technology, particularly in the development of new materials, processes, systems, etc. India is one of the participants in mission innovation (MI) and working on MI's eight innovation challenges (IC8) that are focused on renewable and clean hydrogen [29]. Many Indian institutions and researchers are constantly working on improvement in hydrogen production, its storage, and transport which will in the future help to increase the utilization of hydrogen. Similarly, many organizations and researchers are working on developing more economical and efficient fuel cells to promote their commercial use for the transport sector. Indian organizations are also promoting many demonstration projects. Well-developed fuel cell technology and locally produced green hydrogen will be key players to decarbonize the Indian transport sector by replacing the current petroleum-based vehicle engines.

The demand incentives provided by the governments in the purchase of BEVs are playing a key role in their acceptance all over the world. FC-based electric two-wheelers and three-wheelers are not an affordable option over their other competitors mainly because of their high initial cost. However, as shown using TCO analysis in this work, FC-based four-wheeler passenger vehicles and buses can be a better alternative for heavy and long-range of travel even with the very high initial cost. The hydrogen-based FC electric buses can become a game-changer in the Indian public transport sector. If subsidies are given, the demand incentives under the FAME II scheme and subsidies by the local government in combination will reduce the purchase price of FCEVs significantly. This will make FCEVs more cost-competitive with all other types of vehicles and will help to again reduce the cost of ownership for these vehicles.

Acknowledgment

The authors would like to thank both Mr. Yash Tambi and Mr. Sudhir Sarawat whose regular support made this work possible.

References

[1] D. Workman, Crude oil imports by country 2019. WorldstopexportsCom 2020. http://www.worldstopexports.com/crude-oil-imports-by-country/. (Accessed 23 March 2021).

[2] United Nations Framework Convention on Climate Change (n.d.). The Paris Agreement. https://unfccc.int/process-and-meetings/the-paris-agreement.

[3] R.K. Shrivastava, S. Neeta, G. Geeta, Air pollution due to road transportation in India: a review on assessment and reduction strategies, Environ. Res. 8 (1) (2013) 69.

[4] S.K. Yadav, M. Sahay, A Study on Automobile Industry Growth in India and its Impact on Air Pollution. J. Adv. Res. Dynam. Control Syst. 2015(17):1991–9.

[5] K.J. Warner, G.A. Jones, The 21st century coal question: China, India, development, and climate change, Atmosphere (Basel) 10 (2019) 1–17.

[6] Climate Action Tracker (2020, December 17). Decarbonising the Indian transport sector: pathways and policies. https://climateactiontracker.org/publications/decarbonising-indian-transport-sector-pathways-and-policies/.

[7] GoI. Initiatives | Ministry of Road Transport & Highways, Government of India n.d. https://morth.nic.in/initiatives. (Accessed 9 November 2021).

[8] D. Fedrigo-Fazio, P. Ten Brink, S. Bassi, J. Emond, T. Lucas. Green economy: what do we mean by green economy. (United Nations Environment Programme) UNEP: Nairobi, Kenya. 2012.

[9] A.M. Tursunalievna, "Green" economy: the essence, Principl. Prosp. 6 (2017) 451–453.

[10] Ministry of Power and New & Renewable Energy (n.d.). Power Sector at a Glance All India. https://powermin.gov.in/en/content/power-sector-glance-all-india.

[11] N.C. Giri, R.C. Mohanty, P.D. Scholar, Accelerating India's energy sector to sustainable sources, Hum. Health Benefits Plant Bioact. Compd. 10 (2020) 18066–18076.

[12] N.K. Sharma, N. Tyagi, V. Rana, Renewable energy scenario in India: a current status, Int. J. Appl. Eng. Res. 14 (2019) 150–154.

[13] Indian Railways. Ministry of Railways (Railway Board) n.d. https://indianrailways.gov.in/railwayboard/view_section.jsp?lang=0&id=0,1,304,366,532. (Accessed 9 November 2021).

[14] I. Staffell, D. Scamman, V. Abad, P. Balcombe, P.E. Dodds, P. Ekins, et al., Environmental science the role of hydrogen and fuel cells in the global energy system, Energy Environ. Sci. 12 (2019) 463–491.

[15] B.K. Chaturvedi, A. Nautiyal, T.C. Kandpal, M. Yaqoot, Projected transition to electric vehicles in India and its impact on stakeholders, Energy Sustain. Dev. 66 (2022) 189–200.

[16] V.M. Medisetty, R. Kumar, M.H. Ahmadi, D.V.N. Vo, A.A.V. Ochoa, R. Solanki, Overview on the current status of hydrogen energy research and development in India, Chem. Eng. Technol. 43 (2020) 613–624. https://doi.org/10.1002/ceat.201900496.

[17] R. Rath, P. Kumar, S. Mohanty, S.K. Nayak, Recent advances, unsolved deficiencies, and future perspectives of hydrogen fuel cells in transportation and portable sectors, Int. J. Energy Res. 43 (2019) 8931–8955.

[18] Ritchie H. The price of batteries has declined by 97% in the last three decades: our world in data 2021. https://ourworldindata.org/battery-price-decline. (Accessed 19 June 2021).

[19] A. Vijaykumar, P. Kumar, P. Mulukutla, O. Agarwal (March 2021). Procurement of Electric Buses: Insights from Total Cost of Ownership (TCO) Analysis. WRI, Bangalore. https://wri-india.org/sites/default/files/WRI_EBus_Procurement_Commentary_FINAL_0.pdf.

[20] S. Saxena, A. Gopal, A. Phadke, Electrical consumption of two-, three- and four-wheel light-duty electric vehicles in India, Appl. Energy 115 (2014) 582–590.

[21] F. An, D. Santini, Assessing tank-to-wheel efficiencies of advanced technology vehicles, SAE Technical Paper (2003).

[22] U.S. Department of Energy (n.d.). Fuel Properties Comparison. https://afdc.energy.gov/fuels/properties.

[23] Battelle Memorial Institute, Manufacturing cost analysis of PEM fuel cell systems for 5- and 10-kW backup power applications, US Dep Energy, 2016, p. 124.

[24] I. Dincer, Green methods for hydrogen production, Int. J. Hydrogen Energy 37 (2011) 1954–1971. https://doi.org/10.1016/j.ijhydene.2011.03.173.

[25] Ministry of Heavy Industries and Public Enterprises Government of India. National Automotive Board (NAB) 2020. https://fame2.heavyindustry.gov.in/. (Accessed 13 June 2021).

[26] motoroids.com. HyAlfa, the world's first hydrogen fuelled 3-wheeler launched at Auto Expo Motoroids 2012. https://www.motoroids.com/news/hyalfa-the-worlds-first-hydrogen-fuelled-3-wheeler-launched-at-auto-expo/. (Accessed 23 November 2021).

[27] World Bank. 2021. Electrification of Public Transport: A Case Study of the Shenzhen Bus Group. © World Bank. https://doi.org/10.1596/35935. Creative Commons Attribution CC BY 3.0 IGO.

[28] Electric buses procurement in India: Indian cities got the viable rates 2017:1–14 by The International Association of Public Transport (UITP, from the https://en.wikipedia.org/wiki/French_language, French: Union Internationale des Transports Publics).

[29] Mission Innovation. IC8: renewable and clean hydrogen: mission innovation n.d. http://mission-innovation.net/our-work/innovation-challenges/renewable-and-clean-hydrogen/. (Accessed 9 November 2021).

Overview on application of hydrogen

Joydev Manna
Hydrogen Energy Division, National Institute of Solar Energy, Gwal Pahari, Gurugram, Haryana, India

7.2.1 Introduction

Jaques Charles (1783) utilized hydrogen for the first time in a hydrogen balloon to propel himself and a companion over 36 km at a height of around 550 m. Early in the 20th century, three significant discoveries based on hydrogen's chemical potential led to the widespread use of hydrogen. These three discoveries were as follows: (1) hydrogenation (in 1897, the French chemist Paul Sabatier discovered that the addition of hydrogen to carbon molecule in the presence of nickel is possible); (2) the Haber–Bosch process (in 1910, the two scientists discovered that it is possible to synthesize ammonia using nitrogen and hydrogen); and (3) hydrocracking (in the 1920s, it was discovered that lighter liquid fossil fuels can be produced by adding hydrogen to heavier liquid fossil fuels) [14]. As of now, hydrogen is mostly used in industrial sectors and a bit of hydrogen is used in transport sectors [3]. Although, it is envisaged that hydrogen could be used in other sectors such as power sectors, and for household applications such as heating, cooling, and cooking. In the transport sector, a minimum amount of hydrogen is used for fuel cell vehicle (FCV) applications in only a few countries. Therefore, the application of hydrogen can be categorized as:

(a) Use in the industrial sector
(b) Use in power sector
(c) Use in transport sector
(d) Use in household applications

The following section of this chapter elaborates on the above application of hydrogen.

7.2.2 Use of hydrogen in the industrial sector

Hydrogen is used in numerous industrial sectors for a wide range of purposes. Its chemical properties (such as its reducing properties, high

Towards Hydrogen Infrastructure: Advances and Challenges in Preparing for the Hydrogen Economy.
DOI: https://doi.org/10.1016/B978-0-323-95553-9.00006-6

401

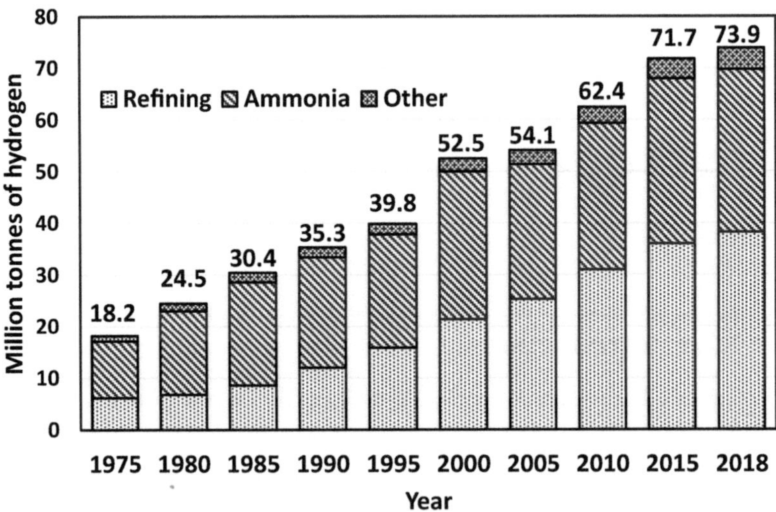

Figure 7.2.1 Global hydrogen demand in various industrial sectors during 1975–2018. *(From: IEA).*

reactivity, etc.) are what make up most of its uses. Most processes that use hydrogen as a reactant involve hydrogenation, which involves adding hydrogen to molecules or cleaving molecules, or removing heterogeneous elements (e.g., sulfur and nitrogen, etc.).

The chemical and petroleum industries employ hydrogen as a reactant in large quantities. Presently, hydrogen is considered an essential element in petrochemical, fertilizers, and some other industries. Since 1980, the demand for hydrogen gas has roughly tripled, reaching a peak of 120 million tons in 2018, of which two-thirds were pure hydrogen and one-third were a mixture of hydrogen and other gases. Growth of global hydrogen demand over a period of about four decades (1980–2018) is shown in Fig. 7.2.1. Refining industries consume around 51.7% of the total hydrogen uses, followed by ammonia production at 42.6% and others at around 5%. Several environmental restrictions such as changes in crude, environmental regulations, for example, sulfur limits for diesel, an emission limit of NOx and SOx, percentages of aromatic and light hydrocarbons in gasoline, etc., are likely to cause a significant growth in the utilization in petroleum processing and it has been estimated that hydrogen demand in refineries will grow by 7% and reach 41 MMT by 2030.

In ammonia synthesis, around 31 MMT of hydrogen was consumed in 2018. Most of the synthesized ammonia is used to produce fertilizers (urea, ammonium nitrate/phosphate, etc.). The vast majority of hydrogen is presently produced for on-site consumption by the industry. Steam reforming of methane (SMR) or partial oxidation (POX) of hydrocarbons followed by a water gas shift reaction are the two main methods of producing hydrogen for these applications [4]. The following sections give a brief account of the major applications of hydrogen by the different industries.

7.2.2.1 Petroleum refinery and petrochemical industry

Several industrial operations are carried out in petroleum refineries to turn crude oils into valuable products including jet fuel, diesel, petrol, kerosene, liquified petroleum gas (LPG), etc. The term "hydro-processing" is used to describe a variety of procedures that are carried out in petroleum refineries using hydrogen. Hydro-processing reactions can be divided into two types: hydrocracking and hydrotreating. Hydrocarbons (often heavy gas oils) are simultaneously cracked and hydrogenated in the hydrocracking process to provide refined fuels with smaller molecules and higher H/C ratios. High yields of kerosene and diesel are produced via hydrocracking. In the hydrotreating process, hydrogen is utilized to hydrogenate sulfur and nitrogen molecules, which ultimately results in their removal as H_2S and NH_3 gas. The sulfur removal process is also known as hydro-desulfurization (HDS). The petroleum refineries' need for hydrogen for hydro-processing has been rising continuously. The main causes of this are the increasingly stringent emission and product specification regulations, the rising demand for lighter, hydrogen-rich products like gasoline, and the necessity for refiners to increase their profit margins by processing crudes of lower quality. Broadly, hydrogen is used in refineries for the following processes:

(a) removal of sulfur, halides, oxygen, nitrogen, and metals,

(b) saturation of olefins and aromatics,

(c) isomerization,

(d) de-cyclization, and

(e) cracking.

The hydrogen generation unit (HGU) produces hydrogen in a refinery using steam reforming of natural gas (NG), LPG, naphtha, or by POX process. Some hydrogen is also generated in refineries as a byproduct of other operations. Hydrogen can also be supplied to refineries using a pipeline from some other source. The amount of hydrogen needed for a specific refinery process depends on the quality of the crude and the number of heteroatoms (sulfur, nitrogen, etc.) that need to be removed. A schematic of a typical HGU is shown in Fig 7.2.2.

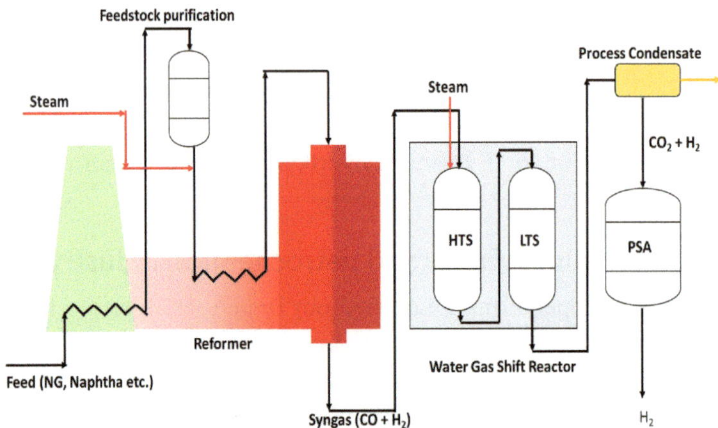

Figure 7.2.2 Schematic diagram of a typical hydrogen generation unit.

Hydrogen is extensively used in the petrochemical industry for the production of many chemicals such as acetic acid, butanediol, tetrahydrofuran, butyraldehyde, hexamethylene diamine, cyclohexane, etc. In the manufacture of polymers like polypropylene, hydrogen is also employed to regulate the molecular weight of the polymer. Recently hydrogen is also used in polymer recycling where molten plastic is hydrogenated to create lighter molecules.

7.2.2.2 Fertilizer industry

The fertilizer industry is heavily dependent on ammonia which is produced by reacting nitrogen and hydrogen. Haber–Bosch process is the name of the industrial method used to produce ammonia. In this method, high pressure (150–250 bar) and high temperature (400–500°C) conditions are used to infuse hydrogen and nitrogen in a 3:1 ratio into a reaction vessel containing a specific catalyst [7]. The most used catalyst for this process is iron (Fe) promoted potassium oxide, calcium oxide, silicon dioxide, and aluminum oxide (Al_2O_3). The mixed gases are often cycled through multiple catalyst beds with cooling in between to increase conversion efficiency. A schematic of the ammonia synthesis process sequence is shown in Fig 7.2.3. An overall conversion efficiency of 95–97% is achievable in the current ammonia production plants. Nitrogen is collected from the air by purifying it, and the hydrogen needed for ammonia manufacture is often produced from fossil fuels (NG or naphtha) through the reforming process. The generated ammonia is used to make nitrogen-based single-nutrient fertilizers like urea

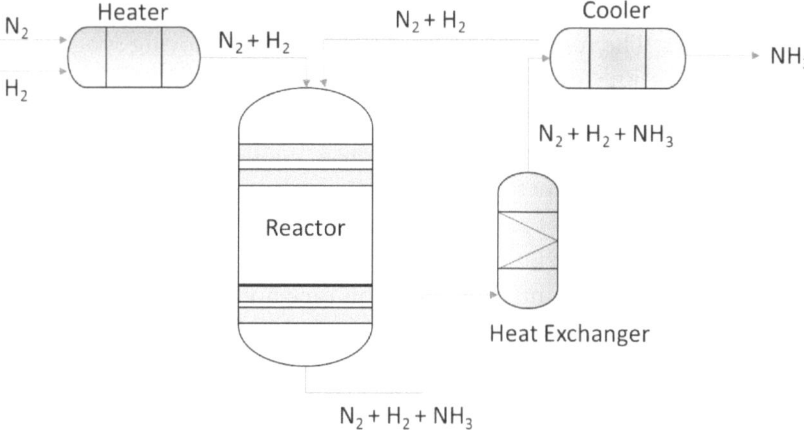

Figure 7.2.3 Schematic process sequence of ammonia synthesis process using Haber–Bosch method.

[CO(NH$_2$)$_2$] and other ammonium nitrate salts, as well as ammonium nitrate (NH$_4$NO$_3$). These fertilizers are further combined with phosphorus (P) and/or potassium (K) based fertilizers [11].

The Haber–Bosch ammonia synthesis method uses more than 1% of the energy produced globally and emits more than 400 MT of CO$_2$ every year. Ammonia output is expected to rise from its present level of 180 MT/year to 270 MT/year by the year 2050. Therefore, CO$_2$ emission will also increase with the increase in ammonia production. A long-sought scientific interest is in the process development for green ammonia production utilizing green hydrogen at mild temperature and pressure conditions. The production of green ammonia by electrochemical, photochemical, and chemical looping methods is more environmentally friendly but comes with additional technical difficulties that have not yet been resolved. Another viable option is to opt for carbon capture and sequestration (CCS) where CCS will be retrofitted into the existing ammonia plants as well as new ammonia synthesis plants will be designed with an integrated CCS facility.

7.2.2.3 Methanol production

Industrially, methanol is produced from syngas using a fixed bed reactor in presence of catalysts (copper on alumina and zinc oxide support). Feedstock used for syngas production is generally natural gas, coal, oil, or biomass.

In recent years, research is focused on the production of methanol by the direct reaction of hydrogen and carbon dioxide (CO$_2$). This is mainly attractive as it offers the removal of atmospheric CO_2 and turns it into a useful product. The methanol production method employing syngas differs from the procedure using pure CO_2 and H_2 in a number of ways. CO and CO_2, along with numerous additional light and heavy-weight co-products, are frequently entrained in syngas methanol synthesis processes together with the final methanol product. The co-products are the result of a more complex series of reactions that start when the catalytic surface and the three reactant gases interact. Another major challenge is to discover a suitable catalyst that can make the reaction thermodynamically favorable so that the energy required for this reaction could be reduced. Palladium and copper-based catalysts are found to be highly active towards this reaction and methanol could be produced at 180–250°C with a conversion efficiency of about 24%. The methanol production process could be expressed using the following three reactions (Eqs. 7.2.1–7.2.3):

$$CO_2 + 3H_2 = CH_3OH + H_2O \qquad \Delta H = -49.16 \, kJ/mol \, CO_2 \quad (7.2.1)$$

$$CO_2 + H_2 = CO + H_2O \qquad \Delta H = -41.22 \, kJ/mol \, CO_2 \quad (7.2.2)$$

$$CO_2 + 2H_2 = CH_3OH \qquad \Delta H = -89.4 \, kJ/mol \, CO_2 \quad (7.2.3)$$

Multiple sets of reactors are utilized to increase the conversion efficiency because the overall CO_2 to methanol conversion efficiency is low. For increased effectiveness, hydrogen is also recycled into the reactor. To completely eco-friendly methanol manufacturing, it is also anticipated to employ CCS.

7.2.2.4 Vanaspati production

Hydrogen has been utilized extensively to reduce the amount of unsaturation in fats and oils and to convert unsaturated fats to fully or partially saturated oils and fats. Hydrogenated vegetable oils such as vanaspati ghee are produced using this process.

The hydrogenation process is performed in a heated tank under vacuum at around 140–250°C. A process diagram used for this purpose is shown in Fig. 7.2.4. In general, palm oil, sunflower seed, or olive oil is used as oil feed for this process. Mechanical stirring is often applied to maintain a uniform temperature throughout the reaction chamber. A catalyst (mostly nickel-based solid catalyst) and the hydrogen gas (at 2.7–4 bar pressure) are introduced in the reaction chamber as the temperature stabilizes. As the hydrogenation reaction is exothermic, no additional heating is required and

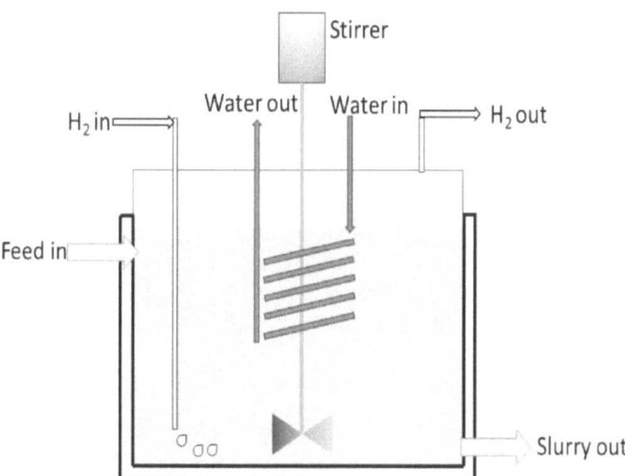

Figure 7.2.4 Hydrogenation process during vanaspati production.

as soon as the reaction starts, cooling is applied. The mechanical stirring process is continued till the end of the reaction. Finally, the hydrogenated oil mixture is extracted, and filtration methods are used to filter out the catalyst. Hydrogen is often used to improve the activity of the nickel catalyst as it may from surface oxide layers.

7.2.2.5 Metal industry

Industrially, tungsten is extracted from its ore (WO_3) using the hydrogen reduction method. Carbon is not used for the reduction of WO_3 as it can form stable tungsten carbide. In this process, an excess amount of hydrogen feed is passed through the powdered tungsten ore at high temperatures (550–850°C) as shown in Eq. (7.2.4). The generated steam is carried out with the excess hydrogen gas.

$$WO_3 + 3H_2 = W + 3H_2O \tag{7.2.4}$$

It has been envisaged that direct reduction of iron and other metals such as copper is possible by using hydrogen as a reducing agent instead of carbon. A Swedish steel plant (Luleå Steel) intends to implement a hydrogen reduction method for iron production. Trials of the pilot plant are on and are likely to be completed by 2024. The iron and steel industry contributes 5% of global CO_2 emissions and the use of hydrogen in place of carbon will provide avenues for decarbonization of this sector, which is very hard to decarbonize.

7.2.2.6 Float glass industry

Hydrogen is a common reducing agent used in the production of float or plate glass. Float glass is a unique kind of glass with a very smooth,

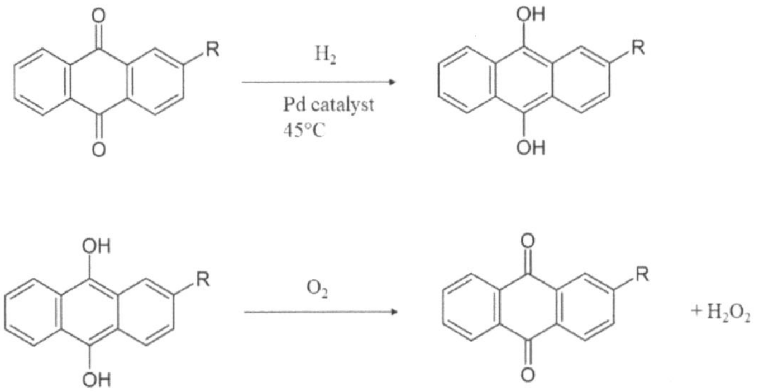

Figure 7.2.5 Anthraquinone process for industrial hydrogen peroxide production.

homogeneous structure, and superior optical qualities. This qualifies it for a wide range of uses, including glass windows, solar panels, LCD screens, and car windscreens. Float glass is made using a procedure that involves floating the liquid glass over a bed of liquid metal and letting it cool. Large glass panes with controlled thickness are produced successfully with this method. Molten glass at a temperature of about 1540°C is poured onto a molten tin bat, which serves as a level template for the glass to disperse over and eventually harden. The tin bath is kept in an environment composed of 90% nitrogen and 10% hydrogen to prevent the molten tin from oxidizing. Tin oxidation causes glass' surface to develop defects. By interacting with any oxygen, hydrogen ensures that the oxygen that causes tin to oxidize is removed. Huge quantities of nitrogen and hydrogen gases are needed to prevent tin from oxidizing.

7.2.2.7 Hydrogen peroxide production

Hydrogen peroxide (H_2O_2) is used in many applications such as (1) sterilizing agent in clinics and hospitals; (2) strong oxidizing agent in chemical synthesis; (3) bleaching agent for hair, clothes, and teeth; and (4) antiseptic to clean wounds, cuts, and other damaged tissue portions.

Industrially, hydrogen peroxide is manufactured using a cyclic process known as the anthraquinone process (Fig 7.2.5). Anthraquinone mainly acts as a hydrogen carrier and reacts with oxygen to produce hydrogen peroxide. This process consists of three steps: hydrogenation, filtration, and oxidation.

In the first step, hydrogen and anthraquinone react in presence of a palladium (Pd) catalyst at 45°C to form anthrahydroquinone. The hydrogenation stage is needed to be carefully controlled to avoid the formation of by-products due to the over-hydrogenation of the anthraquinone rings.

If any trace of Pd catalyst is present in the solution, it will decompose the produced H_2O_2 in later stages, and consequently, yields will be reduced. Therefore, In the second step, the catalyst is taken out.

At last, anthrahydroquinone is oxidized using air/oxygen, forming H_2O_2, and releasing the anthraquinone. The produced H_2O_2 is isolated from the solution using a liquid–liquid extraction column. Demineralized water (DM) is used for the extraction process. The generated crude H_2O_2 is then purified and distilled to increase the concentrations. Proper stabilizers are used to avoid unwanted decomposition.

Recent studies are focused on the synthesis of H_2O_2 directly from hydrogen and oxygen. Bimetallic materials consisting of palladium are found to be effective in the direct synthesis of H_2O_2 from hydrogen and oxygen.

7.2.2.8 Thermal power plant's generator cooling

In thermal power plants, electrical generators are cooled using hydrogen gas. Due to its low density, high thermal conductivity, high specific heat capacity, and cleaner-than-air nature, hydrogen offers many benefits as a coolant. In general, hydrogen used for power plant cooling is at a pressure of around 4 bar. Hydrogen absorbs heat from power generating winding enclosure. The amount of hydrogen used per day is a function of the capacity of the power generator and the condition of the hydrogen seals. The higher-capacity power generators require hydrogen to be maintained at high pressure for effective cooling. Higher pressure results in a higher leakage rate and thereby requiring replenishment of more hydrogen to maintain the pressure for optimal heat transfer. Another issue that is important is maintaining the purity of hydrogen above a particular level from safety considerations. The hydrogen required for cooling of generators could either be brought in cylinders or it could be produced on-site using electrolyzers.

7.2.2.9 Space applications

In space applications, hydrogen is being used as fuel for decades. National Aeronautics and Space Administration (NASA), Indian Space Research Organisation (ISRO), and other space agencies are using liquid hydrogen as a fuel for their spacecraft. In the spacecraft, mainly liquid hydrogen is used. However, for other applications such as flight acceptance tests, shock tunnel tests, meteorological balloons, and sintering processes compressed hydrogen gas (at around 120–300 bar pressure) is used.

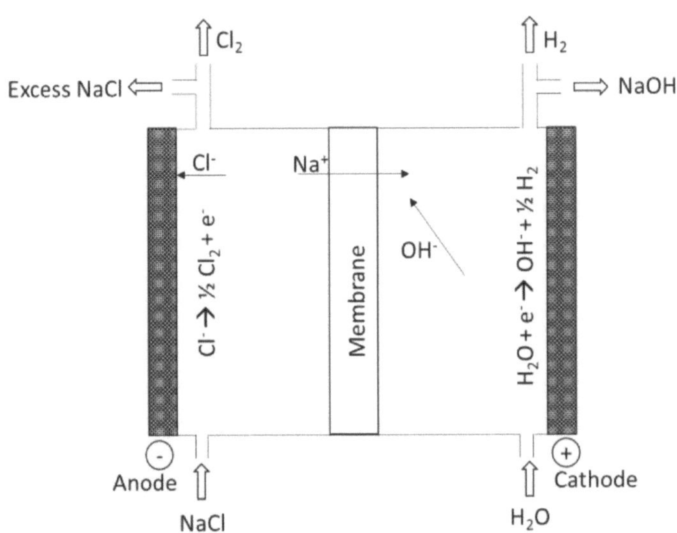

Figure 7.2.6 Schematic diagram of membrane cell used in chlor-alkali industry.

NASA has used hydrogen gas as a rocket fuel for years. The Centaur, Apollo, and space shuttle missions have given NASA extensive knowledge regarding the safe and efficient handling of hydrogen. For instance, each shuttle flight's rocket engines need around 500,000 gallons (around 1900 m^3) of cold liquid hydrogen, with storage boil off and transfer processes using an additional 239,000 gallons (around 900 m^3) of hydrogen.

7.2.2.10 Chlor-alkali industries

One of the oldest industries, chlor-alkali, is crucial in delivering chemicals to other manufacturing industries, including textiles, pulp and paper, alumina, detergents and soaps, pharmaceuticals, etc.

The primary product of the chlor-alkali industry is caustic soda, with chlorine and hydrogen being by-products in the ratio of 1.00:0.89:0.025. Caustic soda can be produced using any one of the three electrochemical cell technologies: (1) diaphragm cell, (2) mercury cell, and (3) membrane cell. The mercury cell technology has been phased out due to its low efficiency and environmentally pollutive nature. The membrane cell technology is the latest development of a clean and energy-efficient method of production of caustic soda. In a membrane cell, the brine solution is fed into the anode half of the system and chlorine is separated from the brine when the electric current is passed through the anode (Fig. 7.2.6). The caustic solution is dispersed along the cathode through the permeable membrane, and 30–35% caustic soda and hydrogen is obtained at the cathode shell.

Both hydrogen and chlorine are useful gases that have uses across a variety of industries. There are several industrial applications for chlorine, including the production of chlorine-based plastics like PVC, solvents like chloroform, and bulk commodities like bleached paper products. Applications for chlorine can be found in the dye, textile, medicinal, antiseptic, pesticide, and paint industries. Hydrochloric acid, chlorinated paraffin wax, organic and inorganic chemicals, chloromethane, vinyls (PVC), and the paper industry are all made with chlorine as well. PVC is one of the chlorine derivatives that has grown quickly, driven by numerous consumer industries. Almost 40% of the world's chlorine supply is used to produce PVC. On the other hand, most of the hydrogen generated in chlor-alkali industries is utilized to synthesize various downstream chemicals, such as Hydrochloric Acid (HCl), or as fuel in boilers.

7.2.2.11 Other uses of hydrogen in industrial sectors

In the semiconductor industry, hydrogen is employed as an oxygen scavenger. The devices are shielded from oxygen damage by hydrogen gas.

Hydrogen gas is also used in welding such as the welding of tungsten and refractory metals using the atomic hydrogen welding (AHW) procedure. Two metal tungsten electrodes are held in a hydrogen-shielding environment during this arc welding operation.

Hydrogen is used in a number of chemical analysis techniques, including gas chromatography and atomic absorption spectroscopy.

7.2.3 Power sector

The use of hydrogen in the power sector for the production of electricity using fuel cells, traditional generators, hydrogen-fired gas turbines, and combined-cycle gas turbines has enormous potential [5]. A gas mixture containing 3–5% hydrogen could be used in the gas turbine technology now in use. A successful demonstration of using a maximum of 30% hydrogen gas mixture in a modified gas turbine was performed by Mitsubishi Hitachi Power Systems. The power industry is highly confident that gas turbines entirely run-on hydrogen could be developed by 2030.

Fuel cells (FC) can be utilized to generate power using hydrogen. As of now the current global installed capacity of the fuel cell is around 1.6 GW

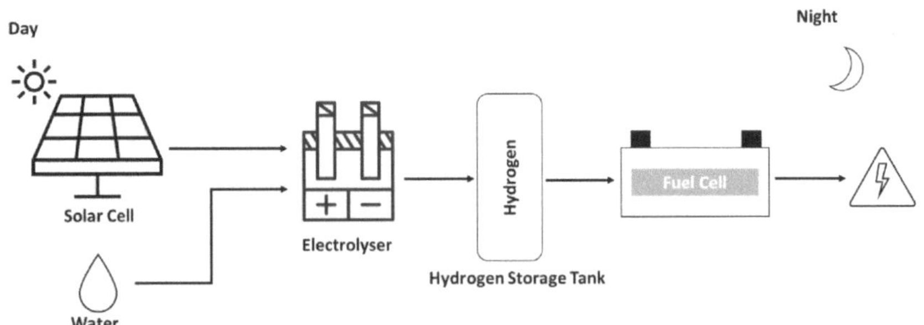

Figure 7.2.7 A typical hydrogen-based microgrid using solar photovoltaic, electrolyzer, and fuel cell.

which has been installed for stationary power generation. The United States, Japan, Germany, and Korea have installed the majority of the stationary FC units. By 2030, Japan wants to have 1 GW of hydrogen-based power generation capacity. Korea intends to have 15 GW of installed FC capacity by 2040. Hydrogen is an excellent energy storage medium with variable renewable power generation systems based on solar and wind and FC using stored hydrogen could be used for power generation, depending on needs. Just to mention the demand for hydrogen, if it were to be used for power generation too using fuel cell technology having efficiencies of about 50%, an aggregate capacity of 1 MW operating with 70% capacity utilization would need about 372 tons of hydrogen per annum. Hydrogen may serve the following application in power sectors:

7.2.3.1 Micro-grid application

As RE options, the idea of a microgrid based on green hydrogen appears to be highly appealing [1]. It will be efficient in mitigating the greenhouse gas (GHG) emission issue and pollution of fossil fuel-based thermal power plants. Among the several options, hydrogen production through electrolysis and use in fuel cells later present an alluring alternative to traditional systems for the production of electricity [17]. These microgrids can be utilized to electrify remote and rural places where it is challenging to provide electricity through a grid connection. A typical microgrid may look similar as shown in Fig 7.2.7. The rural economy is frequently viewed as having its foundation in rural electrification. Energy is needed in rural areas for things like cooking, lighting, irrigation, communication, heating, and other things which will also have positive effects on communication, economic growth, and improved health, education, and farming in rural communities.

7.2.3.2 Energy storage

Hydrogen-based energy storage has been gaining popularity recently because of the large energy storage capacity of hydrogen. It can meet the demand for energy storage throughout a wide timescale range, from short-term frequency regulation of the system to medium- and long-term (seasonal) energy supply and demand balance. In recent years, there has been a sharp rise in the generation of RE resources. Due to their intermittent nature, certain REs, however, rely on the time of year and the season. In order to endure the fluctuation and randomness of such RE sources, the produced electricity must be stored in a reliable manner. Batteries, compressed air energy storage, pumped hydro energy storage, among other options, are just a few of the many energy storage systems currently on the market. Hydrogen offers the advantages of having a huge energy storage capacity, a long storage time, and flexibility in contrast to these. It can absorb excess RE electricity while reducing energy volatility and uncertainty. It can be used to address energy time shifts and seasonal variation [18].

When the amount of electricity produced by RE sources is higher than the demand, the surplus amount could be stored as hydrogen, which can later be utilized to produce electricity or injected into the grid using fuel cells [8]. Particularly, the electricity generated during periods of low demand and low prices could be stored as hydrogen which will further be used to generate electricity during periods of high demand and high electricity prices, reaping the maximum benefits. Moreover, hydrogen can be kept for weeks or months at a time as opposed to the hourly or weekly storage of batteries. Generally, batteries are used in kWh to MWh scale to store energy, and therefore, one must increase the size of the battery systems to attain a higher storage capacity. In contrast, hydrogen storage capacity can reach up to MWh or even TWh due to its high energy density. Along with higher storage capacity, hydrogen can also be used to transfer RE from one place to another as renewable intensity varies with places [20].

7.2.3.3 Grid stabilization services

Electrolyzers and fuel cells can both contribute to ancillary services such as frequency regulation, voltage support, black start, congestion reduction, and lowering of negative price occurrences, etc. [19]. Hydrogen can be used to reduce transmission and distribution line congestion, which might happen as a result of insufficient line capacity, like other forms of energy storage.

For instance, this might occur if the capacity of the export line is exceeded by RE generation. Both RE curtailment and traffic congestion have the tendency to raise market pricing. When this kind of congestion happens, re-dispatching charges are also incurred. This will also assist minimize spending on new, pricey transmission and distribution infrastructure. The second kind of service involves making an effort to lessen the instances of marketplaces experiencing negative prices. Inflexibility in the generation, whether RE or conventional, is the primary cause of these negative pricing. One quick and effective way to boost demand, or a sort of negative generation, is to increase the use of electrolyzers. Reducing or stopping the power output of fuel cells may produce a similar outcome.

Frequency regulation keeps the grid frequency near to its reference value (50 or 60 Hz) by injecting or absorbing power and ensures that supply and demand are balanced. When frequency deviates from the nominal value in a relatively short period of time, frequency containment reserve (FCR), also known as the primary reserve, offers consistent containment of frequency. Frequency restoration reserve (FRR), which has a significantly bigger capacity, is responsible for handling frequency restoration for frequency variations lasting longer than the 30s. Fuel cells and electrolyzers can provide FCR and FRR services by varying their power set point in response to a frequency signal.

Voltage support is another service that hydrogen and fuel cells can provide. The power factor of fuel cells and electrolyzers can be modified based on the local requirements for voltage support similar to electronics-based converters. This is then accomplished by using an inverter or rectifier control to either generate or consume reactive power.

In the event of a blackout, a black start is an additional option. A power plant is normally restarted using a traditional generator, such as a diesel generator. Without any emissions or noise, a fuel cell can likewise accomplish this.

Although other energy storage systems such as battery could be used for these solutions, it has been envisaged that hydrogen has the advantage of having a large energy storage capacity and a quick response time to setpoint changes. Due to the high level of adaptability, hydrogen-based energy storage and power generation system have also the potential to generate more revenue. The majority of the study suggests that the hydrogen economy cannot be profitable without providing a range of services and selling hydrogen for use in other applications.

7.2.4 Transport sector

Currently, electric vehicles (EV) and hydrogen-fueled vehicles (HFV) are the only options with no GHG emissions at the use point as the energy carriers used for their operation can be generated using renewable sources [13]. HFV offer a similar driving range and lower refueling time as available with fossil fuel-based IC engine vehicles (ICEV). The potential of hydrogen in the transport sector is huge and different types of vehicles can be powered by either FC or hydrogen-based ICE technologies. It has been estimated that about 300 MT H_2/year will be needed if all current vehicles are completely replaced by hydrogen economy. Nevertheless, the high cost of hydrogen and FCs and technical/infrastructure challenges associated with hydrogen production, transportation, and storage are inhibiting the introduction of HFVs and their rapid market deployment. Hydrogen could also be used in the maritime sector. Around 2.5% of global energy-related CO_2 emissions are caused by maritime freight. The rail sector is mostly run by electricity and alternative fuels could be used instead of diesel in those portions of the tracks where electrification is difficult. Germany is already using two hydrogen FC trains which can travel 800 km/day on a single refueling and they intend to run 14 such trains by the end of 2021. Austria, France, Japan, and the United Kingdom have plans to deploy hydrogen trains by 2022. For the operation of material handling equipment like forklifts, hydrogen has already emerged as an attractive option in the United States. Hydrogen could also be used in the future for other rail yards and logistics hub machinery. The aviation sector is responsible for around 2.8% of global CO_2 emissions (2017) and it will increase further. Alternative fuels such as biofuels and hydrogen are among the best options to avoid an increase in emissions from this sector. Although feasibility tests and demonstrations have been performed for using hydrogen in small planes, significant research and development (R&D) on refueling, storage infrastructure, and aircraft design are needed to use 100% hydrogen-fueled engines in the aviation sector. Hydrogen-based liquid fuel could be used in the current aircraft with little or no modifications.

7.2.4.1 Status of hydrogen-fueled vehicles

Hydrogen-fueled vehicles can be broadly classified into the following two categories:

a) Internal combustion engine (ICE) based vehicles
b) Fuel cell vehicles (FCV)

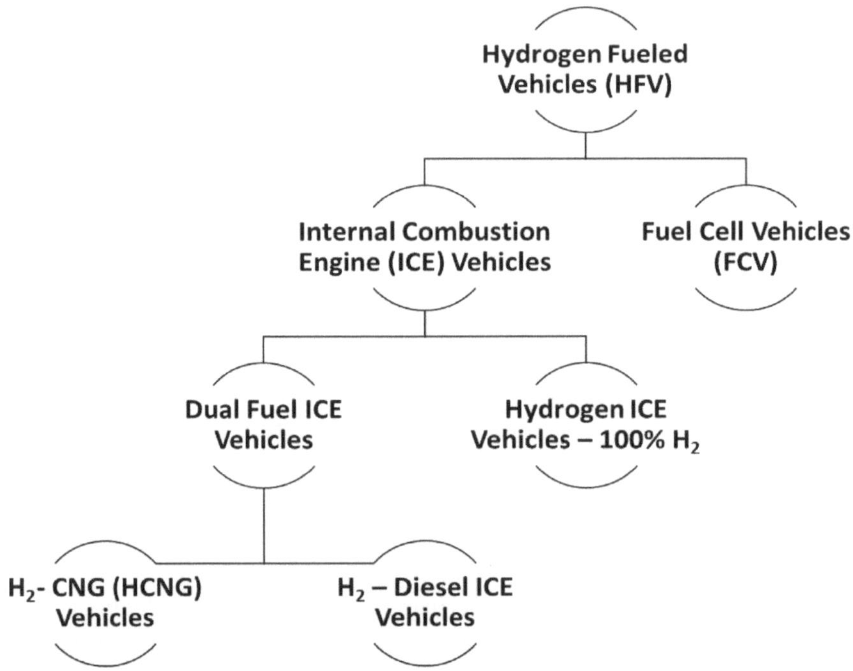

Figure 7.2.8 Classifications of hydrogen-fueled vehicles.

Hydrogen Internal Combustion Engine (HICE) vehicles can further be classified as (1) Dual fuel vehicles and (2) 100% hydrogen-fueled vehicles. In the dual fuel category, it could be either hydrogen–diesel or hydrogen–compressed natural gas (H–CNG), as shown in Fig. 7.2.8.

7.2.4.2 Global HFV scenario

Among different HFVs, FCVs have been preferred globally except for initial efforts made by BMW to develop HICE car some years back. The factors which are responsible for the growth of FCVs over HICE vehicles are high energy conversion efficiencies and no tailpipe emissions. FCVs at the international level have generally experienced the following four phases of development:

- Before 2000: designing the FCV concept, putting the fundamentals to the test, and validating it.
- 2000–2009: FCV R&D along with technology validation and demonstration.
- 2010–2014: early niche-market commercial deployment.

- After 2015: early commercialization (selling to private users in selected areas and introduction of long-haul trucks).

FCVs are EVs without the need for charging batteries. The refueling time of FCVs is about 5 min compared to several hours (in case of slow charging) for an EV. In recent years, some major global automobile companies have taken major initiatives for the commercialization of FC cars. Toyota Motor Corporation developed a Fuel Cell car "Mirai," with a driving range of about 500 km at the end of 2014. Mirai has onboard storage of about 5 kg of hydrogen in a Type IV cylinder at 700 bar pressure. Honda Motor Company and Hyundai Motor Company are two other active players in the FCV market. They each introduced commercial versions of their FCV, the Honda FCX Clarity, and Hyundai ix35 FCEV. Hyundai Motor Corporation unveiled the NEXO FCV in 2018. Apart from road transport, FCs have also been used for propelling an airplane and running a train (Fig. 7.2.9). A four-seat passenger plane named "HY4" launched its first official flight from Stuttgart Airport in September 2016. This was the first hydrogen FC-powered airplane ever built. In the same year, Alstom exhibited "Coradia iLint"—a zero-emission train at Berlin during a trade fair. Germany proposed to launch "Hydrail" soon in its market. Over the past five years, FC forklifts have significantly increased their market share in the US material handling industry.

With the sale of 4000 FCVs in 2018, the global stock reached almost 11,200 units by year-end 2018. The United States, Japan, and European Union are the major markets for fuel-cell cars. By 2030, China plans to deploy 10 lakh FC cars and South Korea has set a target of 6.3 lakh such cars. Japan plans to deploy 500 FCVs in the 2020 Olympics in Tokyo. The vehicles include Toyota "Mirai" and FC bus "Sora" (Fig. 7.2.10) and have set a target of 8 lakh FCVs by 2030. Globally 175 FC buses are operating out of which about 100 are in China [15]. About 28,000 FC-based forklifts are also operational globally. Internationally Toyota Mirai, Honda Clarity, and Hyundai Nexo are being sold in the United States, Europe, Japan, and Korea with average selling prices as given in Table 7.2.1.

Internationally, a typical FC bus costs about US$ 1.2 million. European studies indicate that the cost of FC bus may be in the range of Euro 625,000–1,000,000 depending on order size. Toyota Sora FC bus costs about US$ 945,000. There may be some variation in the specifications of FC buses developed and marketed elsewhere and the one that have been developed in India by Tata Motors Limited (TML). But in terms of initial capital cost, FC buses developed by TML appear to be quite competitive as the unit cost

Figure 7.2.9 Different transport vehicles based on fuel cell (A) HY4 aircraft (B) Hyundai NEXO (C) Coradia iLint, a zero-emission train developed by Alstom.

Figure 7.2.10 "Mirai"—a fuel cell car and "Sora"—a fuel cell bus developed by Toyota.

Table 7.2.1 Price and specifications of available FC cars in the market.

Manufacturers	Model	FC power output (kW)	H₂ storage Maximum pressure (bar)	Maximum weight of H₂ (kg)	Range (km)	Cost (US$)
Toyota	Mirai	114	700	~5	483	66,000
Honda	Clarity FC	103	700	~5.5	355	67,300
Hyundai	Nexo	95	–	~6.3	600	63,000

is in the range of 0.27–0.45 million US$, depending on the order size in comparison to the average cost of 0.06–0.15 million US$ in Europe and the United States.

Purely in terms of the unit cost of HFVs based on the very small order size up to 500 vehicles, these are not competitive vis-à-vis similar diesel and CNG vehicles that are available commercially in the country. Fig. 7.2.11 shows the relative cost of the different sub-systems of FC bus, developed by TML. The relative cost of the sub-systems that include these critical imported systems such as composite cylinders, solenoid valve, hydrogen

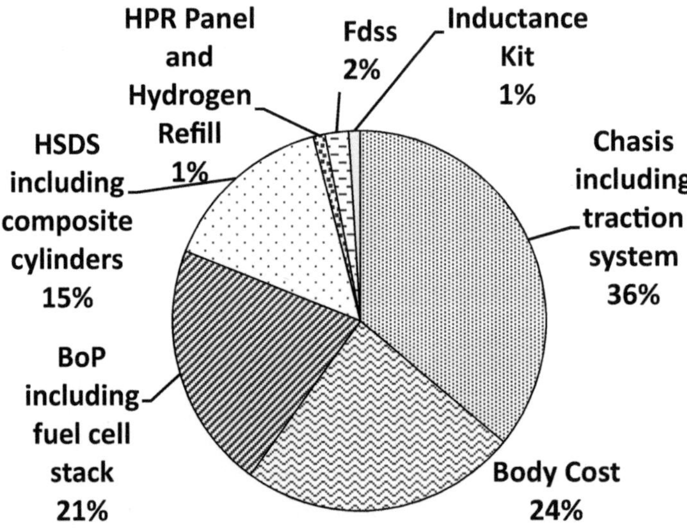

Figure 7.2.11 Relative cost of the different sub-systems of FC bus developed by Tata Motors Ltd. (*BoP*, balance of plant; *HSDS*, H_2 storage and delivery system). *(From: TML).*

injectors, FC stacks, traction motors, etc., as shown in Fig. 7.2.11 is quite significant. However, the potential for bringing down the unit cost of the FC bus may be quite significant if the critical components such as composite cylinders, solenoid valve, hydrogen injectors, FC stacks, traction motors, etc. are produced in the country and the demand for the vehicles can be created through regulatory measures and by providing fiscal and financial incentives like other environmentally sustainable and clean energy vehicle technologies.

For new energy carriers such as electricity and hydrogen that can be used in almost all sectors of the economy, it is better to compare their cost in terms of energy delivered per unit of money (MJ/USD) with other established fossil fuels. Fig. 7.2.12 shows this comparison. It is clear from the figure that the energy content of electricity per unit of money is comparable to petrol, diesel, and methanol. However, in the case of hydrogen depending on its cost (procured from different sources), energy content per unit of money could be more cost competitive compared to petrol, diesel, or methanol.

As per the calorific value of the fuel, the current fuel price, and the mileage of the vehicles that are now on the market or have been produced as prototypes, Table 7.2.2 shows the values for energy consumption and fuel cost for operating the vehicle for 1 km. The advantages of EVs in terms of energy consumption per km and minimal cost per km become quite evident. Certain hydrogen–powered vehicles (HFVs) might be competitive

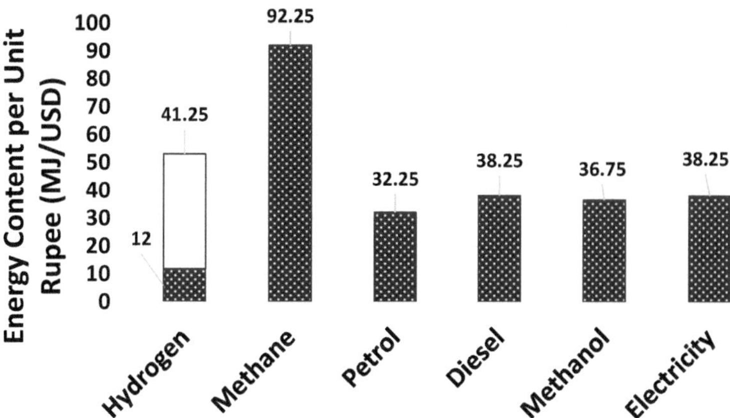

Figure 7.2.12 Comparison of energy content per unit money for different fuels.

Table 7.2.2 Energy consumption and fuel cost for operation of different vehicles for 1 km.

S. no.	Type of vehicle	Energy consumption (MJ/km)	Fuel cost (U.S. cents/km)
1	3-Wheeler—H_2 ICE	1.84	4.15–17.65
2	3-wheeler electric	0.18	0.47
3	Electric car	0.33	0.86
4	Minibus—H_2 ICE	4.80	10.80–45.86
5	FC bus	8.0	18.0–76.4
7	Large bus—CNG	26.32	28.82
8	9 m electric bus	3.6	9.44
13	12 m clcctric bus	4.86	12.76

(*From:* Nouni et al. [12] with permission from Elsevier.)

with comparable fossil-fueled vehicles where hydrogen may be easily accessible and priced competitively at around US\$ 2.5 to 3/kg of H_2. In comparison to low-entry diesel and CNG buses, electric buses are the best in terms of energy consumption per km of vehicle operation. FC buses could only be a good option when hydrogen is priced between US\$2.5–3/kg. However, diesel-fueled ICE vehicles are better compared to HICE vehicles in view of the higher brake thermal efficiency of diesel engines compared to hydrogen engines.

7.2.4.3 Hydrogen infrastructure for transport sector

This section aims to provide a broad understanding of each part of the hydrogen infrastructure needed to refuel HFVs. A hydrogen refueling station

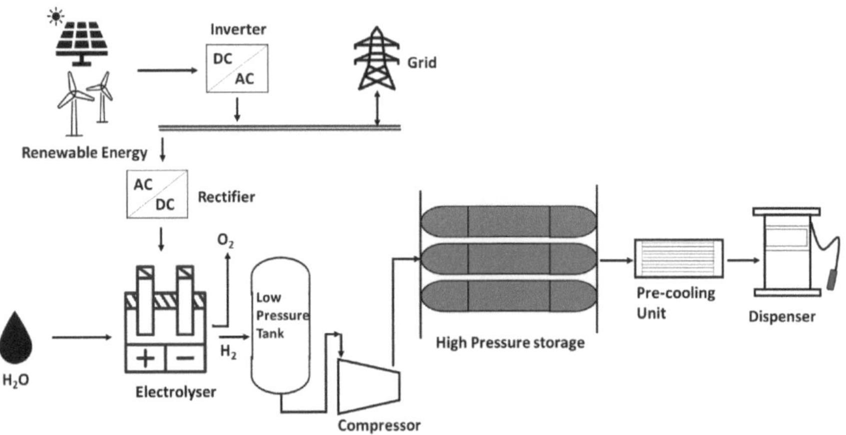

Figure 7.2.13 Schematic diagram of an on-site hydrogen production cum refueling station.

(HRS) can be classified as either on-site or off-site depending on the source of hydrogen. In an off-site HRS, hydrogen is transported from the location where it is produced, whereas in an on-site HRS, hydrogen is generated in the HRS site. The on-site HRS will have an energy source, such as fossil fuels or RE systems like solar, wind, biomass, etc., that may be utilized to produce hydrogen. The method most typically employed at on-site HRS to create hydrogen is water electrolysis. On-grid HRS refers to the station that is connected to the electrical grid, and off-grid HRS or stand-alone HRS refers to the station that is not connected to grid. Fig. 7.2.13 depicts a schematic of an on-site HRS. Electric control units, electrolyzer, and other equipment can all be mounted in a common container. The station gains modularity as a result, which makes it simpler to install, expand, or remove. The capacity of hydrogen production could be increased in the future, depending on the increase in hydrogen demand due to modular design. The additional electrolyzer units could be integrated with HRS. Compressor, buffer storage, and high-pressure storage tanks will be installed in the open within the boundaries of HRS to get rid of any emergency during operation.

Hydrogen used in off-site HRS is created in a separate industrial setting, such as a petroleum refinery, chlor-alkali plant, or in the electrolyzer, and then it is transferred to the HRS through a pipeline, tube-trailer/skid-mounted cylinder bank, or ship. These facilities accept hydrogen at low pressure, compress it, and cool it before transferring it to the vehicle. A schematic of this type of HRS is shown in Fig. 7.2.14.

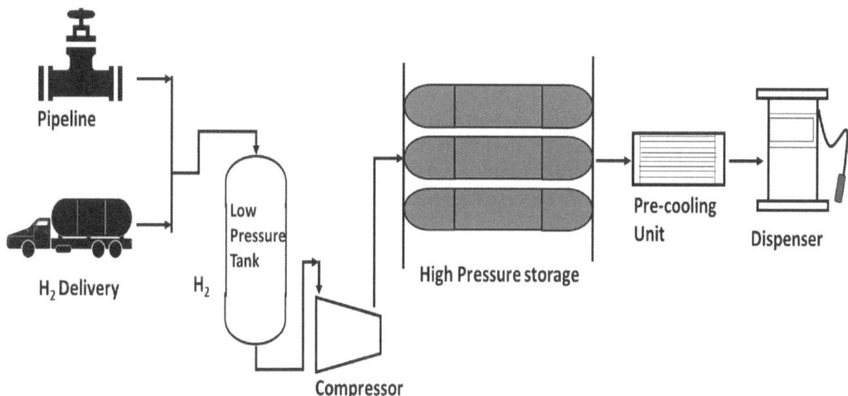

Figure 7.2.14 Schematic diagram of a hydrogen refueling station with delivered hydrogen.

An overall HRS will need a compressor, some storage tank, and a chiller system other than the power supply unit and HGU such as an electrolyzer. Brief on these components is described in the following section.

(a) Power supply

Hydrogen production and associated processes of an HRS are highly energy intensive. Electricity is needed for the operation of the electrolyzer, where hydrogen is produced by splitting water, compression of hydrogen to raise its pressure as hydrogen is the lightest element and for storing enough hydrogen on-board for providing sufficient driving range to the vehicles between refueling it needs to be stored at very high pressure, and cooling along with the operation of other electrical systems of the HRS. The electricity requirement for these processes (electrolysis, compression, and cooling, etc.) can be met by electricity generated by the RE plant (e.g., SPV, wind, biomass, etc.) or from the grid or both RE power generating unit and grid. Grid electricity will have to be necessarily needed if it is decided to operate the electrolyzer at higher capacity utilization factors. However, stand-alone or off-grid electrolysis would be the only option for green hydrogen production. The main components of an SPV power plant are the photovoltaic array and power conditioning unit (inverter and charge controller). PV array produces direct current (DC) electricity from solar energy. To generate the most electricity, the charge controller adjusts the voltage and current of the panels. If the electricity is not used right away, it is then transformed into alternating current (AC) via an inverter and used by AC loads or fed into the power grid. In these situations, the power grid acts as a fictitious battery by delivering electricity during periods of low generation and absorbing generation surpluses during periods of high generation.

(b) Electrolyzer

Although hydrocarbons are now the primary feedstock used to produce hydrogen, there is a rising awareness of the need to employ more RE in the production of green hydrogen. The most efficient way for green hydrogen production is by splitting water through an electrolysis process. Alkaline electrolyzers (AE), proton exchange membrane electrolyzers (PEME), anion exchange membrane electrolyzer (AEME), and solid oxide electrolyzers (SOE) are available options for water electrolysis. Water can be electrochemically converted into H_2 and O_2 in an electrolyzer by applying an external voltage as per Eq. (7.2.5).

$$2H_2O \rightarrow 2H_2 + O_2 \qquad (7.2.5)$$

The type of electrolyzer, materials used in the electrolyzer, and thermodynamic working parameters, such as temperature and pressure, affect the energy consumption or efficiency for hydrogen generation. Amongst different electrolyzer technologies, AEs are the most mature technology for hydrogen production. They are more durable, with 80,000 hours of stack lifetime. Whereas PEME have about 65,000 hours stack lifetime. Since the 1920s, AE has been the dominant water electrolysis technology and is frequently employed for extensive industrial applications. Due to the nonuse of noble metals and reasonably established stack components, AE systems are easily accessible, robust, and exhibit comparatively cheap capital costs. Low operating pressure and current density, however, have a detrimental impact on system size and hydrogen production costs in AE technology. Economies of scale are most likely to be the driving force behind future cost reductions. The parts of electrolyzers can be split into two categories: the stack and the balance of the plant (BoP). The stack includes hydrogen generation as well as a purification and cooling unit. The BoP includes wiring, a control system (PLC), a water treatment unit, a chiller, and a rectifier. The rectifier is required to supply DC current to the electrolyzer. An additional uninterruptible power supply (UPS) unit is incorporated to withstand sudden power supply loss. The entire electrolyzer system is typically contained in a single container. This container is connected to the power and water supply. The water supply is connected to the water treatment system, where it is modified and DM is produced and stored in a tank. DM water is pumped from the tank to another tank where alkali (KOH) is dosed to make the electrolyte solution. In the case of PEME, KOH is not used. The electrolyte solution is pumped to the electrolyzer, where it dissociates into hydrogen and oxygen using an applied voltage. Generated oxygen is usually vented out and hydrogen is collected in the low-pressure vessel. The pressure at which hydrogen exits the electrolyzer is between 5 bars and 10 bars.

(c) Compressor

Hydrogen storage in hydrogen-fueled vehicles is a challenging task. Since hydrogen is the lightest molecule, hydrogen gas has the lowest density: at room temperature and atmospheric pressure, 1 kg of hydrogen gas takes up of about 11 m^3. Hence, an increase in storage density is required for hydrogen storage to be economically viable. Hydrogen could be compressed using a compressor to increase the storage density. FCV and buses have onboard hydrogen storage capability of 700 bar and 350 bar pressure, respectively, to provide a sufficient driving range. Generally, reciprocating compressors are used for hydrogen compression. Sometimes, centrifugal compressors are also used. They raise the hydrogen pressure from the low pressure to the necessary high pressure. During compression, the temperature is increased and it is cooled using a closed-loop cooling system. A cooling pump with a typical capacity of 3/4 HP is used to circulate a water–propylene glycol mixture which is used as a coolant. Finally, high-pressure storage tanks can be used to store pressurized hydrogen.

(d) Storage tanks

In general, hydrogen is stored after production/delivery at relatively low pressure and then compressed to high pressure such that it can be successfully delivered to the vehicle's tank. Therefore, two types of tanks are used to store hydrogen at two different pressures. A buffer hydrogen vessel is used to ensure a constant gas flow to the inlet of the compressor. The buffer storage tank will have the same hydrogen pressure as the electrolyzer discharge pressure. The high-pressure tank will be connected to the compressor discharge point, and it ensures a proper supply of hydrogen to the vehicle tank through the dispenser.

(e) Dispenser and precooling unit (PCU)

A dispenser with an accurate connecting hose is necessary to refuel a vehicle. A specially designed dispenser is used for hydrogen refueling. Two types of hoses are available namely, H35 and H70 to refuel a vehicle with 350 and 700 bar pressures, respectively. Precooling unit is not necessary for 350 bar refueling. However, during 700 bar refueling the temperature of the gas inside vehicle tanks increases due to the Joule–Thomson effect. It is recommended that while refueling, the hydrogen temperature in the storage tank of vehicle should not rise above 85°C. Direct refueling may cause the gas to overheat depending on the initial temperature, so hydrogen must be cooled before refueling. A significant amount of energy is needed to precool the hydrogen before refueling. The dispensing unit does not consume any power when it is not dispensing any hydrogen.

7.2.4.3.1 Cost for HRS system

Compression cost mainly depends on the power consumption of the system. The compressor's output pressure, flow rate, and inlet pressure all affect its power consumption. Cost can increase with increase in the operating pressure. Efficiency and cost both are higher for reciprocating compressor compared to a similar centrifugal compressor. Cost of compressor could vary in the range US\$ 650–6600/kW depending on the size of the compressor. An establishment cost is needed for the first time for a compression unit. Another cost component is concerning the cost of the power input (in kWh) to compress 1 kg of H_2. Energy consumed for compressing hydrogen up to 400 bar to allow dispensing at 350 bar is estimated in the range of 2–4 kWh/kg. The operation and maintenance cost for gauge checks, lubrication and inspections is around US\$ 2000–3000/year [6].

Type III and IV hydrogen storage tanks are suitable for on-board hydrogen storage in fuel cell buses, trucks, and HFVs. In these cylinders, expensive carbon-fiber composite material is used to provide strength to withstand the high pressure of the hydrogen gas. These cylinders are 70% lighter than steel cylinders, safe, and durable with a lifespan of around 20 years [9]. However, the use of carbon fiber composites increases the cost of hydrogen storage tanks.

In Japan, around 100 HRSs have been installed to refuel 2000 numbers of FC cars and two hydrogen buses as of 2018. The initial capital expenditure (CAPEX) for HRSs were around US\$ 4.5–5.4 million with an operational expense of around US\$ 450,000. The Japanese government has allocated around US\$ 106 million for the development of hydrogen infrastructure along with additional subsidies on FCVs. The government has provided around 50% incentives on the CAPEX of HRS and consequently, the cost of installation of HRS became around US\$ 2.5 million [10]. In North America and Europe, a number of HRSs have been installed by various developers (Table 7.2.3). It can be noted that the cost of HRS is highly dependent on the average daily capacity as well as station technology. The cost of generated hydrogen may also vary with the station development cost. It is in the range of 10–33 US\$/kg of H_2. In the United Kingdom and Japan delivered cost of hydrogen was reported as US\$ 10–13/kg of H_2 in the recent past.

Table 7.2.3 List of developed hydrogen refueling stations in the United States with their cost and specifications [6].

Station developer	Number of station funded	Station technology	Average capacity (kg H$_2$/day)	Total cost (million US\$)
First element	19	Delivered H$_2$ gas	180	2.05
Air products	10	Delivered H$_2$ gas	180	1.93
Linde	7	Delivered liquid H$_2$	350	2.78
HyGen	3	Onsite electrolysis	130	3.25
Air Liquide	2	Delivered H$_2$ gas	180	3.26
ITM power	1	Onsite electrolysis	100	2.73
H$_2$ frontier	1	Onsite electrolysis	100	4.61
HTEC	1	Onsite electrolysis + delivered H$_2$ gas	140	3.25
Ontario CNG	1	Onsite electrolysis	100	2.51

7.2.5 Household applications

Natural gas is the current source for cooking and heating in buildings and a huge amount of gas is consumed every year. For use in buildings, hydrogen might be included into the current natural gas infrastructure. Local district energy networks could also use hydrogen for cooling or heating their building networks. In fact, the complete use of hydrogen in household applications such as cooking, heating, cooling, etc. could significantly reduce environmental impact, caused by current use of fossil fuels (mostly NG). Different type of systems using hydrogen for space heating could be adopted. The usual choices include a hydrogen-powered boiler, a hydrogen-powered fuel cell with an additional hydrogen boiler for cold periods, a hybrid heat pump with 100% renewable energy and an additional hydrogen boiler for cold spells, or any other combination of these systems [16]. Many studies have been performed for the space heating using various options and it has been reported that the cost of heating would be quite higher than the current NG-based heating system [2]. This is mainly due to the current cost of hydrogen production (electrolyzer system), the higher cost of fuel cells as well as higher cost of RE electricity.

7.2.6 Conclusion

Rapid industrialization at the global level has not only created huge demand for energy sources but has also contributed to climate change and global warming. Growing energy demand has largely been met by increasing dependence on fossil fuels, which are principal sources of increasing CO_2 emissions. Nearly 70% of greenhouse gases are contributed by thermal power plants, transport sector, construction, and fugitive emissions. Realizing the consequences of environmental degradation and deteriorating air quality across a vast geographical stretch globally, there is increasing awareness to undertake concrete actions to mitigate them. With a view to decelerate the growth of fossil fuels, several measures have been taken in the last few decades, including the development and deployment of renewable energy sources that also help in ensuring energy security in countries that are dependent on imported energy sources. Renewable energy can provide a sustainable solution for mitigating climate change and worsening air quality-related issues, apart from reducing dependence on hydrocarbons and thereby offering energy security. For decarbonizing the industrial and transport sector, energy carriers like hydrogen are viewed as promising option.

Currently, the industrial sector is the main consumer of hydrogen. The two industries that use hydrogen the most widely are those that produce ammonia and refine petroleum. Hydrogen is used to produce lighter fuels such as diesel, gasoline, and other petroleum products through hydrocracking and hydro-treating processes of heavier fuels obtained during the refining of crude oil. Hydrogen is also used for sulfur removal in petroleum refineries. Other industries such as metal extraction, production of float glass, steel, methanol, synthetic fuel, cooling of generator, food industries, semiconductor production, and spacecraft propellant also uses hydrogen gas. In food industries, it is used for converting unsaturated fats like butter, margarine, etc. to saturated fats. Hydrogen has been widely used in the semiconductor industries for manufacturing LEDs, photovoltaic panels, and other electronic components due to its efficient etching and reducing properties. However, most of the hydrogen used in the industrial sector is produced from fossil fuels such as NG, coal, etc. The hydrogen production methods such as reforming, and gasification produce a huge amount of GHGs and are harmful to our environment.

Hydrogen is an attractive energy vector for transport applications as it is a clean, better alternative of EVs and secures energy dependency. Extensive R&D efforts for the development of both ICE as well as FCV technologies over the last several decades have led to the introduction of FCV in recent years. Hydrogen fuel cells have been used in different kinds of vehicles for roads, rails, waterways, and airways. Fuel cell-based forklifts have made a favorable impact and are considered a better option compared to electric mobility for long-haul vehicles.

In the last few years, a resurgence in hydrogen-related activities, especially for green hydrogen production and its utilization across the different sectors of the economies in many countries has been witnessed. Hydrogen can be produced in a large scale without any emission by electrolysis process using RE electricity from solar PV, wind, etc. Hydrogen produced using RE is considered a clean energy carrier as the by-product is only water after burning it in the air or using it in fuel cells for electricity generation. With the increasing penetration of variable RE-based power-generating sources such as solar and wind in the energy mix of various countries, the global discourse is focused on decarbonization by reducing the consumption of hydrocarbons and substituting them by green hydrogen. It has been envisaged to replace hydrogen production methods in industrial sectors and to use green hydrogen. Other sector such as power, household, and transport sectors could also use hydrogen to replace fossil fuels and are considered as emerging options for using green hydrogen.

In power sector, hydrogen has been envisaged as an energy storage medium for long term to balance the power grid. Renewable energy-based hydrogen generation system energy-based can be employed as a microgrid at remote places where electric transmission is not feasible or not commercially viable. Recently, fuel cell systems are being employed in hospitals, telecom sectors and data centers to provide a continuous supply of electricity. It has also been envisaged that hydrogen could also be used to replace CNG, LPG for household applications such as heating, cooling, and cooking purposes and feeding of hydrogen in natural gas streams has already been started in a few countries as a small step in this direction.

References

[1] A. Alzahrani, S.K. Ramu, G. Devarajan, I. Vairavasundaram, S. Vairavasundaram, A review on hydrogen-based hybrid microgrid system: topologies for hydrogen energy storage, integration, and energy management with solar and wind energy, Energies 15 (2022) 7979.

[2] C. Baldino, J. O'Malley, S. Searle, Y. Zhou, A. Christensen, Hydrogen for heating? Decarbonization Options for Households in the United Kingdom in 2050, ICCT, 2020.

[3] J.O. Bockris, The hydrogen economy, in: J.O. Bockris (Ed.), Environmental Chemistry, Springer, Boston, MA, 1977.

[4] R.S. Cherry, A hydrogen utopia? Int. J. Hydrogen Energy 29 (2004) 125–129.

[5] G. Crabtree, M. Dresselhaus, M. Buchanan, The hydrogen economy, Phys. Today (2004) 39–44.

[6] A.J. Cornish, Hydrogen Fuelling Station Cost Reduction Study, LLC, Lakewood, CO, 2011.

[7] J. Guo, P. Chen, Catalyst: NH_3 as an energy carrier, Chem 3 (2017) 709–714.

[8] House of Commons Science and Technology Committee, The role of hydrogen in achieving net zero, Fourth Report of Session 2022–23. https://publications.parliament. uk/pa/cm5803/cmselect/cmsctech/99/summary.html# [Last accessed on 15th May, 2023].

[9] B.D. James, et al., 700 bar type IV H2 pressure vessel cost projections, Department of Energy Physical-Based Hydrogen Storage Workshop: Identifying Potential Pathways for Lower Cost 700 Bar Storage Vessels, USCAR, Southfield, MI, 2016.

[10] B.D. James, 2019 DOE hydrogen and fuel cells program review: hydrogen storage cost analysis (ST100), in: Strategic Analysis Presentation, 2019, US DOE.

[11] J. Manna, P. Jha, R. Sarkhel, C. Banerjee, A.K. Tripathi, M.R. Nouni, Opportunities for green hydrogen production in petroleum refining and ammonia synthesis industries in India, Int. J. Hydrogen Energy 46 (77) (2021) 38212.

[12] M.R. Nouni, P. Jha, R. Sarkhel, C. Banerjee, A.K. Tripathi, J. Manna, Alternative fuels for decarbonisation of road transport sector in India: options, present status, opportunities, and challenges, Fuel 305 (2021) 121583.

[13] A.M. Oliveira, R.R. Beswick, Y. Yan, A green hydrogen economy for a renewable energy society, Curr. Opin. Chem. Eng. 33 (2021) 100701.

[14] R. Ramachandran, R.K. Menon, An overview of industrial uses of hydrogen, Int. J. Hydrogen Energy 23 (7) (1998) 593–598.

[15] SAE China, Hydrogen fuel cell vehicle technology roadmap (English version), Strategy Advisory Committee of the Technology Roadmap for Energy Saving and New Energy Vehicles, Society of Automotive Engineers of China, 2016. Accessed from https://en.sae-china.org/a3967.html [Last accessed on 20th May 2023].

[16] O. Sandri, S. Holdsworth, J. Hayes, N. Willand, T. Moore, Hydrogen for all? Household energy vulnerability and the transition to hydrogen in Australia, Energy Res. Soc. Sci. 79 (2021) 102179.

[17] A. Serna, F. Tadeo, I. Yahyaoui, J.E. Normey-Rico, Business background analysis for a controlled hydrogen-based microgrid, in: The 14th International Workshop on Advanced Control and Diagnosis (ACD 2017) at: Bucharest, Romania, 2017.

[18] L. Valverde, F. Rosa, C. Bordons, Design, planning and management of a hydrogen-based microgrid, IEEE Trans. Ind. Inf. 9 (3) (2013) 398–1404.

[19] Y. Xiang, H. Cai, J. Liu, X. Zhang, Techno-economic design of energy systems for airport electrification: a hydrogen solar-storage integrated microgrid solution, Appl. Energy 283 (2021) 116374.

[20] M. Yue, H. Lambert, E. Pahon, R. Roche, S. Jemei, D. Hissel, Hydrogen energy systems: a critical review of technologies, applications, trends and challenges, Renew. Sustain. Energy Rev. 146 (2021) 111180.

Fuel Cell applications: Portable-domestic-distributed-mobility

M Shaneeth
Vikram Sarabhai Space Centre, ISRO, Thiruvananathapuram, Kerala, India

7.3.1 Introduction

The world is confronting the daunting task of liming potential damages due to global warming [1]. It has long been confirmed that the root cause of the problem lies in the uncontrolled and increasing emission of carbon due to various activities of humans in their quest for progress [2]. This has led to a serious imbalance in the natural carbon cycle on our planet, leading to net emission of carbon into the atmosphere and increasing carbon dioxide levels. Concerted efforts are already underway across the world to decarbonize all activities of mankind towards addressing the problems that transcend all walks of life. In this connection, urgent amendments are required in our way of life. There are certain specific technological interventions that are very promising as well. The whole efforts in this direction can be broadly classified as given below:

1. Recycle and reuse of materials and goods.
2. Waste-to-energy.
3. Lower energy and materials-intensive alternate processes and lifestyles; by enhancement in efficiency and reduction in wastage of both goods and energy.
4. Sustainable/greener feedstocks in manufacturing.
5. Sustainable generation and utilization of energy with the lowest possible carbon emission.

Renewable energy resources-based sustainable energy solutions are the way forward the world has now reconciled. This has caused a significant acceleration in the deployment of renewable energy technologies worldwide, in recent times [3,4]. In this context, hydrogen has emerged as a powerful single vector to reduce the carbon emission owing to energy and transportation sectors, as a cleaner fuel/energy carrier, as well as several hard-to-decarbonize sectors such as steel, fertilizer, cement industries, as a cleaner feedstock, addressing points (3) and (4) listed above [4,5]. Hydrogen produced through potentially low carbon intensive or carbon neutral or carbon negative methods such as renewable energy powered water electrolysis or biomass gasification or waste-to-hydrogen processes form a cleaner fuel substitution for coal or hydrocarbon fuels in energy and transport. It also forms a cleaner feedstock, in place of hydrogen from carbon-intensive processes such as coal gasification or steam methane reforming in vogue in industry for manufacturing ammonia, fertilizer, cement, and so on. Hydrogen also helps store renewable energy on grid scale which is very crucial in adopting highly variable renewable energy, solar, and wind, on very large scales.

It is a well-known fact that hydrogen is not readily available in nature. It requires to be produced from hydrogen-abundant compounds such as water, biomass, or even coal or methane. These are energy–intensive processes and a significant amount of energy requires to be invested to generate clean fuel hydrogen. It is especially so for generating hydrogen from water and biomass, the process associated with a much lesser carbon footprint and the choice for the future. Hence, it is important that, the methods by which energy is harnessed back from hydrogen works at highest possible efficiency so that loss of energy in the entire cycle is minimum and the cycle can be sustained. Conventional energy conversion process involving IC engines operate at relatively low-efficiency levels, owing to multiple energy conversion steps and also due to the fact that, inherently, they work on heat transfer-based cycles. Also, they operate at very high-temperature levels leading to the generation of NOx making the whole process less clean. Direct energy conversion process involving electrochemical reactions gains importance and fuel cells are considered in this context [6,7].

Fuel cells are devices that convert the chemical energy of fuels, mostly Hydrogen gas, directly to electric energy and water. They operate at efficiency levels much higher than conventional engines. Such high conversion efficiency becomes possible since fuel cells operate on electrochemical reactions. In fuel cells, fuel and oxidizers are not allowed to react together, as in conventional chemical reactions. Instead, through an electrochemical reaction, the net available energy output of the reaction, Gibbs free energy, gets converted directly to electricity. This stands in contrast to the conversion of energy output into heat as it occurs in conventional chemical reactions. Fuel cells enable the production of electricity directly from fuels in a single-step process, without any other additional intermediate energy conversion steps such as thermal-to-mechanical and mechanical-to-electrical. A comparison of fuel cells vis-à-vis conventional engines for the generation of electrical energy is depicted in Fig. 7.3.1. A direct result of this is the fact that a hydrogen fuel cell car would run for almost 2–3 times more distance per kg of hydrogen than its IC engine counterpart.

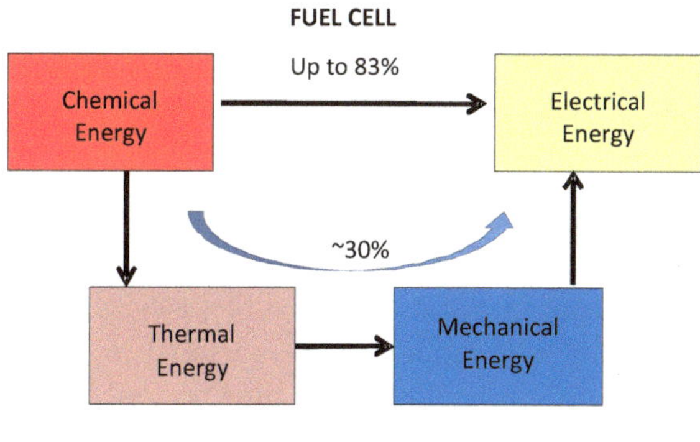

Figure 7.3.1 Electrical energy generation pathway: fuel cell versus conventional engine.

Further, there are fuel cells that can work on multiple fuels, namely, natural gas, biogas, methanol, hydrogen, etc. efficiently. More importantly, since fuel cells work on the direct energy conversion principle based on electrochemical reactions, undesirable side products are avoided and they enable reduced air pollution and carbon emissions and no pollutants such as NOx. In view of the above, fuel cells, in general, need a lesser amount of water for unit power generation as compared to conventional power generation.

Hydrogen-based electric power systems have been used for onboard electric power in several space missions. Fuel cells were the primary power system onboard for many pioneering missions of NASA, viz., Gemini, Apollo, and Space Shuttles [8]. Since then, the technology has evolved considerably and has emerged as a potentially robust system for replacing conventional engines of automobiles, locomotives, airplanes, diesel generators, and so on. In this regard, fuel cell systems compete with batteries with a specific advantage over batteries in terms of driving range and recharge time which is equal to that of present-day petrol and diesel vehicles.

7.3.2 Types of fuel cells and their applications

There are different types of fuel cells: classified based on the materials of construction and operating temperature. The following table, Table 7.3.1, provides the details [9].

Table 7.3.1 Types of fuel cells.

Fuel cell type	Charge carrier ions	Operating temperature
Proton exchange membrane fuel cell (PEMFC)	H^+ ions	<80°C
Phosphoric acid fuel cell (PAFC)	H^+ ions	150–220°C
Alkaline fuel cell (AFC)	H^- ions	250°C
Molten carbonate fuel cell (MCFC)	CO_3^{2-} ions	600–700°C
Solid oxide fuel cell (SOFC)	O^{2-} ions	800–1200°C

Among the fuel cells listed above, PEM fuel cell was the fuel cell type used first time in any real application; it was used in Gemini spacecraft. However, technological superiority of the time favored alkaline technology for further use. Thus, Apollo and Space Shuttle missions operated on alkaline fuel cell systems (Bacon fuel cell) [8]. Fast paced innovations in the last 3 decades towards transport applications, which needed combination of features, namely, low operating temperature, highest power density, and fast start-up/down capability, brought PEM FC back into the limelight again for all portable uses including household, motor transport, and space. There are thousands of fuel-cell cars, buses, and trucks on the road worldwide. There are also boats, aircraft, and trains working with hydrogen fuel cells which are in operational/demonstration mode.

SOFC, MCFC, and PAFC are the other types of fuel cells that are also finding wide applications these days. Many upcoming distributed power generation systems of several hundred kW to MW range, operating across the world as part of smart cities and clean energy initiatives, are based on fuel cell systems and are SOFCs. These also have flexibility with regard to fuels. Several such power systems are reported to be powering corporate offices and other important establishments in Silicon Valley and other places in the United States and elsewhere which include those generating energy from biogas produced at site or nearby landfills [10–13]. Some of the biggest fuel cell power plants in the world in terms of power rating, 50 MW and 78 MW, which are functional today are based on PAFC and are in South Korea [14]. MCFCs of several MW capacities are also operating at several places across the globe.

Each fuel cell has its own design methodology, performance characteristics, and applications areas. Among the different types, PEM fuel cell is one of the most popular types today and these are considered for portable and automobile applications.

In this chapter, it is indented to limit detailed discussions to PEM fuel cells while major application scenario of other major types is also mentioned.

7.3.3 Fuel cell fundamentals

A fuel cell power system consists of a (1) power generator, (2) reactant storage systems, and (3) balance-of-plant. The system can be configured either as a single unit consisting of all the three subsystems in a single unit or can be maintained separately in a distributed fashion, depending on the application and operating power level. Power generator is where fuel gets converted into electric power and fuel cell electrode stack forms the power generator. Gas storage systems store and supply the reactant gas (fuel). Power plant elements consist of all elements to facilitate required gas flow from gas storage into stack, thermal and power management, and overall control and monitoring of the operation of the systems for the indented application. Following block diagram (Fig. 7.3.2) depicts a typical fuel cell power system.

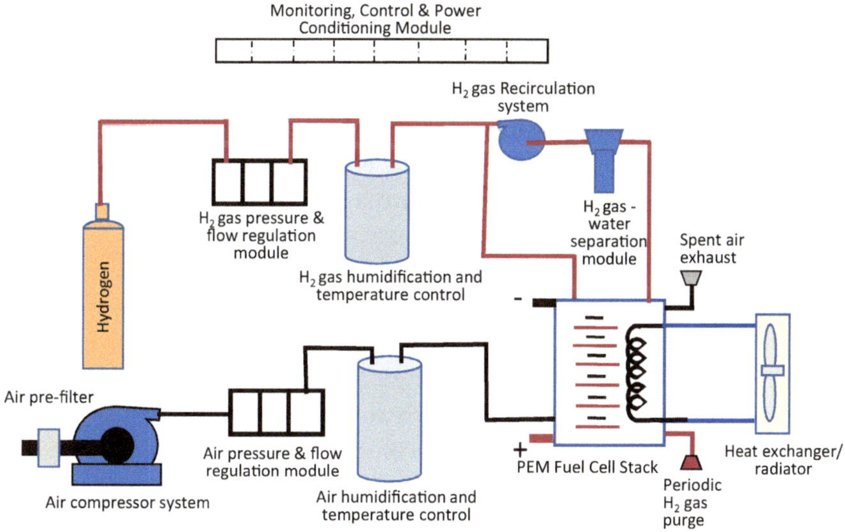

Figure 7.3.2 Typical fuel cell system: block diagram.

A fuel cell stack involves one or more fuel cell in it. Each fuel cell is an electrochemical cell having its own electrodes and electrolyte. A stack consists of multiple cells connected together. They are arrays of tens or hundreds of electrodes. In the electrodes of each cell, a fuel, typically hydrogen, and an oxidizer, pure oxygen gas or oxygen present in air, react electrochemically to form water, releasing electricity and heat along with.

In a typical single PEM fuel cell, a planar polymer film that has the ability to conduct proton (H^+ ion) works as the electrolyte and it is called a proton exchange membrane or polymer electrolyte membrane (PEM). They are normally provided with catalyst coating on either side and are sandwiched between two carbon-based porous sheets, which has the role of gas diffusion. This sandwich assembly of polymer electrolyte membrane and catalyst layer, along with porous carbon-based gas diffusion layers on either side is called a membrane electrode assembly (MEA). In each of one the MEAs, the catalyst layer-porous layer assembly on one side of the polymer membrane wherein oxygen reacts is called a cathode and the catalyst layer-porous layer assembly on the other side of the polymer membrane wherein hydrogen reacts is called anode. Such MEAs are stacked one over the other with intervening fluid distribution plates; one such unit which is placed between two plates, constitutes a cell. These plates are so designed and arranged as to enable distribution of hydrogen gas on one side of the catalyst coated polymer membrane and oxygen gas on the other side. Also, a coolant is distributed over all the plates by providing channels on the back side of the plates that distribute gases to electrodes. Generally, each gas is provided with a single entry to the stack and exist from the stack; the gases are effectively routed and distributed inside the stack to ensure uniform distribution into each plate and over the surface of each electrode. Voltage and current from a fuel cell stack are decided by the number of electrodes and size of electrodes, respectively. This construction scheme makes fuel cells highly modular and, hence, fuel cells can be constructed and used from mW to kW and MW, without any disparity in efficiency. A schematic of a single fuel cell is provided in Fig. 7.3.3.

When reactant gases, H_2 and O_2, are fed to a fuel cell, the terminals of a single cell start exhibiting a voltage to the tune of 1 V. As we draw current from the cell, the voltage starts to decrease. If the current is held constant, cell voltage also will remain constant as long as required gases are fed to the system. Typical performance characteristics of a single cell are depicted in Fig. 7.3.4. Consumption of the gases will be in proportion to the current drawn. Normally, gases in a predefined excess amount will be fed

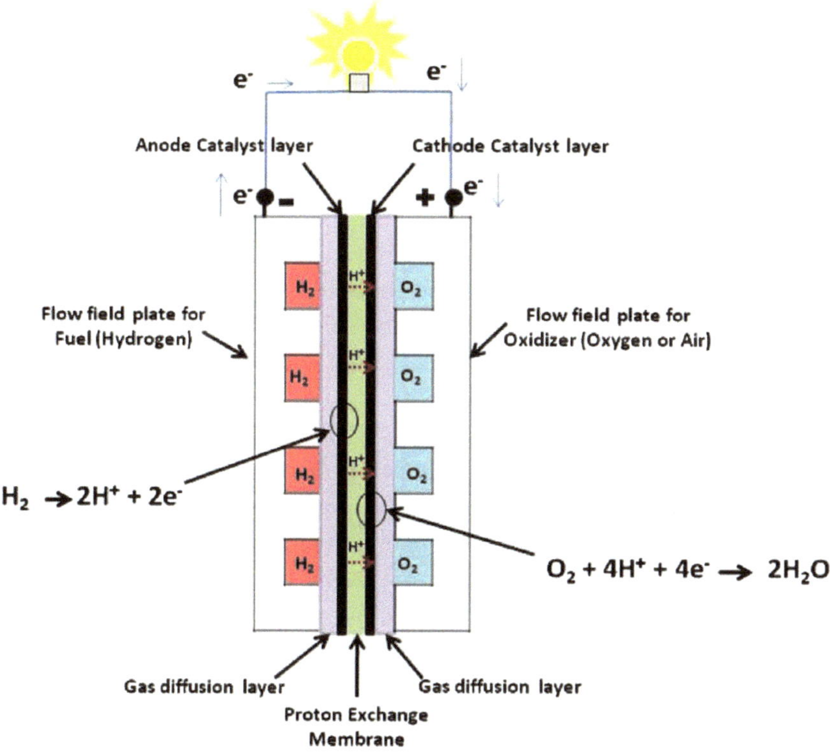

Figure 7.3.3 Single fuel cell schematic.

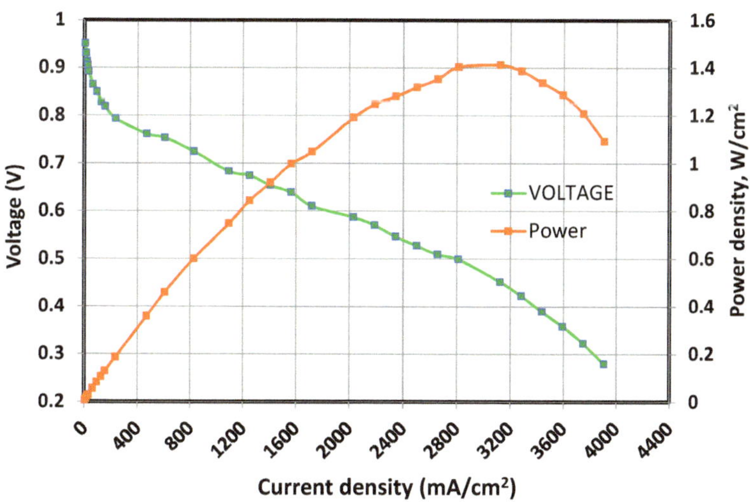

Figure 7.3.4 Typical performance characteristics of a single fuel cell.

to the cell/stack. The gases required to produce the indented power output through the electrochemical reaction will be consumed and the balance gases will come out of the cell which can be recirculated.

Regarding the remaining elements of a fuel cell power system, with regard to reactant storage units, especially for hydrogen, there are different types of storage systems. Those generally used are high-pressure gases, cryogenic fluids, or metal hydrides. Selection of the type of reactant storage is dependent on the quantum of reactants to be stored and the type of application, constraints, etc. Air forms the cathode reactants in fuel cells except those meant for specific applications, viz., space crafts and submarines/underwater vehicles. High-power systems generally use special air compressor systems for distributing air into fuel cells.

The third part of a fuel cell system, balance-of-plant, typically consists of reactant distribution systems consisting of pressure regulators, valves, tubing, etc., gas and air humidification systems, thermal management systems, and monitoring and power management and control systems. In some commercially available high-temperature fuel cell power plants, a fuel reformer that converts fuels like CNG, biogas, etc. into hydrogen before being fed to the fuel cell stack, also forms part of the balance of the plant.

7.3.4 Evolution of fuel cells

Invention of fuel cell is credited to Sir William Robert Grove based on his publication of "gas voltaic battery" in 1839. Christian Friedrick Schonbein is also reported to have brought out the same idea, during the same time or a little earlier [15]. However, it took more than a century for the invention of the fuel cell to get into a practical system. Francis Thomas Bacon, successfully demonstrated a 6-kW class fuel cell system in 1959 which was basically an alkaline fuel cell. During the same time, NASA considered fuel cells as the alternative to the weak batteries of the time for their upcoming big missions [8,16]. PEM fuel cell systems, developed and supplied by M/s. General Electric, were used in Gemini missions (1964–66). However, based on superior performance characteristics exhibited by Bacon fuel cells, NASA changed over to the improvised Bacon cells for the ensuing big missions, the Apollo. The fuel cell power plant for these missions was manufactured and delivered to NASA by M/s. Pratt and Whitney. Later on, further upgraded alkaline fuel cell was used in Space Shuttles. These space missions provided tremendous publicity for fuel cells to the entire world. However, the technological elements remained to be complex and relatively expensive. This made it difficult for fuel cells to get into a wider use, in spite of the simplicity and attractiveness of the concept and the utility it registered very successfully. Until very recently, fuel cells always remained to be a technology for the future.

During the last decade and more, exponential growth was seen in research and improvisation in fuel cell technology, thanks to the greater awareness worldwide about the serious consequences of global warming and the need to reduce carbon emissions. The idea of generating hydrogen from water using a grid scale of renewable energy of GW level through electrolysis, and thereby storing such a huge amount of energy in the form of hydrogen fuel, forms the basis for this resurgence. The huge quantities of hydrogen so generated can be stored or distributed. Electricity can be generated from this hydrogen whenever and wherever it is required at high efficiency with fuel cell systems. The same hydrogen can be used as fuel in vehicles as well, combining the twin advantages of fuel-filled-drive, as in the present case of petrol engines, and a nonpolluting renewable fuel. The whole concept makes it possible to store and use renewable energy from kW to GW level and utilize it for various energy requirements including vehicles. Modularity and scalability of water electrolyzer and fuel cell systems coupled with exponential growth in renewable energy are the enabling factors in this regard. This opened up a new and potentially green and sustainable way of energy management by infusing renewable energy in every walk of life. Today, there are a number of stationery power systems based on fuel cells which power major cities and corporate offices; also, thousands of vehicles are running worldwide on hydrogen energy, powered by fuel cells.

7.3.5 Potential applications

7.3.5.1 Portable power

Power requirements generally met by present day small battery systems and DGs fall into this category. These include off-grid remote powering solution for field surveillance and survey needs, emergency power in disaster management scenario, backup power for domestic needs and so on. Portable fuel cell systems get favored over DGs and batteries on account of the following reasons:

1. Low operating temperature which does not produce any notable thermal signatures, nearly silent operation, and totally carbon emission and pollution-free power generation render the system more favorable compared to the conventional DGs.

2. Greater energy density gives greater energy storage for the same weight and volume, whenever energy storage demands are relatively high, and instant recharge capability; this provides greater competitiveness to fuel cell system compared to battery-based system.

Typically, the power level ranges from a few tens of watts to a few kW. Such systems are generally configured as compact unit which can be easily carried/moved from one place to the other. Another important common feature is that the gas storage will generally be part of the power system; both gas storage and power generation will be confined into a single unit or both could be movable independently, however, they will be kept together and connected on need. Depending on the type of indented applications and power level and duration of powering involved different configurations outlined above are possible and they are pictorially depicted below in Fig. 7.3.5.

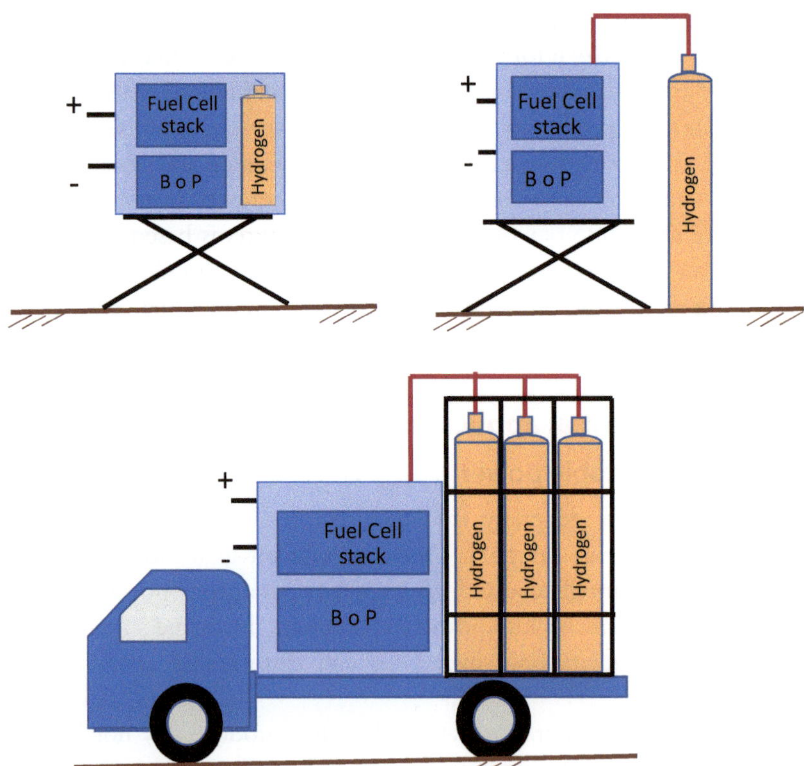

Figure 7.3.5 Different types of portable fuel cell system configurations.

As said before, generally, such systems are used for applications involving comparatively low power requirements, within a few kW. Also, they are typically considered for a finite duration. In view of this, stacks of power of a few tens of watts to a few kW are used in such systems. Metal hydride–based gas storage modules are generally a good choice. The limited quantity of hydrogen which is required to be stored in such systems to meet the finite and generally prefixed duration is one of the reasons behind it. Storage units with such limited hydrogen quantity can be realized within the storage density limits of current metal hydride technology. Also, metal hydride–based systems provide the safest mode of storage for hydrogen. This augurs well for portable systems which could be used in confined environments wherein abundantly safe practices are essential. Wherever repetitive usage or continuous usage is there, periodic replacement of the hydrogen module is convenient and is generally practiced. Further, duty cycles are normally simple with single or multiple start-stops and a complicated nonlinear power profile is normally not there. This significantly simplifies the stack design and enables achieving a prolonged lifetime. Another characteristic feature of such systems would be the simplicity in configuration, especially that involving passive/noncritical air and thermal management schemes and systems controls. Typically, miniature fan or blower-driven air flow serves as both cathode reactants and coolant.

Some of the applications where portable fuel cell systems of the above-said configuration could be beneficial are (1) powering of portable field operation equipment, (2) fixed monitoring systems of scientific, strategic, ecological, or environmental monitoring interest having finite duration, (3) portable emergency recharge systems for battery based applications such as FCEVs, RE connected critical installations such as wireless communication towers, data servers, etc., and (4) temporary or extra power solutions for specific events/programs, in place of DGs, and so on. To meet these, portable power systems could be configured as palm/hand held units, trolley mounted units or vehicle mounted systems.

Portable fuel cell system of 100 W developed and used for remote field survey and a fixed system [17] which was used for powering automatic weather stations on an experimental basis by Vikram Sarabhai Space Centre (VSSC), Indian Space Research Organization (ISRO), are provided in Fig. 7.3.6.

(A)   (B)

Figure 7.3.6 Portable fuel cell power developed by Vikram Sarabhai Space Centre, ISRO used for powering remote field survey instruments (A) and for automatic weather stations (B).

7.3.5.2 Domestic power

Domestic power requirements in industrialized nations or those which are in the process of fast paced industrialization are typically met with electricity from the grid. Even in such situations, there are cases of settlements to which the grid is unable to reach, mostly due to geographical constraints, and also where power outages are frequent. There are communities around the world where the grid has not reached yet. In such settings, the power demand varies from a few hundred of watts to a few kWs typically. Standalone power systems based on various schemes assume importance in such situations. While solar power is the answer in such cases, ensuring uninterrupted power during varying spells of nonsolar hours, due to night time as well as inclement weather, is indeed a challenge. This is due to the fact that, economically sized batteries can store a very limited amount of energy, on one hand, and they require a finite and relatively long recharge time, on the other.

With the hydrogen gas distribution network emerging into a reality much like the present-day "city gas", it is feasible to consider houses getting powered by hydrogen. Houses possessing a fuel cell power system can generate power in accordance with the requirement and in a clean and nearly silent manner. Towards this, hydrogen could be generated from the same locality using solar, wind, or biomass and distributed to multiple users.

Typical domestic power systems will be based on kW class PEM fuel cell stack. City gas like hydrogen gas supply could be hooked to such systems. Alternatively, hydrogen gas cylinder "modules" that house one or more cylinders mounted in it are also used, wherever a centralized gas supply is not available or feasible. These cylinder modules are so designed that they can be easily replaced with a new one, when they are empty, and can be transported for filling. In such domestic power systems, the fuel cell stack is typically designed to meet the standard household duty cycle conditions. A battery–fuel cell hybrid system with a battery in the front end taking up the load directly and the fuel cell charging the battery as well as sharing the load is the most favorable configuration. This will help size both the battery and fuel cell economically; the battery can be sized exclusively to meet the variations in load swiftly and the fuel cell can be optimally configured to operate primarily in charging mode.

The power system mentioned could be connected to the gas distribution network on a continuous basis if there is one available. If not, a gas cylinder module serves the purpose. The size of the gas modules, wherever they are employed, would depend on the energy to be delivered per fill and module replacement frequency. High-pressure or metal hydride gas cylinders are the choice. In order to have an affordable refilling duration of weeks or months, it is often preferred to employ a battery of cylinders connected together, mostly in a single physical framework. The entire module can be replaced periodically. Houses or other buildings, using such systems need to have a separate ventilated area/cabin to house the cylinder module and system.

7.3.5.3 Distributed power system

There are cases involving relatively larger power demand than those described above. These include hospitals, universities, housing colonies/settlements/townships, shopping malls/complexes, sports stadia, theatre complexes, community/cultural centers, commercial centers, cities, etc., each of which can be defined as a set of common activities of their own, typically autonomous in nature. Though they are mostly connected to the grid nowadays, it makes sense to have an independent power system, a centralized power station for those set of activities, to meet their demand owing to various already existing or emerging reasons.

This includes, possibly higher tariffs imposed by authorities concerned for the power consumed by the activities hosted, carbon footprint/greener energy considerations along with economic incentives associated with it, grid access/reliability issues, strategic considerations, and so on. Instances of such distributed power systems powering specific installations, as against getting hooked to a centralized thermal/hydro power station powering an entire district or state, is increasing. Fuel cell-based power systems form a very attractive solution in this regard. This is especially so when there exists feasibility for renewable hydrogen generation/supply of relatively inexpensive and sustainable biomass gasified hydrogen or even biogas. Moreover, when there exist both heat and power requirements, by engineering such solutions appropriately, it would be possible to achieve combined heat and power efficiency level of 80% in such systems.

High temperature fuel cells are generally considered for such installations since they can deliver better quality heat. With higher operating temperature levels and associated improved kinetics, they are potentially more efficient. Also, such power systems are not associated with frequent start stop requirements as seen in portable and domestic systems and hence high temperature systems will not face related issues. The hydrogen gas purity requirements are much less stringent and they have the flexibility to operate on other fuel gases too, potentially facilitating direct biomass gasifier integration into the entire scheme.

Typically, a power level of hundreds of kW to MW is involved in such systems. In view of the advantages mentioned earlier, high-temperature fuel cells, viz., SOFC, MCFC, and PAFC-based systems are used for such applications. Hydrogen, biogas, or syngas from bio gasifiers form the fuels for this. Generally, such power systems are co-installed/operated with fuel cell generation systems. As said before, these power systems deliver the highest efficiency when they are engineered to provide combined heat and power solutions, with hot water or steam output driving various heating applications and electric power output driving other nonheating applications.

There are several such systems varying in power levels operating worldwide. As said before, SOFC systems of several hundred kW operating in Silicon Valley [10–13] and MW class systems of MCFC and PAFC operating in South Korea [14] are well known examples, apart from several more elsewhere. Such distributed power systems based on fuel cells using renewable fuel are proving to be a solution for specific high power consuming segments of autonomous in nature listed earlier to switch to a low carbon footprint and reap benefits due to the same, in the near term.

7.3.5.4 Mobility

Application of fuel cells in mobility is one of the most attention grabbing, most awaited, and possibly most challenging. This is in view of multiple reasons including technical readiness of fuel cell technology for mobility, readiness of hydrogen gas storage technology, scale of the infrastructure essential and the associated economic and time scale considerations, preparedness of relevant rules and regulation and its enforcements, and complexity of the application involving dynamic environment and cost. The points are briefly explained below:

i. Fuel cell technology readiness:

    a. Performance: Mobility involves highly demanding performance and durability requirements due to the performance, especially the duty cycle characteristics of vehicle propulsion systems. Hence fuel cell stacks need to meet very high standards of performance. Some of the specific aspects along with the respective United States Department of Energy (US DOE) targets (ultimate), in a consolidated fashion, are given in the following table (Table 7.3.2).

    b. Features: The fuel cell power system requires to be within the weight and volume limits which is permissible in the envelope of various

Table 7.3.2 United States Department of Energy targets for polymer electrolyte membrane fuel cell systems and stacks [18].

Features	Ultimate target
Performance (0.8 V)	300 mA/cm^2
Peak energy efficiency	70%
Start–stop cycles	5000 cycles
Automotive drive cycle	8000 h
Specific power	2000 W/kg stack 850 W/kg system
Stack power density	2500 W/L stack 650 W/kg system
Low temperature nonassisted start	−30°C
Low temperature assisted start	−40°C
Cold start up time (No load to 50% load)	30 s @ −20°C 5 s @ +20°C
Cost	15$ (stack) 30$ (system)

vehicles. Implication of this is more pronounced in the case of 2 and 4 wheelers where available space to accommodate fuel tank and engine/power system is extremely small. This imposes severe constraints in the design and realization of stack, balance-of-plant, and gas storage modules. The entire system has to be realized without compromising the facilities which are available in the vehicles that the public is used to.

ii. Hydrogen gas storage: A key aspect in ensuring acceptability of hydrogen fuel cell vehicles is the driving range that it can deliver in a single fuel fill. It requires to be competitive with conventional vehicles and also should be having distinct advantage compared to the best battery EVs already available and newer versions that are on the anvil. This is directly related to the maximum quantity of hydrogen that can be stored onboard a vehicle. The issue is very acute in the case of smaller vehicles like two wheelers three wheelers, cars and SUVs where the available space for accommodating storage systems is very constrained; whereas, buses have enough space and it is not of a big concern. Storing in gaseous state is the most accepted method today due to various advantages it offers, especially the fact that hydrogen is a gas under ambient conditions and other modes of storage require extra energy and special storage conditions. However, storing in a gaseous state requires that hydrogen needs to be stored under very high-pressure conditions of the order of 700 bar to contain the volume of required quantity of hydrogen; typically, over 5 kg in the available space in a sedan, for example. Such high-pressure storage needs to be achieved with minimum weight and volume of the container for which polymer liner carbon composite wrapped type four cylinders are essential. These cylinders require to be leakproof with respect to hydrogen gas and shall have long cycle life. The technology of such cylinders is currently available with very limited agencies around the world. The technology available requires more field trials and is not yet in a mass manufacturing mode. Further, the design of cylinders and their mounting shall be in accordance with the safety and operating standards of automobiles since they are subjected to highly dynamic conditions and environments.

iii. Infrastructure: Seamless operation of hydrogen fuel cell vehicles require delivery of pure hydrogen in abundance, at multiple places along the routes in which they are expected to ply. Towards this, the production of hydrogen gas, its storage, and distribution to different sites for dispensing into vehicles and dispensers, all in optimal capacities and in numbers are very essential. This implies that it is required to put a fuel infrastructure based on hydrogen in parallel to the existing automobile fuel infrastructure based on petrol and diesel well in place for the hydrogen vehicles to flourish.

Magnitude of this can be gauged by estimating the requirements for a typical Indian metropolitan city with a registered vehicle fleet in the range of 5–10 million, in which most of the major cities do figure. In order to operationalize a reasonable hydrogen car density to the tune of 10% of the total vehicles, multiple dispensers in different geographical locations of the city are needed. For this small beginning itself, hydrogen production, distribution and dispensing requirements would be in the range of 175–350 tons/day (assuming avg. 1000 km/month/vehicle and 100 km/kg H_2). Since clean hydrogen is the need of the hour for decarbonization requirements, renewable power would be required. Hence, installed solar power capacity to the tune of 2–4 MW/city will be needed to begin with, at the rate of 50 kWh/kg H_2. Further increase in vehicle density will call for proportional increase in all elements of the infrastructure. The quantum of change is massive in terms of cost involved as well as establishing necessary supply chain of relevant elements viz., fuel cells, electrolyzers, gas storage systems, distribution elements, etc. However, it is to be noted that introduction of fuel cell vehicles will start with new vehicles only in limited numbers, in tens and hundreds, and not as a fraction of the number of existing vehicles. This number will be increasing gradually over a decade or longer period. The fuel infrastructure can develop in parallel, in a mutually enabling business model.

iv. Safety: Generally, safety standards and regulations assume significant importance in the context of all types of hydrogen fuel cell-based systems [19]. These are discussed in detail under a dedicated section on Safety. However, there are specific reasons due to which hydrogen fuel cell systems used in mobility require greater attention.

Mobility requires that the system employed possesses significant vibration tolerance. Typical stack involves several hundred meters of sealing and they need to be robust enough to avoid possible susceptibility. Similarly, systems possess numerous gas connections and joints. With hydrogen being the medium involved, such joints deserve special attention with respect design, material conditions, and joint design.

High pressure cylinder based gas storage is the other specific reason. Based on the quantity of gas to be stored and the need to minimize the weight of such cylinders, polymer–liner-based type 4 composite cylinders are the primary choice. Ensuring that hydrogen leak and permeation rates are within the stringent limits under the dynamic environmental conditions that vehicles are expected to be exposed is a major requirement. Further, possibility of accidents and potential damages such cylinder could suffer give rise to safety concerns. Since very high-pressure levels of 350 ksc and 700 ksc are involved, specific vehicle-level design considerations are needed to ensure the safety of such pressure under various conditions. This includes maintaining specific distances from the front and back ends of the vehicle as well as sides. In addition, pressure vessels mounting schemes also assume significant importance. Moreover, being a high-pressure vessel, hydrogen storage vessels onboard will have a finite cycle life. This makes it necessary to track the number of pressure and de-pressurization cycles, or in other words the number of gas filling, and replace the tank accordingly. A mechanism will need to be in place in accordance with the specified life of the vehicle and specifically the pressure vessel used in it.

Further, automobiles undergo very dynamic environment, on one hand, and they are personally owned and operated, on the other. In view of this, abundant safety margins are built into the design to take care of all possible unsafe conditions and they are tested and demonstrated. Moreover, being in the technology introduction stage, performance of the systems is very carefully monitored during the ongoing field trials and pilot demonstrations worldwide, in various driving conditions and environments.

Apart from the well-known applications in cars and buses, the use of fuel cells for aircraft propulsion, marine vessels viz., boats, submarines and ships, trains and, space crafts also share similar requirements. They are good number of past and present examples for each of these applications worldwide. With regard to application of fuel cells in cars and buses, all major vehicle manufactures around the world are having programs and they very promising models already on the road, either on a demonstration and evaluation basis or in commercial mode.

v. Standards, codes, and regulations: Automobiles being highly dynamic in nature and personally owned and operated and have personal and public safety risks due to the very nature of their use, and need to meet specific technical capabilities, and set of guidelines. Also, it is essential that they conform to specific laws enacted by governments and safety-oriented rules and procedures adopted by authorities concerned. So is the case with hydrogen fuel cell systems. It requires elaborate trials and validation and continuous monitoring whenever a new technology is introduced in them. This includes technical definitions/benchmarks and guidelines for designers, manufactures, and operators/users in the form of standards, laws adopted by governments for enforcement as codes and rules and procedures as regulations, covering all aspects of hydrogen-based vehicles. Major aspects in this regard include,

1. defining/treating hydrogen as a fuel, as against being considered as an industrial feedstock and its generation, storage, distribution, dispensing, and utilization/handling,
2. fuel cells and other auxiliary systems that form part of the power system onboard,
3. vehicle level modifications, and
4. performance and maintenance vis-a-vis the existing mechanisms and practices with respect the conventional systems.

It is necessary to certify compliance of the design of various new devices, equipment, and systems, arising out of introduction of the new technological elements that form part of the above said transformation, such as fuel cell stack and systems, with respect to respective standards. Fuel cell stack involves both high energy fuel and high voltage levels and hence need its own unique design and safety standards. Standards for many of the emerging technological elements including fuel cell stack are in evolving stage globally. It is also essential to put in place rules and guidelines for operations, repair, maintenance, and ending of service and laws for mandatory compliance. Further it is imperative to work out mechanisms to enforce the regulatory frame work.

A specific case in this regard is the recertification of a vehicle involved in an accident. Being a high-pressure hydrogen vessel-based system along with high-voltage electrical systems including fuel cell stack and battery pack, specific criteria required to be in place to address various performance, and safety aspects involved. This includes, standardization of design of storage system, fuel cell systems and battery pack and their mounting, classification of the magnitude of the impact from the accident and revalidation tests and analysis required, on one hand, and enforcing mechanisms essential to implement the same, on the other.

Almost all nations having significant hydrogen-based activities are in the process of finalizing relevant rules, regulations, and standards in this regard. International organizations such as Global Technical Regulations (GTR), International Electrotechnical Commission (IEC), International Organization Standardization (ISO), and Institute of Electrical and Electronic Engineers (IEEE) are working in these areas. In addition, US-based agencies viz. Society of Automotive Engineers (SAE), CSA Standards (CSA), Underwriters Laboratory (UL), American Society for Mechanical Engineers (ASME), American Petroleum Institute (API), Compressed Gas Association (CGA), National Fire Protection Association (NFPA), Japanese agency, Japanese Standards Association (JSA), Korean Agency, Korean Standards Association (KSA), Chinese agency, Standardization Administration (SAC) involving National Technical Committee for Hydrogen Energy and National Technical Committee for Fuel Cell and Flow battery and Indian agency, Automotive Industry Standards Committee (AISC) and many others agencies from different countries are putting up a significant amount of work towards developing relevant standards. Details and updates regarding these are available in websites of the agencies concerned; information in a consolidated form is available from www.fchea.org and www.h2tools.org. The standards, codes, and regulatory framework are getting evolved and updated based on emerging local experiences and research findings. Since automobile manufacturing and application are no longer localized, there are significant efforts ongoing worldwide to harmonize the standards, rules, and regulations to avoid conflicts, if any, and ensure consistency.

7.3.6 Safety considerations for hydrogen fuel cell applications

Hydrogen is associated with significant safety concerns possibly due to the memories of infamous Hindenburg incident [20]. The specific incident in question can be understood to be due to a combination of poor design and unwarranted operating conditions rather than usage of hydrogen, per se. However, there are specific characteristics of hydrogen which necessitate safety considerations at all levels, including material, design, manufacturing, operation, and maintenance. Following are some the critical properties in this regard [21].

- Extremely low ignition energy of hydrogen, 0.02 mJ.
- Very wide flammability limit, 4.1–74.8% in air.
- Low molecular size and fast diffusing characteristics.

- Hydrogen embrittlement.
- The need to contain hydrogen at very high-pressure levels (350–750 ksc) or extremely low temperature conditions (20 K).

The low ignition energy implies that, simply by eliminating source of ignition, it would not be practically possible to avoid possible fire or explosion due to hydrogen. Even static electricity on our body would be able to initiate ignition if hydrogen is present in adequate concentration in the surroundings. Hence efforts are required to avoid such concentration build up. Providing abundant ventilation wherever hydrogen is used is one of the mostly followed practice to ensure safety.

Flammability limits refer to the concentration level of hydrogen in air within which hydrogen can get ignited. Outside this limit, even if a source of energy is present, it will not get ignited. Hence, safe practices with hydrogen heavily rely on this characteristic and all efforts are generally focused to avoid any accumulation of hydrogen even if there is a leakage.

Fast diffusing characteristics is both a bane and a boon; it is a bane since leak-proofing of hydrogen-based systems, such as fuel cells, electrolyzers, mechanical joints, etc. are quite challenging and they are susceptible to give higher leak rate with time. In view of this, high standards of leak proofing is essential, quality of materials required for sealing is critical and periodic leak rate assessment and corrections are mandatory. This assumes significance for typical EV class fuel cells since they involve sealing length of more than 1500 m. Fast diffusing characteristics enable leaked gas to spread and thin out extremely fast. The fact that density of hydrogen gas is much less compared to air, adds to the advantage. Due to buoyancy effect, hydrogen does not get accumulated anywhere and immediately rises up under ambient conditions. Hence, under fully ventilated conditions, hydrogen dissipate fast helping to avoid participating and contributing to a fire, if any. Whereas, almost all contemporary liquid and gaseous fuels remain at the place of leakage and sustain the flame, in case of fire, and participate in the fire and contribute to the damage. In view of this, twin strategies involving abundant natural ventilation, in places where hydrogen systems are used, and high order leak-proofness of hydrogen systems ensure safety.

Hydrogen embrittlement is the phenomena by which hydrogen molecules easily diffuse into materials that are exposed to hydrogen gas, especially metals used as the construction material for high pressure containers. Such infusion of hydrogen molecules in metal matrices would lead to drastic reduction in ductility of the material. This results in unexpected and abrupt failure of such materials under load, often giving rise to catastrophic situations. It is essential to select materials that are not prone to Hydrogen embrittlement such as aluminum alloys, viz. AA6061, when considering the material of construction.

Hydrogen being a gas under normal pressure and temperature conditions, it requires to be stored under very high-pressure conditions or cryogenic conditions in liquefied form or in hydrides. Loss of integrity of a high pressure vessel or a cryogenic tank could lead to fuel gushing out in massive quantity. Specific safe practices are employed in the design, manufacturing, mounting, and operation of such tanks to rule out any such eventualities.

7.3.7 Innovations in fuel cells and cost reduction

Towards large scale induction of fuel cells in different applications mentioned above, the major hurdles today are the cost and durability. Though fuel cells possess unassailable record for delivering cleaner energy at high efficiency, it is not competitive in terms of cost and it does not last as much duration as its conventional counterpart. When fuel cell is considered to replace conventional systems in an application, these pose major challenges. Strategies to address the above issues are contradictory, on one hand, and impact other attributes of fuel cells, on the other. For example, cost reduction of fuel cell entails significant reduction in the noble metal catalyst loading while the same could potentially bring down the achievable durability and can limit efficiency. Lower efficiency can enhance consumption fuel and hence operational cost.

In view these, major innovations in fuel cells are directed towards achieving the following in a balanced way.

- Enhancement of energy conversion efficiency.
- Size reduction to achieve high power density.
- Simplification of the system configuration.
- Enhancement of life.
- Cost reduction.

7.3.7.1 Efficiency

Maximizing power output from unit mass of fuel is one of the crucial factors in reducing operational cost and enhancing competitiveness and affordability of fuel cell. The efficiency of converting embedded chemical energy of fuels to electrical energy requires that hydrogen fuel consumption to generate unit energy output is as close to thermodynamic value as possible.

In this regard, the following factors are important in the case of a PEM fuel cell:

- Electrodes shall exhibit highest possible open circuit voltage, close to thermodynamic voltage, and the gap between voltage-on-load and open circuit voltage shall be minimal.
- Electrode stack needs to operate with theoretical or near theoretical gas flow rates for which loss of H_2 gas due to parasitic mechanisms are minimal.
- Active components of balance-of-plant viz. pumps, gas re-circulators, humidifiers, etc. shall work at maximum efficiency with minimal power consumption.

It is very essential to have a very robust sealing scheme for the cells to prevent H_2 gas crossing over to air compartment and thereby affecting open circuit voltage. Cross flow reduces efficiency in two ways; gas that crosses over is not available for electrical energy production and, by forming mixed voltages on the other side, it contributes to reduction in cell voltage thereby impacting efficiency. Effectiveness of the sealing design, its implementation, and its durability are very important in this regard. With sealing length running into several hundred meters, susceptibility for leak is very high and critical attention is needed for sealing of large stacks. Overall stack design, assembly scheme and features incorporated to ensure effectiveness of the sealing over time are all major aspects requiring attention.

Since H_2 gas can permeate through the proton conducting membrane which separates the anode and cathode compartments, the type of the membrane has a major role in deciding the extent of such permeation. So is the case with H_2 permeation blocking species in membrane, either added separately or built-in, which can limit H_2 crossing. While thin membranes can minimize resistive losses in performance, as explained below, it could contribute to enhanced cross permeation and hence open circuit loss, unless specific design elements are brought in.

The net electrical output from a fuel cell electrode is sum of the energy available with the fuel and the losses. Broadly, it can be explained as given below:

$$\text{Net electrical output} = \text{Enthalpy}_{H_2} - \text{Loss}_{\text{thermodynamic}} - \text{Loss}_{\text{kinetic}}$$
$$- \text{Loss}_{\text{ohmic}} - \text{Loss}_{\text{concentration}} - \text{Loss}_{\text{miscellaneous}}$$

Among these, kinetic, Ohmic and, concentration losses can be contained through appropriate material and component level design and operation strategies. This includes,

- improvements and innovation with regard to catalysts and catalyst layer structures,
- improvements in membranes conductivity,
- flow filed structures,
- interfacial and interconnection conductivity, and
- operating conditions viz., temperature, RH and reactant stoichiometry.

Loss under miscellaneous includes gas leak, gas cross permeation, etc. This results in direct loss of gas which reduces gas utilization. Further, it interferes with power generation by affecting voltage. Hence, leaking/permeation of gas from fuel cell system contributes to reduction in efficiency require to be effectively minimized through appropriate materials selection, design innovations and process quality control and assurances.

For the total system, the net electrical output is also decided by the power balance-of-plant consumes. It is essential to have a check on the efficiencies of various devices which form part of the power system and that their power consumption is within predefined limits. An inefficient pump, blower, compressor, etc. can seriously compromise overall efficiency of fuel cell systems and derail the efforts taken to have a high efficiency fuel cell stack.

7.3.7.2 Size reduction and systems simplification

Size reduction and simplification of the system require that deliverable power-to-mass and power-to-volume ratio of the fuel cell stack shall be as high as possible. Fuel cell electrodes and stack components of low thickness and small size are required. Towards this, the components of the stack shall be capable of high-power density operation and high electrical and thermal power flux. High mass activity catalyst and electrodes form the key in this regard. New catalyst materials, higher conductivity membrane, coating that provides high interfacial conductivity for plates, advanced flow structures, etc. are the contributing factors. These enable operating electrodes at higher power densities thereby reducing the size of electrodes and components. Flow field plates are a significant weight and volume contributing element and they shall be light and compact. Material selection, type of design of flow channels considered, manufacturing techniques that can be adopted, etc. are deciding factors in this regard. Plates based on graphitic materials or stainless steel or titanium with varying plate configuration and manufacturing methods are in vogue. They often have related cost implications as well.

Fuel cell system balance-of-plant shall also be as compact and simple as possible. Stack producing minimal heat output is essential so that related thermal systems can be small. Miniature and more efficient balance-of-plant elements viz., pressure regulators, thermal management system elements, power management controllers, system structure, etc. and high storage density reactant storage systems form the essential focus areas in this regard.

7.3.7.3 Durability

Regarding life of fuel cell system, it is the degradation of materials used, such as, catalyst, proton conducting membrane, sealing materials, separator plates, etc. which limit the life of fuel cell stack. Hence, new derivatives of the materials in use or totally novel materials are being tried to enhance life. There are multiple degradation mechanisms in PEM fuel cells and degradation of catalyst and sealing are the two most critical elements in this regard [22–25]. Various modifications in catalyst materials and catalyst layer structure are being pursued worldwide to minimize or slow down the degradation [26–39]. Particle size and structure optimization, new chemistry for catalyst and support, etc. are some of the widely attempted strategies. New variants and new materials are being attempted to enhance life of sealing materials.

Diagnostics of fuel cell stack is another crucial area which is constantly getting evolved. It is desired to diagnose degradation, if any, length of life and health of a fuel cell stack in operation periodically. The most reported method in durability studies involve posttest destructive analyses. However, the same will not be useful for a fuel cell stack in service and in-situ tools are needed in this regard. There are good number of studies which look into different protocols, techniques, and equipment towards this [40-49]. As fuel cells get into more and more applications, including EVs, the in-situ diagnostic techniques with the help of portable equipment are getting increasingly attractive and essential. This will help probing the heath of fuel cell stack in greater detail, apart from the inference from obvious performance aspects, during routine maintenance activities itself rather easily and not so expensive way. When this scheme gets matured to operational level, it is expected to provide a big boost to various applications of fuel cells mentioned earlier.

7.3.7.4 Cost

Cost reduction is indeed one of the highest priorities and existential challenge in the area of fuel cells. Being a new technology, future of fuel cell application is squarely dependent on its acceptance by masses which is a direct function of cost. Cost analysis and cost predictions for fuel cells are reported mostly for EV class fuel cell stacks. Approximately 50% of the cost due to fuel cell power systems onboard vehicle is due to fuel cell stack [50]. Mass production of fuel cell stacks is one of the most accepted routes to get its cost reduced. Manufacturing scale to the tune of one million is very essential to possibly achieve the ultimate cost target of $15/kW for stacks [18,51]. Reaching such ambitious targets also requires that technology is indeed at the level of ultimate technical targets [52]. There are multiple stack and component level technical targets which also has the potential to reduce the cost. Some of the important targets in this regard are listed below in Table 7.3.3. The targets are specified as those which are better than or equal to that mentioned for the year 2020 vide the cited reference.

High power density operation of electrodes, to the tune of >1000 mW/ cm^2, is one the key technical targets. This is required to reduce the size and weight of the stack to accommodate it in vehicles. This indirectly helps to reduce size of various components, including membrane electrode assemblies (MEAs) and flow field plates, going into the stack and hence the quantity of the material consumed for each of it. In fact, for EV class fuel cell stack manufactured under one million scale, around 60% of the cost of stack is expected to be contributed by the MEAs. This is primarily due to the expensive materials involved, such as precious Pt based catalysts and

Table 7.3.3 Some of the major United States Department of Energy targets having direct cost reduction potential [52].

Features	Target
Electrode power density	≥ 1000 mW/cm^2
Pt group metal content	≤ 0.125 g/kW
Pt group metal loading	≤ 0.125 mg/cm^2
Catalyst mass activity	≤ 0.44 A/g
Bipolar plate areal specific resistance	≤ 0.01 Ωcm^2
Membrane area specific proton resistance (at max. operating temperature &water partial pressure of 40–80 kPa)	≤ 0.02 Ωcm^2
Bipolar plate cost	≤ 3 $/kW$_{net}$[a]
Membrane cost	≤ 20 $/m^{2}[a]
Membrane electrode assembly cost	≤ 14/kW$_{net}$[a]

[a] Estimated for manufacturing scale of 5×10^5 stack/year.

special proton conducting membrane. High processing cost of MEAs is also a contributing factor.

In order to address the high cost of MEAs, apart from the strategy to reduce the size of MEAs through high power density operation, the other research direction is to explore lower cost materials, in place of the expensive ones in use today. Fundamental molecular level designing and tuning are very essential in this regard and alternate low cost materials are being pursued. Search for nonnoble metal catalyst, MoF based catalyst, etc. are best examples in this connection However, such materials are yet to reach the level of performance and durability of Pt based catalyst and fall short of the electrode level or stack level performance targets. The other approach is to reduce the amount of expensive catalyst material per kW of power. This is targeted to be achieved with new chemistries and new catalyst layer designs that are capable of delivering higher mass activity, Ampere per gram of catalyst (A/g), thereby enabling to achieve low catalyst loading in electrodes. This way, catalyst consumption can be drastically reduced. It is to be noted that, this also to be done without compromising the performance level since it will affect other technical targets regarding size, efficiency, durability, etc.

Further, reduction in stack area specific resistance (ASR) by means of advanced interface coating on plates help minimizing ohmic voltage drop. Along with it, pushing regimes of mass transport to ultra-high current density levels through novel flow structures help boosting the specific power density of stack significantly and enable reducing stack size and thereby cost due to materials.

Processing of electrodes, plate, sealing, and stack is expensive due to the requirements of special and often custom designed processes and nonconventional machineries which are required. Large scale production is a strategy considered to reduce the cost. Towards this, mass production-oriented methods are being increasingly adopted, among various methods that are in vogue covering processing of electrodes, separator plate fabrication, application of sealing, and stack assembly. For example, there are multiple methods, viz., spray coating, decal method, die coating, screen printing, along with other lesser known methods, which are being used in the processing of MEAs. In the case of flow field plates, injection or compression molded graphite-based plates, formed or stamped metal foils, machined graphite or metal-based plates, etc. are there. So is the case with surface coating of plates for which noble metal electroplating or cladding, Ti nitriding, carbiding, etc. are pursued. It is essential to choose the appropriate manufacturing oriented method, among the various options that are available, very judiciously. This way, cost due to processing activities is expected to be reduced to the targeted level when millions of stacks are produced. To summarize, multipronged approach is needed to reduce the cost.

7.3.8 Conclusion

Fuel cells have emerged as the clean power source in several applications, covering portable, domestic, distributed, strategic, and mobility. Presently, fuel cell systems of few watts to MW power are powering variety of applications across the world. Vehicles, cars, trucks, buses, in thousands, and multiple train services, are in operation worldwide today based on fuel cells. There are also reports of powering, in operation mode or demonstration mode, submarines, boats, ships, and aircrafts with fuel cells.

Some of most celebrated space missions of the past, moon missions and space shuttles, were powered by fuel cells. Production and operation of high temperature fuel cells of kW and MW power is understood to be in commercial mode now. However, in the case of PEM fuel cells, the same appears to be true only with low power systems, in the range of a few hundred watts to a few kWs, and these are suitable for typical low-end applications. The technology for highly compact, durable, cost effective, and efficient PEM fuel cells of tens of kW to 100 kW needed for automobiles/buses, are found to be still limited to a very few agencies around the globe, as this chapter is written. Globally, very largescale manufacturing of EV class fuel cell is yet to begin and cost targets are expected be achieved with mass manufacturing.

Readiness of fuel infrastructure based on renewable hydrogen and biomass derived hydrogen form the key in widespread use of fuel cells. Robust quality, performance and safety standards, and regulatory framework are very essential in proliferation of this technology further. Significant amount of work has already been completed and all reputed standardization agencies around the world are working on the same. With climate actions gaining higher priority and renewable energy growing exponentially, fuel cells which provide twin advantages of higher efficiency and cleaner energy are gaining greater thrust. Fuel cells are expected to replace the existing petrol and diesel engines in numerous applications in the near future.

References

[1] IPCC, Climate Change 2022: Impacts, Adaptation, and Vulnerability, in: H.-O. Pörtner, D.C. Roberts, M. Tignor, E.S. Poloczanska, K. Mintenbeck, A. Alegría, M. Craig, S. Langsdorf, S. Löschke, V. Möller, A. Okem, B. Rama (Eds.), Contribution of Working Group II to the Sixth Assessment Report of the Intergovernmental Panel on Climate

Change, Cambridge University Press, Cambridge, UK and New York, 2022, p. 3056. https://doi.org/10.1017/9781009325844.

[2] J. Hansen, D. Johnson, A. Lacis, S. Lebedeff, P. Lee, D. Rind, G. Russell, Climate impact of increasing atmospheric carbon dioxide, Science 213 (1981) 957–966.

[3] I. Dincer, C. Zamfirescu, Sustainable Energy Systems and Applications, Springer Verlag, New York, 2011.

[4] C.J. Winter, Hydrogen energy: abundant, efficient, Clean: A debate over the energy-system change, Int. J. Hydrogen Energy 34 (2009) S1–S52.

[5] M.F. Hordeski, Alternative Fuels: The Future of Hydrogen, The Fairmont Press, Lilburn, GA, 2007.

[6] S. Basu (Ed.), Recent Trends in Fuel Cell Science and Technology, Springer, New York, 2007.

[7] N. Rajalakshmi, K.S. Dathathreyan, Catalyst layer in PEMFC electrodes: fabrication, characterisation and analysis, Chem. Eng. J. 129 (2007) 31–40.

[8] Fuel Cells for Space Science Applications, AIAA–2003–5938 (NASA/TM—2003-212730), 2003.

[9] R.P. O'Hayre, S.W. Cha, W. Colella, F.B. Prinz, Fuel Cell Fundamentals, John Wiley & Sons, Inc., New York, 2006.

[10] Equinix installs 1 MW bloom unit at Silicon valley data centre, Fuel Cell Bulletin 2015 (6) (2015) 6. https://doi.org/10.1016/S1464-2859(15)30151-6.

[11] Bloom Energy SOFC power Adobe's San Jose headquarters, Fuel Cell Bulletin 2010 (11) (2010) 5–6. https://doi.org/10.1016/S1464-2859(17)30047-0.

[12] Apple-Bloom Energy Server Phase 2 Fuel Cell System, US, 2021. https://www.power-technology.com/marketdata/apple-bloom-energy-server-phase-2-fuel-cell-system-us/.

[13] K. Fosberg, Fuel cell systems provide backup power in telecom applications, Fuel Cell Bulletin 2010 (12) (2010) 12–14. https://doi.org/10.1016/S1464-2859(10)70361-8.

[14] World's largest hydrogen fuel cell power plant, Fuel Cell Bulletin 2021 (11) (2021) 5–6. https://doi.org/10.1016/S1464-2859(21)00602-7.

[15] W. Vielstich, H.A. Gasteiger, A. Lamm (Eds.), Handbook of Fuel Cells: Fundamentals, Technology and Applications, John Wiley & Sons Ltd, New York, 2003.

[16] Francis Thomas Bacon by K.R. Williams in biographical memoirs, The Royal Society, 1993. https://royalsocietypublishing.org/. (Accessed 28 August 2021).

[17] Indian patent application No. 201941048564, for "An autonomous, silent and portable fuel cell system and a fuel cell stack therefor".

[18] DoE technical targets for PEM fuel cell systems and stacks for transportation applications, Office of Energy Efficiency and Renewable Energy, Department of Energy, U.S. https://www.energy.gov/eere/fuelcells/doe-technical-targets-fuel-cell-systems-and-stacks-transportation-applications. (Accessed 27 October 2022).

[19] 2015 Safety, codes and standards section, multi-year research, development, and demonstration plan page 3.7.28 (Fuel Cell Technologies Program Multi-Year Research, Development and Demonstration Plan – Section 3.7 Hydrogen Safety, Codes and Standards. https://www.energy.gov/sites/default/files/2015/06/f23/fcto_myrdd_safety_codes.pdf.

[20] The Hindenburg, Before and After Disaster, Britannica. https://www.britannica.com/story/the-hindenburg-before-and-after-disaster.

[21] Safety standards for hydrogen and hydrogen systems, Office of Safety and Mission Assurances, NASA, NSS: 1740.16, 1997.

[22] W. Vielstich, A. Lamn, H.A. Gasteiger (Eds.), Handbook of Fuel Cells: Advances in Electrocatalysis, Materials, Diagnostics and Durability, Vol. 5 and 6, Joh Wiley and Sons, New York, 2003.

[23] W. Vielstich, A. Lamn, H.A. Gasteiger (Eds.), Handbook of Fuel Cells: Fundamentals, Technology and Applications, Vol. 3, John Wiley and Sons, New York, 2003.

[24] M.M. Mench, E.M. Kumbur, T.N. Vezirolglu (Eds.), Polymer Electrolyte Fuel Cell Degradation, Academic Press, Elsevier, Boston, 2012.

[25] H. Wang, H. Li, X.-Z. Yuan (Eds.), PEM Fuel Cell Failure Mode Analysis, CRC Press, Taylor & Francis Group, New York, 2012.

[26] S. Koh, P. Strasser, Electrocatalysis on bimetallic surfaces: modifying catalytic reactivity for oxygen reduction by voltammetric surface dealloying, J. Am. Chem. Soc. 129 (2007) 12624–12625.

[27] Z. Chen, M. Waje, W. Li, Y. Yan, Supportless Pt and Pt-Pd nanotubes with high durability and activity towards oxygen reduction reaction for PEMF, ECS Trans. 11 (2007) 1301–1311.

[28] M.R. Tarasevich, V.A. Bogdanovskaya, E.N. Loubnin, L.A. Reznikova, Comparative study of the corrosion behavior of platinum-based nanosized cathodic catalysts for fuel cells, Prot. Met. 43 (2007) 689–693.

[29] P. Yu, M. Pemberton, P. Plasse, PtCo/C cathode catalyst for improved durability in PEMFCs, J. Power Sources 144 (2005) 11–20.

[30] X. Wang, W. Li, Z. Chen, M. Waje, Y. Yan, Durability investigation of carbon nanotube as catalyst support for proton exchange membrane fuel cell, J. Power Sources 158 (2006) 154–159.

[31] N. Rajalakshmi, H. Ryu, M.M. Shaijumon, S. Ramaprabhu, Performance of polymer electrolyte membrane fuel cells with carbon nanotubes as oxygen reduction catalyst support material, J. Power Sources 140 (2005) 250–257.

[32] T. Ioroi, H. Senoh, S.-i. Yamazaki, Z. Siroma, N. Fujiwara, K. Yasuda, Stability of corrosion-resistant magnéli-phase Ti4O7-supported PEMFC catalysts at high potentials, J. Electrochem. Soc. 155 (2008) B321–B326.

[33] H. Chhina, S. Campbell, O. Kesler, Ex situ evaluation of tungsten oxide as a catalyst support for PEMFCs, J. Electrochem. Soc. 154 (2007) B533–B539.

[34] N. Rajalakshmi, N. Lakshmi, K.S. Dhathathreyan, Nano titanium oxide catalyst support for proton exchange membrane fuel cells, Int. J. Hydrogen Energy 33 (2008) 7521–7526.

[35] A. Bharti, S. Muliankeezhu, G. Cheruvally, Pt–TiO$_2$ nanocomposites as catalysts for proton exchange membrane fuel cell: prominent effects of synthesis medium pH, J. Nanosci. Nanotech. 18 (4) (2018) 2781–2789.

[36] H. Chhina, S. Campbell, O. Kesler, High surface area synthesis, electrochemical activity, and stability of tungsten carbide supported Pt during oxygen reduction in proton exchange membrane fuel cells, J. Power Sources 179 (2008) 50–59.

[37] A. Bonakdarpour, K. Stevens, G.D. Vernstrom, R. Atansoski, A.K. Schmoeckel, M.K. Debe, J.R. Dahn, Oxygen reduction activity of Pt and Pt–Mn–Co electrocatalysts sputtered on nano-structured thin film support, Electrochim. Acta. 53 (2007) 688–694.

[38] P. Strasser, S. Koh, T. Anniyev, J. Greeley, K. More, C. Yu, Lattice-strain control of the activity in dealloyed core–shell fuel cell catalysts, Nat. Chem. 2 (2010) 454–460.

[39] V.R. Stamenkovic, B. Fowler, B.S. Mun, G. Wang, P.N. Ross, C.A. Lucas, N.M. Markovic, Improved oxygen reduction activity on Pt3Ni (111) via increased surface site availability, Science 315 (2007) 493–497.

[40] O.R. Reid, F.S. Saleh, E.B. Easton, Application of the transmission line EIS model to fuel cell catalyst layer durability, ECS Trans. 61 (23) (2014) 25–32.

[41] D.R. Baker, D.A. Caulk, K.C. Neyerlin, M.W. Murphy, Measurement of oxygen transport resistance in PEM fuel cells by limiting current methods, J. Electrochem. Soc. 156 (9) (2009) B991–B1003.

[42] T.A. Greszler, T.A. Caulk, P. Sinha, The impact of platinum loading on oxygen transport resistance, J. Electrochem. Soc. 159 (9) (2012) F831–F840.

[43] S. Arisetty, X. Wang, R.K. Ahluwalia, R. Mukundan, R. Borup, J. Davey, D. Langlois, F. Gambini, O. Polevaya, S. Blanchet, Catalyst durability in PEM fuel cells with low platinum loading, J. Electrochem. Soc. 159 (5) (2012) B455–B462.

[44] P. Zihrul, I. Hartung, S. Kirsch, G. Huebner, F. Hasche, H.A. Gasteiger, Voltage cycling induced losses in electrochemically active surface area and in H_2/Air: performance of PEM fuel cells, J. Electrochem. Soc. 163 (6) (2016) F492–F498.

[45] X. Ren, P.G. Pickup, Coupling of ion and electron transport during impedance measurements on a conducting polymer with similar ionic and electronic conductivities, J. Chem. Soc. Faraday Trans. 89 (2) (1993) 321–326.

[46] M. Shaneeth, S. Basu, S. Aravamuthan, PEM fuel cell cathode catalyst layer durability: an electrochemical spectroscopic investigation, Chem. Eng. Sci. 154 (2016) 72–80.

[47] M.V. Williams, L. Begg, H. Bonville, H.R. Kunz, J.M. Fenton, Characterization of gas diffusion layers for PEMFC, J. Electrochem. Soc. 151 (2004) A1173–A1180.

[48] J. St-Pierre, B. Wetton, G.S. Kim, K. Promislov, Limiting current operation of proton exchange membrane fuel cells, J. Electrochem. Soc. 154 (2007) B186–B193.

[49] N. Nanoyama, S. Okazaki, A.Z. Weber, Y. Ikogi, T. Yoshida, Analysis of oxygen-transport diffusion resistance in proton–exchange–membrane fuel cells, J. Electrochem. Soc. 158 (4) (2011) B416–B423.

[50] J. Wang, H. Wang, Y. Fan, Techno-economic challenges of fuel cell commercialization, Engineering 4 (2018) 352–360.

[51] DOE workshop on R & D needs for bipolar plates of PEM fuel cell technologies, Strategic Analysis, Southfield, MI, 2017.

[52] Fuel Cell Technologies Office Multi-year research, development, and demonstration plan: section 3.4 fuel cells, U.S. DOE Fuel Cell Technologies Office, 2017. https://www.energy.gov/sites/prod/files/2017/05/f34/fcto_myrdd_fuel_cells.pdf. (Accessed 26 October 2022).

Current and future of hydrogen economy

Current and future of the hydrogen economy

Anandh Subramaniam
Materials Science and Engineering (MSE), Centre for Environmental Science and Engineering (CESE), Sustainable Energy Engineering (SEE), Indian Institute of Technology, Kanpur, Uttar Pradesh, India

8.1.1 Introduction

The hydrogen economy has seen tremendous growth in the past few years; however, serious challenges persist and need to be solved for the widespread use of hydrogen as an alternative fuel. This section, dawning heavily on the content of the other chapters, outlines the current futuristic challenges related to the hydrogen economy. These relate to technological aspects, safety issues, and economic concerns in the path of the expansion of the hydrogen market. Favorable governmental policy and international cooperation are added dimensions, which should be leveraged to give impetus to the growth of the hydrogen economy.

Research and development in the area of hydrogen energy need to be enhanced many-fold, to achieve key breakthroughs in material and systems-level technologies. This requires a concerted and cooperative effort across the globe, along with increased research funding. And for this government organizations and private players not only need to invest heavily, but also to work hand in hand.

Strategies to integrate hydrogen production, storage, and transportation with utilization, keeping in view the economic viewpoint, are important and immediate goals. Public awareness and acceptance of this new paradigm in renewable energy requires hard work by multiple stakeholders (from multiple governmental departments to the average citizens). Standardization of systems and protocols and knowledge dissemination in the area to engineers, researchers, technicians, and end-users would be imperative, given the issues with hydrogen as a fuel. This in conjunction with an effective "marketing strategy" is important for a viable hydrogen economy.

The establishment of hydrogen-related infrastructure permeating the "last mile" and keeping in view the consumer requirements is going to be a major challenge for the future. This will be in addition to the effort required

Towards Hydrogen Infrastructure: Advances and Challenges in Preparing for the Hydrogen Economy.
DOI: https://doi.org/10.1016/B978-0-323-95553-9.00062-5

to convince the manufacturers of equipment like cars to make a switch to hydrogen as a fuel, at least partly. The aforementioned need requires policy changes by the governments' subsidies where needed. The future for the hydrogen economy seems bright, but the seeds of growth need to sow now.

In this chapter, we summarize the essential contents of previous chapters and identify the critical gaps towards the adoption of a widespread hydrogen economy. The detailed assessments along with references can be found in the specific chapters.

Most of the hydrogen used today is in ammonia production (most of which is further used in the fertilizer industry), petroleum refining (in hydrocracking), and methanol production. Hydrogen is also used in the manufacturing sector (glass production, welding, etc.) and food industries (to make hydrogenated vegetable oils). These requirements are expected to increase with time and someday may form an integrated part of the hydrogen economy; but, do not form a focus of the current chapter. However, a point to be kept in view is that hydrogen production has to exceed the requirements of these traditional industries so as to be available for the (newly developing) energy sector.

Often, we will take recourse to vehicular applications of hydrogen as a fuel as a reference point—as this is where the major expansion of the future is expected and is driven by "nil" pollution caused by hydrogen-fuel cell vehicles, which includes a reduction in greenhouse gases. The term "nil" has been placed within quotes as this refers to the lack of pollution at the utilization end and not in the entire cycle, i.e., considerable pollution may occur in other parts of the energy chain if care is not taken. The good news is that cars, train [1], and buses are already running on hydrogen-fuel cell combinations, which serve as demonstrators of the feasibility and as an advertisement for the hydrogen economy.

8.1.2 The hydrogen economy

Transition to green technologies is no longer an "If?," "But," "When?," and "How Soon?." Currently, hydrogen energy has climbed to within the top few spots for alternatives to the fossil fuel-based economy. However, hydrogen energy should be viewed in totality; wherein, cost, governmental policies, standardization, etc. form an integral part of the picture and hence the term "hydrogen economy."

The reasons for the use of hydrogen as a carrier of energy are clear. It has got a higher gravimetric energy density (at 200 bar hydrogen gas has

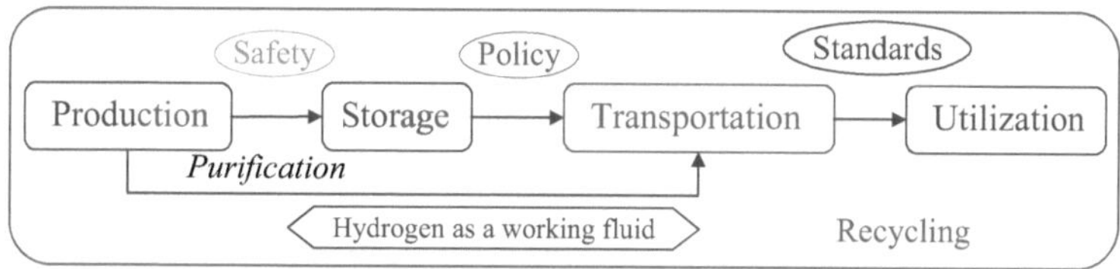

Figure 8.1.1 Components of the hydrogen economy. The color coding is as follows: Green represents a good level of maturity, blue is a medium level, and red low level. It is to be noted that apart from the ones requiring major attention (*red coded*), all other components also need to be improved/addressed. The use of hydrogen as a working fluid is also shown to remind us to keep it in the broader picture of a hydrogen economy.

a gravimetric energy density of 120 MJ/kg, which exceeds that of petrol which has an energy density of 42 MJ/kg) and the product of combustion or use in a fuel cell is water. The energy density (gravimetric and volumetric) of any energy source is one of the most important parameters which plays a key role in many of the applications, e.g., onboard a vehicle.

The key components of a hydrogen economy are related to (Fig. 8.1.1): (1) production, (2) storage, (3) transportation, and (4) utilization. The other important factors to be kept in view are: safety, standards, policy, and recycling. Any shortcomings or lacunae in any of the aforementioned points can severely hamper the realization of the hydrogen economy. As it stands today, each one of the aspects of the hydrogen economy remains a challenge and considerable distance has to be traveled to realize the dream of a widespread hydrogen economy that seamlessly integrates with society.

Most of the clean utilization strategies for hydrogen involve the conversion of chemical energy to electrical energy (or to start with the conversion of electrical energy to chemical energy)—as shown in Fig. 8.1.2. Herein, one of the main challenges to the hydrogen economy comes from the energy flow chains; wherein the inter-conversion of chemical and electrical energy are avoided altogether (e.g., via solar energy to capacitor storage of electrical energy or directly to a very "flexible" grid wherein the storage step is avoided). The battery-based technologies are well established for most parts and hydrogen-based technologies should pitch in strongly in places wherein the battery-based technologies have a weak link (e.g., recharging stations

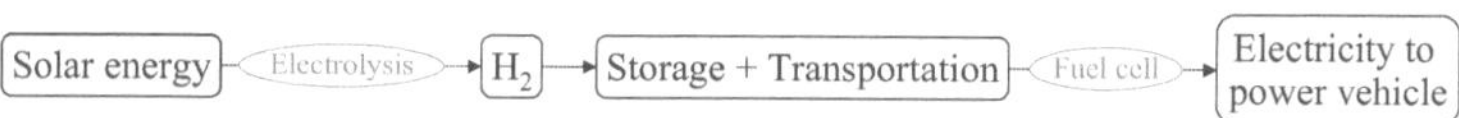

Figure 8.1.2 One possible pathway for the production and utilization of "green hydrogen."

for batteries are scarce and so are hydrogen refueling stations). The time for battery charging is currently much longer than that for gasoline refueling and if we can establish technologies for fast refueling of hydrogen, this may give the hydrogen-based economy a small but finite edge. Additionally, if we can set up hydrogen refueling stations at this critical juncture, the advantage that battery-based systems have in the transportation sector may be mitigated.

As stated earlier, the gravimetric density of hydrogen is good; however, its volumetric energy density is poor. At 200 bar hydrogen has an energy density of 2.1 MJ/L, whilst the value for petrol is 31.5 MJ/L. However, hydrogen outperforms batteries by a huge margin; e.g., Li–ion batteries have an energy density of about 0.5 MJ/kg ($\sim$2 MJ/L). A good energy source is expected to have both a high gravimetric and a volumetric energy density.

The best benchmarking of the hydrogen economy can come from the transportation segment (including earthmovers). Herein, a comparison can be made with other existing and competing technologies, like petrol, diesel, and battery-operated vehicles. A total cost to of ownership (Chapter 7.1 on "Total cost of ownership analysis of fuel cell electric vehicles in India") shows that hydrogen-based vehicles can perform better than vehicles based on other energy sources for larger vehicles (e.g., trucks) and in the long run (10 years of ownership).

The weight of hydrogen-fuel cell-based vehicles increases nominally with range, while the weight of batteries required increases considerably. Currently, batteries are charged with power from the grid, a large fraction of which is produced from nonrenewable sources with high greenhouse gas emissions. This aspect (greenhouse gas emission) is worst for power produced from coal. For hydrogen produced from natural gas, the "well to wheel" efficiency of hydrogen-based vehicles is higher than that of battery-powered vehicles. Certain countries are leading the way in the production of green hydrogen (that is produced from renewable sources only) and in Denmark, all the hydrogen produced is from renewable sources. Germany and other countries are expected to adopt this "green" type of hydrogen production in the coming years. If the power being generated is from a renewable source like solar energy, then batteries outperform hydrogen in the "well to wheel" efficiency as the considerable cost is involved in the storage (packaging) and transport of hydrogen. If "round trip energy efficiency" is used to compare energy stored in batteries versus energy stored in hydrogen (noting that hydrogen energy has to be converted back to electricity via fuel cells), battery technology does much better than hydrogen energy-based technology [2]. However, if a wind turbine is the source of energy and the overgeneration is stored and used, the energy return in this scenario is similar to that for batteries and this is in spite of the lower round-trip efficiency of hydrogen-based energy [2].

Even in the current scenario hydrogen-based technology can perform a vital role and is the preferred choice, like in backup power generation for the health sector where the power may be required sparsely from a backup source.

One important aspect of the use of electric vehicles (battery-based or fuel-cell-based) is the noise factor. Internal combustion (IC) engines cause considerable noise pollution in cities and this aspect is often overlooked but is important from a health perspective (the long-term exposure to noise (and vibration) has deleterious effects on the human system, which is well documented).

Recycling of the components of a hydrogen economy, like the fuel cell, is an important area which is important but lags behind the battery-based technologies and needs considerable attention immediately as this will make the hydrogen economy more competitive.

There is an important part of the hydrogen economy, which is the use of hydrogen not as an energy source but as a working fluid or related to the improvement of the "functional quality" of hydrogen. This includes hydrogen-based cooling systems, thermal storage systems, compressors, and hydrogen purifiers.

8.1.2.1 Value chain and closing the loop

It is imperative that economic analysis is carried out for each of the hydrogen-based technologies, which should ultimately form a closed loop. One such loop is: the reforming of natural gas to produce hydrogen → storage in high-pressure cylinders → transportation in tube trailers → filling hydrogen stations → fueling a car and use in a fuel cell → recycling of the components of a fuel cell (and other components involved in the energy chain). The pollution involved at each stage, including greenhouse gas emissions during production and utilization, needs to be kept in view and accounted for in the balance sheet. The economic and environmental sense of a given hydrogen–based loop can be used to convince policymakers to provide subsidies and other favorable policy incentives. Countries leading the foray into the hydrogen economy need to form consortia to pool these resources, which can then expand further and lead to global coverage.

8.1.3 Hydrogen production

Hydrogen in spite of being the most abundant element in the universe is scarce in free form on earth and this is the main issue for the use of hydrogen as a fuel. In combined form hydrogen has two main sources: hydrocarbons and water.

The primary energy source for the production of hydrogen can be hydrocarbons (including coal), electricity, biomass, nuclear, solar, wind, and other renewable sources. Hydrocarbons maybe refined to give hydrogen, coal can be gasified to give hydrogen, biological processes (involving bacteria) can convert biomass into hydrogen, renewable power sources (solar, wind, etc.) can be used for splitting water and producing hydrogen. Use of nuclear power for the splitting of water is often not preferred (for "obvious reasons"). Most fuel cells require pure hydrogen and hence if the process of production of hydrogen gives any other gases or impurities then the gas needs to be purified before being stored. The key question in the context of the use of power sources like solar and nuclear is: "Why not directly supply the power to the grid (as usually done) and why should we use the power to produce hydrogen (knowing fully well that any energy conversion is lossy)?". The two primary reasons are that for onboard applications we need stored energy (which cannot be tapped from the grid while the vehicle is running, but not on a track), and energy sources like solar and wind are not a constant function of time and when there is excess production we need to store the energy for use (and convert back to electricity when the load exceeds the production). One more reason is that many places may be remote and maybe off the grid.

Currently, most of the hydrogen is produced from fossil fuels like natural gas and coal. Further, most of the hydrogen produced is via steam reforming using natural gas. Electrolysis of water is another option; but forms an attractive option only if the power is supplied by renewable sources such as solar, wind, or hydro-power. The goal is to produce all the hydrogen from renewable energy sources. The efficiency of electrolyzers is not more than 80% and the hydrogen produced by this route is costlier than that produced from natural gas. Many of the electrolyzers use platinum and the search is on for the best alternative catalyst. These aspects (increase in efficiency and re-placement of platinum with inexpensive alternatives) need urgent addressal.

Take home message

Given that multiple industries compete for the hydrogen produced unless the big challenge to produce hydrogen from green sources is addressed immediately, the widespread use of hydrogen as a fuel will remain a dream.

8.1.4 Hydrogen storage and transportation

Once hydrogen has been produced there are two options for its translocation to the point of utilization: storage in tanks and transport via land or sea or use of pipelines. For onboard and certain other applications (like in isolated and remote areas), storage in tanks is the only viable option (though a hybrid scheme with pipelines and tanks can be envisaged). If the quantity of hydrogen produced is small and removed from the main grid of pipelines, then storage in tanks is again the only viable option.

The obvious choice for hydrogen storage is in cylinders (e.g., onboard a vehicle) and the other obvious choice for transport is via pipelines. Unfortunately, both these choices come with issues that have motivated researchers to come up with alternatives. Hydrogen storage cylinders, even when filled upto 700 bar pressure have a low volumetric energy density of 5.6 MJ/L (albeit hydrogen has a high gravimetric energy density of 142 MJ/kg). This low volumetric energy density of hydrogen in gaseous form, along with the safety issues associated with storage of hydrogen in cylinders at high pressures (e.g., during accidents) and the energy associated with pressurization makes hydrogen storage in cylinders, makes us think of alternatives to this method. The critical temperature for hydrogen is $\sim$33K ($\sim$ $-240°$C) and this implies that liquefaction of hydrogen cannot be achieved by pressurization alone, i.e., the gas has to be cooled to low temperatures and this implies that liquid storage of hydrogen is also not attractive (given the cost of cooling to low temperatures).

The transport of hydrogen can be done in cylinders in either compressed gas (in tube trailers) or in liquefied form. Both these methods have attendant issues and further are associated with the emission of greenhouse gases during transportation. Compression of gas expends a considerable amount of energy and so does cooling to cryogenic temperatures. One aspect pointed out in the chapter on the transportation of hydrogen is that existing pipelines for natural gas can be utilized by blending natural gas with hydrogen. The shortcoming with this method is that the maximum blend recommended is usually less than 10% volume (and further this blend may be highly unsuitable for use in most fuel cells). If the demand is low and the distances are small then tube trailers are the preferred method, whilst in the opposite scenario (high demand and long distances) pipelines or cryogenic cylinders are preferred.

Solid state storage, in or on a material, albeit its advantage in terms of safety and high volumetric density, has not come of age yet. Most materials offer hydrogen gravimetric densities lower than 2 MJ/kg and volumetric density lower than 4 MJ/L (e.g., for MgH_2 the values are 2.8 MJ/kg and 4.0 MJ/L) and often suffer from high/low temperature or pressure of operation. The goal here is to develop systems based on inexpensive materials which can: store more than 6.5 MJ/kg (and 4.7 MJ/L) at the system level (not just material level), absorb and desorb around 50°C, withstand more than 1500 cycles, store at less than 12 bar pressure, deliver at pressures greater than 5 bar and fill at a rate of 1.6 g/s.

Take home message

A lot of monitory loss occurs in this section of the value chain. This occurs in infrastructure costs (e.g., laying pipelines) and process costs (e.g., compression of gas). Storage in materials offers a bright future. This weak link needs to be enabled via both policy and research support.

8.1.5 Hydrogen utilization

Had hydrogen been useable with existing IC engines with a matching (or better) efficiency compared to petrol-based IC engines, it would have given a big boost to the adoption of hydrogen as a fuel and would also have been a good alternative to compressed natural gas (CNG) as a fuel. Since this is not the case (the key issue being premature ignition in the combustion chamber, which occurs due to the lower ignition energy, wider flammability range, and shorter quenching distance of hydrogen), the best alternative to use hydrogen as a fuel is to use it in conjunction with a fuel cell.

Typically, five different types of fuel cells are identified: polymer electrolyte membrane (PEM), alkaline, phosphoric acid, molten carbonate, and solid oxide. The one based on polymer electrolyte has been the preferred one for onboard applications. Commercial models like the Toyota Mirai are propelling the demonstration of the functionality of the technology on road, which is important for widespread acceptability. The fuel cell volume power density in this vehicle (Toyota Mirai) is 3.1 kW/L.

The key issue with fuel cells is the cost of fuel cells which uses platinum as the catalyst in the electrodes. Battery-powered vehicles outperform fuel cells based vehicles in the short-distance and smaller vehicle segments (scooters/motorbikes or more than 250 km/day). Fuel cells are best suited currently for buses and trucks. This is due to the fact that with the increasing capacity of the battery which is required for longer distances, the added battery weight consumes considerable power by itself. The key challenge today remains the availability of hydrogen fueling stations, wherein battery charging stations enjoy an advantage. The scenario is being changed slowly and only locally in regions like California and Germany. The distinct advantage that hydrogen enjoys is the short refueling time as compared to the charging time of batteries. This big advantage should be used to leverage the widespread use of hydrogen for onboard applications. Some recent work has suggested that platinum can be replaced with cheaper alternatives [3] and hence make fuel cells competitive.

For stationary applications, other types of fuel cells (non–PEM) may be used. In this segment (stationary applications) and for grid-level storage batteries enjoy the overall advantage due to the low round-trip efficiency of fuel cells. The typical efficiency of fuel cells is around 50%, although this is much higher than that for IC engines ($\sim$25%), massive improvements are required on this front.

Take home message

The price of fuel cells remains one of the biggest challenges in the utilization of hydrogen. Technologies that enable minimal or nil use of the precious metal platinum need to be developed. It is not clear if it is the end of the road for IC engines based on hydrogen or will it rise to be a challenger for fuel cell-based vehicles for onboard applications (and will it be worth the effort to develop IC engines based on hydrogen?).

8.1.6 Hydrogen as a working fluid/improvement of its "functional quality"

Some of the material-based systems used for the storage of hydrogen can also serve to use hydrogen as a working fluid to: compress hydrogen, build environment cooling systems using waste heat at about 200°C, and use it for the storage of heat (thermal storage). The functional quality of hydrogen can be improved by either increasing the pressure of the hydrogen or its purity and both of these can be achieved by the use of hydride materials. The use of waste heat requires the design of integrated systems, either with existing technologies or with newer ones being developed. As and when breakthroughs are achieved in hydrogen storage materials, the viability of these applications becomes higher.

Take home message

Other technologies based on hydrogen (not as an energy source) are important as they provide alternate use of the same materials/technologies and conserve energy. These technologies are maturing at a fast pace. The adoption of these technologies can not only hasten the acceptability of the hydrogen economy but also serve as a good torch bearer for the same.

8.1.7 Safety and standards

We need standards for safety, especially with newer technologies or massive scale-up, but safety per say goes beyond standards. This is because standards cannot account for many of the unforeseen circumstances (or local conditions), whilst safety measures should consider these aspects. For example, vehicular accidents are higher in some regions/roads and hence tube trailers travelling in these high-risk zones should evolve their own safety measures.

Hydrogen has been used for more than a century now and hence most of the safety issues have been well laid out and solved. However, any new technology brings with it newer set of issues. For example, hydrogen has not been mass transported in tube trailers or pipelines and hence issues arising therein are yet to be fully understood and addressed. Issues arising during long-term usage of fuel cells with hydrogen as the energy source have also not been fully studied.

Some accidents related to hydrogen have been reported, which occurred either during production or transport (the well-known one being the explosion at OneH2 plant in North Carolina, United States which produces hydrogen—luckily in this accident no lives were lost). Another well-reported accident, noteworthy due to a lack of explosion, happened during the transport of liquid hydrogen in a tanker, which crashed and caught fire on I-71 (highway) in the United States. Apart from the ill effects of such accidents, they give rise to bad press at this nascent stage of the technology, which can further affect governmental aid and policies and hence should be definitely avoided.

The key issues related to the combustion of hydrogen are as follows. Hydrogen is flammable over a very wide range of concentrations in air (4–75%) and it is also explosive over a wide range of concentrations (15–59%) at standard atmospheric pressure. The flammability limits increase with temperature. Hydrogen flames are very pale blue and are almost invisible in daylight due to the absence of soot hence it is hard to detect. The burning speed of hydrogen is in the range of 2.65–3.25 m/s and is nearly an order of magnitude higher than that of methane or gasoline (at stoichiometric conditions). Thus hydrogen fires burn quickly and, as a result, tend to be relatively short-lived. The quenching gap of hydrogen is $\sim$0.064 cm and is approximately three times less than that of other fuels, such as gasoline. Thus, hydrogen flames travel closer to the cylinder wall before they are extinguished making them more difficult to quench than gasoline flames. This smaller quenching distance can also increase the tendency for backfire as the flame from a hydrogen–air mixture can more readily get past a nearly closed intake valve than the flame from a hydrocarbon–air mixture. This implies that hydrogen lines have to be fitted with back flame arrestors.

However, hydrogen is not dangerous compared to other fuels due to the following. For hydrogen, the autoignition temperature is relatively high at 585°C (it is not be noted that though hydrogen has a higher autoignition temperature than methane, propane, or gasoline, its ignition energy at 0.02 mJ is about an order of magnitude lower than fuels and is, therefore, more easily ignitable). This makes it difficult to ignite a hydrogen/air mixture on the basis of heat alone without some additional ignition source. In many respects, hydrogen fires are safer than gasoline fires. Hydrogen gas rises quickly due to its high buoyancy and diffusivity. Consequently, hydrogen fires are vertical and highly localized. When a car's hydrogen cylinder ruptures and is ignited, the fire burns away from the car and the interior typically does not get very hot. On the other hand, Gasoline forms a pool, spreads laterally, and the vapors form a lingering cloud so that gasoline fires are broad and encompass a wide area. Hydrogen emits nontoxic combustion products when burned. Gasoline fires generate toxic smoke.

The key components of a hydrogen safety system are strategically located sensors, alarms, and mitigation equipment and procedures. Many more features of a safe system are listed in the chapter 5.2 on "Hydrogen Safety/Standards." These technologies are well-developed and can be installed with ease. However, system-level planning has to be done before installations, so that fault tolerance is built-in and leaks and repairs can be handled with ease (e.g., if a pipeline develops a leak then the design should be such that section of the pipeline should be isolatable before either replacement or repair).

Multiple national and international Standards Development Organizations are involved in the development of standards. These organizations collaborate closely with both private and public players to establish standards. In recent times codes, standards and regulations are being established for hydrogen systems for transportation, commercial and residential applications. Two key issues persist here: the involvement of multiple organizations in the generation of the codes and standards and the lack of universal standards across nations which can be used as "user manual."

8.1.8 Recycling of components

Needless to point out recycling of key components is an important part of any energy loop (and development of the "circular economy"), which can make the loop truly energy efficient and greener. This is especially true when the technology becomes widespread and there are precious metals involved, like in the case of Li-ion batteries which may have cobalt in addition to Li. Herein comes geopolitical issues related to control of these precious metals and humanitarian issues related to their mining and processing (e.g., most of the cobalt comes from Congo and is some of it is referred to as "blood cobalt"). Some of these issues persist in the recycling stage and put together with the production issue, poses threat to an uninterrupted supply chain—which can jeopardize the very success of a given technology.

Recycling of battery components has reached a certain stage of maturity, arising from their use in the electronics industry, and hence has an advantage over recycling of fuel cells. The lack of widespread use of fuel cells, coupled with the fact that the already installed fuel cells only now reaching the end of their life, implies that the technology for the recycling of fuel cells is not well developed. But now the timer has been set for the development of dedicated recycling technologies for fuel cell components. It is imperative that the new technologies of recycling fuel cells (and other components of the hydrogen energy chain) have to be environmentally friendly as the currently available techniques are not green and this aspect needs to be urgently addressed.

Table 8.1.1 The current level of maturity for widespread use: L—low, G—good, M—currently workable but needs improvement, and P—poor and requires real focus.

Component	Current Level	Comment
Production	L–M	Production of green hydrogen is the best option for a quick adoption of the hydrogen economy.
Storage	L–M	Storage in cylinders may have to co-exist with storage in materials. Serious efforts are required in this front for onboard applications, which can be driver for the hydrogen economy.
Transport	P	The use of dedicated pipelines remains a major challenge. Other modes of transportation are not efficient, especially given the current storage methods.
Utilization	M	Direct combustion in IC engines remains a challenge. The cost of fuel cells needs to be reduced and their efficiency needs to be increased for application in a widespread manner.
Standards	M	"One stop shop" for hydrogen standards needs to be made available, which can serve as a user manual for standards and safety.
Policy	P	There is complete lack of coherent policy (across the globe). Having such policies can prove to be a big driver for the hydrogen economy.
Safety	G	Though safety is not a big issue now, but "zero accidents" may be required to avoid bad press and accidents can hamper the adoption of hydrogen based technologies.
Recycling	P	Batteries have been around for a while and hence the recycling technology is more matured as compared to that for fuel cells. The clock is ticking for developing such technologies for fuel cells.

8.1.9 Concluding Remarks

The hydrogen economy has a bright future in the long run, i.e., when we run out of fossil fuels. In the short and medium term, adoptions of hydrogen energy technologies require immediate action, which is spread across the spectrum of activities and includes: production-related technologies to the creation of uniform standards. Government investments and international cooperation are paramount if the hydrogen economy has to take hold of a higher share of the marketplace. This includes subsidization policies, the creation of hydrogen corridors, hydrogen valleys, and huge monetary investments. One point often overlooked is that some energy technologies do not have a great net benefit in terms of it being green; however, their use will avoid concentrated pollution in areas of high inhabitation. For example, even if the overall level of carbon emission is the same as that of the use of fossil fuel-based vehicles, the use of hydrogen as a fuel can negate the emission levels in cities (especially of toxic gases) and towns where most of the population is concentrated and hence will offer significant health benefits to humans. Table 8.1.1 lists the level of maturity of various components of the hydrogen economy. The "current level" column has to set the tone for urgent improvement in certain areas (marked with an "L").

In the short-term hydrogen energy has to compete with battery-based technologies. The disadvantages of this technology are best seen in the context of vehicles; wherein battery-based vehicles currently have limited travel range, suffer from long charging times, and are sensitive to the temperature of the environment. Albeit the fact that hydrogen-based technologies also suffer from defects, addressing these on a priority basis can give hydrogen energy a strong foothold currently dominated by batteries.

The future of the hydrogen economy is bright, but we have to light the fire now.

References

[1] Coradia iLint, https://www.alstom.com/solutions/rolling-stock/alstom-coradia-ilint-worlds-1st-hydrogen-powered-train. (Note: the train uses batteries in addition to the fuel cell for power during acceleration and to store the energy recovered during braking).

[2] M.A. Pellow, C.J.M. Emmott, C.J. Barnhart, S.M. Benson, Hydrogen or batteries for grid storage? A net energy analysis, Energy Environ. Sci. 8 (2015) 1938. https://doi.org/10.1039/C4EE04041D.

[3] Y. Gao, Y. Yang, R. Schimmenti, E. Murray, H. Peng, Y. Wang, C. Ge, W. Jiang, G. Wang, F.J. DiSalvo, D.A. Muller, M. Mavrikakis, L. Xiao, H.D. Abruna, L. Zhuang, A completely precious metal–free alkaline fuel cell with enhanced performance using a carbon-coated nickel anode, Proc. Natl. Acad. Sci. 119 (13) (2022) e2119883119. https://doi.org/10.1073/pnas.2119883119.

Index

Page numbers followed by "*f*" and "*t*" indicate, figures and tables respectively.

M

N

O

P